SAS® Applications Programming

THE DUXBURY SERIES
IN STATISTICS AND DECISION SCIENCES

Applications, Basics, and Computing of Exploratory Data Analysis Velleman/Hoaglin
Applied Regression Analysis for Business and Economics Dielman
Elementary Statistics, Fifth Edition Johnson
Elementary Statistics for Business, Second Edition Johnson and Siskin
Essential Business Statistics: A Minitab Framework Bond/Scott
Fundamental Statistics for the Behavioral Sciences, Second Edition Howell
Fundamentals of Biostatistics, Third Edition Rosner
Introduction to Mathematical Programming: Applications and Algorithms Winston
Introduction to Probability and Statistics, Eighth Edition Mendenhall/Beaver
Introduction to Statistical Methods and Data Analysis, Third Edition Ott
Introductory Business Statistics with Microcomputer Applications Shiffler/Adams
Introductory Statistical Methods: An Integrated Approach Using Minitab Groeneveld
Introductory Statistics for Management and Economics, Third Edition Kenkel
Making Hard Decisions: An Introduction to Decision Analysis Clemen
Mathematical Statistics with Applications, Fourth Edition Mendenhall/Wackerly/Scheaffer
Minitab Handbook, Second Edition Ryan/Joiner/Ryan
Minitab Handbook for Business and Economics Miller
Operations Research: Applications and Algorithms, Second Edition Winston
Probability and Statistics for Engineers, Third Edition Scheaffer/McClave
Probability and Statistics for Modern Engineering, Second Edition Lapin
SAS Applications Programming: A Gentle Introduction DiIorio
Statistical Methods for Psychology, Second Edition Howell
Statistical Thinking for Managers, Third Edition Hildebrand/Ott
Statistics: A Tool for the Social Sciences, Fourth Edition Ott/Larson/Mendenhall
Statistics for Business and Economics Bechtold/Johnson
Statistics for Business: Data Analysis and Modelling Cryer/Miller
Statistics for Management and Economics, Sixth Edition Mendenhall/Reinmuth/Beaver
Student Edition of Execustat The Developers of Statgraphics
Understanding Statistics, Fifth Edition Ott/Mendenhall

THE DUXBURY ADVANCED SERIES
IN STATISTICS AND DECISION SCIENCES

A First Course in the Theory of Linear Statistical Models Myers/Milton
Applied Nonparametric Statistics, Second Edition Daniel
Applied Regression Analysis and Other Multivariate Methods, Second Edition
    Kleinbaum/Kupper/Muller
Classical and Modern Regression with Applications, Second Edition Myers
Elementary Survey Sampling, Fourth Edition Scheaffer/Mendenhall/Ott
Introduction to Contemporary Statistical Methods, Second Edition Koopmans
Introduction to Probability and Its Applications Scheaffer
Introduction to Probability and Mathematical Statistics Bain/Englehardt
Linear Statistical Models: An Applied Approach, Second Edition Bowerman/O'Connell
Probability Modeling and Computer Simulation Matloff
Quantitative Forecasting Methods Farnum/Stanton
Time Series Analysis Cryer
Time Series Forecasting: Unified Concepts and Computer Implementation, Second Edition
    Bowerman/O'Connell

SAS® Applications Programming:

A Gentle Introduction

Frank C. DiIorio

Family Health International

THE DUXBURY SERIES IN STATISTICS AND DECISION SCIENCES

 **PWS-KENT Publishing Company**
Boston

PWS–KENT
Publishing Company

20 Park Plaza
Boston, Massachusetts 02116

PWS-KENT Publishing Company is a division of Wadsworth, Inc.

SAS is the registered trademark of SAS Institute Inc., Cary, NC, USA.

ISE   ISBN 0-534-98464-9

Library of Congress Cataloging-in-Publication Data
DiIorio, Frank C.
    SAS applications programming : a gentle introduction / Frank C.
DiIorio
        p. cm. – (The Duxbury series in statistics and decision
sciences)
    Includes index.
    ISBN 0-534-92390-9
    1. SAS (Computer program) 2. Statistics—Data processing.
I. Title. II. Series
QA276.4.D55 1991
005.369—dc20

Sponsoring Editor: Michael Payne
Assistant Editor: Marcia Cole
Editorial Assistant: Tricia Schumacher
Production Coordinator and Cover Designer: Robine Andrau
Manufacturing Coordinator: Peter D. Leatherwood
Interior Designer: Catherine L. Johnson
Cover Photo: © Steven Hunt/Image Bank
Typesetter: ETP Services, Inc.
Cover Printer: John P. Pow Co.
Printer and Binder: Courier/Westford

Printed in the United States of America

91 92 93 94 95 — 10 9 8 7 6 5 4 3 2 1

Preface

SAS Applications Programming: A Gentle Introduction presents a thorough introduction to the SAS System for data management, statistical analysis, and reporting. Special features such as the graphics, econometrics, and operations research add-on packages are not discussed. Likewise, some parts of the "Base" product are not covered (these include some utilities and its macro language). But even if readers skim the book or read only select chapters, they should come away with an appreciation of what SAS can do and a solid understanding of how to use it in their work. SAS Institute documentation has long been known for being voluminous and somewhat intimidating. The intent of this book is to make the reader's introduction as graceful and pain-free as possible, hence the subtitle's use of the word "gentle."

Level of Coverage

The philosophy behind the level and scope of treatment throughout the book is to trade breadth for depth. From the novice user's viewpoint, SAS is a complex and potentially troublesome system. This view suggests that it is better to thoroughly explain a relatively narrow portion of the language in detail than to skim over a large part of it. This approach helps the user avoid errors or at least correct them faster. It also develops the user's intuition about how to "push" the System as far as he or she needs it to go, a benefit lacking in most other SAS-related publications.

The user will learn probably 20 percent of the entire capability of the SAS System in this book. It is designed to be the most useful 20 percent, the portion most often used by beginners. The temptation to include "tricks," the esoteric features so beloved by professional programmers, was great. In a book such as this, with its novice audience in mind, such a temptation is best avoided.

Intended Audience

The book's primary audience is SAS end users, students and people in the "real world" who have a basic level of computer literacy and want to learn what SAS can do to meet their data management, analysis, and reporting needs. These people are not professional programmers. They simply need a tool for their work and want to explore how the SAS System can meet their needs.

Another audience is computing professionals who want to develop an appreciation of how SAS works. This group will recognize elements of both third and fourth generation languages in the SAS System. For this group the book tries to emphasize how to decide which "generation" to use (a confusing point for experienced programmers and one novices are happily unaware of).

Text Organization

The book is divided into five parts and twenty-five chapters. Chapter 1 is a brief general introduction to SAS. Part I (Chapters 2 and 3) is an overview of the language. It examines sample programs and their output in detail and presents basic terminology. Part II (Chapters 4 through 6) presents material required to run simple programs. Readers will learn how to read several types of data, sort files, create simple univariate statistics, and produce histograms.

Part III (Chapters 7 through 11) introduces data management topics such as flow of control, bracketing related groups of statements, and other topics related to program design. Part IV (Chapters 12 through 20) is a series of specialized topics. Features such as combining files, using date-oriented data, recoding, custom report writing, and handling special kinds of input data are covered.

Part V (Chapters 21 through 24) discusses statistical and related topics in depth. It covers *n*-way crosstabulations, correlations, *t*-tests, linear regression, analysis of variance, and two-variable plotting. This material is typically covered in an introductory statistics course for business and the social or health sciences.

Other parts of the book include Chapter 25, a brief tour of parts of the SAS System not covered in this book, and appendices. Appendix A presents the collating sequences used by the computers SAS runs on. Appendix B discusses a set of informal rules for good program design. Appendix C contains a series of questions to ask computer center staff about using SAS. Appendix D outlines common tasks and their solutions using four major operating systems. Appendix E presents information about the datasets used in the book's examples. The layout of each dataset and a listing of the actual data are provided. The datasets are available in a machine-readable form from the publisher. Appendix F summarizes changes in the new release of the SAS System (version 6.06) that may be of interest to the reader.

Text Features

Many real-world examples are included throughout to help the reader identify uses for SAS features described in the text. A design feature that will help highlight key points is a series of "Pointers and Pitfalls." These are rules of thumb, caveats, and other important issues. Although not a substitute for reading the text, they help focus attention on fundamental principles of and potential problem areas in using the SAS System.

Most chapters have exercises after their major sections. The exercises are a self-test on the material just presented and often require integration of topics covered in earlier chapters. Answers to the exercises are located in the back of the book.

Using This Book

Not even the most enthusiastic computer buff will want to read this book in order, start to finish. Different types of readers will want to approach the chapters in different order or avoid some altogether. Two general approaches are suggested for two of the most likely groups of readers: students using the book as an adjunct to a statistics course and real-world practitioners and programmers.

Students in Statistics Courses   Chapter 2 presents essential background information about SAS terminology and logic. Students should read this chapter carefully and be comfortable with the examples it discusses. This may be a good time for them to search out help at their site for actually running a simple SAS program and retrieving its output. Appendix C should be skimmed for suggestions about who to contact and what questions to ask.

Chapter 3 is a reference for SAS syntax. Many of the basic elements of the language are described here, some in depth. Students should skim this chapter, reading enough to know to come back to it when the need arises. Chapter 4 describes how to read raw data and make it usable by SAS. The first part of

Chapter 5 is background on SAS procedures and should be skimmed for the same reason as Chapter 3. The rest of the chapter describes two routines that will probably be used often.

Chapters 2 through 5 prepare the reader to perform a surprisingly wide variety of statistical analyses. Chapter 6 describes single-variable statistics and their presentation (bar charts). Chapters 21 through 24 progress from scatterplots and contingency tables to correlation, *t*-tests, regression, and analysis of variance (ANOVA). Chapter 17 has sections on calculating standard scores and rank statistics.

As analysis proceeds and the user needs to create new variables, impose conditions on the datasets, and so on, more advanced tools become necessary. At a minimum, the material in Chapters 7 through 9 should be skimmed. This material provides an idea of SAS's data manipulation capabilities and the background necessary for the more advanced topics of Chapters 10 through 20, should the need arise.

Practitioners and Programmers The conceptual and syntactical orientation provided in Chapters 2 and 3 is particularly important for readers who will use SAS primarily as a programming, rather than a statistical, language. The orientation questions and learning resources identified in Appendix C should be reviewed at this point as well. Once the reader is comfortable with this material, Chapters 4 and 5 should be read. Chapter 6 should at least be skimmed by the "stat-phobic."

Careful reading of Chapters 7 through 13 will provide the basis for learning more advanced topics later in this book. These chapters emphasize SAS in its role as a high-level computer language, a role often overlooked or misunderstood by business-sector practitioners. The material in Chapters 14 through 20 can be read as needed; little or no sequential reading is required. These chapters address such specialized topics as date-oriented data handling, character data manipulation, efficient recoding and grouping techniques, custom report writing, and advanced data manipulation features.

Interested readers from both groups should read Chapter 25 for an overview of parts of the SAS System not discussed in this book. They should also review Appendix B's "style sheet," a series of commonly used programming and design rules.

Acknowledgments

I owe a debt of thanks to many people for their assistance with this book. Billie Sessoms, Durham Technical Institute, suggested that I should consider turning my one-semester SAS course notes into something more substantial. Key among those who encouraged this idea were Elizabeth Martinez (Research Triangle Institute, Research Triangle Park, North Carolina), Sally Muller and Keith Brown (University of North Carolina), and Kenneth Hardy (IRSS). They reviewed my proposal and outline and made thoughtful comments. Michael Shank provided invaluable assistance in the business and legal side of this endeavor. My thanks to Nancy Anderson (Computer Sciences Corporation), who produced the graphics output in Chapter 25. Finally, my parents and in-laws provided vast amounts of encouragement and moral support during the proposal phase and over the two years of writing.

I have had the good fortune to get excellent critical reviews from the following people:

Alfred A. Bartolucci, University of Alabama–Birmingham

Bruce Bowerman, Miami University

Hallett German, GTE Laboratories

L. Eric Hallman, National Institute of Environmental Health Sciences

Robert M. Hamer, Virginia Commonwealth University

Jerry M. Lefkowitz

D. Robert McConnaughey, Computer Sciences Corporation

Paul Marsh, North Carolina State University

Their comments ranged from the insightful to the bombastic to the correct spelling of Uruguay. I listened to all their comments and took most of them to heart. Many of the features that readers may find attractive in this book are, I am sure, an outgrowth of their comments.

A special debt of thanks is due Kenneth Hardy of IRSS. In addition to the initial review of the proposal and outline, he drafted and reviewed the regression and analysis of variance chapters. I only hope that the final product approaches the insight and clarity he brings to the subject.

The staff of PWS-KENT has provided valuable editorial and administrative assistance. Marcia Cole and Tricia Schumacher bore the brunt of the manuscript distribution to the reviewers and patiently instructed me in the intricacies of the publishing process. Robine Andrau, Senior Production Editor, and Michael Payne, The Duxbury Series Senior Editor, provided the right amount of cajoling, encouragement, and cheerleading throughout this project. The book's organization and content are in large part due to their ideas and understanding of the marketplace. Special thanks to Sally Stickney for a superb copyediting effort.

This project revised my definition of patience, highlighted the importance of walking in small steps, and relied heavily on the support of my long-suffering family. My daughter Amelia is as old as this project and so has never known a father who didn't feel like he had something hanging over his head. Her cackles of delight at watching my computer's hard disk light flashing and the constant distractions she provided are memories I will always cherish. Finally, I would like to dedicate this work to my wife, Elizabeth. She eventually abandoned the "When will it be done?" question and focused on providing distraction, solitude, and support with uncanny good judgment. This book would not have been possible without her love and good humor.

Contents

PART THREE
Refining the Program

1 Introduction

Think for a minute about the Swiss Army knife you take on camping trips. It is probably bulky, has clever features (some of which you have never exploited) and is sometimes difficult to use. This book is an introduction to the SAS System, one of the bigger Swiss Army knives of the computer world. The SAS System is a tool that, for better and worse, shares many of the Swiss Army knife's characteristics. It is a very large, complex, capable, and powerful collection of software that enables the user to perform a wide variety of tasks. It is also sometimes viewed as hard to use and, like the Swiss Army knife, has features that are not used in everyday applications.

This book introduces the SAS System and prepares you for using it in a variety of practical everyday situations. Concepts and applications are explained in detail. Real-world examples and thought-provoking exercises supplement the text. Once you begin to understand the logic and uses of the SAS System, you will see why millions of people regard it as an indispensable part of their work.

1.1 About SAS

What Is SAS?

The SAS System (usually referred to in this book simply as "SAS") is an integrated collection of data management, analysis, and reporting tools. These capabilities are important: The *data management* features allow the user to read and combine data files in any conceivable fashion. The *analysis* capabilities extend from simple frequency distributions through complex multivariate techniques. Finally, the *reporting* features give the user the capability of presenting data management and analysis results in virtually any format.

1

The power of the SAS System is that it is *integrated*: data handled by its data management facilities can be used without modification by the analysis and reporting components. This feature allows the user to work quickly and effectively, with minimal concern for data formats and structures. The system is also integrated across different computing environments: a SAS program written on an IBM-compatible microcomputer will run with almost no modification on a mainframe or minicomputer.

Origins

The SAS System roots can be traced to the Institute of Statistics at North Carolina State University, Raleigh, North Carolina, in the late 1960s. The product, then known as the Statistical Analysis System, quickly grew in popularity and capability, mostly in academic circles. In 1976 SAS Institute Inc., a privately held corporation, was formed. To use an appropriate cliché, "the rest is history."

From its low-key campus origins, SAS Institute has grown into the world's largest privately held software company. Continual product-line expansion and diversification of clientele have resulted in SAS products being used by over 2 million people in over 10,000 sites worldwide. Revenues in 1989 approached $200 million. SAS Institute maintains offices and subsidiaries throughout the world, employing approximately 2,000 people. SAS has become the statistical analysis package of choice in IBM mainframe environments, is the de facto standard for computer performance analysis, and has also made strong inroads in minicomputers and microcomputers.

Part of the reason for the product's popularity is SAS Institute's continual outreach to the end user. Technical support is free. A quarterly magazine, *SAS Communications*, keeps users abreast of new products and offers technical pointers. The Institute sponsors an annual international conference of SAS users. At these conferences the Institute demonstrates products, and users present papers and tutorials. The conferences are also a good way to get to know other SAS users in your application or industrial niche and see how they solve problems.

Many regional and special-interest user groups have formed. They often maintain electronic bulletin boards, have periodic meetings, publish newsletters, and offer consulting services. The Institute also conducts a wide range of courses at its Cary, North Carolina, headquarters, its regional offices, and in individual companies. Finally, the Institute publishes many manuals: system-specific guides, references, special-topic tutorials, and other documentation abound.

Anyone who uses even a few of these outreach mechanisms cannot help but develop a feel for SAS and its power. Unfortunately, getting to the point where you feel comfortable with SAS terminology and capabilities often takes more time than you might be willing to invest. Intensive start-up effort is particularly true of the manuals. Making your introduction to the SAS System as productive and gentle as possible is the motivation behind this book.

1.2 Syntax Conventions

The syntax descriptions used in the book follow a few simple rules:

- Brackets, [ ], indicate optional items. If you actually need the item do *not* type the brackets.
- Uppercase letters indicate SAS keywords and should be entered as is. These include names of SAS routines and special data management commands.

The capitals are for illustrative purposes only: you can actually enter them in your program in any mixture of uppercase and lowercase.

- lowercase letters indicate items supplied by you. These include names of datasets, data items, and titles to use in program output.
- A vertical bar, ¦, indicates that a choice must be made. Two or more SAS keywords will be separated by bars. The item you choose should *not* include the vertical bar.
- An ellipsis, . . . , indicates that the item just described may be repeated.

These conventions are illustrated in the description of the SAS routine to produce frequency distributions, the FREQ procedure:

```
PROC FREQ DATA=dset [ORDER=INTERNAL|FORMATTED];
[TABLES var ... [ / MISSING ] ; ]
```

All of the groups of statements that follow are syntactically valid runs of the FREQ procedure:

```
proc freq data=temp01;

proc freq data=fiscal90;
tables dept1 dept3 dept5;

proc freq data=labmast order=internal;
tables testid / missing ;
```

Summary

The SAS System is a complex and versatile collection of software for data management, analysis, and reporting. It can be used just as easily for simple, one-time applications as for complicated production systems. Its roots are in academia, but SAS has spread to have a wide variety of industrial and governmental sectors around the world. The SAS Institute, creators of the SAS System, have developed many outreach features and organizations that ensure that the end-user's development and training needs will be met. This responsiveness to user needs is a key factor in the Institute's rise to one of the largest software companies in the world.

PART ONE

The Basics:

SAS Terminology

and Logic

2 The SAS Environment: Terminology and Basics

Every programming language, SAS included, has a unique approach to problem solving. The way the task to be solved is defined, the tools at the user's disposal, the way the program communicates results, and many other features vary greatly from language to language. As much as some people would like to argue to the contrary, no single design philosophy is "right." Some, however, are more enjoyable to use and more powerful than others.

This chapter describes some of the basic terminology of the SAS system and presents a general picture of its approach to data management and analysis. It gives a feel for how SAS "thinks" and presents several annotated examples. Later chapters will get into the specifics of SAS's approach to data management and analysis. While reading this chapter, pay attention to how SAS programs are organized and *what* is being done rather than the particulars of the syntax and *how* the work is accomplished.

2.1 Building Blocks

All SAS programs are built around a few common elements. The program must read data, process it, and communicate results to the user. This section describes these activities. It will also help you become more familiar with the look and feel of SAS syntax and terminology.

Units of Work

Virtually every SAS program is built around one or both of two basic units of work: the DATA step and procedures. One or more program *statements*, each terminated by a semicolon (;), make up each of these units. As we will see throughout this and later chapters, the power of the SAS System lies in the variety and power of program statements and the free-form arrangement of DATA steps and procedures.

The **DATA step** is the heart of SAS's programming language. It reads data, performs calculations, writes customized reports, and creates new datasets and data fields. The DATA step's commands are flexible, powerful, and usually easy to use. Like most of the SAS System, they are also characterized by sensible defaults: if you do not explicitly ask for an option or feature, SAS will take an appropriate action. There can be as many DATA steps as needed in a SAS program.

Procedures, commonly referred to as PROCs, are the other major component of the SAS System. They are designed to perform statistical, graphics, and presentation tasks with a minimum of user effort. Usually all that is required is specification of the type of analysis (e.g., linear regression), the data in the model (dependent and independent variables), and identification of nondefault features (residual plot, covariance matrix, and so on). Just as DATA steps have intelligent defaults, so do PROCs. Also like DATA steps, there can be as many PROCs as necessary in a single program.

There is no set order of DATA steps and PROCs in a program. The sequence is determined entirely by the needs of the user. A program may be set up in a variety of ways:

- It may contain nothing but DATA steps creating datasets for use in later programs.
- It may begin with a DATA step, followed by a PROC, followed by a DATA step analyzing a SAS dataset created by the PROC.
- It may begin with a DATA step, then analyze the SAS dataset with several different PROCs.
- It may consist entirely of PROCs analyzing data created in previous SAS programs.

Any combination of DATA steps and PROCs is appropriate. The only constraint is that the ordering of these units of work be logical: a PROC analyzing a dataset, for example, should not be placed before the DATA step creating the dataset.

POINTERS AND PITFALLS

SAS processes statements in the order they are entered. Any combination of DATA steps and PROCs can be used as long as they make logical sense. Don't "zag" before you "zig."

Datasets

Two varieties of data are analyzed by the DATA step and PROCs: raw data and SAS datasets. **Raw data** is the initial form of the dataset. It is simply a table of rows and columns, entered at a terminal using a text editor or possibly already created and ready for use on the system. The table's rows represent units of analysis (e.g., people, companies, test results for a patient), and the columns represent different pieces of information about each unit of analysis (e.g., a person's height, weight, and age; a company's gross and net earnings). A listing of part of the raw data used in the sample programs in Section 2.2 is presented in Exhibit 2.1.

This table of raw data has several important features. First, it is rectangular—that is, it has an equal number of items, or fields, of information for each row. Second, it is assumed that a particular item represents the same measurement in

EXHIBIT 2.1   Partial listing of raw data used in Section 2.2 programs

01 Alabama	e	6	112	-27	3893888	77	35.600	60.035	4801	346	5894	26.12	1127
02 Alaska	w	9	100	-80	401851	1	0.400	64.344	6410	858	10193	6.28	386
04 Arizona	w	8	127	-40	2718215	24	53.300	83.832	7614	257	7041	14.49	1360
05 Arkansas	w	7	120	-29	2286435	44	46.900	51.589	3743	323	5614	25.09	629
06 California	w	9	134	-45	23667902	151	33.100	91.295	7592	241	8295	20.30	26387
08 Colorado	w	8	118	-60	2889964	28	53.400	80.619	7189	273	7998	14.12	1007
09 Connecticut	e	1	105	-32	3107576	638	16.100	78.832	6552	218	8511	30.95	5132
10 Delaware	e	5	110	-17	594338	308	53.500	70.636	6644	330	7449	23.62	220
11 DC	e	5	.	.	638333	10181	0.000	100.000	10692	.	8960	4.52	.
12 Florida	e	5	109	-2	9746324	180	38.400	84.261	8048	225	7270	12.61	4650

every row. Third, the blanks separating the fields aid human reading of the data but are not required by the computer. Fourth, not all fields are available for all observations. The DC line (next to the last in Exhibit 2.1) has four fields of periods (.). These indicate to SAS that the field's values were not available. They are, in SAS's vocabulary, *missing data*. Finally, by simply looking at the data we can only guess at their meaning. Each field of data requires more explanation: What does it measure? What is the unit of measurement? When was the measurement taken? The simple table in Exhibit 2.1 is readable but not directly usable by SAS procedures.

The raw data can be used to create the second type of data used by SAS: the SAS dataset. Instructions in the DATA step tell SAS how to read the raw data, how to calculate new items, which rows and columns to save, and what other data management tasks to perform. The result of this manipulation is a **SAS dataset**, a dataset that contains not only the data but also instructions to SAS about what they mean, how to display them, and how they were created. SAS programmers refer to the rows of a SAS dataset as *observations* and the columns as *variables*. The SAS dataset is written with a unique layout and, unlike its raw data equivalent, is not decipherable simply by listing it on a computer terminal or printout.

POINTERS AND PITFALLS

Raw data is not directly usable by SAS procedures. DATA steps are used to convert raw data to SAS datasets. Once the data are in SAS dataset format, they can be analyzed with PROCs.

SAS datasets have many performance and economic advantages over their raw data counterparts. These advantages are discussed in detail in Chapter 10. The key point to remember here is that the SAS dataset is the only type of data that can be used by procedures. Raw data is useful to SAS only in its role as input to a DATA step and its subsequent storage as a SAS dataset: raw data is *never* directly handled by procedures.

Running SAS

So far the units of work of the SAS program (DATA steps and PROCs) and the types of datasets they process (raw and SAS) have been identified. Once you have determined how to read the data and what procedures you want to run, how do you actually run the SAS program?

Several different methods, or execution modes, are usually available. The one(s) used depends on the kind of computer being used. Typically the choice is between batch and interactive program execution modes. In **batch mode** the program is submitted to the computer's operating system, given an execution priority, and run; and the results are returned to the terminal or printer. The user is free to perform other tasks and run other programs while the program is being handled by the operating system.

Interactive mode executes the program immediately, taking control of the user's terminal in the process. Results are displayed at the terminal as they are produced. The cost of this immediacy is that the user cannot perform any other work while the program is running.

Appendix D describes the pros and cons of both modes in greater detail. It also presents examples of programs run in some of the more popular computing environments. Most of the examples in this book do not rely on running a program in either environment, since SAS programs run in an interactive environment will also run with very few changes in a batch environment. You should take the time now to learn how to prepare and execute a job using the preferred method at your computer center. (See Appendix C for more information.)

POINTERS AND PITFALLS

There are many ways to run SAS programs, and the "best" method varies greatly. Become familiar with the preferred method at your installation.

Program Output

Once the program runs, what output is produced? SAS programs usually produce two types of output: the SAS Log and procedure output. The **SAS Log**, usually referred to simply as the Log, is produced whether the program ran successfully or not. It contains a listing of the SAS program, notes about resource usage (computer time and storage), error messages and diagnostics, a brief description of SAS datasets created and raw datasets used, and other notes and warnings caused by the program's execution.

Procedure output takes many forms. It may be statistical tables, charts, and graphs or, in the more sophisticated cases, maps and simulated three-dimensional renderings of data. In all but the most esoteric instances, all procedure output is written to the same portion of the computer printout, disk dataset, or terminal display "window."

It is good programming practice to reconcile the procedure output to the statements in the Log that generated it: sometimes problems in the analysis are not noted in the procedure output file but are addressed in the Log. A "clean" Log will always mean "clean," reliable procedure output.

POINTERS AND PITFALLS

Analytical problems are not always reported in procedure output. Make a habit of reconciling procedure output and the SAS Log.

SAS programs can produce other types of output. Among these are permanent SAS datasets, new raw datasets, and customized reports generated by the DATA step. Later chapters deal extensively with all forms of output. One of SAS's strengths is its ability to produce several kinds of output from a single

DATA step or PROC. Examples in Chapters 13, 17, 19, and 20 emphasize that the "typical" SAS program produces many of these types of output.

Exercises

2.1. Examine the following program. (*Note:* The line numbers are included for reference only—do not enter them when writing a program.)

```
( 1) data one;
( 2) infile 'mast.fin(fy83)';
( 3) input sic $5. earn 5. prof 5.;
( 4) run;
( 5)
( 6) proc print data=one;
( 7) run;
( 8)
( 9) proc means data=one;
(10) run;
```

   a. How many units of work are there in the program? Count DATA steps, procedures, and statements separately.
   b. Realizing that statement syntax and functions have not yet been covered, take a guess at what the program does. This task is simplified if you break it into small steps corresponding to SAS's units of work.
   c. Do you think the program would work as intended if lines 1 through 4 came *after* line 8? Why, or why not?
   d. Rewrite the program the "wrong" way: fit as many statements as possible on a single line, and take out the blank lines separating the procedures and DATA step. Examine the result. It is more compact than the original, but how easy is it to change? to read?

2.2. Assume that the listing in Exhibit 2.1 is typical of the entire dataset and that the columns are always separated by at least one blank. How many columns, or variables, are in the dataset? How could you describe the location of the variables to SAS? (There are several ways to do this.)

2.3. a. Find out how to run SAS at your computer center (refer to Appendix C for hints on how to get help doing this). Once you know how, run this simple program and examine its output. Examine the output at your terminal, then print it.

```
data test;
length message $25;
string = 'This is a test program!';
run;

proc print data=test;
run;
```

   b. Which part of the output was the SAS Log? the procedure output? The Log contains a list of program statements and other information. Identify the different pieces of information supplied by SAS in the Log. Speculate about their meaning and why they might be useful (there is no "right" answer!).

2.2 Putting the Concepts to Work: An Annotated Program

A simple example using the data in Exhibit 2.1 will demonstrate some of the concepts described in Section 2.1. The program reads raw data, performs a few calculations, and then runs some simple procedures. Some errors made along the way illustrate how and where in the Log SAS points out incorrect syntax. Again, keep in mind that the purpose of the example is not to describe SAS syntax but simply to give a feel for how a SAS program looks and behaves. The example may also begin to give you an idea of how to use the Log and output listings.

The Original Program

The complete program is shown in Exhibit 2.2. All units of work (DATA steps, PROCs, and their component statements) are separated by a semicolon (;). Notice how the blanks both within and between lines spread the program text out and make it easy to read. Also notice the use of boldface. The terms in boldface are SAS keywords: items that identify certain options or activities to SAS.

EXHIBIT 2.2   Sample program: read data, print, compute single-variable statistics

---

```
* Simple program to demonstrate SAS's "look and feel" ;

options nodate linesize=130;

data states;
infile 'c:\book\data\raw\states.raw';
input name $ 4-19  hightemp 25-27  lowtemp 29-31
      pop80 popsqmi farmpct pcturb
      @83 pcapinc 5.;
urb_pop  = (pcturb * pop80) / 100 ;
temp_rng = hightemp - lowtemp ;
run;

proc print data=states;
run;

proc means;
run;
```

Let us go through the program in some detail to get a better feel for its organization and statement functions. If you are interested in reading ahead, refer to the chapters noted at the end of each statement's description.

```
* Simple program to demonstrate SAS's "look and feel";
```

This is an optional comment describing the program's purpose. These "header comments" can be quite useful if you are writing a series of programs and want to explain the function and sequence of each. Comments are described in Chapter 3.

```
options nodate linesize=130;
```

OPTIONS is an optional statement that sets one or more characteristics of the SAS session. In this case, NODATE suppresses the date on output listings, and LINE-SIZE instructs procedures to use 130 columns when displaying output. Many other session characteristics are automatically set by SAS and can be altered in the OPTIONS statement. Options are described in Chapter 12.

```
data states;
```

This is the first statement in the DATA step. It names the SAS dataset STATES. DATA statements are introduced in Chapter 4 and covered in more depth in Chapters 10, 11, and 12.

```
infile 'c:\book\data\raw\states.raw';
```

INFILE tells SAS where to find the raw data. It has many different formats and options. The form used here simply encloses the complete name of the dataset within single quotes ('). INFILE is introduced in Chapter 4 and covered in greater depth in Chapters 11 and 19.

```
input name $ 4-19  hightemp 25-27  lowtemp 29-31
      pop80 popsqmi farmpct pcturb
      @83 pcapinc 5.;
```

Once the source of the raw data is known, the DATA step can actually read the data with the INPUT statement. INPUT is the most complex statement in the SAS language and one of the most powerful. It is introduced in Chapter 4 and covered in greater depth in Chapters 11 and 19.

This INPUT statement demonstrates its flexibility with reference to identifying the location and type of data and how they should be read. It uses several different methods of describing the variable names and locations to SAS.

To indicate that the variable NAME contains alphabetic characters, use the dollar sign ($). The statement also indicates that NAME should be read from columns 4 through 19 of the raw data. The variables HIGHTEMP and LOWTEMP are numbers (no $) and are found in columns 25 through 27 and 29 through 31, respectively. The next four variables, POP80, POPSQMI, FARMPCT, and PCTURB, are simply identified: SAS figures out where they are by looking for one or more blanks separating the fields. Finally, PCAPINC is identified by moving to column 83 (@83) and reading five columns (5.).

```
urb_pop  = (pcturb * pop80) / 100 ;
temp_rng = hightemp - lowtemp ;
```

Once the variables from the raw data are read, they can be manipulated in calculations via assignment statements. Assignment statements are described in Chapters 3 and 7.

Two new variables, URB_POP and TEMP_RNG, are created. URB_POP is the number of people in urban areas, based on the percent urban (PCTURB) and the population (POP80). TEMP_RNG is the temperature range, the difference between the high (HIGHTEMP) and low (LOWTEMP) extremes. Both these calculated variables and those read in the INPUT statement will be stored in SAS dataset STATES.

```
run;
```

The DATA step ends with a RUN statement. RUN tells SAS that it should begin to execute, or run, the statements in the DATA step. The DATA step statements are executed for each record in the raw data, in this case, 51 times (once for each state or equivalent).

```
proc print data=states;
run;
```

The next unit of work in the program is a PROC, in this case, PRINT. The program indicates that the SAS dataset STATES should be printed. RUN is then entered to end the procedure and actually start printing. (The term *print* refers to either the display of the data at the terminal or their actual printing on paper.) PRINT has many options, but the one chosen here accepts all default settings. Thus PRINT will make all decisions regarding which variables and observations to print, how to display them, what title to use on the output, and so on. The PRINT procedure is described in Chapter 5.

EXHIBIT 2.3   Sample program Log and output files

```
------------------ SAS Log --------------------------
```

```
NOTE: Copyright(c) 1985,86,87 SAS Institute Inc., Cary, NC 27512-8000, U.S.A.      [1]
NOTE: SAS (r) Proprietary Software Release 6.03
      Licensed to D**2 Systems, Site 12345678.
```

```
   [2]
    1    * Simple program to demonstrate SAS's "look and feel" ;
    2
    3    options nodate linesize=130;
    4
    5    data states;
    6    infile 'c:\book\data\raw\states.raw';
    7    input name $ 4-19  hightemp 25-27  lowtemp 29-31
    8          pop80 popsqmi farmpct pcturb
    9          @83 pcapinc 5.;
   10    urb_pop  = (pcturb * pop80) / 100 ;
   11    temp_rng = hightemp - lowtemp ;
   12    run;
```

```
NOTE: The infile 'c:\book\data\raw\states.raw' is file C:\DATA\RAW\STATES.RAW.    [3]
NOTE: 51 records were read from the infile C:\DATA\RAW\STATES.RAW.
      The minimum record length was 100.
      The maximum record length was 100.
```

```
NOTE: Missing values were generated as a result of performing an operation on missing values.    [4]
      Each place is given by: (Number of times) at (Line):(Column).
      1 at 11:12
```

```
NOTE: The data set WORK.STATES has 51 observations and 10 variables.      [5]
NOTE: The DATA statement used 9.00 seconds.
```

```
   13
   14    proc print data=states;        [6]
   15    run;
```

```
NOTE: The PROCEDURE PRINT used 11.00 seconds.
   16
   17    proc means;        [7]
   18    run;
```

```
NOTE: The PROCEDURE MEANS used 5.00 seconds.
```

```
NOTE: SAS Institute Inc., SAS Circle, PO Box 8000, Cary, NC 27512-8000       [8]
```

```
------------------ Output listing --------------------
```

SAS   **[9]**

[10] OBS	NAME	HIGHTEMP	LOWTEMP	POP80	POPSQMI	FARMPCT	PCTURB	PCAPINC	URB_POP	TEMP_RNG [11]
1	Alabama	112	-27	3893888	77	35.6	60.035	5894	2337695.66	139
2	Alaska	100	-80	401851	1	0.4	64.344	10193	258567.01	180
3	Arizona	127	-40	2718215	24	53.3	83.832	7041	2278734.00	167
4	Arkansas	120	-29	2286435	44	46.9	51.589	5614	1179548.95	149

EXHIBIT 2.3 *(continued)*

5	California	134	-45	23667902	151	33.1	91.295	8295	21607611.13	179
6	Colorado	118	-60	2889964	28	53.4	80.619	7998	2329860.08	178
7	Connecticut	105	-32	3107576	638	16.1	78.832	8511	2449764.31	137
8	Delaware	110	-17	594338	308	53.5	70.636	7449	419816.59	127
9	DC	.	.	638333	10181	0.0	100.000	8960	638333.00	.
10	Florida	109	-2	9746324	180	38.4	84.261	7270	8212350.07	111

```
- - - - - - - - - - - New page of output - - - - - - - -
```

SAS

N Obs	[12] Variable	N	Minimum	Maximum	Mean	Std Dev [13]
51	HIGHTEMP	50	100.0000000	134.0000000	113.8600000	6.4933561
	LOWTEMP	50	-80.0000000	14.0000000	-39.6000000	17.2992863
	POP80	51	401851.00	23667902.00	4442074.61	4699160.11
	POPSQMI	51	1.0000000	10181.00	350.9607843	1421.01
	FARMPCT	51	0	94.8000000	45.2294118	25.6461872
	PCTURB	51	33.7730000	100.0000000	67.5971176	14.9966653
	PCAPINC	51	5183.00	10193.00	7092.80	973.5579083
	URB_POP	51	172734.03	21607611.13	3275504.38	4066544.35
	TEMP_RNG	50	86.0000000	187.0000000	153.4600000	19.8414223

```
proc means;
run;
```

The next unit of work is another PROC, the MEANS procedure. An analysis dataset is not specified, which means that SAS decides which one to use. Since no options are specified, MEANS determines which descriptive statistics to compute and which variables to analyze. The MEANS procedure is described in Chapter 6.

The SAS Log and output from this program are shown in Exhibit 2.3.

The bracketed numbers in the following list refer to marked portions of Exhibit 2.3.

- Messages similar to those at **[1]** are printed at the beginning of every SAS Log. These contain legal and licensing information, and have no bearing on the execution of the program.

- The program statements are reproduced on the Log exactly as they were input to SAS. Notice **[2]** that they are automatically numbered by SAS.

- Several useful pieces of information, or NOTEs, are printed following the DATA step. The raw dataset's characteristics—its name, how many lines were read from it, and the minimum and maximum line lengths read (in this case both are 100 columns long)—are identified in **[3]**.

- Another NOTE **[4]** signals a problem at line 11. As Chapter 7 explains, one or both of the variables in the calculation in line 11 (HIGHTEMP and LOWTEMP) lacked valid values and thus could not be used to compute TEMP_RNG. This is often not an error but simply a fact of life when working with incomplete data. SAS always tells the user about this and other instances when it could not carry out all the calculations requested.

- At the end of the DATA step [5] a NOTE describes the size of the SAS dataset STATES (51 observations, 10 variables) and notes the resources required to execute the DATA step (9.00 elapsed seconds).

- The PRINT [6] and MEANS [7] procedure statements are printed on the Log, along with their resource requirements.

- Finally, at the end of the Log the SAS signoff message is printed [8].

- The default title "SAS" [9] is printed at the top of each page of the output listing.

- The PRINT procedure's output is on the first page of the listing. PRINT automatically numbers the observations [10] and arranges the columns [11] in the order in which the variables were read in the DATA step. (Exhibit 2.3 displays only the first 10 observations.)

- Finally, the MEANS procedure output is printed. MEANS automatically selects all numeric variables for analysis [12]. Notice that the default title "SAS" is still being used even though another PROC's output listing has begun. The statistics for each variable are printed across the page [13]: the number of valid observations for the variable (N), the extreme values (MINIMUM and MAXIMUM), the average (MEAN), and the standard deviation (STD DEV).

"Improving" the Program

Even simple programming can become complicated. Suppose you decide that one of the variable names is not descriptive enough. Also suppose that you want to take advantage of the display and calculating options of the procedures. Exhibit 2.4 demonstrates what happens when one or more syntax errors are made in the program. We make what we think are the appropriate changes to the program, and SAS reacts with the Log shown in Exhibit 2.4.

This result is obviously not what we had in mind. Let us look at the changes and error statements in Exhibit 2.4. As before, the bracketed numbers refer to marked portions of the exhibit.

- Line 11 was changed, trying to make the name of the temperature range variable more descriptive [1]. Unfortunately, the new name violates the eight character limit on SAS names. An ERROR message is printed below the statement.

- Two new statements, FORMAT [2] and LABEL [4], are added to enhance procedure output. FORMAT describes how variables will be printed (dollar signs, commas, number of decimal places, and so on). LABEL associates descriptive text with variables. Some procedures will print both the variable name and the label, thus clarifying the sometimes obscure meaning of a brief, eight-character name. Use of both FORMAT and LABEL is optional. They are described in Chapter 5.

- One of the formats is incorrect [3]. SAS prints an ERROR message describing the problem. Two variables (FARMPCT and PCTURB) are affected, so the message is printed twice.

- The LABEL statement is also incorrect [5]. SAS prints an error message and indicates it expected an equals sign (=).

- Because there were errors in the DATA step, SAS does not create the SAS dataset STATEs [6]. It will continue reading program statements, trying to identify other errors.

- SAS finds an unknown option [NOBS] in the PROC PRINT statement [7]. It prints an error message and indicates that it expected a left parenthesis [(].

EXHIBIT 2.4 SAS Log of program with errors

```
-------------------- SAS Log ----------------------------

NOTE: Copyright(c) 1985,86,87 SAS Institute Inc., Cary, NC 27512-8000, U.S.A.
NOTE: SAS (r) Proprietary Software Release 6.03
      Licensed to D**2 Systems, Site 12345678.

   1    * Simple program to demonstrate SAS's "look and feel" ;
   2
   3    options nodate linesize=130;
   4
   5    data states;
   6    infile 'c:\book\data\raw\states.raw';
   7    input name $ 4-19  hightemp 25-27  lowtemp 29-31
   8         pop80 popsqmi farmpct pcturb
   9         @83 pcapinc 5.;
  10    urb_pop  = (pcturb * pop80) / 100 ;
  11    temp_range = hightemp - lowtemp ;
```

```
ERROR: The variable named TEMP_RANGE contains more than 8 characters.   [1]
```

```
  12    format pop80 urbpop comma11.  farmpct pcturb 2.6  pcapinc dollar8.;   [2]
```

```
ERROR: The decimal specification of 6 must be less than the width specification of 2.   [3]
ERROR: The decimal specification of 6 must be less than the width specification of 2.
```

```
  13    label name     = 'State name'        [4]
  14          hightemp = 'Highest recorded temperature'
  15          lowtemp  = 'Lowest recorded temperature'
  16          pop80    = 'Population, 1980'
  17          popsqmi  = 'Population density, 1980'
  18          farmpct  = '% land devoted to agriculture'
  19          pcturb   = '% urban pop (areas over 2,500 pop)'
  20          pcapinc  = '1980 per capita income'
  21          urb_pop  = 'Urban population, 1980'
```

```
  22          temp_range  'Temperature range'      [5]
```

```
         temp_range  'Temperature range'
                     -------------------
ERROR: Syntax error detected.


  23                ;
NOTE: Expecting one of the following:

=
```

```
  24    run;
```

```
NOTE: The SAS System stopped processing this step because of errors.   [6]
NOTE: The data set WORK.STATES has 0 observations and 10 variables.
```

```
NOTE: The DATA statement used 7.00 seconds.
  25
```

```
  26    proc print data=states nobs;   [7]
proc print data=states nobs;
                     ----
ERROR: Syntax error detected.
```

EXHIBIT 2.4   *(continued)*

```
NOTE: Expecting one of the following:

(

   27    var pop80 popsqmi pcturb urb_pop hightemp lowtemp temp_rng;    [8]
   28    id name;
   29    title 'State-level data used in demo program';

   30    run;
NOTE: The SAS System stopped processing this step because of errors.    [9]
NOTE: The PROCEDURE PRINT used 3.00 seconds.
   31
   32    proc means maxdec=2 n nmiss min max average;    [10]

proc means maxdec=2 n nmiss min max average;    [11]
                                       -------
ERROR: Syntax error detected.


   33    run;
NOTE: The SAS System stopped processing this step because of errors.
NOTE: The PROCEDURE MEANS used 3.00 seconds.
NOTE: SAS Institute Inc., SAS Circle, PO Box 8000, Cary, NC 27512-8000
```

- Several statements [8] are added to the PRINT procedure (VAR, ID, and TITLE). These explicitly request variables to print and specify a more meaningful title for the output.
- Because there were errors in the specification of the PRINT procedure, SAS does not print the dataset [9].
- Some of the options of the MEANS procedure are used [10]. These control the number of decimal places printed and explicitly select the statistics calculated.
- Once again, an error is made in specifying a procedure option [11]. AVERAGE is not the correct option name for the average of a variable.

The Corrected Program and a Note About Errors

The errors made in the program are both common and easy to fix. Notice that sometimes SAS suggests what is amiss: sometimes its diagnosis is correct, other times not. All the program's problems were *syntax* errors. Other errors, those in the program's design or *logic*, are harder to diagnose and fix because SAS did not complain about the program's specification and syntax: logic errors tell SAS to "do the wrong thing," while syntax errors tell SAS to "do the thing wrong." Many examples of both types of errors appear throughout this book.

Once the program is corrected and rerun, the SAS Log and output files are much "cleaner," as shown in Exhibit 2.5.

Let us look at the changes to the program and see how the output differs from the first example.

- The offending variable name has been shortened to a valid length [1]. TEMP_RANGE becomes the more terse, but still meaningful, TEMPRNGE.

EXHIBIT 2.5   Syntactically correct program Log and output

```
-------------------- SAS Log ---------------------------

NOTE: Copyright(c) 1985,86,87 SAS Institute Inc., Cary, NC 27512-8000, U.S.A.
NOTE: SAS (r) Proprietary Software Release 6.03
      Licensed to D**2 Systems, Site 12345678.

   1    * Simple program to demonstrate SAS's "look and feel" ;
   2
   3    options nodate linesize=130;
   4
   5    data states;
   6    infile 'c:\book\data\raw\states.raw';
   7    input name $ 4-19  hightemp 25-27  lowtemp 29-31
   8          pop80 popsqmi farmpct pcturb
   9          @83 pcapinc 5.;
  10    urb_pop  = (pcturb * pop80) / 100 ;
  11    temprnge = hightemp - lowtemp ;      [1]
  12    format pop80 urb_pop comma11.  farmpct pcturb 6.2   [2]  pcapinc dollar8.;
  13    label name     = 'State name'
  14          hightemp = 'Highest recorded temperature'
  15          lowtemp  = 'Lowest recorded temperature'
  16          pop80    = 'Population, 1980'
  17          popsqmi  = 'Population density, 1980'
  18          farmpct  = '% land devoted to agriculture'
  19          pcturb   = '% urban pop (areas over 2,500 pop)'
  20          pcapinc  = '1980 per capita income'
  21          urb_pop  = 'Urban population, 1980'
  22          temprnge = 'Temperature range'    [3]
  23          ;
  24    run;
NOTE: The infile 'c:\book\data\raw\states.raw' is file C:\DATA\RAW\STATES.RAW.
NOTE: 51 records were read from the infile C:\DATA\RAW\STATES.RAW.
      The minimum record length was 100.
      The maximum record length was 100.
NOTE: Missing values were generated as a result of performing an operation on missing values.
      Each place is given by: (Number of times) at (Line):(Column).
      1 at 11:12
NOTE: The data set WORK.STATES has 51 observations and 10 variables.
NOTE: The DATA statement used 11.00 seconds.
  25
  26    proc print data=states n;   [4]
  27    var pop80 popsqmi pcturb urb_pop hightemp lowtemp temprnge;
  28    id name;
  29    title 'State-Level Data for Demo Program';
  30    run;
NOTE: The PROCEDURE PRINT used 9.00 seconds.
  31
  32    proc means maxdec=2 n nmiss min max mean [5] ;
  33    run;
NOTE: The PROCEDURE MEANS used 5.00 seconds.
NOTE: SAS Institute Inc., SAS Circle, PO Box 8000, Cary, NC 27512-8000
```

EXHIBIT 2.5 *(continued)*

```
------------------- Output listing --------------------
```

State-Level Data for Demo Program   [6]

[7]

NAME	POP80	POPSQMI	PCTURB	URB_POP	HIGHTEMP	LOWTEMP	TEMPRNGE
Alabama	3,893,888	77	60.03	2,337,696	112	-27	139
Alaska	401,851	1	64.34	258,567	100	-80	180
Arizona	2,718,215	24	83.83	2,278,734	127	-40	167
Arkansas	2,286,435	44	51.59	1,179,549	120	-29	149
California	23,667,902	151	91.29	21,607,611	134	-45	179
Colorado	2,889,964	28	80.62	2,329,860	118	-60	178
Connecticut	3,107,576	638	78.83	2,449,764	105	-32	137
Delaware	594,338	308	70.64	419,817	110	-17	127
DC	638,333	10181	100.00	638,333	.	.	.

. lines skipped
.

N = 51   [4]
```
- - - - - - - - - New page of output - - - - - - - - -
```

State-Level Data for Demo Program

N Obs	Variable	[8] Label	[9] N	Minimum	Maximum	Nmiss	Mean
51	HIGHTEMP	Highest recorded temperature	50	100.00	134.00	1	113.86
	LOWTEMP	Lowest recorded temperature	50	-80.00	14.00	1	-39.60
	POP80	Population, 1980	51	401851.00	23667902.00	0	4442074.61
	POPSQMI	Population density, 1980	51	1.00	10181.00	0	350.96
	FARMPCT	% land devoted to agriculture	51	0.00	94.80	0	45.23
	PCTURB	% urban pop (areas over 2,500 pop)	51	33.77	100.00	0	67.60
	PCAPINC	1980 per capita income	51	5183.00	10193.00	0	7092.80
	URB_POP	Urban population, 1980	51	172734.03	21607611.13	0	3275504.38
	TEMPRNGE	Temperature range	50	86.00	187.00	1	153.46

- The format problem is corrected [2]. The number 6.2, rather than 2.6, is specified, indicating FARMPCT and PCTURB will print using six columns with two decimal places.

- The label for TEMPRNGE is now valid [3].

- The option for PRINT is corrected [4]. N (not NOBS) requests that SAS display a count of the number of observations and print it at the bottom of the procedure output.

- Variable averages are now correctly requested [5] in PROC MEANS.

- The output listing's title is the more explanatory: "State-Level Data for Demo Program" [6]. Until altered, this will be printed on every page of PRINT's output as well as that of subsequent PROCs.

- The output of PRINT is different [7]. The ID option uses variable NAME as the leftmost column, the VAR statement requests specific variables from the dataset in a specific order, and the FORMAT statement controls the display of the variables. (Notice that POP80, for example, is now printed with commas due to the addition of the COMMA format in the DATA step.)
- The MEANS output is also different [8]. Variable labels are printed along with variable names, thus improving the listing's readability. The statistics differ from the defaults shown in the first example [9]: only those requested in the MEANS statement are printed. Also notice that the MAXDEC option controls the number of decimal places printed (two).

Exercises

2.4. Refer to the program in Exhibit 2.2. The INPUT statement names eight variables. The variables could have been named VAR1, VAR2, . . . , VAR8. What is the advantage of naming them as in Exhibit 2.2?

2.5. Refer to Exhibit 2.3.
  a. Highlight the added information produced by SAS in the Log. Try to interpret the meaning of the NOTEs following each unit of work.
  b. Examine the procedure output. What extra piece of information does SAS add to the dataset listing?
  c. Identify the following items: Colorado's population density (variable name POPSQMI) and DC's per capita income (PCAPINC). Contrast the population densities of DC and Alaska.
  d. Move down to the MEANS procedure output. Two general types of information can be identified—the variable name and statistics for the variable. Highlight the location in the output of each type of information.

2.6. Refer to Exhibit 2.4. Some SAS messages are helpful and others simply indicate that there is a mistake. Just by reading the ERROR messages, can you tell how the problems might be corrected?

2.7. Refer to Exhibit 2.5.
  a. Is the PRINT procedure's output easier to read now than it was in Exhibit 2.3? Why, or why not?
  b. Is the MEANS output more readable here than in the output in Exhibit 2.3? Why, or why not?

Summary

Chapter 2 introduces the two "units of work" common to all SAS programs. The DATA step is the portion of the program that reads data and performs arithmetic calculations and logical comparisons. The other unit of work, procedures or PROCs, is a high-level, compact series of statements that allow the user to rearrange the order of the SAS dataset as well as print, graph, and analyze it.

An extended example of program development is discussed. The two most common forms of program output, the SAS Log and the output listing, are examined as a simple program is run and then refined. Several errors were made during the development and elaboration of the program. The nature of these errors is described, how SAS displayed error messages is shown, and the correct, final program and its output are presented.

This chapter is intended as an overview, giving a feel for how SAS programs are organized and for the appearance of the program's output. Once the flexibility and relative ease of writing SAS programs are appreciated, readers will begin to grasp the tremendous computational power at their disposal.

3 Syntax Basics

If the concepts in Chapter 2 are the programming "workshop," this chapter looks at the workshop's "tools." To use SAS or any other language effectively, you must be familiar and comfortable with its syntax. You might have gleaned some of the rules and conventions from the examples in Chapter 2. Once you have used the SAS language for a while, you will appreciate its power and be able to accomplish a great deal of work with relatively little effort.

Before discussing syntax, a few of this book's biases should be mentioned. First, the presentation of programs reflects an idiosyncratic but reasonably mainstream programming style. It relies heavily on the liberal use of blank space to aid program readability and emphasizes the use of meaningful names for datasets and variables. Be aware that SAS, more so than many other languages, allows a great variety of program designs and philosophies. There are usually many "correct" solutions for a particular task. Only one, however, can be conveyed in a book like this. If you do not find the style comfortable, you can probably devise your own.

Another caveat: remember that the SAS language is rich, varied, and very large. A readable *and* concise description of all its capabilities is not likely to succeed. The syntax and capabilities of the language described here are a fairly small subset of the full power of the SAS System. It is, however, the most useful subset.

The material in this chapter is meant to be part tutorial and part reference. On first reading, it may be too detailed for the computer novice. Perhaps the best way to use the chapter is to skim it once just to get a feel for the syntax, then read it thoroughly when later chapters or your own programming tasks require more detail.

3.1 Statements

A SAS statement is the lowest-level, or smallest, unit of work recognized by SAS. It can be part of a DATA step, part of a PROC, or, less frequently, independent of these units of work.

A statement contains any valid SAS command, and it must end with a semicolon (;). Statements are commonly referred to by their function in the program, which is usually indicated by a keyword that begins the statement. Thus you can talk about using an IF statement, a DATA statement, and so on.

The semicolon indicates to SAS that a statement is ending. This delineation makes it possible to enter multiple statements on a single program line (although the practice is discouraged). It also means that if a statement cannot fit on a single line, it can be continued on one or more subsequent lines.

Within a statement, blank space and even blank lines can be inserted to improve legibility. Statements can be entered in uppercase, lowercase, or a mixture of the two. Although mixed case is appealing at first blush, it is generally better to avoid it since many program and text editors become confused during searches and changes when case is not consistent.

POINTERS AND PITFALLS

Statements should be aesthetically pleasing to the eye and syntactically correct to SAS's "eye." Blank space should be used to highlight program structure and logic.

The four groups of statements in Exhibit 3.1 are valid, equivalent, and stylistically different ways to accomplish the same task. The extra effort required to write the last example (alignment of names, indentation, and so on) makes the program easier to read, increases the chance of catching errors before they are made, and lends itself to easy modification.

EXHIBIT 3.1   Simple comparison of programming styles

```
DATA ONE;INFILE DATAHERE;INPUT NAME $ 1-10 GRADE $ 11 SALARY 12-20;

data one;infile datahere;input name $ 1-10 grade $ 11 salary 12-20;

data one;
infile datahere;
input name $ 1-10 grade $ 11 salary 12-20;

data one;
infile datahere;
input name     $  1 - 10
      grade    $ 11
      salary     12 - 20
      ;
```

Two Special Statements: RUN and ENDSAS

Two SAS statements, RUN and ENDSAS, warrant special attention. Entering RUN; at any point in the program tells SAS that a unit of work (DATA step or PROC) has ended. SAS will stop reading program statements and begin to execute the unit of work. The ENDSAS; statement executes the unit of work, then terminates the SAS session.

The RUN statement is required for some uses of procedures discussed in this book (PLOT, REG, GLM, ANOVA, and DATASETS). It is good programming practice, however, to enter RUN whenever a unit of work ends: it is a clear indication to SAS that the DATA step or PROC specification is complete.

POINTERS AND PITFALLS

Regardless of whether you are in batch or interactive mode, a RUN statement is always used to mark the end of a DATA step or PROC. Highlight use of the ENDSAS statement with a comment (see next section).

Exercises

3.1. Refer to the four DATA steps displayed in Exhibit 3.1. How many statements are there in each? Which do you think would be the easiest to write? to read? to alter?

3.2. How many different types of statements are there in the following program? (Don't worry about the *function* of the statements; just focus on the *type* of statement.)

```
data one;
infile datain;
input x y z ;
* Create TYPE, based on different values of all variables
read;
if x > 2 then type = 'h';
if y < 3 then type = 'l';
if z >= 4 then type = 'm';
;;
```

3.3. Count the number of statements in the sample program used in Exercise 3.2.

3.2 Comments

No SAS program *requires* comments. However, they are so important to the programming process that they deserve special attention. Comments, as their name suggests, are notes, within the SAS program, on general or particular aspects of the program's design. When writing long programs, using someone else's program, or returning to a project after an absence, the rationale for a program's design and quirks may be elusive. Comments satisfy the need for memory-jogging and guidance.

Syntax

Both embedded comments and comment statements are available. **Embedded comments** take the following form:

```
/* text of comment */
```

They can occur at the beginning, the end, or within a single statement. Any text between the /* and */ is ignored by SAS, effectively treated as a blank. **Comment statements** take the following form:

```
* text of comment ;
```

Like other SAS statements, they start with a keyword (*) and end with a semi-colon (;). Programs may use both forms of comments. The only restriction on their use is that in IBM MVS and DEC VAX systems embedded comments should not begin in column 1 of a program line.

Uses

Comments provide a record of what went on in a program and how it took place. The "what went on" portion is often handled in a longer comment at the beginning of the program. It indicates authorship, development dates, purpose, any prior or subsequent programs to be run, datasets used and created, and other general features. Style and content for these comments vary greatly. A sample "header" comment is shown in Exhibit 3.2. Notice that it is a *single* comment: it begins with an asterisk and ends with a semicolon.

EXHIBIT 3.2   Sample "header" comment

```
*----------------------------------------------------------------
|
| \mort\rate3.sas
|
| First written July 12, 1988
|
| Prior program(s): \mort\setupdb.sas
| Subsequent programs(s): \mort\graph.sas <optional>
|
| Read mortality data, construct cohort-specific rates (broken
| down by age, sex, race).
|
| Modification history
|
| Person    Date    Changes
| ACD       8/3/88  Add race cohort, adjust first DATA step and
|                   sort
| JMP       10/1/88 Performance tuning: eliminate unnecessary
|                   variables, rewrite recoding using user-
|                   written formats.
|
*-------------------------------------------------------------;
```

Other common locations for comments are before and within PROCs and DATA steps. Just as the comment described the function of the entire program, so can it document the purpose of a PROC or DATA step. This is illustrated in Exhibit 3.3.

Comments *within* DATA steps and PROCs are used to explain how something was done. These are usually a mixture of embedded comments and comment statements. Their use is demonstrated in Exhibit 3.4. In this example the

EXHIBIT 3.3   Comments used to describe DATA step and procedure

```
*--- Read trade data, create surplus/deficit indicator. Keep only
     those observations from underdeveloped nations. ;
data underdev;
... rest of DATA step follows ...

*--- Frequencies only for vars created in the data step. ;
proc freq;
... rest of PROC follows ...
```

EXHIBIT 3.4   Comments used to describe DATA step logic

```
data underdev;
set  wrldbank;
if exp > imp              /* Current quarter surplus */
   or expavg4 > impavg4   /* 4-qtr average surplus */
   then surplus = 1;
   else surplus = 0;

* Underdeveloped criteria determined by committee meeting,
  July 3, 1988. ;
if pcapinc < 1000 or brthrate > 10 then output;
run;
```

embedded comment explains how a variable was calculated. The comment statement explains the rationale for selecting observations to be output. Notice that if the embedded statement is rewritten as a comment *statement* (as in Exhibit 3.5), SAS would detect bad syntax. The * indicating the beginning of the ill-fated comment is interpreted as part of the IF statement, and the semicolon at the end of the comment ends the IF statement. The next statement type (OR . . .) is unknown to SAS and causes an error.

EXHIBIT 3.5   Inappropriate use of comment statement

```
if exp > imp              * Current quarter surplus ;
   or expavg4 > impavg4   * 4-qtr average surplus ;
   then surplus = 1;
   else surplus = 0;
```

A final use of both forms of comments is to "turn off" groups of statements. Suppose you write a program that has several PROCs. You are satisfied with output from some of these but still have to fine-tune others. You do not want to rerun the finished PROCs, nor do you want to delete them from the program. A solution is to use an embedded comment as in Exhibit 3.6. The FREQ and MEANS PROCs have all been embedded within two comments. They are said to be "turned off," or "commented out." This is a useful, if not tidy, way to avoid having the computer do extra work.

EXHIBIT 3.6 "Commenting out" units of work

```
    /****************************************
proc freq;
... rest of PROC ...

proc means;
... rest of PROC ...
    *****************************************/

proc reg;
... rest of PROC ...
```

POINTERS AND PITFALLS

Use comments to document your work. Do not overcomment: too many comments cause visual clutter and become counterproductive. Be sure that a change to the program is accompanied by parallel changes in comments.

Exercises

3.4. Examine the following program:

```
* Sample program showing effective use of comments. ;
* Both types of comment formats are used. ;
data one;
infile 'uba.test.data';   /* Sample dataset, NOT the
                            production file */
input id  1 - 8    /* company CUSIP # */
   hi  9 - 12  /* seasonal high value */
   low 13 - 16 /* seasonal low value */
   avg         /* average closing value, fy '90 */
   ;

proc print data=one;
title 'Listing of test data';
run;
```

a. How many comments are in this program? Distinguish them by type: How many are embedded comments? How many are comment statements?

b. Give the program some context and write an appropriate header, or beginning, comment to describe its function. For example, the program might be written as part of homework 3 of Finance 240 and may use the EEC sample data.

c. "Turn off" the PRINT procedure and its associated TITLE statement. Do it twice, once using comment statements, once using embedded comments.

3.3 Variables

A variable is a name assigned to an item of data that occurs in all rows, or observations, of the SAS dataset. Variables have already been used in the programs and program segments in this book. For example, the dataset introduced in Chapter 2 used items named STATES, HIGHTEMP, LOWTEMP, and POP80.

Their necessity and usefulness is easy to grasp: before pieces of raw data may be used in the program, they must be identified to SAS. SAS places a few restrictions on their names:

- Their length cannot exceed eight characters.
- The name must start with an alphabetic (a through z) character or an underscore ("_"; *not* a minus sign, "-").
- The name can contain only alphabetic characters (a through z), underscores, and numerals (0 through 9).

Examples of legal and invalid variable names are shown in Exhibit 3.7.

EXHIBIT 3.7 Legal and invalid variable names

```
Legal Names

squaremi
sales88
_88
_zipcode

_


Invalid Names

squaremil      Too many characters (max is eight)
88sales        Begins with a number (only a-z, _)
zip code       Blank is illegal character (only a-z, _, 0-9)
```

One of the most effective ways to write clear, readable programs is to select sensible variable names. Names like "a," "widget," and "var1" are quick to code, acceptable to SAS, and sometimes even humorous. They do not, however, convey any information. If variables are named something meaningful, like SAT0688, SAT1188, and SAT0389, their meaning will be grasped (and later recalled!) more readily. This so-called mnemonic naming style is a cornerstone of good programming practice.

There is a middle ground in this variable-naming continuum. Try to think of an eight-character name for the survey question, "Management's attitude toward subordinates' resistance to introduction of fax machines." You probably can't. Even if you can, it is likely that you will not be able to recall its meaning in later programs. Situations like this suggest compromise: if the item comes from a questionnaire, it makes better sense to name the variable so that it corresponds to the question. Q23A might, in this case, convey more information than a tortured mnemonic. The LABEL statement, discussed in Chapter 5, provides a partial solution to this naming dilemma.

POINTERS AND PITFALLS

Use names that suggest the variable's content. If its meaning cannot be adequately summarized in SAS's eight-character limit, try to cross-reference the variable: question numbers on a survey are an example of this method.

Variable Lists

Often in DATA steps or PROCs you will want to refer to a group of variables at a time. You might want to print them, perform calculations on them, drop them from a dataset, or graphically display them. SAS provides a simple and compact means to refer to such variable lists. Before these lists can be used, however, the concept of variable order in a SAS dataset must be understood.

Whenever a SAS dataset is created or additional variables are added to an existing SAS dataset, SAS creates variables in a particular order. The order is determined by how the variables were first made known to SAS. Consider the following program segment:

```
input x1 1    x2 2-3    x3 4-7    name $ 9-14    type $ 15
      y 17    z 18      x4 19-22 ;
sumx1_3 = x1 + x2 + x3 ;
x4 = x4 / 2;
```

The variable order here is X1, X2, X3, NAME, TYPE, Y, Z, X4, and SUMX1_3. When SAS read these statements it encountered X1, then X2, and so on through SUMX1_3. SAS is not concerned with the *type* of statement (INPUT, assignment, and so on) where it found the variables. The important characteristic is the *order* in which the variables are found.

With this in mind, let's go back to variable lists. SAS provides a family of shorthand methods to refer to a dataset's variables. These are summarized in Exhibit 3.8.

EXHIBIT 3.8   Variable list styles and meanings

---

(1) VARn - VARx	VARn through VARx, where n and x are integers and x is greater than n. Only selects those variables starting with VAR. All of VARn, VARn+1, and so on through VARx must exist.
(2) VARn -- VARx	All variables between VARn and VARx, even those not starting with VAR
(3) VARa -- VARb	All variables between VARa and VARb
(4) VARa - numeric - VARb	All numeric variables between VARa and VARb
(5) VARa - character - VARb	All character variables between VARa and VARb
(6) _character_	All character variables
(7) _numeric_	All numeric variables

A common feature of these lists is that if they identify a variable that is not in the dataset, SAS will generate a warning or, worse, an error. For example, if a list specified all character variables between X1 and X5 and there were none in that range, SAS would print a warning in the Log, saying "no variables found in specified range."

Using the program segment from Exhibit 3.8 as a starting point, let's look at how we can write each of the variable lists outlined in Exhibit 3.9 and elaborate a bit on each.

List Style 1: VARn--VARx.   This form assumes a common prefix of VARn and VARx. It also assumes that *each* variable implied by the list actually exists: if the list is SCORE1-SCORE10 and variable SCORE7 was never created, SAS will stop the program with an error. Examples are presented in Exhibit 3.9.

EXHIBIT 3.9   Variable list style VARn--VARx

Variable list	Refers to variables
x1 - x3	x1, x2, x3
x1 - x4	x1, x2, x3, x4
x4 - x1	*** error *** starting suffix must be less than ending suffix. 4 is not greater than 1.
x01 - x3	*** error *** Even though x01 looks like x1, it is, in fact, a different variable, and so produces an error. If x01 did exist, the list would be legal.

List Style 2: VARn--VARx.   List style 2 is similar to list style 1 in that the beginning and ending variables in the list must be in the dataset. The difference is that the "--" indicates that *all* variables found between VARn and VARx should be included in the list. Notice that since this form does not place any requirements on intermediate variables, the absence of SCORE7 in list style 1's discussion would *not* cause an error. Examples are presented in Exhibit 3.10.

EXHIBIT 3.10   Variable list VARn--VARx

Variable list	Refers to variable
x1 -- x3	x1, x2, x3
x1 -- x4	x1, x2, x3, name, type, y, z, x4
x3 -- x4	x3, name, type, y, z, x4

List Style 3: VARa--VARb.   List style 3 is the more general case of list style 2 in that it does not require a common prefix (i.e., the VAR in VARn and VARx). The only requirement is that VARa be defined in the dataset prior to VARb. Examples are presented in Exhibit 3.11.

EXHIBIT 3.11   Variable list VARa--VARb

Variable list	Refers to variables
x1 -- x4	x1, x2, x3, name, type, y, z, x4, sum1_3
x3 -- sum1_3	x3, name, type, y, z, x4, sum1_3
z -- name	*** error *** variable "z" was defined after "name"

List Style 4: VARa-NUMERIC-VARb.   List style 4 is a constrained form of list style 3. The list references only the *numeric* variables between VARa and VARb (numeric variables are discussed in Section 3.4). Neither VARa nor VARb has to be numeric. Examples are presented in Exhibit 3.12.

EXHIBIT 3.12   Variable list VARa – numeric – VARb

Variable list	Refers to variables
x1- numeric - sum1_3	x1, x2, x3, y, z, x4, sum1_3
name - numeric - x4	y, z, x4
name - numeric - type	*** Warning! *** there are no numeric variables between name and type

List Style 5: VARa-CHARACTER-VARb.   List style 5 is another constrained form of list style 3. Only the *character* variables between VARa and VARb are referenced by the list (character variables are discussed in Section 3.4). Neither VARa nor VARb has to be character. Examples are presented in Exhibit 3.13.

EXHIBIT 3.13   Variable list VARa – character – VARb

Variable list	Refers to variables
x3- character - y	name, type
name - character - x4	name, type
name - character - type	name, type
y - character - type	*** Warning! *** there are no character variables between y and type

List Style 6: _CHARACTER_.   One of the simpler lists to write, _CHARACTER_, refers to *all* character variables in the dataset. No beginning and ending variables are needed. "_CHAR_" is not a valid abbreviation.

List Style 7: _NUMERIC_.   Another quickly written list, _NUMERIC_ refers to *all* numeric variables in the dataset. No beginning and ending variables are needed. "_NUM_" is not a valid abbreviation.

POINTERS AND PITFALLS

Variable lists are an extremely useful and compact method for referring to a dataset's variables. Their effective use relies on your knowledge of variable *order*. Be certain of the dataset's variable order. Use the CONTENTS procedure (Chapter 10) if you are not sure.

Variable lists are one of the nice touches that make SAS capable of doing a great deal of work with relatively little coding effort. By the same token, when variable lists are misused, SAS produces obtuse and rather puzzling error messages. Another form of list abuse arises when the list refers to more variables than were anticipated and a procedure produces voluminous amounts of output. Being confident of variable order is important. A reliable method for determining this order, PROC CONTENTS, is presented in Chapter 10.

Exercises

3.5. Identify legal and illegal variable names. If illegal, briefly state which rule the name violates.

```
-
---------
--------
_1st
area code
_Count
-count
prev_pregs
measure1
1on1
```

3.6. Assign variable names to the following items:
   a. APGAR score measured at 1 minute old
   b. Gross sales of MegaHealth vitamins, fiscal year 1990
   c. Question 2 on a survey instrument: respondent's child's age when first contracted measles
   d. Question 15 on a survey instrument: respondent's age

3.7. Refer to the dataset descriptions in Appendix E. Select a dataset, review its variables' meaning (the "Label" column), and assign variable names based on this information. (These names should differ from those already used!)

3.8. Suppose the following statements appear in a SAS program. (*Note:* The $ following a variable name indicates a character variable. All other variables are numeric.)

```
data test2;
input x1 1 x2 2 x3 3 x4 $ 5-12 x5 $ 13-14 x6 $ 16;
avg = (x1 + x2 + x3) / 3;
```

   a. How many variables are identified in these statements?
   b. What is the order of the variables?
   c. Which variables are identified by the following variable lists? If the list is not valid, state why.

   (1) x1-x6
   (2) x1-numeric-x6
   (3) x1-character-x6
   (4) _character_
   (5) _numeric_
   (6) _numeric
   (7) x1-numeric-avg
   (8) x4-numeric-x6
   (9) start--end
   (10) x01--x06

3.4 Data Types

All data processed by SAS DATA steps and PROCs are one of two data types: numeric or character.

Numeric

Numeric data are amounts or concepts that can be represented by numbers and that will be manipulated in an arithmetic expression. Dates, industry rankings, salaries, state codes, yes/no indicators, and class averages are examples of numeric data. SAS does not have to be told whether a number is an integer (a whole number) or real (a number with a fractional component). All numbers are handled with the assumption that there may be a fraction. Chapter 14 presents a variation on numeric types useful for handling variables containing date-oriented data.

Character

Character data are less restricted in scope. *Anything* that can be put into the raw datasets described in Chapter 2 is valid character data. Names, addresses, quotations, and numbers are examples of character data. A character variable can be from 1 to 200 columns long. Any name may be used for a SAS character variable: you cannot tell from a variable's name whether it is character or numeric.

Character data are the only instance in the SAS System where text is sensitive to upper- and lowercase and blanks. The following list displays eight legitimate ways to represent a person's name:

```
George Bailey
George  Bailey
 George Bailey
GEORGE BAILEY
george Bailey
 George  Bailey
GEORGE  BAILEY
george bailey
```

Even though all of Mr. Bailey's names look the same to the human eye, SAS's "eye" is more exact. Each name is unique (this may explain why Mr. Bailey gets duplicate pieces of mass mailings). Reading and handling character data can be tricky. Chapter 18 is devoted to the topic.

Choosing Data Types

The choice of data type is usually straightforward. If you will be running correlations between salary and years employed, your only choice would be to store the data as numbers. Likewise, character data is the only choice if you are putting together a mailing list. Further, some PROCs will accept only a particular type of data, so the choice of data type is dictated.

There are gray areas, however. Consider the following situation. A questionnaire item is scored on a 1 to 5 scale. It will be used in frequency tables and charts and in selecting subpopulations of the dataset. The first instinct is to call it numeric, since it "looks" like a number. Now ask whether any of the variable's anticipated uses involves any form of arithmetic. For example, will it be added or multiplied? If the answer is no, the variable should be stored as character.

Another way to make the choice is to ask whether the data could be coded as "a,b,c,d,e" rather than "1,2,3,4,5" and not lose their information value. If there would not be any loss, choose character data.

POINTERS AND PITFALLS

Do not assume that a data item that "looks" numeric should be read as a numeric variable. Most items that will not be used in arithmetic operations can be safely read as character data.

Why not just call a number a number? There are two reasons. First, a conscious and reasoned selection of data type encourages planning of analyses and manipulations. Second, SAS uses fewer machine resources to handle character data. For small programs such an efficiency consideration yields negligible performance improvements. As program and dataset sizes grow, however, such data selection issues become more important.

Exercises

3.9. A dataset will have the following variables. Choose a data type for each. Justify your choice.
   a. Name of the state (e.g., North Carolina)
   b. State postal code (NC)
   c. State FIPS code (37)
   d. Region of the country (1 = North, 2 = South, . . .)
   e. Population, 1980
   f. Population, 1990
   g. Percent population change, 1980–1990
   h. State motto ("First in Flight")

3.5   Missing Data

Sooner or later most researchers encounter incomplete, or missing, data. A respondent did not answer all the questions on your questionnaire, a firm was too small to have payroll figures made public so the Bureau of Labor Statistics withheld information, and so on.

The absence of information is not the only cause of missing data values. The cause may also be structural. For instance, in states with same-party-only primaries, crossover votes would not exist: the count of Democrats who voted for Republicans would be missing. Likewise, an analysis of hospital admissions data would not produce a meaningful count of men who were admitted with a reason code of "delivery."

Dataset design may also be the source of missing values. You may want to allow a maximum of five test scores in your dataset, and so create variables SCORE1 through SCORE5. This means that anyone taking fewer than five tests will have one or more missing values. In most business and research applications a few missing data values are tolerable. You should be aware of how SAS handles such data.

Recall the depiction of a SAS dataset as a matrix of observations and variables, or rows and columns. If the dataset has no missing values, its appearance is that of a uniformly filled-in series of numbers and characters. A dataset with missing values, on the other hand, appears moth-eaten: "holes" are present instead of legitimate values. The important concept is that a missing value is caused by the *absence* of data for a particular variable in a given observation. A zero (0) for a variable value is *not* missing but simply a small, valid, "present" value. For example, a person could be paid up on a credit card and have a balance of $0. If another person did not have a credit card, the balance variable could not be measured and would thus be missing.

POINTERS AND PITFALLS

Missing data indicate the absence of data for an observation's value of a variable. A value of zero (0) is not a missing value.

Reading Missing Values

Reading missing values is usually not problematic. As described in Chapter 4, one style of reading raw data is telling SAS to go to a column of the data, read one or more columns, and store the result in a numeric or character variable. If SAS is told to read a numeric variable and encounters blanks or a period (.) in the columns, it assumes the variable is missing. Think of the missing value "." as a number with nothing either to the right or left of the decimal point (i.e., "missing values" on both sides). If SAS reads blanks for a character variable, it assumes a missing value. Notice that since any character is fair game for character data, a period (.) would be read as a legitimate, nonmissing character value.

Calculations with Missing Values

SAS follows a simple, but important, rule when it finds a missing value in a calculation: the "target" variable is also set to missing. Consider the program segment in Exhibit 3.14. There are "holes" in the data: only the first line of data has entries for all three variables. When the program attempts to add TERM, CD, and PASSBOOK to create TOT_SAVE, SAS writes a Log similar to Exhibit 3.15. SAS tried to add missing values in line 3. Each time it attempts to use a variable with a missing value, SAS notes the statement's line number and increases its count of "number of missing." Only the first line of data in Exhibit 3.14 produces a nonmissing value of TOT_SAVE.

At the end of the DATA step, SAS prints a NOTE indicating where it created missing values. This may be simply an informative message, not an error! It

---

EXHIBIT 3.14   Program using data with missing values

```
data accounts;
input term 1-5 cd 6-10 passbook 11-15;
tot_save = term + cd + passbook;
cards;
1000 12000   350
 500         1000
             500
1400  5000
run;
```

EXHIBIT 3.15   Log showing NOTE about missing values

```
------------------- SAS Log ----------------------

  1     data accounts;
  2     input term 1-5 cd 6-10 passbook 11-15;
  3     tot_save = term + cd + passbook;
  4     cards;
```
```
NOTE: Missing values were generated as a result of performing an operation
      on missing values.
      Each place is given by: (Number of times) at (Line):(Column).
      4 at 3:12
```
```
NOTE: The data set WORK.ACCOUNTS has 4 observations and 4 variables.
NOTE: The DATA statement used 9.00 seconds.
```

---

can be very useful when diagnosing longer programs and often explains peculiar PROC output.

POINTERS AND PITFALLS

Carefully examine Log messages about the number and location of missing values. A missing value count equal to the number of observations in the dataset usually suggests that you read the raw data incorrectly.

Just as you can create a missing value inadvertently, you can create one deliberately. At the beginning of this section, an example of payroll figures being withheld by the Bureau of Labor Statistics was given as a reason for missing data. Consider what would happen if these missing data were coded as −9. Obviously a payroll cannot be $−9: you want this value to become missing. The conversion from a value that in fact is missing to a SAS missing value is covered in Chapter 7.

SAS's Assumptions about Missing Data

SAS makes some assumptions about missing data. Most PROCs assume that observations with one or more missing values in the analysis variables should be excluded from the analysis. A message indicating the number of excluded observations is usually displayed in the procedure output. If you decide that the missing value is a valid value of a variable, some PROCs will let you take special options forcing them to remain in the analysis.

The DATA step also makes an important assumption about the data. Each time execution loops to the top of the DATA step to read from the raw data, SAS sets most variables to missing (exceptions are discussed in Chapter 19). It assumes you want to start with a "clean slate" for each observation. Most variables are initially empty, or missing. It is up to your calculations and data reading statements to assign values to the variables. Throughout the book you will see how this and other default assumptions can be overridden.

Exercises

3.10. How many missing values are in the following list of variable values (commas separate each number):

```
1.0, ., , .1, ., ., 9, 0.1, -9, .
```

3.11. Refer to the DATA step building dataset ACCOUNTS. If the missing values were replaced by zeroes, how would the output in Exhibit 3.14 be affected? Why would you convert the .'s to 0's? Why would you leave the .'s as is?

3.12. A survey instrument has valid responses of "Don't know" (coded as −9) and "Refused to answer" (coded as −8). Is this a situation where you would want to enter the numbers as missing? Why, or why not?

3.6 Constants

Many applications require a calculation in which one or more terms have fixed, or constant, values. The value of the constant does not change from observation to observation as the SAS dataset is built. An example of this was shown in

Exhibit 2.2, where a constant value of 100 was used as a divisor in a calculation. Constants assist calculation of variables and are not stored in SAS datasets. Just as there are two SAS variable data types, so are there two types of constants: numeric and character.

Numeric Constants

As the percent urban example in Exhibit 2.1 showed, a numeric constant is all or part of an expression written like a number would be written. The only restriction on its entry is that numbers cannot be written with commas, dollar signs, or blanks. Scientific notation is acceptable, and decimal points for integers are not required. Some valid numeric constants are shown in Exhibit 3.16.

EXHIBIT 3.16　Valid numeric constants

```
1000
1000.
1000.0
1e3
10e2
2.3
-2
-2.7
3e-1
```

Notice that the first four constants are equivalent. The method of entry is a matter of convenience and preference. Numeric constants are restricted in size to about plus or minus 72E15, the same as numeric variables.

Character Constants

Character constants can contain any combination of alphabetic, numeric, and other characters. They must be enclosed by a pair of single quotes (') or quotation marks ("). A single quote in a constant enclosed by single quotes is represented by entering a *pair* of single quotes(' '; not the quotation mark, "). Like character variables, their length ranges from 1 to 200 characters. Also like their variable counterparts, they are case and spacing sensitive. Exhibit 3.17 illustrates valid character constants and highlights the blank and case-sensitive nature of character data.

EXHIBIT 3.17　Valid character constants

```
'High res'
'High   res'
"High   res"
'True'
'TRUE'
'Isn''t present'
"Isn't present"
'1'
'---> Moderate <---'
' '
''
" "
```

Exercises

3.13. Which pairs of constants are equal in value?
   a. 2 2.0
   b. −1.70 −1.7
   c. '−8' −8
   d. "eieio" 'eieio'
   e. 'tricky' 'tricky'

3.14. Which constants are not valid?
   a. 2,000
   b. 'Region 1'
   c. '2,000'
   d. 'Region 1"
   e. ""
   f. −"2,000"
   g. " ' "

3.7   Expressions

At this point we can choose data types and distinguish variables from constants. With these concepts we can create new variables and make decisions about DATA step execution based on variable and constant values. Examples of both activities were illustrated in the sample program in Section 2.2. Both activities required the use of arithmetic and logical expressions, which are discussed in this section.

The importance of expressions will become clear in the rest of the book. Nevertheless, it is helpful to see some examples right away. The sample uses below are followed by descriptions of the instructions SAS carried out:

```
totscore = verbal + math ;
```

Add variables VERBAL and MATH, and store the result in the variable TOTSCORE.

```
if income > expenses then surplus = gross - outlay ;
```

Compare the variables INCOME and EXPENSES. If the value in INCOME is greater than the value in EXPENSES, then subtract the value of OUTLAY from the value of GROSS, and store the result in the variable SURPLUS.

```
repay = prin + (prin * .12 * yrs) ;
```

Multiply the variable YRS by the constant .12, then multiply that product by the variable PRIN. Add this result to the variable PRIN. Store the result in the variable REPAY (this is the formula for total loan repayment at 12% with principal PRIN and term YRS).

```
if sex = 'M' & grade > 8 then eligible = 1;
```

If the value of SEX is 'M' and the value of variable GRADE is greater than 8, then store a 1 in variable ELIGIBLE.

```
if mpg * capacity > 400 then range = 'long';
```

Multiply the values of variables MPG and CAPACITY. If the result exceeds 400, then assign a value of 'long' to the variable RANGE.

The variety and power of SAS expressions cannot be overemphasized. Their syntax is simple when the needs are straightforward, and very complex when calculations are difficult or lengthy. The pattern of even the most complex expressions, however, boils down to a succession of "variable or constant"--"operator"--"variable or constant."

Expression Syntax

Both arithmetic (numeric manipulation) and logical (true/false evaluation) operators appear in this general form:

```
{ variable   }               { variable   } ]
{ constant   } [ operator  { constant   } ]
{ expression }               { expression } ]
```

Expressions can be part of other expressions. Operators join subordinate expressions and enable comparison and manipulation of values. Operators come in three varieties: arithmetic, comparison, and logical. The examples used below emphasize numeric operations; using character data is discussed in Chapter 18.

Arithmetic Operators.   These are used in situations that require an arithmetic operation on any mixture of variables and constants. Some of SAS's arithmetic operators and their uses are illustrated in Exhibit 3.18.

EXHIBIT 3.18   Arithmetic operators

Operator	Arithmetic operation	Examples
**	Exponentiation	x**.5
		x**y
		x**-2
*	Multiplication	x * 8
		z * a
/	Division	rate / 1.06
		gallons / miles
+	Addition	x + y
-	Subtraction	x - y

If the operator is asked to work with a missing value, the result of the operation will be missing (recall the discussion of missing values in Section 3.5). If, however, SAS attempts to perform a mathematically undefined operation, such as division by zero, it prints a warning in the Log. Exhibit 3.19 is typical of a program with such potential problems.

SAS indicates where the zero-divide was detected, displays the offending line of raw data, dumps the value of each variable at the time of the error, and tells at the bottom of the Log how many times this error occurred. Notice that the SAS dataset was still created.

Comparison Operators.   While arithmetic operators *create* values, comparison operators, as their name suggests, *compare* them. They are useful for taking actions based on the relationship between two variables or constants of a similar data type. The comparison could tell SAS how, or whether, to create a variable (refer to the example program and its use of the IF, THEN, and ELSE statements in Section 2.2). This decision making is covered in detail in Chapter 8. Exhibit 3.20 lists comparison operators and their meaning.

EXHIBIT 3.19   Log indicating calculation problems

```
-------------------- SAS Log --------------------------

    1      data one;
    2      input radius 1-2 scale 4;
    3      circumf = 3.14159 * radius**2;
    4      adjusted = circumf / scale ;
    5      cards;
```
┌──┐
│ ERROR: Division by zero detected at line 4 column 12.│ ◄────────── Zero-divide error
└──┘   message

┌──┐
│ RULE:----+----1----+----2----+----3----+----4----+----5----+----6----+----7---│ ◄──────┐
│   8   5 0   │
└──┘

┌──┐
│ RADIUS=5 SCALE=0 CIRCUMF=78.53975 ADJUSTED=. _ERROR_=1 _N_=3   │
└──┘
```
NOTE: Missing values were generated as a result of performing an operation on     Raw data line that
      missing values.                                                             created the error
      Each place is given by: (Number of times) at (Line):(Column).
      3 at 4:12
      4 at 3:11
```
┌──┐
│ NOTE: Mathematical operations could not be performed at the following places.│
│   The results of the operations have been set to missing values. │
│   Each place is given by: (Number of times) at (Line):(Column). │
│   1 at 4:12   │
└──┘
```
NOTE: The data set WORK.ONE has 5 observations and 4 variables.

Summary of mathematical errors in the DATA step

---

**EXHIBIT 3.20   Comparison operators**

| Operator | Meaning |
|----------|---------|
| = | **Equal**. The values of the variables/constants are equal. |
| ^= | **Unequal**. The values of the variables/constants are not identical. Note: Finding the "not" key (the "^") can be a challenge on some terminals. On many keyboards it is a Shift-6 ("uppercase" 6). Verify the location with someone who has a similar keyboard. |
| > | **Greater than**. The value of the variable/constant on the left of the > is greater than the value on the right. |
| >= | **Greater than or equal**. The value of the variable/constant on the left of the >= is greater than *or equal to* the value on the right. |
| < | **Less than**. The value of the variable/constant on the left of the > is less than the value on the right. |
| <= | **Less than or equal**. The value of the variable/constant on the left of the >= is less than *or equal to* the value on the right. |

How does SAS perform these comparisons? It simply takes the values (either variables or constants) on either side of the operator and evaluates whether the condition is met. If, for example, the condition is $A > 4.5$, SAS would examine the contents of variable $A$ and compare it to the constant 4.5. If $A$ is greater than
```

4.5, the condition is met, or "true." If *A* is less than or equal to 4.5, the condition is not met, or "false." Internally, SAS uses the values 1 and 0 to represent "true" and "false," respectively. Some uses of comparison operators are presented in Exhibit 3.21.

EXHIBIT 3.21   Use of comparison operators

Assume that an observation in a SAS dataset has the following values:

| state = 'PA' | coast = . | ocean = 'AT' | tidal = 89 |

Expression	Evaluation
ocean = 'AT'	True. The value of variable OCEAN is equal to the constant 'AT'.
ocean = 'at'	False. Variable OCEAN is stored in the SAS dataset as uppercase. The constant 'at' is in lowercase.
state ^= 'PA'	False. The value of variable STATE is equal to the constant 'PA'.
state ^= 'MA'	True. The value of variable STATE is not equal to the constant 'MA'.
coast > 100	False. Variable COAST is missing (.), so will be treated as smaller than any numeric variable or constant.
coast = .	True. The value of COAST is missing.
tidal > 100	False. The value of variable TIDAL is less than 100.
coast < tidal	True. COAST's missing value is, by definition, smaller than any nonmissing value.

Three final notes about using comparison operators are important. First, although it is possible in some cases to compare different data types (i.e., compare numeric and character variables), it is not good programming practice. These comparisons often do not work properly and are usually the result of a misunderstanding on the user's part about the data types of the variables.

Second, numeric missing values "compare low." Numeric missing values are always smaller than any nonmissing value. Character missing values *usually* compare low. This equivocation is required since it is possible, though not easy, to read data that are comparatively smaller than a blank. Fortunately, this is the sort of data usually handled only by professional programmers.

Third, comparing numeric values for equality sometimes produces unexpected results. Due to the way computers perform arithmetic and store numbers, data that *appear* to be integers may actually be treated as slightly larger or smaller. Thus a test for equality of two apparent integers may produce an unexpected "false."

Logical Operators.   The comparisons in Exhibit 3.21 were useful, simplistic, and not representative of the kinds of decision making "real life" programs have to make. Actions are often based on two or more comparisons. These situations require using **compound expressions**. SAS provides "logical operators" (sometimes referred to as Boolean operators) to combine comparisons. Recall the true/false nature of the comparisons just described while examining Exhibit 3.22.

EXHIBIT 3.22   Logical (Boolean) operators

Operator	Meaning
&	*Both* comparisons joined by the operator must be true. Can also be written as AND.
|	*Either* expression joined by the operator must be true. Some keyboards have two keys that look like this symbol. You should select the unbroken vertical line in this case. Can also be written as OR.
^	The expression is *not* true. This operator is usually used as a prefix for a simple or compound expression. It can also be written as NOT.
IN	The variable on the left of the operator is IN a list of parenthetical values to the right of the operator. Examples: REGION IN ('NE','W','S'), GRADE IN (1,2,4,8).

Logical operators are usually used in the same decision-making situations that comparison operators are used. Their objective is to help SAS boil down complex expressions into a simple true/false evaluation. SAS converts the true/false value to an internal numeric value of 1 or 0, respectively. This feature can be useful for making arithmetic expressions more compact. Exhibit 3.23 illustrates the use of logical operators.

EXHIBIT 3.23   Use of logical operators

Assume that an observation in a SAS dataset has the following values:

state = 'PA'   coast = .   ocean = 'AT'   tidal = 89

Expression	Evaluation
ocean = 'AT' & state = 'PA'	True. Both expressions surrounding the "&" are true.
ocean = 'AT' | state = 'PA'	True. For the OR condition to be true, at least one expression must be true. In this case, both conditions are true.
ocean = 'AT' | state = 'NY'	True. At least one of the conditions was met. STATE was *not* 'NY', but OCEAN *was* 'AT'.
tidal > 100 | coast > 250	False. Neither condition was true.

Complex Expressions

The expressions in Exhibit 3.23 were an improvement over simple comparisons: We were able to define conditions based on the evaluation of two or more comparisons. If we want even more flexibility, then we should be able to join together an arbitrary number of comparisons. That is, instead of the one or two comparisons used up to this point, we should be allowed to join three, four, or however many are needed to express our decision criteria. The same freedom should be allowed for constructing arithmetic expressions. We should be able to handle many numbers.

The problem inherent in joining comparisons and performing long arithmetic calculations is one of the order in which the operations are carried out. Consider the following expressions:

```
a**2 / b + c**2 * d - 2

a > b & c > d | e <= f
```

When SAS evaluates these expressions, does it ask what order to do the operations? raise *a* to the power of 2? then subtract 2 from *d*? or then add *b* and *c*, then square the result? compare *a* and *b*? or first compare *d* and *e*? Because blank space does not have any effect on how SAS reads its statements, no clues about order can be gleaned by looking at just the two lines. Some rules are needed.

SAS provides simple mechanisms to build a logical condition or arithmetic expression of any complexity. The first form is to give priority to certain operators [arithmetic: **, *, /, +, - comparison: NOT(^), AND(&), OR(¦)], with similar operators evaluated left to right in the expression. This "operator hierarchy" is simple to grasp conceptually. However, it is rare to find an experienced programmer, much less a novice, who commits the hierarchy to memory.

The second way to be sure the expression will be evaluated as intended is simpler and has visual appeal. Simply enclose the variables and their operators in parentheses. This ensures that the expression in the parentheses will be evaluated as you intended. Expressions within expressions can also be enclosed in parentheses (so-called nested expressions). The most deeply nested expressions are evaluated first. Redundant parentheses cause visual clutter and have no effect on a statement's evaluation. When part of the expression falls outside the parentheses, the hierarchy of operators mentioned in the preceding paragraph is used.

POINTERS AND PITFALLS

Use parentheses to emphasize logic and override the default order of expression evaluation.

The only restriction on the use of parentheses is that they stay "balanced": each left parenthesis should have an accompanying right parenthesis. Many examples of this explicit ordering of evaluation and execution appear throughout this book. For now, Exhibit 3.24 demonstrates how the two expressions above can be reinterpreted using parentheses. Notice how the use of parentheses clarifies the order of operations. This order can greatly affect the results of the calculation or comparison!

Expression Shorthand. SAS offers a compact notation for some forms of comparisons. Suppose you wanted to write an expression that would test for the variable AGE being in the range 21 through 45. You would test for AGE greater than 21 *and* AGE less than 45, or you could write it more compactly by placing the variable name between the operators and their comparisons (the AND is implied). The compact form of this and other expressions is illustrated in Exhibit 3.25.

EXHIBIT 3.24   Impact of parentheses on expression evaluation

---

The numbers beneath the operators indicate the order in which the expression will be evaluated. Assume that the variables for an observation have the following values:

a = 3   b = -2   c = 4   d = 0   e = 1   f = 2

Expression	Result
a**2 / b + c**2 * d - 2 1 3 5 2 4 6	-6.5
a**(2/b) + c**(2*d) - 2 3 1 5 4 2 6	-.66667
a**(2/(b+c**2)) * (d-2) 5 3 2 1 6 4	-2.3398
(a**2 / (b+c))**2 * (d-2) 2 3 1 5 6 4	-40.5
a > b & c > d	e <= f 1 3 2 5 4	true
(a>b & c>d)	e <=f 1 3 2 5 4	true
a>b & (c>d | e<=f) 4 5 1 3 2	true

EXHIBIT 3.25   Compact comparisons

Expression	Meaning
21 < age < 45	AGE greater than 21 and less than 45.
high > score > low	SCORE less than HIGH and greater than LOW.
low < score < high	Same as above, endpoints and signs reversed.
60 > temp > 90	TEMP less than 60 and greater than 90. A badly defined expression better written as temp < 60	temp > 90.

Exercises

3.15. An observation has variables with the following values:

```
ID = 'SH'
GROUP = 'C'
SCORE1 = 3
SCORE2 = 6
SCORE3 = 5
SCORE4 = .
```

Determine the results of the following expressions.
a. score1 + score2 + score3
b. (score1 + score2 + score3) / 3

    c. score1 − score2 − score3

    d. (score1 / score2) − score3

    e. score1 + (score2 * score3)

    f. score4 * score3 * score2

    g. id = 'sh'

    h. id = 'SH'

    i. id = 'SH' & group = 'c'

    j. id = 'SH' ¦ group = 'c'

    k. score1 + score2 + score3 = 14

    l. score1 + score2 + score3 ^= 14

    m. id = 'SH' ¦ (group = 'C' & score1 = 3)

    n. 3 < score1 < 5

    o. 3 <= score1 < 5

Summary

Chapter 3 introduces much of the syntax and many of the programming guidelines used throughout this book. SAS program statements, which are the smallest "pieces" of the DATA steps and procedures in any SAS program, may be written using virtually any style that suits the user. Comments and the reasons for their use are tools that provide a powerful means of documenting programs. Items such as program author, date, purpose, special conditions, and the like may all be specified in a comment.

Data items, or variables, and the different forms of listing them were enumerated. SAS variable lists allow very flexible specification of variables to be used in procedures and DATA step operations. The variables may be set up as either *character*, which contain any alphabetic character, number, or special character, or *numeric*, which contain numeric values that will be handled arithmetically later in the DATA step or in a procedure.

The concept of missing data may indicate a problem with the data collection process or may simply be a natural byproduct of the DATA step. In either case, SAS has special rules and notation for handling these values.

After a description of numeric and character constants, which are values that are fixed for the duration of the DATA step, the logic and use of arithmetic and logical expressions were discussed. Arithmetic expressions manipulate variables and constants by adding, subtracting, multiplying, dividing, or raising numbers to a power. Logical expressions produce true/false values based on conditions of variable equality or magnitude. Both forms of expressions can implement very complex logic. They are used not only to calculate new variables but also to determine whether part of a DATA step should be executed for a particular observation. Both these functions of expressions are discussed in greater detail in later chapters.

PART TWO

Simple Tasks,

Simple Statistics:

Reading and Analyzing Data

4 DATA Step I:

Reading the Data

Before doing any work with either the DATA step or the PROCs, you have to describe the raw data to SAS. This chapter presents the basic steps for reading data. It discusses the kinds of data SAS can read, how to tell SAS where the data are located, two methods to describe the layout of the data, and the errors typically made by beginners.

4.1 What Kind of Data Can SAS Read?

SAS can read virtually any kind of data, in any format. It incorporates the best features of two worlds: the low-level language capabilities built into the DATA step and the high-level fourth generation commands. The user can either assume complete control or can give SAS a minimum of information and let it do the work.

Among the files that can be read are those created by such products as SPSS/PC, SPSS/X, dBASE III, Lotus 1-2-3® (WKS, WK1, DIF formats), DB2, IDMS, IMS, ORACLE®, Ingres, VSAM, and BMDP. Any hierarchical or "flat" file in ASCII or EBCDIC format can be read. The data can be stored on disk, tape, or diskettes. Any number of these media in any combination can be handled by a single SAS program.

The format of data can be *fixed*, where the location of data is consistent from record to record, and/or *list*, where the data are more free-form, separated by commas, tabs, and spaces. Such data descriptor features make it possible to describe record layouts with a minimum of effort.

4.2   Naming the Dataset: The DATA Statement

Raw data are read in the DATA step, and the DATA step begins with a DATA statement. This statement does three things:

1. Tells SAS that a DATA step is starting.
2. Names the SAS dataset being created. If a name is not specified, SAS assigns one.
3. Sets variables used in the DATA step to missing values.

The DATA statement appears in the following format:

```
DATA dsname[options] ... ;
```

This DATA statement contains two elements:

1. The *dsname* is the name of a SAS dataset. The name must begin with an alphabetic character (a through z), be a maximum of eight characters, and contain only alphabetic characters and numbers. Dataset names that suggest the contents of the dataset should be chosen: FISCAL88 is more meaningful than TEMP. If *dsname* is not specified, SAS will generate one named DATA*nn*, where *nn* is the number of times so far in this program that SAS had to choose (e.g., DATA1, DATA2).

2. The *options* are special features that allow finer control of *dsname*'s handling. Variables can be renamed, and a subset of variables can be selected for or excluded from processing. Using these options, we can also label the dataset. Options are covered in detail in Chapter 10.

Exercises

4.1. Which dataset names in the following list are not valid? Justify your answers.
  a. a
  b. _a
  c. 1sttrial
  d. fiscal1990

4.2. Choose dataset names for the following collections of data:
  a. SAT scores, other demographic data for incoming freshmen, 1988
  b. State-level data (demographic, social, economic, geographic, other miscellaneous data)
  c. State-level demographic data

4.3. Refer to the descriptions of the example datasets in Appendix E. Select appropriate dataset names other than those already assigned.

4.3   Identifying the Data Source

Any data source, regardless of complexity or storage medium, must be described to SAS before it can be used. This identification is accomplished through the INFILE, FILENAME, and CARDS statements. The INFILE statement links SAS to data sources physically external to the program. The FILENAME statement actually identifies the name and location of the dataset. The CARDS statement signals the beginning of a series of lines of data contained within the program

itself. These internal data lines are often called **in-stream** data. Each of these statements has its own syntax.

The INFILE Statement

INFILE is one of the few SAS statements described in this book that behaves differently in different computing environments. Since its purpose is to link the DATA step to a source outside the program, this chapter emphasizes the SAS end of this link and confines system-dependent particulars to Appendix D. The INFILE statement can appear in two forms:

```
INFILE fileref [options] ;
INFILE '[pathname]filename' [options];
```

Some common applications of the INFILE statement follow:

```
infile temp;

infile 'trials:[pass1.region1]feb91.raw';
```

In the INFILE description, three elements are important:

1. The *fileref* identifies an external file whose name has been assigned a FILE-NAME statement. The special *fileref* CARDS can be used when data are in-stream (follow the DATA step) and special INFILE *options* are required.

2. The *pathname* and *filename* are an alternative form of file specification that explicitly identify the external file to process in the DATA step. The *pathname* specifies the path, or subdirectory, of the file. The default is the path current when interactive SAS was started. Batch SAS has no default for *pathname*.

3. The *options* refers to special features that allow control over where within the file to begin and end reading, which columns will be read, and the like. These topics are beyond the scope of this chapter but are discussed at length in Chapters 11 and 19.

FILENAME Statement

The FILENAME statement links the DATA step(s) to an external file (it is not available in IBM MVS environments in version 5). It has the following format:

```
FILENAME fileref '[pathname]filename';
```

Some common applications of the FILENAME statement follow:

```
filename x 'd:\sample.dat';

filename test '[]usa90.rawdata';

filename eec 'ubaacs.tariff.eec.ver1';
```

In the FILENAME statement, *fileref*, *pathname*, and *filename* have the same meaning as in the INFILE statement. The choice to use a FILENAME and an INFILE or simply to identify the file directly in the INFILE statement is usually a matter of taste. If an external file will be used by more than one dataset, however, using FILENAME is probably the better option since the file name has to be specified only once. The FILENAME statement may be placed anywhere in the SAS program prior to or within the DATA step using it. It is usually a good idea, however, to place it near the beginning of the program, before any DATA steps that use it.

CARDS Statement

The CARDS statement signals the beginning of in-stream data. It has two forms:

```
CARDS;
CARDS4;
```

CARDS may also be entered as either DATALINES or LINES. CARDS4, DATA-LINES4, and LINES4 are also equivalent.

Since the statement tells SAS that the DATA step's raw data begins with the next line, CARDS and its variations signal the end of DATA step statements. No DATA step statements for the current DATA step may follow the CARDS statements.

The CARDS4 statement and its variants should be chosen over CARDS when in-stream data contains semicolons (;). Using CARDS4 requires that a data line beginning with four semicolons immediately follows the data.

If CARDS or its variants are used, SAS detects the end of the data by a semicolon in a program line. It is good practice to follow data with a RUN statement or simply ;. The line containing the terminating semicolon must be placed immediately after the last line of data. Blank lines after the data are interpreted as more (and missing) data.

Exercises

4.4. You have stored your data in a file on your microcomputer and want to analyze it with SAS. The file name is C:\TEST.RAW.
  a. Write FILENAME and INFILE statements to make the file available to the DATA step.
  b. Follow the same instructions as in part **a**, but use a single INFILE statement (no FILENAME).

4.5. Which of the following program arrangements is correct? The points of ellipses (. . .) indicate all or part of a statement that is not germane to the exercise. If a program is not valid, explain why.

```
a. filename in ... ;
   data ... ;
   infile in;
b. filename indata ... ;
   data ... ;
   infile datain;
c. data ...;
   infile cards;
   input ...;
   datalines;
   ...
d. data ...;
   input ...;
   cards;
   ...
```

4.4 Describing the Data: The INPUT Statement

By now SAS knows *what* to name the SAS dataset and *where* the data are located. The final detail is to describe *how* to move through the raw data and pick out the dataset's variables. This data description takes place in the INPUT statement.

How the INPUT Statement Works

The INPUT statement tells SAS to read one or more lines from the data source identified by the CARDS or INFILE statement. The process usually entails the following steps:

1. The contents of the line are copied into the computer's memory.
2. SAS moves to a column indicated by INPUT and reads to the end of the data field.
3. SAS converts the data to a SAS numeric or character variable and stores the SAS-formatted data in memory.
4. The process is repeated for each variable indicated in the INPUT statement.
5. Once all variables are read, they can be written to the dataset specified in the DATA statement.

INPUT actually initiates the reading of the data: INFILE, CARDS, and DATA simply tell SAS where to look for it and what to name it. You do not have to read every field of raw data into your SAS dataset: you may only be interested in a few variables even though the raw dataset contains hundreds of items. Conversely, you can read the same field more than once, even using different data types. Later examples in this section demonstrate using a field of raw data for more than one variable. This use is a simple, powerful, and often-overlooked capability of SAS.

POINTERS AND PITFALLS

INFILE or CARDS tells SAS *where* to find the data. INPUT tells SAS *how* to read the data and to start moving it from the raw dataset into the computer's memory.

Syntax

Only the simplest forms of INPUT are covered here. They deal only with well-organized, orderly datasets. In hapters 11 and 19 the complexity of the dataset layouts and, in turn, the INPUT statement is increased. For now, consider these forms of the INPUT statement:

```
INPUT var $ start [- end] ... ;
INPUT var [modifier] start [- end] [.dec] ... ;

INPUT var [modifier] ... ;
INPUT var $ ... ;
```

Some common applications of the INPUT statement follow:

```
input name $ 1-20 grade 22 status $ 24;

input city $ state $ zip $;

input month day year low_temp hi_temp;

input loc $ 1-4 span 8-17 .1  rank;
```

In the syntax description, the first pair of statements describes fixed-column input and the second illustrates the syntax for the simpler list input. In these descriptions, each component is important:

- The *var* is a variable name.
- The $ indicates *var* is character. If not specified, SAS assumes it is numeric.

- The *start* is the column number in the raw data to begin reading *var*.
- The *end* is the column number in the raw data to end reading *var*. If *var* is only one column wide, *end* is optional.
- The *dec* is the number of implied decimal places in the data field. If decimal points are actually entered in the data, they override this specification. (Exhibit 4.2 illustrates this process for a number of different situations.)
- The *modifier* warns SAS that *var* may contain invalid data and tells SAS how to react to this bad data. Two modifiers, ? and ??, are discussed in Section 4.5.

Fixed versus List Format. In the INPUT statement forms above, notice that the first pair specifies column locations and the second pair does not. This is the essential difference between fixed and list input, respectively. Most data supplied by government agencies and processed in commercial environments is fixed format. Since list input only requires the data fields be separated by one or more blanks, it has the advantage of being better suited for quick entry of small amounts of data. List input has the drawback of being unable to skip fields: If you are interested in the fifth field of data, you must read the preceding four as well.

Mixed Format Styles. Another feature of INPUT styles is that they can be mixed. It is perfectly acceptable to read a line of data in fixed format for a few columns, read the next variable using list INPUT, then go back to fixed format, possibly rereading fields at the beginning of the record or even reading the data from right to left! The INPUT statement is extremely flexible: you do not have to confine yourself to a single format style.

Flexible Input Formats. The final point about this introduction to the INPUT process is that it is, indeed, only an introduction. If you find the syntax restrictive and feel the potential is limited, refer to Chapter 11 for an illustration of some intermediate-level capabilities.

POINTERS AND PITFALLS

Take advantage of the INPUT statement's power and flexibility. Remember that different format styles may be legally combined in a single INPUT statement.

Exhibit 4.1 presents a series of legal INPUT statements.

Implied and Actual Decimal Places

One of the most confusing aspects of the conversion of raw data to SAS variables is the insertion of decimal places. The DEC specification tells SAS to insert a decimal point *only* if none is actually entered in the data! If this "no decimal point" condition is true, SAS inserts the decimal even if it must insert leading zeroes to satisfy the number of positions coded in the INPUT statement.

If there *are* decimal points in the data, they override the DEC specification. If no DEC specification is used, the decimal places are kept. Exhibit 4.2 illustrates these distinctions.

EXHIBIT 4.1   Simple INPUT statements explained

```
input a b c ;
```
Read numeric variables A, B, and C, separated by one or more blanks.

```
input code 1-4 code_1 1
      code_2 2-4 ;
```
Read CODE in columns 1 through 4, then reread it, picking out only column 1 for CODE_1 and the rest of the field for CODE_2.

```
input a 1-7 .2 b c ;
```
Read A in columns 1 through 7, inserting a decimal place two places from the left if one is not coded in the data. Then, starting in column 8, use list format to assign values to B and C.

EXHIBIT 4.2   Impact of decimal points in raw data

Assume that variable TEST was read with the INPUT statement

```
input x 1-5 .2 ;
```

Stored in Raw dataset as:	Store in SAS dataset as:
1	.01
12	.12
123	1.23
1234	12.34
12345	123.45
1.	1.00
1.2	1.20
12.	12.00
123.4	123.40

4.5   Sample Program for Reading a Dataset

This section presents a program to read a simple raw dataset and discusses some of the problems that might be encountered when reading it. Data are collected for all states with ocean and/or tidal shorelines (a more complete description of the dataset is found in Appendix E). The general outline of the state's coast is measured in miles, as is the tidal shoreline (places subject to tides do not have to border on the ocean to qualify). The first few records of the dataset are shown in Exhibit 4.3. The columns and their meanings are described in Exhibit 4.4.

One of SAS's strong points is its willingness to accept alternative specifications for the same task. Exhibit 4.5 illustrates several equivalent and valid INPUT statements. Keep in mind that each form answers SAS's questions of *what* variable will be read, *where* it will be found, and *how* it should be read (character or numeric).

EXHIBIT 4.3   First lines from COASTAL dataset

```
ecat   me    228    3478
ecat   nh     13     131
ecat   ma    192    1519
ecat   ri     40     384
```

EXHIBIT 4.4   Layout of COASTAL dataset

Field	Type	Start column	End column	Decimal places
Coast-ocean identifier	char	1	4	
Coast	char	1	2	
Ocean	char	3	4	
State postal code	char	7	8	
General coastline	num	9	14	0
Tidal coastline	num	15	21	0

EXHIBIT 4.5   Alternative INPUT statements for COASTAL data

```
input id $ 1-4 coast $ 1-2 ocean $ 3-4 state $ 7-8
      gencst 9-14 tidalcst 15-21;

input id $ 1-4 coast $ 1-2 ocean $ 3-4 state $ 7-8
      gencst 9-14 .0 tidalcst 15-21 .0;

input tidalcst 15-21 gencst 9-14 state $ 7-8 ocean $ 3-4
      coast $ 1-2 id $ 1-4;

input id $ 1-4 ocean $ 3-4 coast $ 1-2 state $ 7-8
      gencst tidalcst ;
```

Notice the following features of the INPUT statements in Exhibit 4.5:

- The columns used for ID are reread, forming two other variables (COAST and OCEAN).

- A space was not needed to separate the fields of data. COAST ended in column 2 and OCEAN began in column 3. GENCST ended in column 14 and TIDALCST began in column 15. As long as SAS knows where to look for the variables, it does not matter whether there are blanks or commas separating the fields or even whether they overlap.

- The .0 decimal place indicator in the second example is not needed. If the decimal point is present in the data, the fractional portion will be saved.

- The data can be read in any order: left to right (first two examples) or right to left (third example).

Bad Data, Bad Directions

What if life is not this simple? Or what if it really is this simple? There are still ways to complicate matters. Two types of errors are often made when reading even the simplest data: bad data and bad directions.

Bad Data.   Most datasets are initially beset by the presence of invalid values. The letter "O" may be used for the number "0", gender may be entered as "M" or "F" but described in the dataset's documentation as "0" or "1", and so on. To show how SAS reacts to these types of situations, the data used earlier are deliberately contaminated. The new data are illustrated in Exhibit 4.6.

EXHIBIT 4.6   "Problem" data

```
ECat   me    228   3,478
ecat   nh     13     131
ECat   ma    192   1,519
ecat   ri     40     384
```

There are several problems with the data in Exhibit 4.6:

- The tidal coastline variable contains commas. SAS can handle such input data, but not with the descriptor (TIDALCST 15-21) that was used.
- The last value of general coastline contains a letter "O" instead of a number "0".
- The coast identifier is entered sometimes as uppercase EC and sometimes as lowercase ec. As far as SAS is concerned, these are two distinct levels of the COAST variable and will similarly force ID into unintentionally distinct categories.

If the first INPUT statement is used to read the abbreviated, corrupted dataset, we would see a SAS Log resembling that in Exhibit 4.7.

EXHIBIT 4.7   SAS Log for problem data

```
------------------- SAS Log ----------------------------

   1     data coastal;
   2     input id $ 1-4    coast $ 1-2    ocean $ 3-4    state $ 7-8
   3          gencst 9-14    tidalcst 15-21 ;
   4     cards;

NOTE: Invalid data for TIDALCST in line 5 15-21.
RULE:----+----1----+----2----+----3----+----4----+----5----+----6----+----7---
   5 ECat   me    228  3,478
ID=ECat COAST=EC OCEAN=at STATE=me GENCST=228 TIDALCST=. _ERROR_=1 _N_=1

NOTE: Invalid data for TIDALCST in line 7 15-21.
   7 ECat   ma    192  1,519
ID=ECat COAST=EC OCEAN=at STATE=ma GENCST=192 TIDALCST=. _ERROR_=1 _N_=3

NOTE: Invalid data for GENCST in line 8 9-14.
   8 ecat   ri    40     384
ID=ecat COAST=ec OCEAN=at STATE=ri GENCST=. TIDALCST=384 _ERROR_=1 _N_=4

NOTE: The data set WORK.COASTAL has 4 observations and 6 variables.
```

NOTE for each observation with invalid data

Exhibit 4.7 indicates that SAS tries to suggest what went wrong. The Log points out problems in several areas:

- There are three NOTEs about "Invalid data for" a variable. They indicate which lines and columns of the data caused the problem (line 5, columns 15–21, for example).
- Each time a data line contains bad data, it is displayed on the Log exactly as it was read by SAS. The first time a NOTE is printed on a page, a column RULE is displayed above the offending data line to help you locate the data more easily.
- The line below the raw data listing displays the values of all the observation's variables, invalid or not. Notice that in the first such display TIDALCST has been set to a missing value ("."). Whenever SAS cannot read a value, it sets the variable to missing.
- The dataset has the correct number of observations (4) and variables (6), but three of the values in this data matrix will be missing due to data problems.
- Finally, notice that SAS did *not* complain about the dual entry COAST as "EC" and "ec".

To suppress all or part of these messages, the ? and ?? **format modifiers** can be used for numeric variables. The ? modifier suppresses the message about invalid data. The ?? modifier suppresses the message about invalid data *and* the raw data and variable listing. Thus ?? makes even "dirty" data look "clean" to the reader of the SAS Log.

The format modifiers are used in Exhibits 4.8 and 4.9. Consider using them only when you want to avoid long Logs, when you are already aware that something may be wrong with your data, and when you are willing to let SAS set these values to missing. SAS supplies these diagnostic tools as aids: use the ? and ?? modifiers only when you are comfortable with the presence of bad data and SAS's default reaction to it.

EXHIBIT 4.8   Use of the ? modifier

```
------------------- SAS Log ----------------------------

   1    data coastal;
   2    input id $ 1-4    coast $ 1-2    ocean $ 3-4    state $ 7-8
   3          gencst ? 9-14    tidalcst ? 15-21 ;
   4    cards;
RULE:----+----1----+----2----+----3----+----4----+----5----+----6----+----7---
   5 ECat   me   228  3,478
ID=ECat COAST=EC OCEAN=at STATE=me GENCST=228 TIDALCST=. _ERROR_=1 _N_=1
   7 ECat   ma   192  1,519
ID=ECat COAST=EC OCEAN=at STATE=ma GENCST=192 TIDALCST=. _ERROR_=1 _N_=3
   8 ecat   ri    40    384
ID=ecat COAST=ec OCEAN=at STATE=ri GENCST=. TIDALCST=384 _ERROR_=1 _N_=4
NOTE: The data set WORK.COASTAL has 4 observations and 6 variables.
NOTE: The DATA statement used 9.00 seconds.
```

> ? modifier suppresses NOTEs but still prints data line and variable values.

Bad Directions.   Dirty data are not the only complication. Bad programming directions can also thwart clean, effective use of the data. If variable location, scale, and/or type are misspecified, SAS may or may not catch the error. Even

---

EXHIBIT 4.9   Use of the ?? modifier

```
------------------- SAS Log ----------------------------

   1    data coastal;
   2    input id $ 1-4    coast $ 1-2    ocean $ 3-4    state $ 7-8
   3          gencst ?? 9-14    tidalcst ?? 15-21 ;
   4    cards;
NOTE: The data set WORK.COASTAL has 4 observations and 6 variables.
NOTE: The DATA statement used 8.00 seconds.
```

?? modifier suppresses NOTEs, raw data listing, variable listing.

---

if the data are correct, telling SAS to read numeric data in a column filled with character data results in the same types of messages seen in the preceding section.

To illustrate how inaccurate INPUT specifications can affect the way SAS reads data, we will return to a partial listing of the COASTAL dataset. Suppose we mistakenly used this INPUT statement:

```
input id $ 1-3    coast $ 1-2    ocean $ 3    state    7-8
      gencst 9-13    tidalcst 14-21 ;
```

Notice that OCEAN's width is now one column, STATE is numeric rather than character, and TIDALCST begins a column early (14 rather than 15). The SAS Log for the program appears in Exhibit 4.10.

In Exhibit 4.10 SAS encountered invalid data for STATE and TIDALCST in *each* observation. Characters ("me", "nh", and so on) were found for numeric STATE. Each TIDALCST value began with a number in column 14, then had a series of blanks before encountering numbers again. All were set to missing values. Such uniformly bad reactions to an INPUT statement almost always mean bad instructions rather than bad data. Notice also that SAS could not detect the misspecification of OCEAN as a one-column-wide character variable. The error here is particularly insidious since it is not really an error per se. We would probably notice the mistake if we printed the dataset and noticed that OCEAN was "shorter" than expected. This type of mistake, if undetected, can lead to peculiar results from analyses. By taking only the first of two letters, the AT's (ATlantic) and the AR's (ARctic) are inadvertently combined into the single category "A": the gap between Alaska and the Lower Forty-Eight suddenly vanishes!

Exercises

4.6. Refer to the descriptions of the LIFESPAN and NATLPARK datasets in Appendix E. Select one of these for this exercise.
   a. Write a list style INPUT statement to read the dataset. You can use the variable names used in Appendix E.
   b. Write an INPUT statement using column input.
   c. Write an INPUT statement that uses a combination of both column and list style specifications.

4.7. The following INPUT statement successfully read a raw dataset:

```
INPUT ID Q1-Q3 AVG 12-15 .1;
```

Write several lines of the raw dataset that would conform to this input specification. The actual values do not matter. What *is* important is their location in the input record.

EXHIBIT 4.10   Incorrect INPUT directions for the COASTAL data

```
------------------- SAS Log -------------------

 1    data coastal;

 2    input id $ 1-3   coast $ 1-2   ocean $| 3 |   state  7-8  ◄───────    Should be 3-4

 3           gencst 9-13   tidalcst |14-21|;  ◄───────    Should be 15-21

 4    cards;
```

```
NOTE: Invalid data for STATE in line 5 7-8.
NOTE: Invalid data for TIDALCST in line 5 14-21.
```
```
RULE:----+----1----+----2----+----3----+----4----+----5----+----6----+----7---
 5 ECat  me   228   3478
ID=ECa COAST=EC OCEAN=a STATE=. GENCST=22 TIDALCST=. _ERROR_=1 _N_=1
```
```
NOTE: Invalid data for STATE in line 6 7-8.
NOTE: Invalid data for TIDALCST in line 6 14-21.
```
```
 6 ecat  nh    13    131
ID=eca COAST=ec OCEAN=a STATE=. GENCST=1 TIDALCST=. _ERROR_=1 _N_=2
```
```
NOTE: Invalid data for STATE in line 7 7-8.
NOTE: Invalid data for TIDALCST in line 7 14-21.
```
STATE and TIDALCST are invalid in *every* observation.
```
 7 ECat  ma   192   1519
ID=ECa COAST=EC OCEAN=a STATE=. GENCST=19 TIDALCST=. _ERROR_=1 _N_=3
```
```
NOTE: Invalid data for STATE in line 8 7-8.
NOTE: Invalid data for TIDALCST in line 8 14-21.
```
```
 8 ecat  ri    40    384
ID=eca COAST=ec OCEAN=a STATE=. GENCST=4 TIDALCST=. _ERROR_=1 _N_=4
```
```
NOTE: The data set WORK.COASTAL has| 4 observations |and 6 variables.

NOTE: The DATA statement used 10.00 seconds.
```

4.8. The following line of data is read from a raw dataset (the column ruler is not part of the data):

```
----+----1----+----2----+----3
ICD174.0 86 3005 450000 156.8
```

A correct INPUT statement follows:

```
input codetype $ 1-3 code_num 4-8  year 10-11
    incid 12-16 base 17-23  comprate 25-29;
```

Which of the following INPUT statements will not work correctly? Explain your answer.

a. input codetype $ code_num 4-8 year incid 12-16
      base comprate;

b. input codetype $ 1-3 code_num 4-8 .1 year
      incid base comprate;

c. input codetype $ 1-3 comprate 25-29 base 17-23
      incid 12-16 year 10-11 code_num 4-8;

d. input comprate 25-29 year 10-11 incid base
      code_num codetype $ 1-3;

4.6   Complete Examples of the DATA Step

Four complete examples of the DATA step are presented in this section. The first creates a SAS dataset from an external source, the second from in-stream data. The remaining examples illustrate a common error and its results. The complete SAS Log is shown for each example. The examples are included to give a feel for how the SAS Log appears for a "clean" run through a raw dataset. The data are read from a file named PRECIV stored in the directory \DATA\RAW. The first five lines of the data are shown in Exhibit 4.11.

EXHIBIT 4.11   First lines of presidential dataset

Washington	fed	1732	va	1789	57	67
Adams, J.	fed	1735	ma	1797	61	90
Jefferson	dem/rep	1743	va	1801	57	83
Madison	dem/rep	1751	va	1809	57	85
Monroe	dem/rep	1758	va	1817	58	73

Key features and options used in the examples are summarized in the following list:

Exhibit	Features/Options
4.12	Read an external file
4.13	Read in-stream data
4.14	Misplaced INPUT and INFILE statements
4.15	Single-statement DATA step

External Dataset

This program creates a SAS dataset PRES containing information about the 16 antebellum U.S. presidents. The SAS Log is presented in Exhibit 4.12.

EXHIBIT 4.12   Create a SAS dataset from an external file

```
------------------ SAS Log -----------------------

NOTE: Copyright(c) 1985,86,87 SAS Institute Inc., Cary, NC 27512-8000, U.S.A.
NOTE: SAS (r) Proprietary Software Release 6.03
      Licensed to D**2 Systems, Site 12345678.

  1    data pres;
  2    infile '\data\raw\preciv';
  3    input name $ 1-20 party $ 21-29 born 31-34  bornst $ 37-38
  4         inaug 43-46   ageinaug 51-52    agedeath 56-57
  5         ;
  6    run;
NOTE: The infile '\data\raw\preciv' is file C:\DATA\RAW\PRECIV.

NOTE: 16 records were read from the infile  C:\DATA\RAW\PRECIV.
      The minimum record length was 57.
      The maximum record length was 57.
NOTE: The data set WORK.PRES has 16 observations and 7 variables.
NOTE: The DATA statement used 9.00 seconds.
NOTE: SAS Institute Inc., SAS Circle, PO Box 8000, Cary, NC 27512-8000
```

INFILE identifies an external dataset.

SAS specifies the full name of the dataset (note the addition of the disk drive–C).

Summary of external dataset size, record lengths

In-Stream Data

This example reads data comparing the cost of mailing a first-class letter in several industrialized countries. Note that an INFILE CARDS; statement could have been added after the DATA statement with exactly the same results. The SAS Log is reproduced in Exhibit 4.13.

EXHIBIT 4.13   Create a SAS dataset from in-stream raw data

```
------------------- SAS Log --------------------------

NOTE: Copyright(c) 1985,86,87 SAS Institute Inc., Cary, NC 27512-8000, U.S.A.
NOTE: SAS (r) Proprietary Software Release 6.03
      Licensed to D**2 Systems, Site 12345678.

   1      data postcomp;
   2      input country $ 1-19 currency $ 20-29 units 32-38
   3           us_equiv 44-47 ;
   4      cards;
```

Fourteen data lines follow CARDS statement.

```
NOTE: The data set WORK.POSTCOMP has 14 observations and 4 variables.

NOTE: The DATA statement used 8.00 seconds.
NOTE: SAS Institute Inc., SAS Circle, PO Box 8000, Cary, NC 27512-8000
```

Misplaced Statements

If the order of the INFILE and INPUT statements in Exhibit 4.12 is switched by mistake, SAS would produce the Log in Exhibit 4.14. Remember that in the DATA step SAS carries out instructions in the order that they are found. We told it *what* to read (the INPUT statement) before telling it *where* to find it (INFILE). SAS detected the error but did not assume anything: it defined the seven variables in the INPUT statement, read the INFILE, noted the problem, and created a dataset with zero observations since it could not read any data.

EXHIBIT 4.14   Misplaced INPUT and INFILE statements

```
------------------- SAS Log --------------------------

NOTE: Copyright(c) 1985,86,87 SAS Institute Inc., Cary, NC 27512-8000, U.S.A.
NOTE: SAS (r) Proprietary Software Release 6.03
      Licensed to D**2 Systems, Site 12345678.

   1      data pres;
   2      input name $ 1-20 party $ 21-29 born 31-34  bornst $ 37-38
   3           inaug 43-46   ageinaug 51-52   agedeath 56-57
   4           ;
   5      infile '\data\raw\preciv';
   6      run;

ERROR: INPUT statement executed before INFILE statement.

NAME=  PARTY=  BORN=. BORNST=  INAUG=. AGEINAUG=. AGEDEATH=. _ERROR_=1 _N_=1

NOTE: The data set WORK.PRES has 0 observations and 7 variables.

NOTE: The DATA statement used 9.00 seconds.
NOTE: SAS Institute Inc., SAS Circle, PO Box 8000, Cary, NC 27512-8000
```

Misplaced Program

What if the INPUT and INFILE statements were inadvertently deleted from the program in Exhibit 4.14? Such a mistake is more common than you might expect. When editing long programs and moving blocks of text (e.g., DATA steps), users sometimes forget to move the *entire* block. Exhibit 4.15 shows how SAS reacts to this minimalist DATA step.

EXHIBIT 4.15   Single-statement DATA step

```
------------------ SAS Log ----------------------------

NOTE: Copyright(c) 1985,86,87 SAS Institute Inc., Cary, NC 27512-8000, U.S.A.
NOTE: SAS (r) Proprietary Software Release 6.03
      Licensed to D**2 Systems, Site 12345678.

    1     data pres;                                          No DATA step
    2     run;                                                statements

NOTE: The data set WORK.PRES has  1 observations  and  0 variables.

NOTE: The DATA statement used 7.00 seconds.
NOTE: SAS Institute Inc., SAS Circle, PO Box 8000, Cary, NC 27512-8000
```

Summary

Chapter 4 describes the mechanics of reading raw data and creating a SAS dataset. The DATA statement names the dataset being created. Several statements help SAS identify the location of the raw data being read. INFILE, FILENAME and CARDS may be used to specify the raw data source. The INPUT statement provides the instructions for actually reading the data. These instructions tell SAS which columns are used by a variable, the name of the variable, and whether the variable is character or numeric.

The INPUT statement is the most complex and powerful in the SAS language. This chapter describes only its most basic forms, points out how SAS reacts to invalid data, and illustrates the impact of incorrect column and data type specifications. As shown in the example programs, SAS can sometimes appear to work correctly even if it is supplied with incorrect descriptions of the data. Caution and double-checking of results are very important.

5 Using the Dataset: Introduction to PROCs

This chapter describes commonly used procedures. It begins by looking at features that are common to all PROCs and then discusses the PRINT and SORT procedures. This background provides the basis for many data management and analysis tasks. The same caution noted at the beginning of Chapter 3 is also appropriate here: the material covered in the first section is more reference than tutorial. Some readers, particularly computer novices, may find it easier to skim the first section and return to read it in detail when the statements actually have to be used.

5.1 Common Features of Procedures

PROCs perform specialized tasks and thus have unique options and statements. One of SAS's strengths is its consistency of syntax among disparate tasks and the procedures' ability to use many similar statements. This section describes the conventions, options, and statements that are common to nearly all the PROCs used in this book. Discussions of the individual PROCs in later chapters will note which common features may be used.

DATA=: Specifying the Dataset to Use

Most PROCs discussed in this book must process a SAS dataset. The name of the dataset is conveyed either *implicitly* or *explicitly*, via the DATA option in the PROC statement.

Implicit Specification. A PROC statement without an explicit SAS dataset reference causes SAS to use the most recently created dataset. This is sometimes referred to as the *rule of last use*. Consider the program in Exhibit 5.1. The PRINT

EXHIBIT 5.1   Illustration of rule of last use

```
data fiscal88;
[DATA step statements]
run;

data fiscal89;
[DATA step statements]
run;

proc print;
run;
```

procedure processes the SAS dataset FISCAL89 because it was created most recently before the PROC statement. This dataset selection rule applies to all SAS procedures.

Explicit Specification.   To use FISCAL88 in the PRINT procedure, the default must be overridden by using the DATA option of the PROC statement, as shown in Exhibit 5.2.

EXHIBIT 5.2   Using the DATA option in PROC to override rule of last use

```
data fiscal88;
[DATA step statements]
run;

data fiscal89;
[DATA step statements]
run;

proc print data=fiscal88;
run;
```

Many experienced SAS programmers prefer to enter the DATA option even if the rule of last use would point to the required dataset. Although implicit selection may be adequate in a small program, it may create confusion and ambiguity in long programs, following DATA steps that create multiple datasets, or following PROCs that create datasets (covered in Chapter 17). As a rule, then, you should develop the good habit of explicitly specifying the SAS dataset to avoid potential pitfalls.

POINTERS AND PITFALLS

Do not regard dataset specification in PROCs as an option. Treat it as a requirement: explicit references remove potential for confusion, especially in long or complex programs.

VARIABLES: Specifying the Variables to Use

The VARIABLES statement, abbreviated as VAR, tells SAS which variables to use in the PROC. Some of the procedures discussed later in this book use other statements, but VAR is the most common method of specifying variables. Its syntax is shown below:

```
VARIABLES var ... ;
```

Some common applications of the VARIABLES statement follow:

```
var avgtemp;

variables _numeric_;

var score1-score20 name id;
```

In the VARIABLES statement, *var* is any combination of variables and variable lists. If the statement is not present, the PROC uses all *appropriate* variables: PROC MEANS and PROC UNIVARIATE, for example, use only numeric variables, while PROC PRINT uses all variables. Exhibit 5.3 illustrates the use of VAR statements with PROCs.

EXHIBIT 5.3   Using VARIABLE statement with PROCs

```
print data=fiscal88;
var _numeric_ bankname;
run;

proc means data=temp;
var score1-score10;
run;
```

Reordering Variables.   When used with some PROCs, VAR can determine the order in which variables are used. For example, if an INPUT statement read variables BANKNAME and CITY (character) and QTR188, QTR288, QTR388, and QTR488 (numeric), and you wanted to print the numerics first, followed by BANKNAME, you could use the first VAR statement in Exhibit 5.3.

FORMAT: Controlling Display of the Variables

The FORMAT statement allows control of two procedure activities. It tells SAS, "write the values of the following variables this way." *Formats control how the data are displayed; the data themselves are not affected.* When used with a PROC, the formats specified are in effect only while the procedure is running. Formats that come with the SAS System are described in detail in Chapter 9. Chapter 10 demonstrates how formats can be permanently associated with SAS dataset variables.

The second activity controlled by the FORMAT statement is more subtle. In procedures where variables are used for grouping, formats can determine how the grouping of the data contained in the variables takes place. This use of formats is extremely powerful and a bit tricky. It is discussed in Chapter 15.

Syntax.   The syntax of the FORMAT statement is illustrated below:

```
FORMAT vars [format] ... ;
```

Some common applications of the FORMAT statement follow:

```
format score1-score20 5.3 avg;
```

```
format name $20. style $3. basis1-basis5;
```

The FORMAT statement has two elements:

1. The *vars* specifies variables and/or variable lists using a format. Each group of variables must identify all character or all numeric variables.

2. The *format* is optional. It specifies the name of the format to use with *vars*. The *format* must begin with a dollar sign ($) if *vars* are character. All formats must contain a period (.) to help SAS distinguish them from variable names. SAS selects a default format if *format* is not specified. Constant reliance on these defaults, however, usually means you are not taking full advantage of SAS's display and grouping capabilities.

More than one FORMAT statement can be used with a PROC. If more than one FORMAT is assigned to a variable, the one closest to the end of the PROC's statements is used.

Example.   Formats can be used with the presidential data from Chapter 4. A PRINT procedure is shown in Exhibit 5.4.

EXHIBIT 5.4   Using a FORMAT statement with a PROC

```
proc print data=pres;
format party $1.    ageinaug 2.    born century. ;
run;
```

The FORMAT statement gives SAS the following instructions for printing the variables.

- Only the first column of the character variable PARTY (format $1.). The format is part of the SAS System's library of formats.

- AGEINAUG using two columns with no decimal places. The 2. format is also part of the SAS System library.

- The BORN variable using a special format not found in the SAS System library. CENTURY is a "user-written" format that customizes data display according to user-specified standards. User-written formats are discussed in Chapters 14 and 15.

LABEL: Annotating Variable Names

Eight characters is sometimes not enough to communicate the meaning of a variable. The LABEL statement enables variable name annotation. Most PROCs print the label information along with the variable name.

Like FORMAT, LABEL's impact is felt only when the procedure is running. Also like FORMAT, there are ways to permanently attach its descriptive information to a SAS dataset. No leading dollar sign is needed to distinguish character variables from numerics: the syntax is identical for both data types.

Syntax.   The syntax of the LABEL statement is straightforward:

```
LABEL var = 'text' ... ;
```

Some common applications of the LABEL statement follow:

```
label name = "Respondent's name"
      avg  = '4-yr average rating'
      ;
```

The LABEL statement has two components:

1. The *var* is a single variable name (*not* a variable list).
2. The *text* is a descriptive label for *var*. It may not exceed 40 characters and must be enclosed in a pair of single quotes (') or quotation marks ("). If a single quote is part of *text* enclosed in single quotes, enter two consecutive single quotes. Any character is valid as *text*.

Any number of LABEL statements may appear in the same PROC. If more than one LABEL is assigned to the same variable, the one closest to the end of the PROC's statements is used.

Example. An example of the LABEL statement using the antebellum presidential data is shown in Exhibit 5.5.

EXHIBIT 5.5   LABEL statement for PRES dataset

```
label born    = 'Year born (4-digit)'
      name    = "Last name (initials where needed)"
      party   = 'Party''s name at time of inaug.'
      ;
```

ATTRIB: Combining Formats and Labels

Formats and labels may be assigned to variables in a single statement. ATTRIB allows you to specify the format and/or label of a variable next to each other. This makes a variable's characteristics immediately understood: both its display format and description are beside the variable name. The disadvantage of AT-TRIB's syntax is that you cannot use variable lists. This point is discussed at greater length in Chapter 9.

Syntax. The ATTRIB statement's syntax is shown below:

```
ATTRIB [var [FORMAT=format] [LABEL='text']] ... ;
```

Some common applications of the ATTRIB statement follow:

```
attrib name format=$20 label="Respondent's name"
       avg format=5.2 label="4-year rating average";
```

The ATTRIB statement has three elements:

1. The *var* is a variable name (*not* a variable list).
2. The *format* is a format specification. It is identical in form to the formats specified in the FORMAT statement.
3. The *text* is a descriptive label for variable *var*. It is identical in form to the labels specified in the LABEL statement.

Any number of ATTRIB statements may appear in the same PROC. If more than one FORMAT or LABEL is assigned to the same variable, the one closest to the end of the PROC's statements is used.

TITLE and FOOTNOTE: Annotating Procedure Output

SAS procedure output can be enhanced by including titles and footnotes. Title lines are printed at the top of each page of output, footnotes at the bottom. The syntax of both features is presented below:

```
TITLE[n] ['text'];
FOOTNOTE[n] ['text'];
```

Some common applications of TITLEs and FOOTNOTEs follow:

```
title1 'Average Scores: Districts 1 and 2';
title2 'Data current as of 23FEB91';

title1;

footnote 'Program producing this output in TRY1.SAS';
```

The TITLE and FOOTNOTE statements have two components:

1. The *n* is a number from 1 to 10 indicating the number of the title or footnote line. The default value is 1.

2. The *text* is the text of the title or footnote. The rules about use of single quotes and quotation marks are identical to those of the LABEL statement. Omitted *text* or that with a null (") or blank (' ') value prints an empty TITLE or FOOTNOTE.

Unless told otherwise, SAS centers the titles and footnotes. Control over the centering is discussed in Chapter 12. If you do not specify any titles, the default (not surprisingly) is "SAS" or "the SAS System."

Stacking TITLEs and FOOTNOTEs. SAS stacks titles and footnotes: changing TITLE number *n* deletes titles with a number higher than *n*. If, for example, a program uses TITLE1, TITLE2, and TITLE3 statements and TITLE2 is changed, TITLE1 is still in effect but TITLE3 is lost. The same logic also applies to FOOTNOTEs. Stacking is demonstrated in Exhibit 5.6.

──────────

EXHIBIT 5.6   TITLE stacking and replacement

───

```
proc print data=pres18;
title 'US Presidents';
title2 '18th Century';
run;

proc print data=pres19;
title2 '19th Century';
run;

proc print data=pres;
title 'All Antebellum US Presidents';
run;
```

───

The title lines for the printing of the first dataset are "US Presidents" and "18th Century." Overriding the second title when printing the second dataset keeps "US Presidents" as the first title line and "19th Century" as the second. The third PRINT changes TITLE1 to "All Antebellum US Presidents." This deletes the second title line.

BY: Processing in Groups

PROCs usually process the SAS dataset as a whole; that is, they print, chart, and calculate statistics using all observations in the dataset. The BY statement runs the procedure separately for each value, or level, of the variables in the statement. This process is sometimes referred to as **BY-group processing**.

Syntax. The BY statement may be written two ways:

```
BY [DESCENDING] sort_key ... ;

BY sort_key ... NOTSORTED ;
```

Some common applications of the BY statement follow:

```
by descending date descending time;

by dept name;

by type notsorted;
```

The BY statement has three elements:

1. The *sort_key* is a variable (*not* a variable list) in the dataset by which the dataset is ordered. If no options are specified elsewhere in the statement, SAS assumes *sort_key* is in ascending, or low-to-high, order from observation to observation: any observation's value of *sort_key* is greater than or equal to the preceding observation's value.

2. *DESCENDING* indicates that the next *sort_key* is found in descending, or high-to-low, order from observation to observation in the dataset: any observation's value of *sort_key* is less than or equal to the preceding observation's value.

3. *NOTSORTED* indicates that BY-group processing should take place on grouped, rather than sorted, data; that is, the data are not necessarily in ascending or descending order, but you are willing to begin a BY-group whenever any *sort_key* in the BY statement's variable list changes value. This is illustrated in Exhibit 5.7, Example 1.

Example. Exhibit 5.7 illustrates distinctions between the different styles of observation ordering.

BY-Group Procedure Appearance. Most procedures display output differently when BY-group processing than when processing the entire dataset at once. Some PROCs draw a line across the page that identifies the BY-group being processed. Others begin a new page of output when a new level of the BY-group begins. Do not think something has gone awry simply because the listing looks different from "normal" (no BY-group) output.

Observations Out of BY-Group Order. SAS prints an error message and halts program execution if a BY statement misspecifies the dataset's order. Suppose we PRINT the first dataset in Exhibit 5.7 with a BY variable of DEPT. A partial SAS Log appears in Example 5.8.

We could rearrange the raw data, moving the lines into the correct order with a text editor, then rerun the SAS program. There is a much quicker and more reliable means of reordering—the SORT procedure. PROC SORT is discussed in Section 5.2.

EXHIBIT 5.7   Comparison of different BY-variable arrangements

Example 1: initial, unordered dataset. Could process this using BY DEPT
           NOTSORTED; since the records are out of DEPT sort order (i.e.,
           not alphabetical), but all like-valued DEPTs are next to each
           other.

NAME	DEPT
Adams	MKT
Spencer	FIN
Johnson	FIN
Young	MFG
Adams	MFG

Example 2: BY NAME; dataset sorted by employee name.

NAME	DEPT
Adams	MKT
Adams	MFG
Johnson	FIN
Spencer	FIN
Young	MFG

Example 3: BY DEPT; dataset sorted by department.

NAME	DEPT
Spencer	FIN
Johnson	FIN
Young	MFG
Adams	MFG
Adams	MKT

Example 4: BY DEPT NAME; dataset sorted by department, then by name within
           each department.

NAME	DEPT
Johnson	FIN
Spencer	FIN
Adams	MFG
Young	MFG
Adams	MKT

Example 5: BY DESCENDING NAME; names sorted in reverse alphabetical order.

NAME	DEPT
Young	MFG
Spencer	FIN
Johnson	FIN
Adams	MFG
Adams	MKT

**FREQ: One
Observation Counting
as Many**

Sometimes an observation represents more than one occurrence of a variable.
The FREQ statement identifies a variable whose value represents the number of
times the observation should be counted in an analysis. For example, survey data
may have a value by which the observation should be weighted, based on the

EXHIBIT 5.8   Improper BY specification

```
-------------------- SAS Log ----------------------------

    1      proc print data=emps;
    2      by dept;
    3      run;
ERROR: Data set WORK.EMPS is not sorted in ascending sequence. The current
       by-group has DEPT = MKT and the next by-group has DEPT = FIN.
NOTE: The above message was for the following by-group:
      DEPT=MKT
NOTE: The SAS System stopped processing this step because of errors.
NOTE: The PROCEDURE PRINT used 7.00 seconds.
```

respondent's age, gender, and race. Such values allow calculation of population totals based on sample values. A frequency count of the survey's variables *without* a FREQ statement reflects distributions within the *sample*. Frequencies *using* a FREQ statement are distributions of the *population* the survey supposedly represents. The frequency value is *not* a weight. See the end of the WEIGHT statement's discussion for a comparison of FREQ and WEIGHT variables.

Syntax.   The syntax of the FREQ statement is shown below:

```
FREQ var ;
```

Two things are important in this statement:

1. The *var* indicates the frequency variable. The procedure treats each observation as if it had actually occurred *var* times in the dataset.

2. The *var*'s value must be at least 1 and is truncated (uses only the integer portion) if fractional. Values that are either missing or less than 1 exclude an observation from the analysis.

WEIGHT: Relative Weights

Unlike the variable used in the FREQ statement, a WEIGHT variable does not affect the observation count of the analysis. WEIGHT is often used in linear models procedures when observations' distributions (variances) differ. In simple, univariate statistics the individual observation's contribution to the dataset's statistic is proportional to its contribution to the sum of the weight variable across the entire dataset. Exhibit 5.9 illustrates weights and compares them to the FREQ statement variable.

Syntax.   The syntax of the WEIGHT statement is shown below:

```
WEIGHT var;
```

In this statement, *var* is the weight variable. The value each observation contributes to the PROC's analysis is equal to its value of *var* over the sum of all *var*s used in the analysis. The *var* must be a positive numeric variable. If not, the observation will be excluded from the analysis.

FREQ versus WEIGHT.   The distinction between FREQ and WEIGHT is both significant and sometimes difficult to grasp. Most importantly, FREQ usually changes the number of observations used in the analysis. This, in turn, affects the statistics and their significance tests. WEIGHT, however, does not alter the

EXHIBIT 5.9 FREQ-WEIGHT comparison

Analysis Variable	FREQ var.	WEIGHT var.
5	2	1.0
10	3	1.5
20	2	2.5
15	1	5.0

	Count	Sum	Mean	Variance
No FREQ or WEIGHT	4	50	12.5	41.66
Using FREQ	8	95	11.87	32.56
Using WEIGHT	4	145	14.50	65.83

observation count but will affect the statistics. Exhibit 5.9 shows how some simple univariate statistics (counts, sums, averages, and variances) are affected by the statements' use.

Exercises

5.1. Refer to the postal rate dataset used in Exhibit 4.13. Variable COUNTRY is a country name, CURRENCY the name of the currency, UNITS the local currency amount to send a first class letter, and US_EQUIV the equivalent cost in 1987 U.S. dollars.

    a. Write VARIABLES statements to meet the following requirements:

        (1) Identify the numeric variables.

        (2) Identify the character variables.

        (3) Identify the variables for country name and U.S. dollar equivalent.

        (4) List, in order, the variables for country, U.S. dollar equivalent, local currency, and local units.

    b. Write a FORMAT statement associating the format $20. with COUNTRY and CURRENCY, 5.2 with US_EQUIV, and COMMA7. with UNITS. For now, do not worry about the meaning of the formats; just concentrate on getting the syntax correct.

    c. Write a LABEL statement to assign labels to each variable in the dataset.

5.2. There are four procedures in this program. What are the TITLEs and FOOTNOTEs that will be used with each one?

```
proc means data=mort8088;
title1 'Mortality rates, 1980-88';
footnote1 'Program in dataset \m80s\rates.sas';
run;

proc print data=mort9092;
title2 'Preliminary rates for 1990-2';
run;

proc freq data=demog;
title2;
footnote2 'Background data (see \m80s\demog)';
run;
```

```
proc print data=demog;
title 'Demographic profiles';
footnote '';
```

5.3. A dataset has two variables, REGION and INDEX, arranged as in the following list. (The numbers in parentheses are included for use in answering the questions that follow. They are not part of the dataset.)

```
    REGION INDEX
(1) e     2.0
(2) e     2.0
(3) e1    3.1
(4) s     4.2
(5) w     8.1
(6) w     8.0
(7) n     8.0
```

How would the dataset be arranged if it met the following criteria (just list the order of observations using the observation numbers):
a. by index
b. by region index
c. by descending region
d. by descending region index
e. by descending region descending index

5.2   Rearranging the Dataset: PROC SORT

Many situations call for rearranging the order of the observations in a dataset. For example, a list of names must be in alphabetical order, scores must be listed in descending order, or an analysis must be repeated for different levels of one or more stratifying variables (as in BY-group processing). Combining two or more SAS datasets may require sorting by one or more variables common to the datasets. The SORT procedure is a simple way to satisfy the requirements of these, and other, situations.

Syntax

The syntax for the SORT procedure is outlined here:

```
PROC SORT [DATA=datain] [OUT=dataout] [NODUP];
BY byvarlist;
```

Some common applications of the SORT procedure follow:

```
proc sort data=random out=ordered;
by grade name descending gmat;
```

```
proc sort data=reuse;
by area bin;
```

PROC SORT has four elements:

1. The *datain* identifies the input, unsorted SAS dataset. If the *DATA* option is not entered, the most recently created dataset is used.

2. The *dataout* identifies the output, sorted dataset. If it is not entered, the dataset specified by the *DATA* option is replaced by the sorted dataset. This replacement, the default handling of the data, is usually called an *in-place* sort.

3. *NODUP* eliminates duplicate observations from the output dataset. Values for all variables in the observations, not just the BY variables, must be identical for observations to be deleted. *NODUP* is not available in all implementations of SAS.

4. The *byvarlist* identifies one or more character or numeric variables whose values determine the order of the sorted dataset. See the description of the BY statement for more details. There is no practical limit to the number of variables in *byvarlist*.

Usage Notes

Keep the following rules in mind when sorting in SAS:

- Missing values always sort low. Thus a blank (missing character value) is "smaller" than an "a", "a " is smaller than "ab", and the numeric missing value "." is smaller than any number, negative or positive.

- SORT does not produce printed output. The SAS Log contains a message about the number of observations in the output dataset and, if NODUP was used, the number of duplicate observations that were eliminated.

- The order of the sorted dataset varies from one computer environment to another, since each computer has its own idea of which values are large and which are small. For example, on minicomputers the number zero is considered smaller than the letter A, while on IBM mainframe systems the opposite is true. Appendix A summarizes these two major collating sequences.

- If the dataset's order is acceptable, it can be used directly with BY-group processing, bypassing the SORT. Sorting an already sorted dataset does not cause any harm: it is just not necessary.

- When two or more variables in the input dataset have identical values of the BY variable(s), they are included in the output dataset in the order they were found on input. If, for example, observations 2, 30, and 45 had the same value of a BY variable, they would be written to the dataset as three consecutive observations, 2 coming first, then 30 and 45.

POINTERS AND PITFALLS

Before sorting, determine if the sort is really necessary: are the data already in the order you require? Can groups of data be processed with the NOTSORTED option?

Be sure you understand the output dataset's order, especially when processing with two or more BY variables. List some observations of the output dataset to be sure the sort worked correctly. (The PRINT procedure can be used for this. It is described later in Section 5.3.)

Examples

Key features and options used in this section are summarized in the following list:

Exhibit	Features/Options
5.10	Separate input and output datasets
5.11	In-place sort; NODUP

The dataset SALES is first sorted by ascending values of the variable PROD-. UCT. The sorted dataset, SALES2, is shown in Exhibit 5.10.

EXHIBIT 5.10   Separate input and output datasets

```
------------------- Program listing -------------------

proc sort data=sales out=sales2;
by product;
run;

------------------- Listing, SALES2 -------------------

PRODUCT    REGION     GROSS

 00100       NE       34800
 00100       SW       45890
 00100       MA       39050
 00110       MA       18730
 001A9       NE      434000
 00201       MA      754000
 A0034       NE      128000
 A0034       SW       90000
```

> Smallest values of PRODUCT come first in the sorted dataset.

In Exhibit 5.11 the dataset SALES is sorted by two variables. The output dataset has ascending values of REGION and descending values of GROSS within each REGION. We suspect that there may be duplicate records in the dataset and

EXHIBIT 5.11   An "in-place" sort, eliminating duplicate records

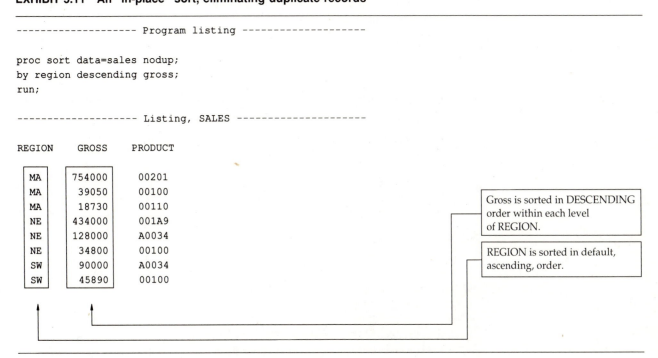

```
------------------- Program listing -------------------

proc sort data=sales nodup;
by region descending gross;
run;

------------------- Listing, SALES -------------------

REGION     GROSS     PRODUCT

  MA       754000     00201
  MA        39050     00100
  MA        18730     00110
  NE       434000     001A9
  NE       128000     A0034
  NE        34800     00100
  SW        90000     A0034
  SW        45890     00100
```

> Gross is sorted in DESCENDING order within each level of REGION.

> REGION is sorted in default, ascending, order.

use NODUP to eliminate the duplicates. The sort is in-place, overwriting the original version of SALES.

Exercises

5.4. Write SORT procedure statements to satisfy the following requirements:
   a. Sort dataset RESULT80 in-place. Arrange it in ascending order by variable LAB_ID.
   b. Follow the instructions in part **a**, but write the sorted dataset to RES80SRT.
   c. Sort STATES by POP_GRP within REGION. That is, POP_GRP should be arranged in low-to-high order within each level of REGION, also sorted low to high. Eliminate duplicate records in the dataset.

5.5. What, if anything, is wrong with the following SORT statements?
   a. `proc sort nodup data=master out=; by area;`
   b. `proc sort out=sortlab;`

5.3   Listing the Dataset: PROC PRINT

One of the most useful applications for a dataset is a simple listing. This section describes the PRINT procedure. It can list all or just a few of the variables, "dress up" the display, and calculate subtotals and totals for numeric variables.

The listing can be an end in itself. Typically, however, it is one of the first steps down the data management and analysis road. It provides a quick visual check that the data appear correct, thus helping to prevent later confusion and misleading results.

POINTERS AND PITFALLS

Use PRINT liberally. Listing all or part of the dataset is a simple and effective means of diagnosing problems with the data (too many missing values, incorrect input formats, and so on).

Syntax

The syntax for the PRINT procedure is described below. These statements are unique to PRINT:

```
PROC PRINT [DATA=dataset] [N] [U] [D] [LABEL]
           [SPLIT='char'] [NOOBS] ;
[ID idvars;]
[SUM sumvars;]
[SUMBY sumbyvar;]
[PAGEBY pagebyvar;]
```

Some common applications of the PRINT procedure follow:

```
proc print data=sales;
id district;
sum gross net discount profit;
by area;

proc print data=maillist split='*' n u;
id zipcode;
```

The PRINT procedure has the following components:

1. The *dataset* identifies the dataset to print. If the *DATA* option is not entered, the most recently created dataset is used.

2. *N* prints an observation count at the end of the listing. If BY-groups are used, it also prints the number of observations at the end of each BY-group.

3. *U* requests uniform display of each page/screen of the listing. By default, SAS adjusts the width of the columns (but not their order) from page to page, depending on the size of variable values for that page. This option sometimes requires significant additional computer resources.

4. *D* double-spaces output. If not specified, SAS prints observations with no blank lines separating observations (single spaces).

5. *LABEL* requests variable labels as column headings. If not specified or if specified and a variable label is blank, SAS uses the variable name as the column header. This option usually increases the size of the output file, especially if the *U* option is also specified.

6. *SPLIT='char'* requests variable headings as column labels, just like *LABEL*. *SPLIT*, however, tells SAS to display the label in up to three lines. The location of the line splits is determined by the single character *char* in the text of the label.

7. *NOOBS* suppresses the column of observation numbers on each line of output.

8. The *idvars* identifies one or more variables to print at the beginning of each line of output for the observation. It also suppresses the column of observation numbers printed at the beginning of each line of output. The *ID* statement is especially useful when there are many variables to print, forcing SAS to split an observation across several lines of output. Properly chosen *idvars* in this situation will simplify identification of observations.

9. The *sumvars* identifies numeric variables to add across all observations in the dataset.

10. The *sumbyvar* requests a subtotal whenever *sumbyvar*, a BY statement variable, or any other variable to the left of the *sumbyvar* in the BY statement variable list changes value. The subtotal is printed for variables specified in the *SUM* statement. If *sumbyvar* is not specified, subtotals are printed when *any* BY variable changes.

11. The *pagebyvar* requests a new page/screen of the output file whenever *pagebyvar* or any variable to the left of *pagebyvar* in the BY statement variable list changes value. If not specified, pages and screens end only when there is no more room to display more lines.

Statements common to all PROCs that can be used with PRINT include VARIABLES, FORMAT, LABEL, ATTRIB, TITLE, FOOTNOTE, and BY. These statements were described earlier in this chapter.

Usage Notes

PROC PRINT has the following features:

- The VARIABLES statement determines the order (from left to right) in which variables not in the ID or BY statements are displayed. If not specified, all variables in the dataset will be printed in the order they were defined to SAS.

- If PAGEBY or SUMBY is used, a BY statement must be present.

- If a BY statement is used, PRINT will display the name and value of the BY variable(s) only when the value of one of the BY variable(s) changes.

Exhibits 5.14 and 5.15 illustrate how the appearance of the output is altered when BY-group processing is used.

- The SUM and ID statements do not have to use variables specified in the VARIABLES statement. This is illustrated in Exhibit 5.13.

Examples

Key features and options used in this section are summarized in the following list:

Exhibit	Features/Options
5.12	Use all defaults
5.13	VAR, ID, SUM, and FORMAT statements
5.14	BY statement; LABEL and N options
5.15	Interaction of BY and ID statements

EXHIBIT 5.12   Use PRINT defaults to display the data

```
------------------- Program listing --------------------

proc print data=coastal;
title 'Take all defaults';
run;

------------------- Output listing ---------------------
```

                        Take all defaults   1

OBS	COAST	OCEAN	STATE	GENCST	TIDALCST
1	ec	at	me	228	3478
2	ec	at	nh	13	131
3	ec	at	ma	192	1519
4	ec	at	ri	40	384
5	ec	at	ct	.	618
6	ec	at	ny	127	1850
7	ec	at	nj	130	1792
8	ec	at	pa	.	89
9	ec	at	de	28	381
10	ec	at	md	31	3190
11	ec	at	va	112	3315
12	ec	at	nc	301	3375
13	ec	at	sc	187	2876
14	ec	at	ga	100	2344
15	ec	at	fl	580	3331
16	ec	gu	fl	770	5095
17	ec	gu	al	53	607
18	ec	gu	ms	44	359
19	ec	gu	la	397	7721
20	ec	gu	tx	367	3359
21	wc	pa	ca	840	3427
22	wc	pa	or	296	1410
23	wc	pa	wa	157	3026
24	wc	pa	hi	750	1052
25	wc	pa	ak	5580	31383
26	wc	ar	ak	1060	2521

Variables are listed in their order in the dataset.

OBS, an observation counter, is produced automatically by the procedure.

Exhibit 5.12 uses all defaults for options, variables, and display formats.

Exhibit 5.13 uses the VAR, ID, SUM, and FORMAT statements. The variables to print are specified (VAR statement). The variable COAST is used to identify each observation (ID statement). All numeric variables are added (SUM

EXHIBIT 5.13   Use VAR, ID, SUM, and FORMAT statements

```
------------------- Program listing -------------------

proc print data=coastal;
var ocean state gencst tidalcst;
id coast;
sum _numeric_;
format gencst tidalcst comma7.;
title 'Use VAR, ID, SUM, and FORMAT statements';
run;

------------------- Output listing --------------------
```

Use VAR, ID, SUM, and FORMAT statements   1

COAST	OCEAN	STATE	GENCST	TIDALCST
ec	at	me	228	3,478
ec	at	nh	13	131
ec	at	ma	192	1,519
ec	at	ri	40	384
ec	at	ct	.	618
ec	at	ny	127	1,850
ec	at	nj	130	1,792
ec	at	pa	.	89
ec	at	de	28	381
ec	at	md	31	3,190
ec	at	va	112	3,315
ec	at	nc	301	3,375
ec	at	sc	187	2,876
ec	at	ga	100	2,344
ec	at	fl	580	3,331
ec	gu	fl	770	5,095
ec	gu	al	53	607
ec	gu	ms	44	359
ec	gu	la	397	7,721
ec	gu	tx	367	3,359
wc	pa	ca	840	3,427
wc	pa	or	296	1,410
wc	pa	wa	157	3,026
wc	pa	hi	750	1,052
wc	pa	ak	5,580	31,383
wc	ar	ak	1,060	2,521
			=======	========
			12,383	88,633

VAR statement replaces default column order and restricts variables which will be printed.

Sum statement is used to add numeric variables.

ID statement is used to replace OBS.

statement). GENCST and TIDALCST and their sums are printed with commas (FORMAT statement).

    Exhibit 5.14 uses the BY statement and the LABEL and N options. A BY statement is used to separate the listing by levels of the variable COAST. Labels rather than names are used to identify the variables (LABEL option and statement). A count of the number of observations printed is produced (N). Notice that since BY groups are used, the count for each group plus the grand total is printed.

EXHIBIT 5.14   BY-group processing; LABEL, N options

```
------------------- Program listing --------------------

proc print data=coastal split=' ' n;
var ocean state gencst tidalcst;
label tidalcst = 'Detailed outline' ;
sum _numeric_;
by coast;
format gencst tidalcst comma7.;
title 'Use VAR, FORMAT, BY, and SUM statements';
title2 'Use TIDALCST label for column header, print # of obs';
run;

------------------- Output listing ---------------------

                Use VAR, FORMAT, BY, and SUM statements
            Use TIDALCST label for column header, print # of obs
```

```
------------------------------- COAST=ec --------------------------------
```

BY-group processing forces different display of the data.

OBS	OCEAN	STATE	GENCST	Detailed outline
1	at	me	228	3,478
2	at	nh	13	131
3	at	ma	192	1,519
4	at	ri	40	384
5	at	ct	.	618
6	at	ny	127	1,850
7	at	nj	130	1,792
8	at	pa	.	89
9	at	de	28	381
10	at	md	31	3,190
11	at	va	112	3,315
12	at	nc	301	3,375
13	at	sc	187	2,876
14	at	ga	100	2,344
15	at	fl	580	3,331
16	gu	fl	770	5,095
17	gu	al	53	607
18	gu	ms	44	359
19	gu	la	397	7,721
20	gu	tx	367	3,359
			-------	--------
COAST			3,700	45,814

SPLIT option tells PRINT to use labels as column headers.

N = 20

N option in PROC statement displays counts for each BY-group

EXHIBIT 5.14 *(continued)*

```
-------------------------------- COAST=wc ---------------------------------

                                           Detailed
        OBS    OCEAN    STATE    GENCST     outline

        21      pa       ca        840       3,427
        22      pa       or        296       1,410
        23      pa       wa        157       3,026
        24      pa       hi        750       1,052
        25      pa       ak      5,580      31,383
        26      ar       ak      1,060       2,521
                                -------    --------
       COAST                      8,683      42,819
                                =======    ========
                                12,383      88,633
```

> N = 6
> Total N = 26 ◄────── N option in PROC displays BY-group and total counts.

Exhibit 5.15 is the same as Exhibit 5.14 except that an ID statement specifying the same variable used in the BY statement is added. Notice how the appearance of the output differs from Exhibit 5.14: PRINT displays only the first occurrence of the BY-group variable and does not insert dashed lines across the page.

Exercises

5.6. Write PRINT statements for the following scenarios.

    a. Print RATES90. Add all numeric variables. Identify each line of an observation's listing by variables YRDIAG and YR_SURV. Print a count of the number of observations at the bottom of the listing.

    b. Print dataset SALES. Begin each REGION on a new page. Do not print the automatic observation numbers PRINT usually displays. Double-space the output, and ensure that all pages are printed with an identical format.

    c. Print HOOP using variable labels as column headings. The split character is an asterisk (*). Identify each observation by variables TEAM and YEAR.

    d. Print the most recently created dataset. Just display the numeric variables, adding them for each level of the variable NAME.

5.7. What, if anything, is wrong with the following PRINT statements (each example contains all statements used for the PROC)?

    a.
```
proc print;
sumby region;
```

    b.
```
proc print split='//';
label cohort = 'Cohort/effect';
```

    c.
```
proc print count data=distrib;
sumby sector;
by region market;
```

EXHIBIT 5.15   **BY-ID statement interaction**

```
------------------ Program listing ------------------

proc print data=coastal split=' ' n;
var ocean state gencst tidalcst;
id coast;
label tidalcst = 'Detailed outline' ;
sum _numeric_;
by coast;
format gencst tidalcst comma7.;
title 'Use VAR, ID, FORMAT, BY, and SUM statements';
title2 'Use TIDALCST label for column header, print # of obs';
run;

------------------ Output listing ------------------
```

```
          Use VAR, ID, FORMAT, BY, and SUM statements                    1
          Use TIDALCST label for column header, print # of obs


                                              Detailed
    COAST    OCEAN    STATE    GENCST    outline

    ec         at       me       228       3,478
               at       nh        13         131
               at       ma       192       1,519
               at       ri        40         384
               at       ct         .         618
               at       ny       127       1,850
               at       nj       130       1,792
               at       pa         .          89
               at       de        28         381
               at       md        31       3,190
               at       va       112       3,315
               at       nc       301       3,375
               at       sc       187       2,876
               at       ga       100       2,344
               at       fl       580       3,331
               gu       fl       770       5,095
               gu       al        53         607
               gu       ms        44         359
               gu       la       397       7,721
               gu       tx       367       3,359
    -----                       -------    --------
    ec                            3,700     45,814

                           N = 20
```

COAST is used as both the ID and BY-group variables.

EXHIBIT 5.15 *(continued)*

wc	pa	ca	840	3,427
	pa	or	296	1,410
	pa	wa	157	3,026
	pa	hi	750	1,052
	pa	ak	5,580	31,383
	ar	ak	1,060	2,521
-----			-------	--------
wc			8,683	42,819
			=======	========
			12,383	88,633

```
              N = 6
       Total N = 26
```

Summary

Chapter 5 discusses the use of SAS datasets in procedures and reviews statements and options common to nearly all procedures described in this book. The DATA-option specifies the dataset to be used in the procedure. The VARIABLES statement indicates the variables to use, the FORMAT statement provides directions for the display of the data in the output listing, the LABEL statement allows annotation of a variable with a character string up to 40 characters long, and ATTRIB combines the functions of FORMAT and LABEL.

The TITLE and FOOTNOTE statements allow up to ten lines of descriptive text to be placed at the top and bottom of the output listing's pages. The BY statement enables processing of the SAS dataset in groups rather than as a whole. The FREQ and WEIGHT statements provide the means to allow an observation in the dataset to contribute, or weight, more or less than other observations, based on the value of a specified variable.

Two simple and useful procedures, SORT and PRINT, were also introduced. SORT rearranges the order of a dataset's observations, and PRINT displays the dataset's observations in the output listing. Both procedures have options to control the arrangement and presentation of their output.

6 Descriptive Statistics

One of the basic tasks of commercial and academic research is calculating descriptive statistics. Measures of central tendency and dispersion, histograms, and frequency distributions give a feel for how the data behave and can identify potential problems in the coding and entry of the data. A histogram, for example, can reveal a population's departure from normally distributed data. A frequency table may reveal invalid values of a variable, thus suggesting potential problems in the data.

Descriptive statistics also have an application not strictly related to research. Simple measures such as minimum, maximum, and counts of missing and nonmissing values are useful data management tools. A minimum or maximum value beyond expected, acceptable values is an indication that some observations have been misread or bad values entered. An unexpectedly large count of missing values for a variable may suggest data collection or entry problems.

POINTERS AND PITFALLS

Descriptive statistics have data handling as well as analytical applications. Do not assume that means, sums, and the like are simply the first steps in statistical analysis.

This chapter discusses four procedures used to generate descriptive statistics: MEANS, UNIVARIATE, FREQ, and CHART. The chapter does not discuss why a particular statistic may be useful or how it is calculated. Rather, it emphasizes how to generate the statistic of interest. The bivariate and multivariate aspects of these PROCs are addressed in Chapters 21 and 22.

6.1   Univariate Statistics: PROC MEANS and PROC UNIVARIATE

MEANS and UNIVARIATE Compared

Both the MEANS and UNIVARIATE procedures readily compute means, standard deviations, minimum and maximum values, and other single-variable measures of central tendency and dispersion. Before describing the syntax and use of each, the PROCs' similarities and differences are briefly summarized.

Similarities.   Both PROCs omit missing values when calculating most statistics. Each uses the same formulas. They print a "." (missing value) if the number of available observations is too small for a meaningful statistic. An observation count of 1, for example, is adequate for the minimum and sum of a variable but insufficient and meaningless for variance.

Finally, both procedures implicitly allow paired comparison *t*-tests. Suppose there is a pair of related variables in each observation, possibly pre- and post-treatment measures. Simply performing *t*-tests (discussed in Chapter 22) on each variable does not test for significant change in the scores. Instead, a new variable in a DATA step that represents the *difference* between the scores must be created. The *t*-test measures on the "difference" variable in MEANS and UNIVARIATE provide a test for the change in the measures being significantly different from zero.

Differences.   UNIVARIATE's output usually uses one page per variable. Most of its statistics are computed automatically: you do not have to request calculation of means, variances, and so on. A statement (ID) is available that allows identification of observations containing the minimum and maximum values in the dataset.

Output from the MEANS procedure is much more compact, sometimes taking only one line per analysis variable. It computes relatively few default statistics. When a nondefault measure is required, *all* the statistics MEANS should compute must be specified. MEANS has the ability to stratify the analysis without requiring the dataset to be in sort order. This feature instructs MEANS to compute the statistics for each distinct level of one or more variables (specified in the CLASS statement). If stratification is required, MEANS with a CLASS statement will usually run much faster than UNIVARIATE using BY-group processing.

Features of MEANS and UNIVARIATE Compared

Exhibit 6.1 compares the capabilities of the MEANS and UNIVARIATE procedures. The availability and default status of various features are noted. Entries in the MEANS "Available" column are keywords used to request the statistic in the MEANS statement.

PROC MEANS

The syntax for MEANS is described below:

```
PROC MEANS [DATA=dataset] [MAXDEC=nplaces] [FW=width]
            [stat ...];
    [CLASS group_var ...;]
    [VAR analysis_var ...;]
```

Some common applications of the MEANS procedure follow:

```
proc means data=scholar;
class state;
var gpa sat_m sat_v act;


proc means data=expense sum mean fw=8 maxdec=2;
var mort food daycare debt;
```

EXHIBIT 6.1   Comparison of MEANS and UNIVARIATE features

Statistic/Feature	PROC MEANS Available	Default	PROC UNIVARIATE Available	Default
Number of nonmissing observations	N	yes	yes	yes
Number of missing observations	NMISS		yes	yes
Total number of observations			yes	yes
Mean	MEAN	yes	yes	yes
Median			yes	yes
Mode			yes	yes
Standard deviation	STD	yes	yes	yes
Variance	VAR		yes	yes
Minimum	MIN	yes	yes	yes
Maximum	MAX	yes	yes	yes
Range	RANGE		yes	yes
Uncorrected sum of squares	USS			
Corrected sum of squares	CSS			
Covariance	CV			
Skewness	SKEWNESS		yes	
Kurtosis	KURTOSIS		yes	
Student's t			yes	yes
Probability of non-0 t	PRT		yes	yes
Upper quartile (75th percentile)			yes	yes
Interquartile range			yes	yes
Percentiles: 1, 5, 10, 90, 95, 99			yes	yes
Signed rank statistic			yes	yes
Kolmogorov statistic			yes	yes
Shapiro-Wilk statistic			yes	yes
Box plots			yes	
Stem-and-leaf plots			yes	
Normal probability plot			yes	
Test for H0: normal distribution			yes	

```
proc means data=cohort1 n nmiss sum mean;
class control;
```

The MEANS procedure has the following components:

1. The *dataset* indicates the name of the dataset to be analyzed. If the *DATA* option is not entered, the most recently created dataset is used.

2. The *nplaces* specifies the number of decimal places to use when printing results. It may range from 0 to 8.

3. The *width* specifies the number of columns to use when printing the statistics. The default is 12.

4. The *stat* is a univariate statistic to compute for all variables. The *stat* values and their meaning are summarized in Exhibit 6.1. If not specified, MEANS usually prints the number of nonmissing observations (*stat* value *N*), the mean (*MEAN*), standard deviation (*STD*), minimum (*MIN*), and maximum (*MAX*).

5. The *group_var* specifies stratifiers for the analysis. MEANS computes the requested statistics for each combination of *group_var* values. *CLASS* is optional and has no defaults. If not entered, MEANS summarizes over the entire dataset. The *group_var* may be any combination of numeric and char-

acter variables. The analysis dataset does *not* have to be sorted by *group_var*. The CLASS statement is available in only version 6 of the SAS System.

6. The *analysis_var* is any combination of numeric variables and variable lists. The *VAR* statement identifies variables to use in the analysis. If it is not entered, all numeric variables are analyzed.

Statements common to all PROCs that can be used with MEANS include FORMAT, LABEL, ATTRIB, TITLE, FOOTNOTE, BY, and FREQ. For a complete description of these statements, see Chapter 5.

Usage Notes

PROC MEANS has the following features:

- Both variable names and LABELs, if any, are printed along with their statistics. If the output line width is insufficient, the variable labels may not be printed.

- If statistic specifications are missing in the PROC statement, at least N, MEAN, STD, MIN, and MAX are calculated. If *any* statistics are specified in the PROC statement, *only* those statistics will be printed.

- Both the CLASS and BY statements may appear in the same execution of MEANS. This prints the *group_vars* breakdown for each level of the BY variable(s).

- MEANS cannot print statistics for some variables and not others. All statistics are printed for all variables.

- Statistics and variables print in the order in which they were specified in the PROC and VAR statements.

Examples

The examples in this section use the coastal shoreline dataset introduced in Chapter 5. The dataset is described in detail in Appendix E. Key features and options used in this section are summarized in the following list:

Exhibit	Features/Options
6.2	Use all defaults
6.3	MAXDEC; labels; user-specified statistics
6.4	CLASS statement

EXHIBIT 6.2   MEANS, using all defaults

```
proc means data=coastal;
title 'Take all defaults';
run;
```

------------------- Output listing begins -------------------

                         Take all defaults

N Obs	Variable	N	Minimum	Maximum	Mean	Std Dev
26	GENCST	24	13.0000000	5580.00	515.9583333	1118.06
	TIDALCST	26	89.0000000	31383.00	3408.96	5965.25

Default statistics ←

Analysis variables, by default, are all numeric variables.

Exhibit 6.2 uses all defaults: statistics, variables, and formats.

Exhibit 6.3 controls the decimal places and specifies the statistics to be printed. The number of decimal places printed is controlled by MAXDEC. Specific statistics are requested using N, NMISS, SUM, MEAN, STD, VAR, MIN, and MAX. Variable labels are used to make output more readable.

EXHIBIT 6.3   Control decimal places; specify statistics to print

```
proc means data=coastal maxdec=2
     n nmiss sum mean std var min max;
label gencst = 'General outline'
      tidalcst = 'Detailed outline'

title 'Override default statistics, control decimal places';
title2 'Use LABEL statement';
run;
```

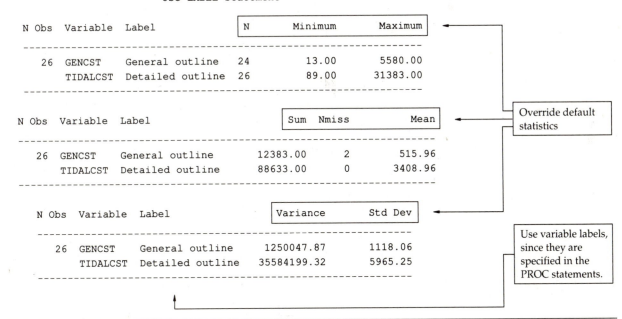

```
------------------- Output listing begins -------------------

         Override default statistics, control decimal places
                    Use LABEL statement

   N Obs   Variable   Label           N      Minimum      Maximum

   -----------------------------------------------------------
     26   GENCST    General outline   24      13.00       5580.00
          TIDALCST  Detailed outline  26      89.00      31383.00
   -----------------------------------------------------------


   N Obs   Variable   Label              Sum   Nmiss        Mean

   -----------------------------------------------------------
     26   GENCST    General outline   12383.00     2       515.96
          TIDALCST  Detailed outline  88633.00     0      3408.96
   -----------------------------------------------------------


    N Obs   Variable   Label             Variance    Std Dev

   -----------------------------------------------------------
      26   GENCST    General outline    1250047.87   1118.06
           TIDALCST  Detailed outline  35584199.32   5965.25
   -----------------------------------------------------------
```

Override default statistics

Use variable labels, since they are specified in the PROC statements.

Exhibit 6.4 uses the CLASS statement. The number of decimal places to print (MAXDEC option) for the SUM and MEAN of the analysis variables is specified. A CLASS statement with two stratifying variables, OCEAN and LENGTH, is used. The procedure calculates the statistics for every nonmissing combination of these variables' values.

PROC UNIVARIATE   The syntax for UNIVARIATE is described below:

```
PROC UNIVARIATE [DATA=dataset] NORMAL
                      PLOT ;
   [ID id_var;]
   [VAR var_list;]
```

EXHIBIT 6.4   CLASS statement with two stratifiers

```
data coastal;
infile '\book\data\raw\coast.raw';
input coast $ 1-2 ocean $ 3-4 state $ 7-8
      gencst 9-14 tidalcst 15-21;
if          gencst > 250  then length = 'Long ';
   else if gencst > 0    then length = 'Short';
run;

proc means data=coastal maxdec=2 sum mean;
label gencst = 'General outline'
      tidalcst = 'Detailed outline'
      ;
class ocean length;
title 'Override default statistics, control decimal places';
title2 'Use LABEL and CLASS statements';
run;
```

```
------------------ Output listing begins ------------------

Override default statistics, control decimal places
Use LABEL and CLASS statements
```

OCEAN	LENGTH	N Obs	Variable	Label	Sum	Mean
ar	Long	1	GENCST	General outline	1060.00	1060.00
			TIDALCST	Detailed outline	2521.00	2521.00
at	Long	2	GENCST	General outline	881.00	440.50
			TIDALCST	Detailed outline	6706.00	3353.00
	Short	11	GENCST	General outline	1188.00	108.00
			TIDALCST	Detailed outline	21260.00	1932.73
gu	Long	3	GENCST	General outline	1534.00	511.33
			TIDALCST	Detailed outline	16175.00	5391.67
	Short	2	GENCST	General outline	97.00	48.50
			TIDALCST	Detailed outline	966.00	483.00
pa	Long	4	GENCST	General outline	7466.00	1866.50
			TIDALCST	Detailed outline	37272.00	9318.00
	Short	1	GENCST	General outline	157.00	157.00
			TIDALCST	Detailed outline	3026.00	3026.00

Number of observations in each CLASS variable level

CLASS variables

Some common applications of PROC UNIVARIATE follow:

```
proc univariate data=round2 plot;

proc univariate data=clinics;
id clinname;
by region;
```

The UNIVARIATE procedure has the following elements:

1. The *dataset* indicates the name of the dataset to be analyzed. If the *DATA* option is not entered, the most recently created dataset is used.

2. *NORMAL* computes tests of the hypothesis that the variable comes from a normal distribution.

3. *PLOT* prints stem-and-leaf, box, and normal probability plots. The appearance of these plots depends on the number of distinct levels of the analysis variable.

4. *ID* specifies *id_var*, a variable to identify the largest and smallest values of the analysis variables. Up to eight characters or significant digits of this character or numeric variable are displayed. If the *ID* statement is not entered, UNIVARIATE uses observation numbers to assist identification.

5. The *var_list* is any combination of numeric variables and variable lists. The *VAR* statement identifies variables to use in the analysis. If it is not entered, all numeric variables are analyzed.

Statements common to all PROCs that can be used with UNIVARIATE include FORMAT, LABEL, ATTRIB, TITLE, FOOTNOTE, BY, and FREQ. These statements are described in detail in Chapter 5.

Usage Notes

PROC UNIVARIATE has the following features:

- Unlike MEANS, UNIVARIATE does not give many choices about which statistics are calculated. This may result in the production of large amounts of unneeded statistics.

- UNIVARIATE uses at least one page of output per variable. It is very easy to use large amounts of paper with a small UNIVARIATE program!

Examples

Key features and options used in this section are summarized in the following list:

Exhibit	Features/Options
6.5	Use all defaults
6.6	PLOT option; LABEL and ID statements
6.7	BY-group processing

Exhibit 6.5 takes defaults for statistics, options, and variables.

Exhibit 6.6 (see page 96) requests plots (PLOT option), uses VAR labels (LABEL statement), and specifies a variable to assist identification of extreme values (ID statement).

Exhibit 6.7 (see page 98) is similar to Exhibit 6.6, but BY-group processing is added. Since the NOTSORTED option was specified, the analysis takes place for groups of OCEAN, displayed in the order in which they were found while UNIVARIATE was passing through the dataset. Only OCEAN levels "at" and "gu" are shown.

EXHIBIT 6.5   Default UNIVARIATE output

```
proc univariate data=coastal;
title 'Take all defaults';
run;
```

-------------------- Output listing begins --------------------

Take all defaults

UNIVARIATE PROCEDURE

| Variable=GENCST | ◄── | Analysis variable |

Moments

N	24	Sum Wgts	24
Mean	515.9583	Sum	12383
Std Dev	1118.055	Variance	1250048
Skewness	4.377855	Kurtosis	20.34983
USS	35140213	CSS	28751101
CV	216.6949	Std Mean	228.2221
T:Mean=0	2.260773	Prob>|T|	0.0335
Sgn Rank	150	Prob>|S|	0.0001
Num ^= 0	24			◄──────── Number of observations not equal to zero (0)

Quantiles(Def=5)

100% Max	5580	99%	5580
75% Q3	488.5	95%	1060
50% Med	189.5	90%	840
25% Q1	76.5	10%	31
0% Min	13	5%	28
		1%	13
Range	5567		
Q3-Q1	412		
Mode	13		

Extremes

Lowest	Obs	Highest	Obs
13(	2)	750(	24)
28(	9)	770(	16)
31(	10)	840(	21)
40(	4)	1060(	26)
44(	18)	5580(	25)

Missing Value	.
Count	2
% Count/Nobs	7.69

◄──────── Missing value information; printed only if a variable actually has missing values

EXHIBIT 6.5 *(continued)*

- - - - - - - - - - - - - - - - - New page - - - - - - - - - - - - - - - - -

Take all defaults

UNIVARIATE PROCEDURE

Variable=TIDALCST ← ─────────────────────────────────── Analysis variable

Moments

N	26	Sum Wgts	26
Mean	3408.962	Sum	88633
Std Dev	5965.249	Variance	35584199
Skewness	4.437806	Kurtosis	21.24676
USS	1.1918E9	CSS	8.896E8
CV	174.9873	Std Mean	1169.882
T:Mean=0	2.913937	Prob>|T|	0.0074
Sgn Rank	175.5	Prob>|S|	0.0001
Num ^= 0	26		

Quantiles(Def=5)

100% Max	31383	99%	31383
75% Q3	3359	95%	7721
50% Med	2432.5	90%	5095
25% Q1	618	10%	359
0% Min	89	5%	131
		1%	89

Range	31294
Q3-Q1	2741
Mode	89

Extremes

Lowest	Obs	Highest	Obs
89(	8)	3427(	21)
131(	2)	3478(	1)
359(	18)	5095(	16)
381(	9)	7721(	19)
384(	4)	31383(	25)

Exercises

6.1. Refer to Exhibit 6.2. If you did *not* take all the default options (as the exhibit did), how would you write the PROC and its associated statements?

6.2. Write MEANS and UNIVARIATE statements for the following situations. You can choose MEANS or UNIVARIATE when possible.

    a. Dataset SALES has to have means, sums, standard deviations, the number of missing observations, and the number of nonmissing observations calculated for each level of DISTRICT. All numeric variables should be analyzed.

    b. Stem-and-leaf plots of variables NET and GROSS are needed. Both variables are in dataset SALES.

EXHIBIT 6.6   PLOT option; LABEL and ID statements

```
proc univariate data=coastal plot;
label tidalcst = 'Detailed outline' ;
var tidalcst;
id state;
title 'Request plots, use VAR, ID, and LABEL statements';
run;
```

------------------- Output listing begins -------------------

Request plots, use VAR, ID, and LABEL statements

UNIVARIATE PROCEDURE

Variable=TIDALCST   Detailed outline

Moments

N	26	Sum Wgts	26
Mean	3408.962	Sum	88633
Std Dev	5965.249	Variance	35584199
Skewness	4.437806	Kurtosis	21.24676
USS	1.1918E9	CSS	8.896E8
CV	174.9873	Std Mean	1169.882
T:Mean=0	2.913937	Prob>|T|	0.0074
Sgn Rank	175.5	Prob>|S|	0.0001
Num ^= 0	26		

Quantiles(Def=5)

100% Max	31383	99%	31383
75% Q3	3359	95%	7721
50% Med	2432.5	90%	5095
25% Q1	618	10%	359
0% Min	89	5%	131
		1%	89
Range	31294		
Q3-Q1	2741		
Mode	89		

Extremes

Lowest	ID		Highest	ID	
89	(pa	)	3427	(ca	)
131	(nh	)	3478	(me	)
359	(ms	)	5095	(fl	)
381	(de	)	7721	(la	)
384	(ri	)	31383	(ak	)

EXHIBIT 6.6 *(continued)*

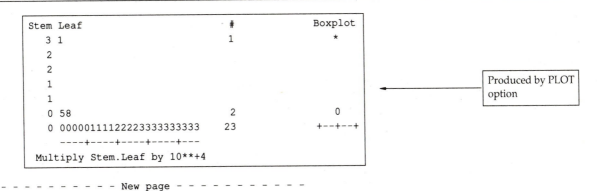

```
Stem Leaf                          #        Boxplot
   3 1                             1           *
   2
   2
   1
   1
   0 58                            2           0
   0 00000111122223333333333      23      +--+--+
     ----+----+----+----+---
Multiply Stem.Leaf by 10**+4
```

Produced by PLOT option

- - - - - - - - - - - - - New page - - - - - - - - - - -

Request plots, use VAR, ID, and LABEL statements

UNIVARIATE PROCEDURE

Variable=TIDALCST   Detailed outline

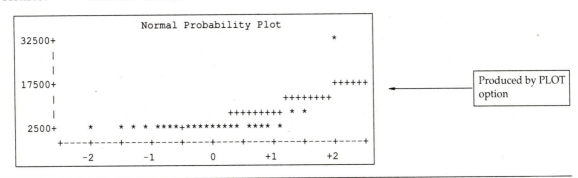

```
                     Normal Probability Plot
   32500+                                       *
        |
        |
   17500+                                    ++++++
        |                              ++++++++
        |                   ++++++++++ *  *
    2500+    *    * * * ****+********* ****  *
        +----+----+----+----+----+----+----+----+----+----+
            -2        -1        0        +1        +2
```

Produced by PLOT option

Exercises
(continued)

c. The five highest and lowest values of NET and GROSS have to be identified. Both variables are in dataset SALES. Each value should be identified by the observation's value of SALEPER.

d. Create a one-page, compact listing of information from the SALES dataset. List number of nonmissing observations for each analysis variable, the minimum, maximum, range, and average. Analyze variables NET and GROSS.

6.3. What, if anything, is wrong with the following MEANS statements (assume that the variables are actually in the dataset)?

a. `proc means maxdec=.2 standard variance count;`
   `class district;`
   `var net gross;`
   `format gross $6. net ;`

b. `proc means data=sales pctile;`
   `class saleper;`
   `variables net gross district saleper;`

c. `proc univariate;`
   `class district;`
   `id saleper product;`

EXHIBIT 6.7 BY-group processing

```
proc univariate data=coastal;
label tidalcst = 'Detailed outline' ;
var tidalcst;
id state;
by ocean notsorted;
title 'Use VAR, ID, LABEL, and BY statements';
run;
```

```
------------------- Output listing begins -------------------
```

```
         Use VAR, ID, LABEL, and BY statements
```

```
-------------------------------- OCEAN=at --------------------------------
```
← BY-group level

```
                    UNIVARIATE PROCEDURE
```

```
Variable=TIDALCST        Detailed outline
```

```
                        Moments
```

N	15	Sum Wgts	15
Mean	1911.533	Sum	28673
Std Dev	1318.686	Variance	1738933
Skewness	-0.19425	Kurtosis	-1.69929
USS	79154455	CSS	24345060
CV	68.98577	Std Mean	340.4833
T:Mean=0	5.614177	Prob>|T|	0.0001
Sgn Rank	60	Prob>|S|	0.0001
Num ^= 0	15		

```
                    Quantiles(Def=5)
```

100% Max	3478	99%	3478
75% Q3	3315	95%	3478
50% Med	1850	90%	3375
25% Q1	384	10%	131
0% Min	89	5%	89
		1%	89
Range	3389		
Q3-Q1	2931		
Mode	89		

```
                      Extremes
```

Lowest	ID	Highest	ID
89	(pa)	3190	(md)
131	(nh)	3315	(va)
381	(de)	3331	(fl)
384	(ri)	3375	(nc)
618	(ct)	3478	(me)

ID statement specified STATE values should be used to identify extreme values.

EXHIBIT 6.7   *(continued)*

- - - - - - - - - - - - - - - - New Page - - - - - - - - - - - - - - - - - -

Use VAR, ID, LABEL, and BY statements

```
------------------------------- OCEAN=gu ------------------------------------
```
← Next BY-group level

UNIVARIATE PROCEDURE

Variable=TIDALCST   Detailed outline

Moments

N	5	Sum Wgts	5
Mean	3428.2	Sum	17141
Std Dev	3106.055	Variance	9647575
Skewness	0.450129	Kurtosis	-1.29311
USS	97353077	CSS	38590301
CV	90.60307	Std Mean	1389.07
T:Mean=0	2.467982	Prob>|T|	0.0691
Sgn Rank	7.5	Prob>|S|	0.0625
Num ^= 0	5		

Quantiles(Def=5)

100%	Max	7721	99%	7721
75%	Q3	5095	95%	7721
50%	Med	3359	90%	7721
25%	Q1	607	10%	359
0%	Min	359	5%	359
			1%	359

Range	7362
Q3-Q1	4488
Mode	359

Extremes

Lowest	ID	Highest	ID
359	(ms)	359	(ms)
607	(al)	607	(al)
3359	(tx)	3359	(tx)
5095	(fl)	5095	(fl)
7721	(la)	7721	(la)

6.2 Counting Categories: PROC FREQ

Another common data management and analysis task is producing frequency tables. These tables count the number of occurrences of a variable's values or categories, such as, how many times does the value "CA" occur for variable STATE? How are individual values of SCORE distributed?

You may also be interested in the *cumulative* distribution of the variable's values. This distribution is the number and percentage of total observations that are included up to and including a particular category. For example, how many observations are classified as 'AK,' 'AL,' and 'AR' by the time category 'CA' is reached? What percent of SCORE values are accounted for by category 3? What level of income group accounts for 50 percent of the population?

This section discusses the use of PROC FREQ to produce frequency distributions. FREQ can also be used for multiway tables and measures of association. Use of FREQ for these instances is discussed in Chapter 22.

Syntax

FREQ's syntax is described below:

```
PROC FREQ [DATA=dataset]
          [ORDER=FREQ|DATA|INTERNAL|FORMATTED];
[TABLES vars ... [ / [MISSING] [NOCUM]] ; ]
```

Some common applications of PROC FREQ follow:

```
proc freq data=struc;


proc freq data=cohorts;
tables group_a group_b / missing;
```

PROC FREQ has the following elements:

1. The *dataset* is the name of the dataset to be analyzed. If the *DATA* option is not entered, the most recently created dataset is used.

2. *ORDER* controls the presentation of the table's categories of nonmissing values. *ORDER=FREQ* prints the variable's most frequently occurring category at the top of the table and the least frequent at the bottom. *ORDER=DATA* lists categories in the order they are found when FREQ reads the dataset. *ORDER=INTERNAL* arranges categories in their internal SAS representation from low values to high. *ORDER=FORMATTED* arranges the categories according to their display formats (see Chapters 14 and 15 for a discussion of formats). *INTERNAL* is the default.

3. *TABLES* specifies analysis variables. The format of *vars* is identical to that of a VAR statement's variable list: any combination of numeric and character variables and variable lists may be entered. If *TABLES* is not entered, all variables in the dataset are included in the analysis.

4. The / tells FREQ that the *TABLES* list ends and that one or more options follows. It is required if either *MISSING* or *NOCUM* is specified.

5. *MISSING* treats missing values of *vars* as nonmissing when calculating percentages and cumulative counts. By default, FREQ prints only the frequency of missing values.

6. *NOCUM* suppresses printing of cumulative frequencies and percentages. Only the individual category counts and percent of total are displayed.

Statements common to all PROCs that can be used with FREQ include FORMAT, LABEL, ATTRIB, TITLE, FOOTNOTE, and BY. For a complete description of these statements, see Chapter 5. Notice that the VAR statement is not recognized by FREQ and is not a substitute for the TABLES statement.

Usage Notes

PROC FREQ has the following features:

- Even though the TABLES statement is optional, it is good practice to specify it. Without a TABLES specification, FREQ produces a distribution for *every* variable in the dataset. This often creates useless calculations and vast amounts of printout. Consider the following scenario: a dataset contains employee names, salaries, and salary grades. FREQ considers employee names and *each* salary as a separate category in the frequency table! If TABLES were omitted, a listing for even a small dataset could easily produce hundreds of pages of meaningless figures. Usually only variables with ordinal or nominal scales are used with FREQ. Formats, particularly user-written ones (discussed in Chapters 14 and 15), may be used to group interval and ratio-scaled data.

- Multiple TABLE statements may be used (see Exhibit 6.10). This is helpful when an option is needed for some variables but not for others.
- FREQ is sensitive to both case and spacing when it creates categories of character variables. Thus "US," "us," and "U S" would be assigned to separate categories even though they look alike.
- FREQ uses only the leftmost 16 positions of character variables when assigning categories. This may lead to unwanted groupings. For example, the following values would all be grouped into a single category ("Very Small Town,"):

```
Very Small Town, IN
Very Small Town, FL
Very Small Town, NC
```

The information that makes each value distinct (IN, FL, and NC) is contained in positions 18 and 19, beyond the 16-position "reach" of FREQ.

- Both system and user-written formats may be used to group or enhance the display of categories. Commas can be inserted into numbers, for instance, and only the first few positions of a character variable can be used to assign categories. A simple use of formats is illustrated in Exhibit 6.10. More advanced uses of formats are discussed in Chapters 15 and 16.

Examples

Key features and options used in this section are summarized in the following list:

Exhibit	Features/Options
6.8	Use display and format defaults; specify TABLES
6.9	Control table order; include missing values
6.10	Multiple TABLES statements, each with different options

Exhibit 6.8 (see page 102) takes all display and format defaults.

Exhibit 6.9 (see page 103) displays categories for all TABLES variables in most to least frequent order (ORDER option in PROC statement). LABEL is used to annotate the output.

Because different options are desired for different variables, Exhibit 6.10 (see page 104) uses TABLES statements. FORMAT is used to redefine levels of OCEAN: $1. uses only one (the leftmost) position of OCEAN when assigning observations to categories.

Exercises

6.4. Write FREQ statements for the following scenarios. The dataset HYDRO has categorical variables LAKEID and DEPTHGRP.
   a. Print a distribution of LAKEID, arranging categories in order of most to least frequent.
   b. Create a distribution of LAKEID. Suppress cumulative distributions.

EXHIBIT 6.8   Use defaults for dislay and format; specify TABLES

```
proc freq data=coastal;
tables coast ocean length;
title 'Take all defaults in PROC, specify vars in TABLE';
run;
```

------------------ Output listing begins ------------------

Take all defaults in PROC, specify vars in TABLE

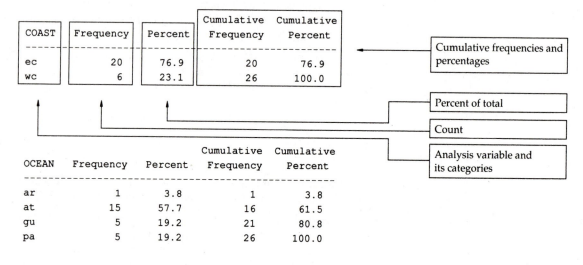

COAST	Frequency	Percent	Cumulative Frequency	Cumulative Percent
ec	20	76.9	20	76.9
wc	6	23.1	26	100.0

Cumulative frequencies and percentages

Percent of total

Count

Analysis variable and its categories

OCEAN	Frequency	Percent	Cumulative Frequency	Cumulative Percent
ar	1	3.8	1	3.8
at	15	57.7	16	61.5
gu	5	19.2	21	80.8
pa	5	19.2	26	100.0

LENGTH	Frequency	Percent	Cumulative Frequency	Cumulative Percent
Long	10	41.7	10	41.7
Short	14	58.3	24	100.0

Frequency Missing = 2

Exercises
(continued)

    c. Print a table of DEPTHGRP, treating its missing values as valid categories.

6.5. What, if anything, is wrong with the following FREQ statements?
    a. `proc freq;`
        `var region style / missing;`
    b. `proc freq order=reverse;`
        `tables / missing nocum;`
    c. `proc freq;`
        `tables missing nocum;`

6.3   Pictorial Displays: PROC CHART

SAS offers a simple way to complement FREQ's tables with more visually oriented output. This section discusses some of the capabilities of the CHART procedure. CHART can produce simple histograms of numeric and character

---

EXHIBIT 6.9   Include missing values in the tables

```
proc freq data=coastal order=freq ;
tables coast ocean length / missing ;
label coast = 'East or West coast'
      ocean = 'Body of water bordered on'
      length = 'Over or under 250 mi. gen. coast'

title 'ORDER=FREQ in PROC, MISSING in TABLES, variable LABELs';
run;
```

------------------- Output listing begins -------------------

ORDER=FREQ in PROC, MISSING in TABLES, variable LABELs

East or West coast   ◄———— Nonblank variable LABELs
                                   improve clarity of the
                                   table.

COAST	Frequency	Percent	Cumulative Frequency	Cumulative Percent
ec	20	76.9	20	76.9
wc	6	23.1	26	100.0

FREQ option displays most
frequent category first.

Body of water bordered on

OCEAN	Frequency	Percent	Cumulative Frequency	Cumulative Percent
at	15	57.7	15	57.7
gu	5	19.2	20	76.9
pa	5	19.2	25	96.2
ar	1	3.8	26	100.0

Over or under 250 mi. gen. coast

LENGTH	Frequency	Percent	Cumulative Frequency	Cumulative Percent
	2	7.7	2	7.7
Short	14	53.8	16	61.5
Long	10	38.5	26	100.0

Missing (blank) values are used
as a valid, nonmissing category
with the MISSING option.

---

variables, where the height or length of each category's bar is proportional to
its occurrence in the dataset. The bars, for example, may count the number of
stocks by their industrial sector. CHART can also create displays whose bars are
the average of a variable for each category of another variable. The bars in these
charts may represent the average number of shares of stock sold by industrial

---

EXHIBIT 6.10 Multiple TABLES, each with different options; FORMAT statement

---

```
proc freq data=coastal order=freq ;
tables coast ocean length / nocum ;
tables length / missing ;
label coast = 'East or West coast'
      ocean = 'Body of water bordered on'
      length = 'Over or under 250 mi. gen. coast'
      ;
format ocean length $1.;
title 'Multiple TABLES, NOCUM and MISSING options';
run;
```

```
------------------ Output listing begins --------------------
```

Multiple TABLES, NOCUM and MISSING options

```
              East or West coast

    COAST    Frequency    Percent
    ---------------------------------
    ec          20         76.9
    wc           6         23.1


           Body of water bordered on

    OCEAN    Frequency    Percent
    ---------------------------------
    a           16         61.5
    g            5         19.2
    p            5         19.2


    Over or under 250 mi. gen. coast

    LENGTH   Frequency    Percent
    ---------------------------------
    S           14         58.3
    L           10         41.7

        Frequency Missing = 2
```

Tables produced by the first TABLES statement: cumulative statistics are suppressed.

```
          Over or under 250 mi. gen. coast

                                  Cumulative  Cumulative
    LENGTH   Frequency   Percent  Frequency    Percent
    ----------------------------------------------------------
                 2         7.7        2          7.7
    S           14        53.8       16         61.5
    L           10        38.5       26        100.0
```

Table produced by the second TABLES statement: use missing value as a valid category.

sector. Bar groups may also be defined, stratifying the analysis by levels of a third variable. The average stock volumes may be grouped by the year they were traded.

CHART has another, more advanced, use. SAS Institute offers a product, SAS/GRAPH, that produces high-quality graphic output (see Chapter 25 for more details). GRAPH procedures allow display of three-dimensional plots, maps, and a more polished form of the CHART procedure. The syntax of the chart procedure for GRAPH and the Base product procedure described in this chapter are nearly identical. This makes CHART an ideal development tool for the more computer-resource-intensive GCHART: you can get the chart contents correct with CHART, then move to GCHART to dress it up with color, bar shadings, and so on.

Using PROC CHART

Although the CHART procedure has many options affecting the appearance and composition of the chart, it is possible to produce useful output with minimal effort. This section presents a simple chart and gives you an idea of what you can change and which options to use. The chart is in Exhibit 6.11.

EXHIBIT 6.11   Default vertical bar chart

```
proc chart data=x.states;
vbar region;
label region = 'US geographic region';
run;

------------------ Output listing begins ------------------
```

FREQUENCY OF [REGION] ⟵────────────────────────── Analysis variable

FREQUENCY

```
       |
       |                                    *****
       |        *****                       *****         *****
       |        *****                       *****         *****
  10  +        *****                       *****         *****
       |        *****        *****          *****         *****
       |        *****        *****          *****         *****
       |        *****        *****          *****         *****       ⟵── Vertical bar orientation
       |        *****        *****          *****         *****
   5  +        *****        *****   *****   *****         *****
       |        *****        *****   *****   *****         *****
       |        *****        *****   *****   *****         *****
       |        *****        *****   *****   *****         *****
       |        *****        *****   *****   *****         *****
       -------------------------------------------------------------
           Midwest     N'east      Pacific    South       West       ⟵── Categories are listed
                                                                           alphabetically.
                      US geographic region  ⟵──  REGION label

                                                                      Number of observations
                                                                      in each bar
```

The dataset's observations are U.S. states and the District of Columbia. The bars represent the distribution of the states by geographic region, the variable REGION. Each bar's height is proportional to its frequency in the dataset. Thus we see the category "South" is present most often (the highest bar), with "Midwest" and "West" tied for second most frequent category. By default, the bars are arranged in ascending order.

Although this chart is informative, it is likely that you will want to enhance its appearance or make it communicate more information. CHART has several groups of features and options that allow you to change the chart to make it more appealing and useful. All the options described in the following subsections are discussed later, in the syntax section.

Inclusion of Bars.   A missing value is not usually considered eligible for display. Sometimes with subgrouping options bars with zero counts are forced into the chart. These defaults are overridden with the MISSING and NOZEROS options.

Order of Bars.   Bars are usually arranged in ascending order, using the values of the analysis variable categories. The ASCENDING and DESCENDING options order the bars by the categories' frequency, mean, or other measure specified by the TYPE option.

Appearance of the Chart.   There are several ways to control the look of the chart. A horizontal or vertical orientation for the chart is selected with the HBAR and VBAR statements. Reference lines are drawn with the REF option. The characters used to draw the bars are specified with the SYMBOL option.

Number of Bars.   Each category of a character variable is represented by a bar. If the variable is numeric, CHART determines an appropriate number of intervals and groups the data. User-written formats (Chapters 15 and 16) and the DISCRETE, LEVELS, and MIDPOINTS options allow more control over the defaults.

Meaning of the Bars.   By default, the bar represents the count of the variable's category in the entire dataset. The TYPE option specifies a different meaning of a bar (average, percentage, and so on), while SUMVAR specifies a variable to use in TYPE's calculation. For example, consider TYPE=MEAN and SUMVAR=SCORE for a chart of variable LAB. The bars represent the average value of SCORE for distinct levels of LAB.

The bar pattern may be repeated for each level of another variable by using the GROUP option. Adding GROUP=VIRUS to the preceding example produces bars that represent the average SCORE for each LAB. This distribution is repeated for each level of VIRUS.

Syntax

The syntax for the CHART procedure is described below. If it seems a bit daunting, keep in mind that you seldom need more than a few options and can even get the job done without using any options!

```
PROC CHART [DATA=dataset];
VBAR|HBAR vars ...
            [/ [MISSING] [DISCRETE] [LEVELS=nlev]
               [MIDPOINTS=midlist] [SUMVAR=svar]
               [TYPE=FREQ|PERCENT|CFREQ|CPERCENT|
                    SUM|MEAN] [NOZEROS]
```

```
[ASCENDING|DESCENDING] [REF=refval]
[SYMBOL='sym']
[vbaropts|hbaropts] ] ;
```

Some common applications of PROC CHART follow:

```
proc chart data=quotas;
vbar dept / sumvar=sales type=mean;

proc chart data=quotas;
hbar dept clerk / group=quarter ref=20000;

proc chart data=quotas;
hbar clerk / type=sum sumvar=net ascending;
```

The CHART procedure has the following options available:

1. The *dataset* is the name of the dataset to be analyzed. If the *DATA* option is not entered, the most recently created dataset is used.

2. *VBAR* requests vertical bar charts for the variables listed in *vars* . . . , while *HBAR* requests horizontal bar charts. The *vars* . . . may be any combination of numeric or character variables and variable lists.

3. The / signals the end of the variable list and the beginning of a list of options. It is required if any option is specified.

4. *MISSING* treats missing values as nonmissing. The default is the exclusion of missings from the display.

5. *DISCRETE* displays each level of an analysis variable in *vars* . . . with a single bar. This option is the default when displaying character data and is useful with numeric data with a limited number of categories. Contrast it with the *LEVELS* and *MIDPOINTS* options described in 6 and 7.

6. The *nlev* specifies the number of *LEVELS* to use as interval midpoints when charting numeric variables. The default is five.

7. The *midlist* specifies values to use as *MIDPOINTS* for numeric variable intervals or as a list of displayable character variable values. The default for numeric variables is five evenly spaced intervals. By default, all levels of character variables are preserved, arranged in low-to-high order. See the "Specifying Midpoints" section later in this chapter for more details.

8. The *svar* specifies a summary variable (*SUMVAR*) other than the chart variables. It is used to compute the bar *TYPE*s described in 9.

9. *TYPE* requests that the chart's bars specify a particular statistic. Choices are *FREQ* (the number of times the value or interval represented by the bar occurred in the dataset), *PERCENT* (the percent of all observations represented by the bar), *CFREQ*, and *CPERCENT* (the cumulative frequency and percentage, respectively, of the categories), and *SUM* and *MEAN* (for numeric variables only, the sum or mean of *vars* . . . or when *SUMVAR* is used, the sum or mean of *svar*).

10. *NOZEROS* suppresses the printing of "empty" bars (those with a category count of zero).

11. *ASCENDING* and *DESCENDING* arranges the bars in low-to-high and high-to-low values of the *TYPE* variable, if specified. If TYPE=MEAN and DESCENDING were specified, the bars would begin with the analysis variable category with the largest mean, followed by the next largest, ending with the category, or bar, with the smallest mean. If neither option is specified, CHART displays bars in either *midlist* or formatted order.

12. *REF* draws a reference line through the bars at *refval*. When *TYPE* is *PERCENT* or *CPERCENT*, *refval* should be between 0 and 100. Otherwise a *refval* within the expected range of the data is used. If the *REF* option is not specified, no lines are drawn.

13. *SYMBOL* uses the single character *sym* rather than an asterisk (*) when displaying bars.

14. The *vbaropts* and *hbaropts* are options that may only be used with vertical or horizontal charts, respectively.

Both the VBAR and HBAR options have some unique limitations, which are described in the following subsections.

Special VBAR Option.   The NOSPACE option may be used only in a VBAR statement. NOSPACE permits CHART to eliminate blank space between bars. If it is not specified and CHART cannot fit all bars on a single page, it will reorient the display as a horizontal chart.

Special HBAR Options.   The following options may be used only in an HBAR statement:

- NOSTAT suppresses the printing of bar frequency statistics to the right of the display. If not specified, the bar's frequency and percentage and the cumulative frequency and percentage are printed.

- FREQ, PERCENT, CFREQ, and CPERCENT individually request printing of the bar's frequency, percentage, cumulative frequency, and cumulative percentage, respectively.

POINTERS AND PITFALLS

When starting to use a new PROC, it is usually easiest to take as many defaults as possible and add options gradually in later runs. This strategy is particularly true with complicated charts.

Usage Notes

PROC CHART has the following features:

- Any number of VBAR and HBAR statements may accompany a single PROC CHART statement.

- Horizontal bar charts have something of an advantage over their vertical counterparts since they display the chart *and* the same types of information (frequency, percent, and so on) provided by PROC FREQ. This gives a visual feel for the data as well as the actual numbers.

- The format of the charts depends in part on the dimensions of the page or screen. There are "sensible" defaults for these settings. See Chapter 12 for more details.

Specifying Midpoints

Midpoint lists for *numeric* variables can be specified in several ways, as illustrated in Exhibit 6.12. Notice that the points do not have to be evenly spaced. The TO and BY specifications create a uniformly spaced midpoint beginning with a

EXHIBIT 6.12   Methods for specifying interval midpoints

```
MIDPOINTS = 5 10 15 20
MIDPOINTS = 5 10 40 80 100
MIDPOINTS = 5 20 TO 40 BY 10      /* points are 5, 20, 30, 40 */
MIDPOINTS = 1 TO 20 BY 2          /* points are 1, 3, ... 17, 19 */
MIDPOINTS = 12 to 1 by -1         /* points are 12, 11, ... 2, 1 */
```

number and ending at another number after counting in units of the BY variable or, by default, one.

Midpoint lists for *character* variables rearrange the default order of bar presentation or display a subset of the data. Exhibit 6.13 shows two possible MID-POINTS statements for values of STATE in a dataset of New England states. The default is presented first, followed by a more north-to-south ordering.

EXHIBIT 6.13   MIDPOINTS rearranging character variable bar order

```
Default:   MIDPOINTS =  'CT' 'MA' 'ME' 'NH' 'RI' 'VT'

Reordered: MIDPOINTS = 'ME' 'NH' 'VT' 'MA' 'RI' 'CT'
```

Continuing with the New England example, Exhibit 6.14 selects only northern New England states: MIDPOINTS specifies only "northern" values of STATE. The display excludes observations with values outside this range.

EXHIBIT 6.14   MIDPOINTS restricting chart content

```
MIDPOINTS = 'ME' 'NH' 'VT'
```

Examples

Key features and options used in this section are summarized in the following list:

Exhibit	Features/Options
6.11	Default VBAR
6.15	Default HBAR
6.16	Specify MIDPOINTS correctly
6.17	Specify MIDPOINTS incorrectly
6.18	SUMVAR
6.19	SUMVAR, TYPE
6.20	Interval-scale variable, default MIDPOINTS
6.21	Ordinal-scale variable, default MIDPOINTS
6.22	DISCRETE
6.23	DISCRETE, GROUP
6.24	DISCRETE, GROUP, NOZEROS
6.25	LEVELS, SUMVAR, TYPE, GROUP
6.26	SUMVAR, TYPE, ASCENDING, SYMBOL, REF

Exhibit 6.15 illustrates the default layout for horizontal bar charts. Exhibit 6.11, used earlier in this section, presented default vertical bar charts. The frequencies and percentages at the right of the table may be suppressed with the NOSTAT option.

EXHIBIT 6.15   Default HBAR

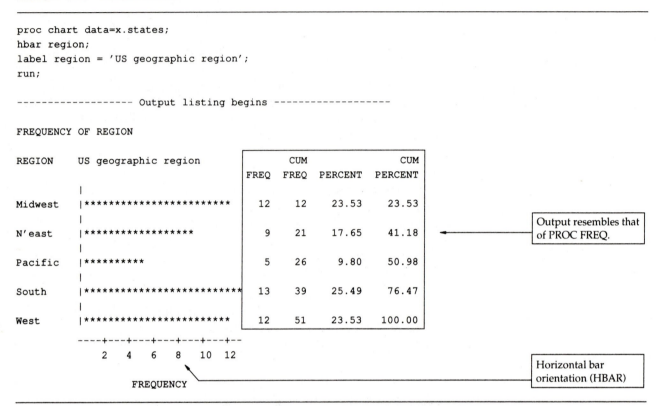

```
proc chart data=x.states;
hbar region;
label region = 'US geographic region';
run;


------------------ Output listing begins ------------------


FREQUENCY OF REGION


REGION    US geographic region
```

REGION	US geographic region		FREQ	CUM FREQ	PERCENT	CUM PERCENT
Midwest	|***********************		12	12	23.53	23.53
N'east	|*****************		9	21	17.65	41.18
Pacific	|*********		5	26	9.80	50.98
South	|*************************		13	39	25.49	76.47
West	|***********************		12	51	23.53	100.00

```
          ----+---+---+---+---+---+--
             2   4   6   8   10  12
                  FREQUENCY
```

Output resembles that of PROC FREQ.

Horizontal bar orientation (HBAR)

Exhibit 6.16 (see page 111) uses the MIDPOINTS option with a character variable to control which levels of REGION are included in the chart. Notice that MIDPOINTS also controls the order of the selected bars.

In Exhibit 6.17 (see page 112) we forget how the dataset specifies the northeastern states and use a midpoint of "Northeast" rather than "N'east." CHART cannot find any observations with this value of REGION and prints a bar with a count of zero. This empty bar could be suppressed with the NOZERO option.

In Exhibit 6.18 (see page 113) the meaning of the bar is changed. Rather than represent a count of the analysis variable as in the previous exhibits, each bar represents the sum (default TYPE statistic) of CRIMERT for each level of REGION.

Since the sum of rates is not a very useful statistic, we decide to override the default value of TYPE and request that the bars represent the MEAN crime rate for each level of REGION. Exhibit 6.19 (see page 114) contains the modified chart.

In Exhibit 6.20 (see page 115) we allow CHART to compute the number and width of the intervals for a numeric analysis variable. The bars represent the number of states whose percent of female-headed households are living in poverty. The numbers below the bars are the midpoints of the intervals selected by CHART.

EXHIBIT 6.16   Correct MIDPOINTS specification

```
proc chart data=x.states;
vbar region / midpoints="N'east" 'South' 'Midwest';
label region = 'US geographic region';
run;

------------------ Output listing begins -------------------
```

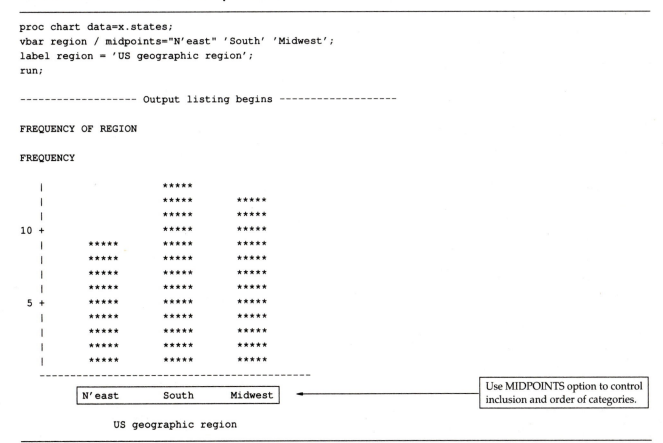

Variable DEN_GRP is numeric, with integer values of 1 through 4. Exhibit 6.21 (see page 116) shows the chart created when no options are taken. CHART cannot distinguish an ordinal numeric like DEN_GRP from an interval one such as PCTFPOV (used in Exhibit 6.20). The interval selection and midpoints obscure the DEN_GRP's values.

In Exhibit 6.22 (see page 117) we clarify the chart created in Exhibit 6.21. The DISCRETE option tells CHART to treat each numeric value of DEN_GRP as a separate category; that is, do not compute intervals and midpoints, just accept the data as is. We now see the distribution we probably expected in the first place!

In Exhibit 6.23, (see page 118) pose the question, How are the density groups represented by DEN_GRP distributed by values of EASTWEST? Use the GROUP option to repeat the analysis in Exhibit 6.22 for each level of variable EASTWEST. Notice that DEN_GRP level 1 does not occur in the East, nor does DEN_GRP value 4 occur in the West. The zero count bars are printed since CHART creates bars for all possible values of DEN_GRP for each value of EASTWEST.

Exhibit 6.24 (see page 119) uses the NOZEROS option to eliminate the bars with values (in this case, frequencies) of zero.

The interpretation of the bars can become complicated as the number of options used increases. In Exhibit 6.25 (see page 120) each bar represents average unemployment benefit categories grouped by geographic region. Notice the different format of the horizontal charts when the GROUP option is used.

EXHIBIT 6.17   Incorrect MIDPOINTS specification

```
proc chart data=x.states;
vbar region / midpoints='Northeast' 'South' 'Midwest';
label region = 'US geographic region';
run;
```

------------------ Output listing begins ------------------

FREQUENCY OF REGION

FREQUENCY

```
     |                 *****
     |                 *****      *****
     |                 *****      *****
10 +                   *****      *****
     |                 *****      *****
     |                 *****      *****
     |                 *****      *****
     |                 *****      *****
 5 +                   *****      *****
     |                 *****      *****
     |                 *****      *****
     |                 *****      *****
     |                 *****      *****
     -------------------------------------------
       ┌─────────┐
       │Northeast│   South      Midwest
       └─────────┘
         ▲
         │    US geographic region
         └─────────────────────────────────
```

A category with no observations: an "empty bar."

Cumulative statistics are eliminated to allow room to print values of the GROUP variable.

Exhibit 6.26 (see page 121) uses the program in Exhibit 6.19 as its starting point and applies some cosmetic changes. The bars are filled with the character *x* rather than an asterisk (*) (the SYMBOL option). A reference line (the REF option) is drawn at 4,000. Finally, the bars are arranged by increasing values of the mean crime rate (the ASCENDING option).

Exercises

6.6. Write CHART statements to meet the following requirements. The analysis dataset is SALES. You can choose the bar orientation (horizontal or vertical) when possible. Variable names are enclosed in parentheses.
   a. The chart's bars represent the average gross sales (GSALE) per product (PROD).
   b. The bars are the average gross sales (GSALE) per product (PROD), grouped by district (DIST).
   c. The bars represent a count of the number of different products (PROD) sold within each region (REGION).
   d. The bars represent the sum of gross sales (GSALE) per product (PROD) within each region (REGION). Suppress bars with zero (0) counts, and arrange the bars in ascending order—smallest sales are printed first.

EXHIBIT 6.18   Use SUMVAR option to change meaning of bars

```
proc chart data=x.states;
vbar region / sumvar=crimert;
label region = 'US geographic region'
      crimert = 'Crime rate per 100,000 pop';
run;


------------------ Output listing begins ------------------

SUM OF CRIMERT BY REGION

CRIMERT SUM


          |                                           *****
          |                                           *****     *****
    6,000 +                                           *****     *****
          |              *****                        *****     *****
          |              *****                        *****     *****
          |              *****     *****              *****     *****
    4,000 +              *****     *****              *****     *****
          |              *****     *****     *****    *****     *****
          |              *****     *****     *****    *****     *****
          |              *****     *****     *****    *****     *****
    2,000 +              *****     *****     *****    *****     *****
          |              *****     *****     *****    *****     *****
          |              *****     *****     *****    *****     *****
          |              *****     *****     *****    *****     *****
          ------------------------------------------------------------
          ^           Midwest    N'east    Pacific   South     West

                          US geographic region
```

Sum of CRIMERT (SUMVAR variable) is represented by each bar.

e. The bars are the number of observations per district (DIST). Display the actual counts beside the chart itself. Fill the bars with the letter 'x' rather than the default asterisk ('*').

f. Follow the instructions in part e, but print bars only for DIST values of 'N,' 'S,' and 'MW,' in that order.

6.7. What, if anything, is wrong with the following chart specification statements?

a. `var district / nozeroes type=sum;`

b. `vbar dist / type=mean sumvar=net group=dist;`

c. `hbar net gross /discrete levels=5;`

EXHIBIT 6.19   Use TYPE to change default bar statistic (to mean CRIMERT)

```
proc chart data=x.states;
vbar region / sumvar=crimert type=mean;
label region = 'US geographic region'
      crimert = 'Crime rate per 100,000 pop';
run;

------------------- Output listing begins -------------------

MEAN OF CRIMERT BY REGION

CRIMERT MEAN
```

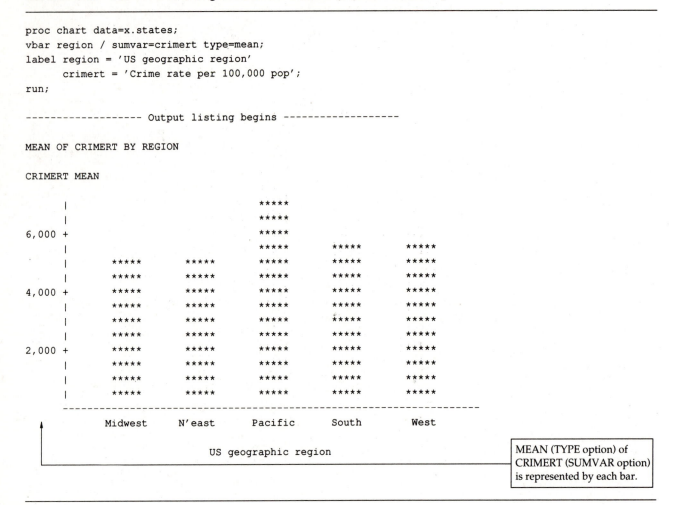

```
         |                          *****
         |                          *****
6,000 +                             *****
         |                          *****     *****     *****
         |          *****   *****   *****     *****     *****
         |          *****   *****   *****     *****     *****
4,000 +          *****   *****   *****     *****     *****
         |          *****   *****   *****     *****     *****
         |          *****   *****   *****     *****     *****
         |          *****   *****   *****     *****     *****
2,000 +          *****   *****   *****     *****     *****
         |          *****   *****   *****     *****     *****
         |          *****   *****   *****     *****     *****
         |          *****   *****   *****     *****     *****
          --------------------------------------------------------
             Midwest     N'east     Pacific     South      West

                       US geographic region
```

MEAN (TYPE option) of
CRIMERT (SUMVAR option)
is represented by each bar.

EXHIBIT 6.20   Use default MIDPOINTS for interval-scale variable

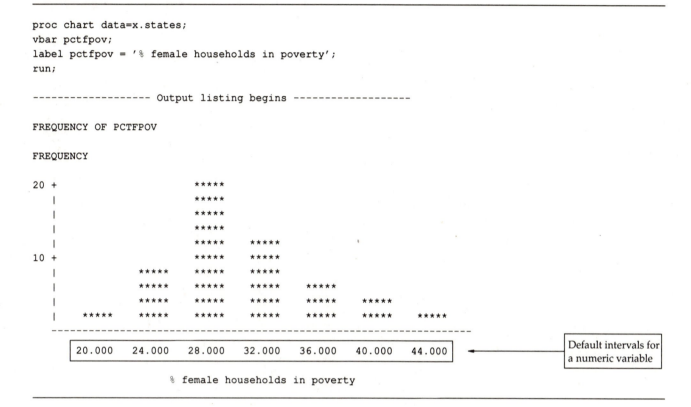

```
proc chart data=x.states;
vbar pctfpov;
label pctfpov = '% female households in poverty';
run;

------------------- Output listing begins -------------------

FREQUENCY OF PCTFPOV

FREQUENCY

20 +                     *****
   |                     *****
   |                     *****
   |                     *****
   |                     *****   *****
10 +                     *****   *****
   |             *****   *****   *****
   |             *****   *****   *****   *****
   |             *****   *****   *****   *****   *****
   |     *****   *****   *****   *****   *****   *****   *****
   --------------------------------------------------------------
```

| 20.000 | 24.000 | 28.000 | 32.000 | 36.000 | 40.000 | 44.000 | ← Default intervals for a numeric variable

% female households in poverty

EXHIBIT 6.21   Use default MIDPOINTS for ordinal-scale variable

```
proc chart data=x.states;
vbar den_grp;
label den_grp = 'Population density group (quartiles)';
run;
```

------------------ Output listing begins ------------------

FREQUENCY OF DEN_GRP

FREQUENCY

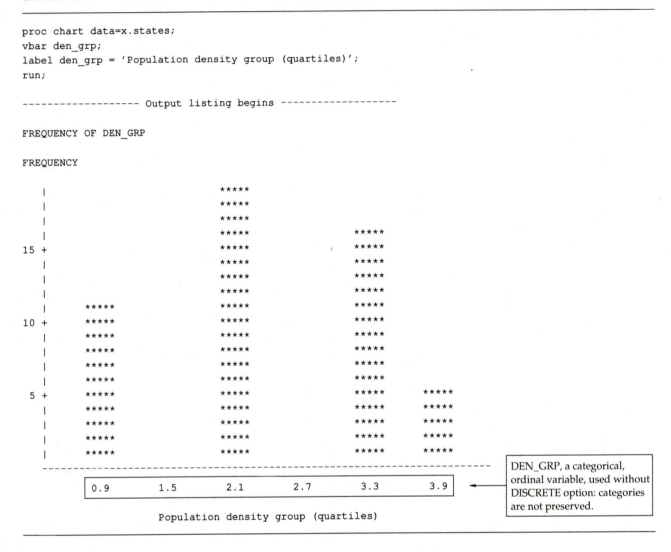

| 0.9 | 1.5 | 2.1 | 2.7 | 3.3 | 3.9 |

DEN_GRP, a categorical, ordinal variable, used without DISCRETE option: categories are not preserved.

Population density group (quartiles)

EXHIBIT 6.22   Use DISCRETE option with an ordinal-scale variable

```
proc chart data=x.states;
vbar den_grp / discrete;
label den_grp = 'Population density group (quartiles)';
run;

------------------- Output listing begins -------------------

FREQUENCY OF DEN_GRP

FREQUENCY

     |                    *****
     |                    *****
     |                    *****
     |                    *****      *****
  15 +                    *****      *****
     |                    *****      *****
     |                    *****      *****
     |                    *****      *****
     |          *****      *****      *****
  10 +          *****      *****      *****
     |          *****      *****      *****
     |          *****      *****      *****
     |          *****      *****      *****
     |          *****      *****      *****
   5 +          *****      *****      *****      *****
     |          *****      *****      *****      *****
     |          *****      *****      *****      *****
     |          *****      *****      *****      *****
     |          *****      *****      *****      *****
     ------------------------------------------------------------
              |    1    |    2    |    3    |    4    |

          Population density group (quartiles)
```

Add DISCRETE option to preserve ordinal variables' categories and avoid calculation of midpoints.

EXHIBIT 6.23   Repeat a chart for each GROUP of variable EASTWEST

```
proc chart data=x.states;
vbar den_grp / discrete group=eastwest;
label den_grp = 'Population density group (quartiles)'
      eastwest = 'East/West Mississippi River indicator';
run;

------------------ Output listing begins ------------------

FREQUENCY OF DEN_GRP GROUPED BY EASTWEST

FREQUENCY
```

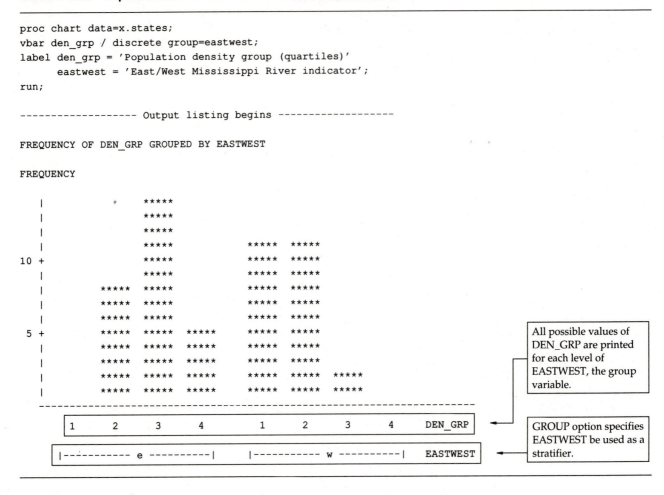

All possible values of DEN_GRP are printed for each level of EASTWEST, the group variable.

GROUP option specifies EASTWEST be used as a stratifier.

EXHIBIT 6.24   Eliminate null (0-count) bars with NOZEROS option

```
proc chart data=x.states;
vbar den_grp / discrete group=eastwest nozeros;
label den_grp = 'Population density group (quartiles)'
      eastwest = 'East/West Mississippi River indicator';
run;

------------------ Output listing begins ------------------

FREQUENCY OF DEN_GRP GROUPED BY EASTWEST

FREQUENCY

     |                 *****
     |                 *****
     |                 *****
     |                 *****             *****     *****
  10 +                 *****             *****     *****
     |                 *****             *****     *****
     |       *****     *****             *****     *****
     |       *****     *****             *****     *****
     |       *****     *****             *****     *****
   5 +       *****     *****     *****   *****     *****
     |       *****     *****     *****   *****     *****
     |       *****     *****     *****   *****     *****
     |       *****     *****     *****   *****     *****     *****
     |       *****     *****     *****   *****     *****     *****
     --------------------------------------------------------------
             2         3         4       1         2         3      DEN_GRP

          |--------- e ---------|     |--------- w ---------|      EASTWEST
```

NOZEROS suppresses printing of empty bars.

EXHIBIT 6.25   Use LEVELS, SUMVAR, TYPE, and GROUP

```
proc chart data=x.states;
hbar pcturb / levels=4 sumvar=unempben type=mean group=eastwest;
label eastwest = 'East/West Mississippi River indicator'
      pcturb  = '% population living in urban areas'
      unempben = 'Monthly unemployment benefits'
      ;
run;
```

```
------------------ Output listing begins ------------------
```

MEAN OF UNEMPBEN BY PCTURB GROUPED BY EASTWEST

```
EASTWEST    PCTURB                                              UNEMPBEN
            MIDPOINT                                  FREQ      MEAN
                    |
e           48.000  |**********************         8         175
            64.000  |***********************        9         182
            80.000  |********************************  8      244
            96.000  |********************************  2      246
                    |
w           48.000  |***********************         5         178
            64.000  |**************************     10         200
            80.000  |***********************         8         190
            96.000  |********************            1         166
                    |
                    --------+-------+-------+-------+-
                        60     120     180     240

                    Monthly unemployment benefits
```

LEVELS option specifies 4 groups. Midpoints are of UNEMPBEN mean values.

EXHIBIT 6.26   Cosmetic changes with SYMBOL, REF, ASCENDING options

```
proc chart data=x.states;
vbar region / sumvar=crimert type=mean ascending symbol='x' ref=4000;
label region = 'US geographic region'
      crimert = 'Crime rate per 100,000 pop';
run;

------------------ Output listing begins ------------------

MEAN OF CRIMERT BY REGION

CRIMERT MEAN
```

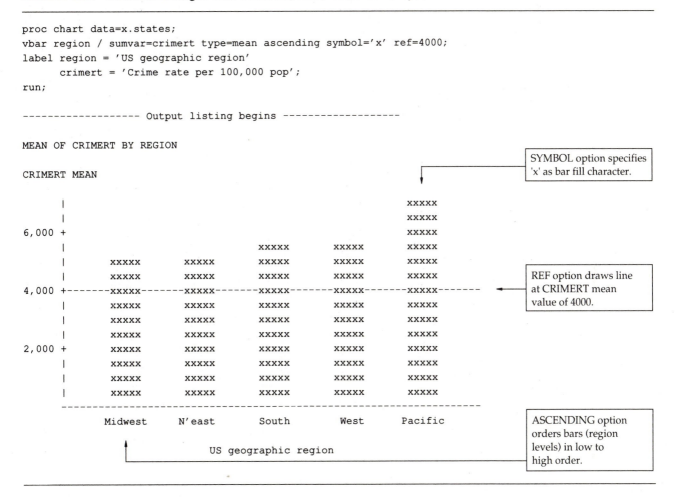

SYMBOL option specifies 'x' as bar fill character.

REF option draws line at CRIMERT mean value of 4000.

ASCENDING option orders bars (region levels) in low to high order.

Summary

Chapter 6 introduces procedures used for single-variable (univariate) statistics. The MEANS and UNIVARIATE procedures produce similar measures of central tendency and dispersion. Both a comparison of the procedures for a variety of statistics and guidelines for choosing between the two are provided. MEANS produces more compact output but does not have UNIVARIATE's scope of statistics and tests.

Frequency distributions are easily produced with the FREQ procedure. Some of FREQ's features are contained in CHART, a graphically oriented procedure. CHART can produce category counts similar to FREQ, while at the same time displaying bars and other graphical devices showing the relative magnitude of a category. CHART's bars may also be adjusted to represent other statistics, such as the mean, sum, cumulative percentage, and count of an analysis variable. CHART's myriad options are illustrated in a lengthy series of proper and inappropriate examples of use.

PART THREE

Refining the Program

7 DATA Step II: Calculations

So far we have seen how to read data, compute simple statistics, and perform other straightforward tasks such as printing the dataset and producing frequency distributions. The sample programs simply read raw data from an external or in-stream file and printed, sorted, and charted the SAS dataset.

This chapter adds some realism to this scenario. It discusses numeric assignment statements: the modification or creation of variables based on their value and/or the values of other variables. The chapter also covers the use of a powerful computational shorthand known as *functions*. Examples throughout the chapter demonstrate the impact of missing values on the calculations. Character variable manipulations are not covered here. They are saved for Chapter 18.

7.1 Numeric Assignment Statements: Syntax, Rules, and Examples

Syntax and Rules

The purpose of an assignment statement is simple: perform one or more arithmetic operations on one or more variables and/or constants, then place the results in a variable (sometimes referred to as a *target* variable). The general syntax for the statement follows:

```
target = expression [arith_oper expression ...];
```

The statement has three important elements:

1. The *target* is the variable being created (no prior assignment or INPUT statements in this DATA step used this variable name) or replaced (the variable was created earlier in the DATA step; it is now being written over).

2. The *expression* is an arithmetic expression manipulating variables and/or constants. For more details, see the discussion of expressions in Section 3.6.

3. The *arith_oper* is an arithmetic operator joining the preceding expression to another expression. Parentheses can be used to enhance the clarity of the joining.

There is no practical limit to the complexity and length of an assignment statement. There are a few reasonable restrictions. First, statement *order* is important: most tasks require that the variables in *expression* have values before they can be used in the expression. Consider the program segment in Exhibit 7.1. The assignment "cart" is put before the INPUT "horse." SAS is asked to manipulate variables that it never heard of. *SAS does not consider this an error.* It simply sets the value of HEIGHT to missing in every observation. If INPUT is placed before the assignment, the results are better: FEET and INCHES can have values when the assignment statement is executed.

EXHIBIT 7.1   Syntactically correct statements out of logical order

```
data teamstat;
height = 12 * feet + inches ;
input name $ 1-20 position $ 22 feet 24 inches 26-27 ;
run;
```

POINTERS AND PITFALLS

SAS is sensitive to *syntax* errors but cannot detect errors in *logic*. Remember that DATA step statements are executed sequentially: values must be assigned to a variable before it can be used in a calculation.

Another restriction is that when SAS encounters a missing variable for *any* variable in *any* expression in the statement, the target variable is set to missing. If a variable is being replaced, the missing value replaces the old, possibly non-missing, value. A common way around this restriction is discussed in Section 7.2.

A word of caution: when replacing variables decide whether you can afford to lose the original value. In some cases the replacement is reasonable. Your initial value may be scaled awkwardly (e.g., thousands of dollars instead of dollars) or in a different unit of measurement than you want (e.g., miles instead of kilometers). Or you may want to adjust the variable by an inflator/deflator (e.g., express as constant dollars). These examples are shown below:

```
dollars = dollars * 1000 ;
trip_len = trip_len * 1.602 ;
gnp = gnp * cpi ;
```

The first two assignments overwrite DOLLARS and TRIP_LEN by multiplying by constants. If you need to get back to the original values, it is a simple matter of dividing by 1000 and 1.602, respectively. GNP, however, may not be as easily recovered if CPI's value is not available in a later DATA step, PROC, or program.

Examples

Exhibit 7.2 illustrates several assignment statements. Notice the use of blank space, parentheses, and multiple lines in more complex expressions. Parentheses are particularly effective in longer assignments. They clarify which operations will be performed first, visually separating distinct portions of the calculation. These and other features do not take long to write and make the statement easy to read. They can even preempt coding errors by highlighting the parallels among different parts of the calculation: the style of the XPROD statement almost guarantees that the correct suffix (1, 2, or 3) will be used.

EXHIBIT 7.2   Sample well-written assignment statements

```
residual = observed - expected ;
interest = prin * rate * time ;
dollars2 = dollark * 1000 ;
loan_val = prin * (1+r)**time ;
xprod = (r1 * d1 * sc1) +
        (r2 * d2 * sc2) +
        (r3 * d3 * sc3) ;
newscore = (score1 * 1.05) + (score2 * .98);
average = (v1 + v2 + v3 + v4 + v5 + v6 +
          v7 + v8 + v9 + v10 + v11 + v12 ) / 12 ;
wgt_avg = (n1*v1 + n2*v2 + n3*v3) / (n1+n2+n3) ;
acres = hectares * 2.471 ;
```

Some of the calculations are straightforward: a variable or constant is multiplied or subtracted from just one other variable. But others are a bit more complex: one (LOAN_VAL) requires using an interest rate formula, while another (AVERAGE) requires entering a dozen variable names. Tedium would quickly turn into irritation if we had to average 20 variables instead of just 12. Numeric functions, discussed in the next section, are an easier, more compact, and foolproof way to do these operations.

Exercises

7.1. Examine the following program segment:

```
input a b c;
b = b * c;
c = b ** 2;
d = c * b;
a = a / 2;
```

a. Which variables are *created* in assignment statements?
b. Which variables are *replaced* in assignment statements?

7.2. An observation has the following values of three variables:

v1 = 2   v2 = 3   v3 = .

What values result from the following operations?
a. v1 + v2
b. v1 + v2 + v3
c. (v1**2) + (v2**2)
d. v3 * 2
e. (v3 = .) + (v2 = 2) [*Hint:* Remember that SAS evaluates expressions as true (0) or false (1).]

7.3. What, if anything, is wrong with the following assignment statements?
  a. `acres = hectares * 2. 471 ;`
  b. `wgt_avg = ((n1*v1)+(n2*v2)_(n3*v3)/(n1+n2+n3) ;`
  c. `hectares = acres * 2.471;`
  d. `residual = (observed - expected) ;`
  e. `newscore = (score1*1.05 + score2*.98) ;`

7.2 Numeric Functions

SAS, like most languages, provides a set of built-in routines that perform common operations on variables and/or constants. These **functions** are extremely powerful and easy to use. They cover a wide range of common scientific, business, and statistical tasks. Just as SAS procedures perform vast amounts of work with just a few statements, functions reduce coding effort in the DATA step. Some of the rules for using functions and some of the commonly used statistical, arithmetic, and financial functions are discussed in this section. Functions for handling character values are covered in Chapter 18.

Basic Components

Functions, with infrequent exceptions, have two components. The first is the name of the function. This tells SAS what you want to do: add, average, take a logarithm, round, and so on. The second part of the function is one or more *parameters* or *arguments* enclosed in parentheses. These tell SAS what variables or constants you want the function to manipulate. Using the function is usually referred to as *calling* it. The value computed is said to be *returned*. Whenever you call a function, you must answer the *what* function and *on which variables* questions. Consider the following use of the MIN function.

```
low_alt = min(ny,chi,rdu,balt);
```

What we want to do is find the minimum (MIN). *On which variables* is answered by a comma-delimited list of variables.

Syntax

The general forms of function calls are illustrated below:

```
funcname(arg1,arg2, ...)
funcname(OF arg1 arg2 ...)
funcname(OF arg_st-arg_end)
funcname(OF arg_a--arg_b)
```

Some common applications of function calls used in assignment statements follow:

```
count = n(of round1-round20);

sumprof = sum(p_trans, p_forest, p_fish, p_mfg);

avg = mean(of vol81q1--vol90q4);
```

Function calls have the following components:

1. The *funcname* is the name of the function.

2. The *arg* is an argument. It can be a constant, variable, or any expression (including other functions) that can be reduced to a numeric result.

3. The *arg_st* and *arg_end* are beginning and ending names in a like-suffixed variable list (see the discussion of variable lists in Section 3.3 for more details). If OF is omitted, SAS has no way of knowing that we want to use a variable list rather than an expression: SAS would interpret v1–v5 as v5 subtracted from v1 instead of v1, v2, and so on through v5.

4. The *arg_a* and *arg_b* are beginning and ending names in a contiguous variable list (see the discussion of variable lists in Section 3.3 for more details). Like the list described in 3, failure to include OF will simply result in subtraction of the intended starting and ending variables.

Examples of Form

The variety and potential complexity of function calls are illustrated in Exhibit 7.3. Notice how some methods combine the different styles of arguments.

EXHIBIT 7.3 Different methods for calling functions

```
sum(a,b,c)
sum(a,b,-c)
var(15,a,b)
std(of v1-v50)
std(of v1 v2 v3)
std(of v1-v3)
min(of init-numeric-fin)
min(mean(of v1-v5), mean(of v6-v10))
sum(a+5, of v1-v10)
```

Rules, Cautions, and Advice

Although functions are flexible and easy to use, they impose a few rules. This section reviews these rules and highlights how functions behave differently from assignment statements.

Parameter Order, Range, and Number. Some functions require parameters in a particular *order*. Others require that the parameter be within a particular *range* of values (e.g., only positive numbers may be evaluated for logarithms). Others may require that only a specific *number* of them be passed. The "Descriptive Statistics Functions" section later in the chapter presents this information when appropriate.

POINTERS AND PITFALLS

When using a function, always be mindful of restrictions on the number and order of its parameters. Do not assume that SAS will produce an error or warning if you pass inappropriate values.

If you do not pass arguments in the right order and number, SAS may or may not produce an error. If the function needs a nonmissing numeric argument and you pass it a missing or character value, a message will be printed in the Log

EXHIBIT 7.4   Log reporting incorrect function arguments

```
------------------- Input dataset ---------------------

DISTANCE    SPEED1    SPEED2
    .         850       875
  1200        400       420
  1140        780        .
     0          .         .


------------------- SAS Log --------------------------

    3     data temp;
    4     input distance 1-5 speed1 7-10 speed2 12-15 ;
    5     log_dist = log(distance);
    6     min      = min(speed1,speed2);
    7     cards;
```

```
NOTE: Invalid argument to function LOG at line 5 column 16.
RULE:----+----1----+----2----+----3----+----4----+----5----+----6----+----7---
   11      0
DISTANCE=0 SPEED1=. SPEED2=. LOG_DIST=. MIN=. _ERROR_=1 _N_=4
   12     run;
```

LOG function will not accept an argument with a missing value.

```
NOTE: Missing values were generated as a result of performing an operation on
      missing values.
      Each place is given by: (Number of times) at (Line):(Column).
      1 at 5:16
      1 at 6:23
```

```
NOTE: Mathematical operations could not be performed at the following places.
      The results of the operations have been set to missing values.
      Each place is given by: (Number of times) at (Line):(Column).
      1 at 5:16
```

This is a summary of invalid operations attempted during the DATA step.

```
NOTE: The data set WORK.TEMP has 4 observations and 5 variables.
NOTE: The DATA statement used 13.00 seconds.
```

and execution will continue. The dataset and Log in such a case are presented in Exhibit 7.4.

Sometimes SAS cannot catch misplaced parameters simply because they are out of order: both parameters may be valid (e.g., nonmissing, numeric, and positive), but you may have passed them in the wrong order. Suppose you want to round variable INCOME to the nearest hundred by using the ROUND function. As we will soon see, ROUND expects the first argument to be the variable to be rounded and the second to be the rounding unit ("round to the nearest . . ."). In Exhibit 7.5 both the correct way (variable ROUND1) and an incorrect *but syntactically valid* way (ROUND2) to perform the rounding are shown.

Missing Values.   Functions that compute descriptive statistics such as counts, sums, variances, or standard deviations use only nonmissing numeric values. These functions return a missing value only if *all* arguments are missing.

This treatment of missing values is a critical difference between functions and the calculations in Section 7.1. A missing value in the v1 + v2 + . . . + v12 example sets the target variable to missing. The SUM function, however, returns the sum of all nonmissing values.

EXHIBIT 7.5   Impact of incorrect parameter order

```
------------------- Program listing -------------------

data inc;
input income 1-5 ;
round1 = round(income,100);
round2 = round(100,income);
cards;
115
250
15
1875
run;

proc print;
run;

------------------- Output listing -------------------
```

INCOME	round(income,100)	round(100,income)
115	100	115
250	300	0
15	0	105
1875	1900	0

Logically incorrect, but syntactically valid: round 100 to the nearest INCOME.

Logically correct: round INCOME to the nearest 100.

Functions versus PROCs.   The functions described in the "Descriptive Statistics Functions" section include some familiar descriptive statistics: MEAN, VAR, STD, SUM, and so on. We encountered these when we reviewed the MEANS and UNIVARIATE procedures. Why do we need to see them again?

POINTERS AND PITFALLS

Functions work *within* an observation, procedures *across* observations.

It is a matter of direction. Procedures compute statistics *across* observations, taking, say, the sum of a variable for all observations in the dataset. Functions, on the other hand, operate *within* observations, calculating the sum of an observation's variables. It is possible to combine the two. The NMISS function, for example, can yield a summary measure for each observation that could then be collapsed across observations by using PROC MEANS.

Frequently Used Numeric Functions

This section describes some frequently used numeric functions. The descriptions are divided into four "families": descriptive statistics, arithmetic, rounding, and logarithmic and exponential. Financial functions are discussed in the next section.

Descriptive Statistics Functions. These functions are identical in name and purpose to the options in the MEANS procedure. In the listing that follows, *arglist* is a list of numeric arguments: any combination of variable lists, individual variables, expressions, and constants is acceptable. References to the number of arguments required refer to the minimum number needed to be syntactically correct. If the correct number of arguments is passed and fewer than the minimum have nonmissing values, the function returns a missing value.

Function Name	Purpose
N([OF] *arglist*)	Number of arguments in *arglist* with nonmissing values.
NMISS([OF] *arglist*)	Number of arguments in *arglist* with missing values.
CSS([OF] *arglist*)	Corrected sum of squares for the arguments in *arglist*. Requires at least two arguments.
CV([OF] *arglist*)	Coefficient of variation for the arguments in *arglist*. Requires at least two arguments.
KURTOSIS([OF] *arglist*)	Kurtosis of the arguments in *arglist*. Requires at least four arguments.
SKEWNESS([OF] *arglist*)	Skewness of the arguments in *arglist*. Requires at least three arguments.
STD([OF] *arglist*)	Standard deviation of the arguments in *arglist*. Requires at least two arguments.
STDERR([OF] *arglist*)	Standard error of the mean of the arguments in *arglist*. Requires at least two arguments.
VAR([OF] *arglist*)	Variance of the arguments in *arglist*. Requires at least two arguments.
USS([OF] *arglist*)	Uncorrected sum of squares of the arguments in *arglist*. Requires at least two arguments.
SUM([OF] *arglist*)	Sum of the arguments in *arglist*.
MEAN([OF] *arglist*)	Mean, or average, of the arguments in *arglist*.
MIN([OF] *arglist*)	Minimum value of the arguments in *arglist*. Requires at least two arguments.
MAX([OF] *arglist*)	Maximum value of the arguments in *arglist*. Requires at least two arguments.
RANGE([OF] *arglist*)	Range of the arguments in *arglist*. This value will always be positive, even if the minimum and maximum are both negative. Requires at least two arguments.

Exhibit 7.6 uses some of these functions. Some observations have complete (no missing) data, some have a few missing values, and one is completely missing. Notice how the functions that require a minimum number of arguments react to the different numbers of missing values. The variables v1 through v5 are

EXHIBIT 7.6   Functions computing single-variable statistics

```
------------------- Output listing --------------------
```

V1 V2 V3 V4 V5	n(of v1-v5)	nmiss(of v1-v5)	css(of v1-v5)	cv(of v1-v5)
1 3 5 7 9	5	0	40.0	63.246
-1 -2 0 . 1	4	1	5.0	-258.199
-1 -2 -3 -4 -5	5	0	10.0	-52.705
. . . . .	0	5	.	.
9	1	4	0.0	.
9 8 . . .	2	3	0.5	8.319
9 8 7 . .	3	2	2.0	12.500
9 8 7 6 .	4	1	5.0	17.213

V1 V2 V3 V4 V5	kurtosis(of v1-v5)	skewness(of v1-v5)	std(of v1-v5)
1 3 5 7 9	-1.2	-4.4452E-17	3.16228
-1 -2 0 . 1	-1.2	5.42169E-17	1.29099
-1 -2 -3 -4 -5	-1.2	4.44523E-17	1.58114
. . . . .	.	.	.
9	.	.	.
9 8 . . .	.	.	0.70711
9 8 7 . .	.	0	1.00000
9 8 7 6 .	-1.2	-4.3657E-17	1.29099

V1 V2 V3 V4 V5	stderr(of v1-v5)	var(of v1-v5)	uss(of v1-v5)	sum(of v1-v5)
1 3 5 7 9	1.41421	10.0000	165	25
-1 -2 0 . 1	0.64550	1.6667	6	-2
-1 -2 -3 -4 -5	0.70711	2.5000	55	-15
. . . . .	.	.	.	.
9	.	.	81	9
9 8 . . .	0.50000	0.5000	145	17
9 8 7 . .	0.57735	1.0000	194	24
9 8 7 6 .	0.64550	1.6667	230	30

V1 V2 V3 V4 V5	mean(of v1-v5)	min(of v1-v5)	max(of v1-v5)	range(of v1-v5)
1 3 5 7 9	5.0	1	9	8
-1 -2 0 . 1	-0.5	-2	1	3
-1 -2 -3 -4 -5	-3.0	-5	-1	4
. . . . .	.	.	.	.
9	9.0	9	9	0
9 8 . . .	8.5	8	9	1
9 8 7 . .	8.0	7	9	2
9 8 7 6 .	7.5	6	9	3

displayed at the beginning of each line. The other column headings show the form of the function call used with those variables.

Arithmetic Functions.   These functions require all arguments to be nonmissing.

Function Name	Purpose
ABS(*arg*)	Absolute value of *arg*.

MOD(*num,div*)	Remainder when *num* is divided by *div*. The *div* cannot be zero.
SQRT(*arg*)	Square root of *arg*. The *arg* must be nonnegative. Other roots are easily calculated by raising to the appropriate fractional power. A cube root, for instance, could be obtained by the following:

```
CUBER = RATE**.3333 ;
```

RANNOR(*arg*)	Generate a normally distributed (mean 0, standard deviation 1) random number. The *arg* is 0, or a five-, six-, or seven-digit odd number.
RANUNI(*arg*)	Generates a random number uniformly distributed between 0 and 1. The *arg* follows the same rules as the RANNOR function.

Exhibit 7.7 uses some of these functions.

EXHIBIT 7.7 Arithmetic functions

```
------------------- Output listing --------------------
```

abs_arg	abs(abs_arg)	mod_num	mod_div	mod(mod_num,mod_div)
-1.0	1.0	10	10.0	0.0
.	.	7	4.0	3.0
8.2	8.2	-2	-1.5	-0.5
13.0	13.0	11	20.0	11.0
-1.2	1.2	11	12.0	11.0

sqrt_arg	sqrt(sqrt_arg)	ranuni(0)	rannor(0)
0	0.0000	0.38404	0.07259
.	.	0.75835	-1.28058
-1	.	0.07702	-0.21960
100	10.0000	0.65434	0.20196
2	1.4142	0.82882	-0.94315

Rounding Functions.   These functions round off values, either by dropping decimal places or by going to the nearest round-off unit.

Function Name	Purpose
CEIL(*arg*)	Rounds off *arg* to the next largest integer (rounds up).
FLOOR(*arg*)	Rounds off *arg* to the next smallest integer (rounds down).
INT(*arg*)	Erases the fractional portion of *arg* (truncates). INT and FLOOR behave identically when *arg* is positive. INT and CEIL behave identically when *arg* is negative.

ROUND(*arg,amount*)   Rounds *arg* to the nearest *amount*. The *amount* is optional. If entered, it must be positive. If omitted, SAS uses a value of 1.

Exhibit 7.8 illustrates the use of these functions. Variable ARG is the original value. Other columns present the results of different function calls.

EXHIBIT 7.8   Rounding functions

```
------------------- Output listing --------------------

         ceil   floor   int   round    round    round    round    round
  ARG   (arg)   (arg)  (arg)  (arg)   (arg,.1) (arg,.5) (arg,2) (arg,10)

 -6.50    -6     -7     -6     -7      -6.5     -6.5      -6      -10
 -6.10    -6     -7     -6     -6      -6.1     -6.0      -6      -10
 -6.80    -6     -7     -6     -7      -6.8     -7.0      -6      -10
   .       .      .      .      .       .        .        .        .
  0.00     0      0      0      0       0.0      0.0       0        0
  0.20     1      0      0      0       0.2      0.0       0        0
  1.50     2      1      1      2       1.5      1.5       2        0
  1.08     2      1      1      1       1.1      1.0       2        0
  1.09     2      1      1      1       1.1      1.0       2        0
  5.00     5      5      5      5       5.0      5.0       6       10
  7.50     8      7      7      8       7.5      7.5       8       10
```

Logarithmic and Exponential Functions.   All functions in this section require positive arguments.

Function Name	Purpose
LOG(*arg*)	Napierian, natural, base e log.
LOG10(*arg*)	Common, base 10 log.
LOG2(*arg*)	Base 2 log.
EXP(*power*)	Raise "e" (approximately 2.71828) to *power*. The *power* can be any value, subject to the limitations of your computer. SAS prints a warning message and returns a missing value if *power* is too large for your computer.

Exhibit 7.9 (see page 136) uses some of these functions. Variable ARG is the original value. Other columns present the results of different function calls.

Financial Functions

SAS provides several families of financial functions to calculate single-period or accumulated depreciation, internal rate of return, and net present value. Its mortgage, savings, and compound-interest functions are goal seeking: the argument of interest is passed as missing and is calculated on the basis of the three nonmissing arguments. This allows you to pose questions such as, "How much do I have to deposit now if I want to save $10,000 after 10 years and the fund yields 7% compounded interest?"

This section discusses how to use these functions. *In all cases, the order in which the arguments are passed is significant.* As was the case with univariate statistics, this book is not intended to be a financial statistics primer. The intent is to illustrate how to get the numbers, not how to interpret them. Financial

---

EXHIBIT 7.9   Logarithmic and exponential functions

```
------------------- Output listing --------------------

     ARG     log(arg)    exp(log(arg))    log10(arg)    log2(arg)

   -1.000        .              .              .             .
    0.000        .              .              .             .
    1.000     0.00000         1.000         0.00000       0.00000
    2.000     0.69315         2.000         0.30103       1.00000
    2.718     1.00000         2.718         0.43429       1.44269
   10.000     2.30259        10.000         1.00000       3.32193
  100.000     4.60517       100.000         2.00000       6.64386
```

---

functions described in this section are available only in version 6 of the SAS System.

Depreciation.   Functions beginning with DACC indicate *accumulated* depreciation, while those beginning with DEP are for a *specific period* in the life of the asset. Five functions are available for each group (function suffix follows in parentheses): declining balance (DB), declining balance converting to straight-line (DBSL), straight-line (SL), sum-of-years' digits (SYD), and a table-driven form that requires entry of depreciation rates for each period (TAB). In no case can the accumulated depreciation exceed the value of the asset.

The DBSL functions differ from their DB and SL namesakes. They select the appropriate depreciation method for a particular interval: the method chosen to calculate a given period's depreciation will be the one maximizing the period's value. Either the declining balance or straight-line method will be used.

The format of the depreciation functions is discussed below:

Function Name	Value Calculated
DACCDB	Accumulated declining balance
DEPDB	Periodic declining balance
	Parameters: *nperiod, depval, life, deprate*
DACCDBSL	Accumulated declining balance converting to straight-line
DEPDBSL	Period declining balance converting to straight-line
	Parameters: *nperiod, depval, life, deprate*
DACCSL	Accumulated straight-line
DEPSL	Periodic straight-line
	Parameters: *nperiod, depval, life*
DACCSYD	Accumulated sum of years' digits
DEPSYD	Periodic sum of years' digits
	Parameters: *nperiod, depval, life*
DACCTAB	Accumulated, from user-specified table
DEPTAB	Periodic, from user-specified table
	Parameters: *nperiod, depval, rate . . .*

The parameters for each function are important:

- The *nperiod* is, for DACC functions, the number of periods for which the accumulated depreciation will be calculated. For DEP functions, it represents a particular period's depreciation. The value may be fractional for all functions except DBSL. If negative, the result is set to zero. For the DB, DBSL, and TAB functions, if *nperiod* is zero, the result will be zero. If *nperiod* exceeds SL's or SYD's *life* or TAB's number of rates, the result is set to *depval*.

- The *depval* is the depreciable value of the asset.
- The *life* is the lifetime of the asset. The unit of measure (years, months, and so on) must agree with the unit used for *nperiod*. The value can be fractional for all functions except DBSL.
- The *deprate* is the *rate* of depreciation, expressed as a fraction (2, 1.5, .75, and so on). A value of 2 indicates a double declining balance.
- The *rate* . . . is a list, or table, of *rate*s used when calculating depreciation. Enter at least as many *rate*s as there are *nperiod*s. DEPTAB does not check the sum of the rates. If it is less than 1., the sum of the periods' depreciation will not equal the value of the asset. Conversely, if it exceeds 1., the sum of the individual periods' depreciation will exceed the value of the asset. The DACCTAB function will stop accumulating the periods' depreciation once the value of the asset has been reached.

Exhibit 7.10 illustrates the use of these functions.

EXHIBIT 7.10　Depreciation functions

	nper		
	1	2	3
	---	---	-----
depdb(nper, 1000, 3, 1.0)	333	222	148
daccdb(nper, 1000, 3, 1.0)	333	556	704
depdbsl(nper, 1000, 3, 1.0)	333	333	333
daccbsl(nper, 1000, 3, 1.0)	333	667	1,000
depdb(nper, 1000, 3, 2.0)	667	222	74
daccdb(nper, 1000, 3, 2.0)	667	889	963
depdbsl(nper, 1000, 3, 2.0)	667	222	111
daccdbsl(nper, 1000, 3, 2.0)	667	889	1,000
depsl(nper, 1000, 3)	333	333	333
daccsl(nper, 1000, 3)	333	667	1,000
depsyd(nper, 1000, 3)	500	333	167
daccsyd(nper, 1000, 3)	500	833	1,000
deptab(nper, 1000, .6, .3, .3)	600	300	300
dacctab(nper, 1000, .6, .3, .3)	600	900	1,000
deptab(nper, 1000, .35, .2, .1)	350	200	100
dacctab(nper, 1000, .35, .2, .1)	350	550	1,000

Other Financial Functions. Several other financial functions are available. Two of these, net present value (NPV) and internal rate of return (IRR), are related. NPV is the difference of the present value of the expected *income* from a project and the expected *costs* of the project. IRR is the discount rate that exactly equates the NPV's expected income and costs (i.e., it is the rate that sets NPV to 0).

Function Name	Purpose
INTRR	Internal rate of return (fraction)
IRR	Internal rate of return (percentage)
	Parameters: *period, cashflow* . . .
NETPV	Net present value (fraction)
NPV	Net present value (percentage)
	Parameters: *rate, period, cashflow* . . .

The parameters of these functions are described below:

- The *rate* is the interest rate per period.
- The *period* is the number of time units of interest. It should agree in measure (year, month, and so on) with rate and *cashflow* . . . . Continuous compounding is assumed when *period* is zero. If *period* is missing, the function returns a missing value.
- The *cashflow* . . . is a series of at least *period* cash flows when *period* is greater than zero. If *cashflow* is an outlay (an investment), it is entered as a negative number. A missing *cashflow* value in the INTRR or RR functions is treated as a zero.

The *rate*, *period*, and *cashflow* must agree in measure: if *rate* represents a yearly rate, then *period* should represent a number of years, and *cashflow* . . . values should also represent yearly values. Exhibit 7.11 illustrates the use of these functions.

EXHIBIT 7.11   Net present value, internal rate of return functions

---

					netpv(rate,1,per1,per2,per3,per4)	intrr(1,per1,per2,per3,per4)	intrr(0,per1,per2,per3,per4)
rate	per1	per2	per3	per4			
1.00000	-100,000	50,000	50,000	120,000	-47,500	0.43314	0.35987
1.00000	-100,000	100,000	10,000	5,000	-46,875	0.12796	0.12041
0.43314	-100,000	50,000	50,000	120,000	0	0.43314	0.35987
0.10000	-100,000	50,000	50,000	120,000	76,935	0.43314	0.35987
0.12796	-100,000	100,000	10,000	5,000	-0	0.12796	0.12041
0.10000	-100,000	100,000	10,000	5,000	2,930	0.12796	0.12041
0.10000	-100,000	50,000	0	120,000	35,612	0.25812	0.22962
0.10000	-100,000	50,000	.	120,000	44,628	0.25812	.
0.00000	-100,000	50,000	50,000	120,000	120,000	0.43314	0.35987
0.00000	-100,000	100,000	10,000	5,000	15,000	0.12796	0.12041

The remaining financial functions are characterized by their "what if" nature: each accepts four arguments, one of them missing (the item of interest). The function evaluates the missing argument subject to the constraints of the three nonmissing ones.

Function Name	Purpose
COMPOUND	Calculations for amounts earning compound interest.
	Parameters: *value, termval, rate, npay*
MORT	Calculations for mortgage loans with equal payments and a fixed interest rate.
	Parameters: *amount, payment, rate, npay*
SAVING	Calculations for savings made in equal amounts and at fixed periods.
	Parameters: *termval, payment, rate, npay*

The parameters of these functions are described below:

- The *value* and *amount* represent starting figures: amount at the beginning of the term (COMPOUND), or amount of the loan (MORT).

- The *termval* represents figures at the end of the period: *value* of the starting amount (COMPOUND), or amount saved (SAVING).

- The *payment* represents periodic outlays: loan payment (MORT), or periodic savings (SAVING).

- The *rate* is the rate per period, expressed as a fraction. This can be entered directly (.015 for an 18% yearly rate) or as an expression (.18/12). The unit of measure for *rate* must agree with that in *npay*. The rate is compounded monthly.

- The *npay* is the number of payments made, entered directly as a number (e.g., 360) or as an expression (30*12).

Exhibit 7.12 illustrates the use of these functions. Notice that when the functions solve for a value of *npay* or rate, the result is expressed in the same units as the nonmissing value. The third MORT example solves for rate, for

---

EXHIBIT 7.12 Other financial functions: COMPOUND, MORTGAGE, and SAVINGS

COMPOUND

value	termval	rate	npay	RESULT
.	20000	.18/12	5*12	8185.92
15000	.	.10/12	120	40605.62
15000	25000	.	7	.0757
12000	15000	.18	.	1.34

MORTGAGE

amount	payment	rate	npay	RESULT
.	750	.11/12	25*12	76521.78
80000	.	.10/12	30*12	702.06
80000	700	.	30*12	.0083
75000	800	.11/12	.	214.98

SAVINGS

termval	payment	rate	npay	RESULT
.	350	.08/12	20*12	207531.
30000	.	.07/12	10*12	172.32
5000	100	.	4.5	1.0792
12000	350	.05/12	.	28.21

example, and expresses it as the monthly rate, since *npay* was specified as a number of months. The third COMPOUND example, on the other hand, returns a yearly rate since *npay* was entered as the number of years (7, instead of 7*12).

Exercises

7.4. This group of questions uses the following dataset:

X1	X2	X3	X4	X5
8	9	10	11	12
.	12	12	16	.
-1	0	1	2	.

For each of the three observations, calculate the following expressions, or state why the operation is not valid.

a. n(of x1–x5)
b. sum(of x1–x5)
c. sum(x1, x2, –x3)
d. mean(x1,x2)
e. mean(of x1–x5, 5)
f. mean(x1–x5, 5)
g. kurtosis(x1, x2)
h. kurtosis(x1–x5)

7.5. In each of the following statements, functions are used to create a variable. Give a brief prose description of what the target variable represents. All statements are syntactically correct.

a. minavg = min(mean(of grp1-grp5),
                                mean(of grp10-grp15));
b. maxmin = max(min(p1,p2), min(p3,p4), min(p5,p6));
c. fewest = min(n(of coh70_1-coh70_10),
                                n(of coh80_1-coh80_10));

7.6. The variable X1 has the following seven values in seven observations:

    -1 0 . 2 3 4 6

Calculate the result of the following expressions:

a. abs(x1)
b. mod(x1, 3)
c. mod(x1, 0)
d. mod(x1, 1)

7.7. The variable RD has the following eight values in eight observations:

    -2.5 -2.4 0 . .2 .5 1.0 2.8

Calculate the result of the following expressions:

a. ceil(rd)
b. floor(rd)
c. int(rd)
d. round(rd)
e. round(rd, 2)
f. round(rd, .5)
g. round(rd, 10)

Summary

Chapter 7 describes the syntax and rules for arithmetic calculations and discusses the use of functions—routines supplied with the SAS System that quickly and efficiently perform common computational tasks. The general rules for using functions are addressed: most functions are sensitive to the order, range, and number of terms, or "parameters," they process. Examples of syntactically incorrect, valid, and logically incorrect functions are shown.

Several "families" of functions are presented. Descriptive statistic functions produce many of the same statistics as the MEANS and UNIVARIATE procedures, the difference being that they operate within an observation (procedures work across observations). Arithmetic functions compute absolute values, remainders, and random numbers. Rounding functions return values closest to the nearest "roundoff unit." Logarithmic functions calculate common, Napierian, and base 2 logarithms. Finally, financial functions calculate depreciation, internal rates of return, and loan repayments. They may be used to perform "what-if" analyses.

8 DATA Step III: Controlling the Flow of Execution

At this point you can read raw data; perform calculations; and print, chart, and analyze numeric data. Chapter 7 introduced some useful real-world complexity in the form of assignment statements. This chapter takes DATA step programming a step further by discussing statements that control execution flow.

Most of the example programs presented so far have simply "dropped through" the DATA step, executing each statement for every observation. What if we want to execute an assignment statement only if a condition is true? What if we want to do several assignments based on a condition? What if we want our output dataset to contain only certain observations (e.g., a particular geographic area, people scoring above a certain test score, and so on)?

SAS has statements that let us accomplish these and other tasks. This chapter discusses IF-THEN, ELSE statements, DO-groups, and OUTPUT, DELETE, and RETURN statements. Their versatility and simple syntax allow conditional statement execution and control which observations will be written to the SAS dataset.

8.1 Conditional Execution: IF Statements

The IF-THEN and ELSE statements (collectively referred to simply as IF statements) do what their names suggest: IF a condition is "true," THEN execute the next statement, otherwise (ELSE) take some other action. The syntax of these statements is illustrated below:

```
IF condition THEN action;
    [ELSE action;]
```

Some common applications of IF-THEN-ELSE statements follow:

```
if avg > 5.0 then type = 'High';
```

```
if          avg > 5.0 & group = '1' then class = 'Over';
    else if avg > 5.0 & group = '2' then class = 'ok  ';
    else                                  class = 'avg.';

if avg > 5.0 then type = 'High';
    else          type = 'Low ';
```

The syntax of IF-THEN-ELSE statements has two required and one optional component:

1. The *condition* is any SAS expression. Any mixture of variables, constants, and expressions (including other IF statements) is allowed. If the condition is "true," the *action* following THEN is taken.

2. The *action* is a SAS statement: assignments, INPUT, IF, INFILE, and other statements discussed in this chapter are allowed here. A null statement (;) is also a valid *action*.

3. *ELSE*, if used, must immediately follow another *ELSE* or *IF* statement. There is no limit to the number of *ELSE* statements allowed. An *ELSE* with an *action* other than another *IF* statement terminates the *IF-THEN-ELSE* sequence.

Legal IF Statements

This section presents examples of valid IF statements. A verbal description of the instructions as understood by SAS accompanies each example. Variables here and throughout the rest of the chapter are from a dataset containing data on the timing and impact of work stoppages in the United States: observations represent calendar years. Pay particular attention to the formatting of these program segments: indentation and liberal use of blank space are especially helpful when writing IF statements.

POINTERS AND PITFALLS

Ensure logical closure of IF statements. Always have an unconditional ELSE statement ending a series of IF statements used for assigning variable values.

Use indentation and blank space to highlight the subordinate statements. This makes the logic of the IF statements easier to follow.

"If variable NSTOP is larger than 267, then assign a value of H to character variable STOPLVL. If NSTOP is less than or equal to 267, do not alter the value of STOPLVL."

```
if nstop > 267 then stoplvl = 'H';
```

"If variable NSTOP is larger than 267, then assign a value of H to character variable STOPLVL. Otherwise, assign a value of L to STOPLVL."

```
if nstop > 267 then stoplvl = 'H';
    else          stoplvl = 'L';
```

"If variable NSTOP exceeds 267 *and* variable WORKERS exceeds 1307, then assign a value of H to character variable STOPLVL. Otherwise, if either variable meets these conditions, assign STOPLVL a value of M. If this condition is also false, place an L in STOPLVL."

```
if          nstop > 267 & workers > 1307 then stoplvl = 'H';
    else if nstop > 267 | workers > 1307 then stoplvl = 'M';
    else                                      stoplvl = 'L';
```

"If variable YEAR is 1969 or earlier, assign a value of 1 to numeric variable EARLY. Otherwise, assign a value of 1 to numeric variable LATE. Implicitly, EARLY will have a missing value for 1970 and later years, and LATE will have a missing value for 1969 and earlier years."

```
if      year <= 1969 then early = 1;
    else                     late  = 1;
```

Invalid or Dubious IF Statements

This section illustrates syntactically invalid, logically suspect, and stylistically poor IF statements. Each example is accompanied by an explanation of the deficiency.

This IF-THEN-ELSE sequence will execute correctly:

```
if nstop > 267 & workers > 1307 then stoplvl = 'H';
 else if nstop > 267 | workers > 1307 then stoplvl ='M';else
 stoplvl = 'L';else stoplvl=' ';
```

However, it is lacking from a stylistic standpoint: the parallelism of the first two IF's is not apparent; the second ELSE is lost at the end of the second line, giving the appearance of an unconditional assignment to STOPLVL in the third line. If this program is hard to read it will also be hard to change. Taking the time to use blank space effectively when the program is first written will save headaches later on.

ELSE must *follow* an IF, not precede it. It is hard to formulate a verbal description of this segment. We normally do not begin a thought by saying "otherwise, we will do such and such." This error is distressingly easy to make when your computer's program editor is new and unfamiliar.

```
    else stoplvl = 'L';
if nstop > 267 then stoplvl = 'H';
```

In this example the first ELSE statement is unconditional and thus ends the IF-THEN-ELSE sequence. This leaves the second ELSE without a corresponding IF. Part of the SAS Log from a program using this bad sequence is reproduced in Exhibit 8.1.

EXHIBIT 8.1   Log reporting poorly structured IF-THEN-ELSE statements

---

```
   1    data workstop;
   2    infile '\data\raw\workstop';
   3    input year nstop workers daysidle;
   4    if        nstop > 267 & workers > 1307 then stoplvl = 'H';
   5        else                               stoplvl = 'L';
   6        else if nstop > 267 | workers > 1307 then stoplvl = 'M';
ERROR: No matching IF-THEN clause.
   7    totwork = workers * 1000 ;
   8    totidle = daysidle * 1000 ;
   9    run;
NOTE: The SAS System stopped processing this step because of errors.
NOTE: The data set WORK.WORKSTOP has 0 observations and 7 variables.
NOTE: The DATA statement used 6.00 seconds.
```

The logic rather than the syntax is flawed in this example:

```
if               nstop > 267 then stoplvl = 'H';
  else if 0 <= nstop < 267 then stoplvl = 'L';
  else                          stoplvl = '?';
```

Notice what happens when an observation's value of NSTOP is 267: the first statement's condition evaluates "false" since it tests NSTOP for being greater than 267. Execution drops into the first ELSE, which tests NSTOP in the range from 0 up to but not including 267. This condition is also "false." Execution continues into the second ELSE, which unconditionally assigns a value of '?' to STOPLVL. Clearly, 267 should be in the 'H' or 'L' levels of STOPLVL. The lack of closure in the IF statement conditions allowed 267 to fall through the cracks. Changing the '>' to a '>=' in the first statement or the '<' to a '<=' in the second will fix this subtle and common problem.

Exercises

8.1. Some observations in a dataset have the following values of X and Y:

```
 X   Y
 .   0
 0   1
-1   0
```

What values of LEVEL are assigned for these observations if only the following groups of statements are used in the DATA step?

a. `if      x > 0 & y > 0 then level = 'p';`
   `  else if x > 0 | y > 0 then level = 'm';`

b. `if x > 0 & y > 0 then level = 'p';`
   `if x > 0 | y > 0 then level = 'm';`

c. `if x > 0 | y > 0 then level = 'm';`
   `if x > 0 & y > 0 then level = 'p';`

d. `if      y > 0 then level = '1';`
   `  else if x < 0 then level = '2';`
   `  else          level = '3';`

e. `if y = 1 & x = 1 then level = 't';`

f. `if y = 1 & x = 1 then level = 'y';`
   `  else          level = 'n';`

8.2. Write the IF-THEN-ELSE statements required to meet the following conditions and assign the appropriate values to variable C:

Condition	Result
a. a > 2 and b = 3	c = 1
a > 2 or b = 3	c = 2
other	c = .
b. b < 4 and a from 1 to 5, inclusive	c = 0
b**2 > 16 and a from 1 to 5, but not 3	c = 1
c. b at least 15	c = 1
others	c = 0
d. b at least 15	c = 1
others, except b = 0	c = 0

8.2 Bracketing Related Statements: DO-Groups

So far only a single statement has been executed after a "true" IF condition. What if you want to do several things following a "true" condition? What if you want to assign 0's and 1's to both EARLY and LATE in the "If variable YEAR is 1969 . . ." example on page 145? One approach is shown in Exhibit 8.2.

EXHIBIT 8.2   Two variables calculated using only IF-THEN statements

```
if year <= 1969 then early = 1;
   else               early = 0;
if year > 1969 then late = 1;
   else               late = 0;
```

In this example IF statements are written for each calculated variable (EARLY and LATE). Such brute force repetition becomes tedious and error-prone if many actions are dependent on the evaluation of an IF statement. SAS enables bracketing, or grouping, any number of statements using **DO-groups**. These statements are either executed or skipped as a *group*. This section discusses their syntax and usage.

Syntax

The syntax of the simplest form of the DO statement, the DO-group, is straightforward:

```
DO;
[statements]
END;
```

Some common applications of DO-groups follow:

```
if type = 'H' then do;
   rate = rate * 1.2;
   scale = 32;
   end;
   else do;
        rate = rate * 1.1;
        scale = 30;
        end;

if type = 'H' then do;
   end;
```

DO begins a DO-group; the *END* statement closes it off. The statements in a DO-group are sometimes referred to as the group's *range*. When used with IF statements, DO and END ensure that the entire group's statements will be executed only when the IF statement's condition is "true."

POINTERS AND PITFALLS

When coding many DO-groups in a DATA step or when enclosing DO-groups within other DO-groups, take care that each DO has a corresponding END. It is very easy to lose an END in a long program!

Groups can be contained within other groups, a useful feature called *nesting*. Each group, however, must terminate with a separate END statement. There is no practical limit to the amount, or *levels*, of nesting. If a program has many nesting levels, however, it is likely that there is a simpler way to express the underlying logic.

Examples

Analysis and data management needs determine the work done in the DO-groups. The examples in this section concentrate on assignment statements. Notice how the use of indentation and blank space in the examples clarifies the beginning and end of each group.

Restructure the awkward recoding of the EARLY and LATE variables used earlier.

```
if year <= 1969 then do;
   early = 1;
   late  = 0;
   end;
   else do;
        early = 0;
        late  = 1;
        end;
```

Create ELECYEAR for every observation, then create some variables dependent on the value of YEAR. EARLYW will be missing in observations with year > 1969, while LATERW will be missing in observations with year <= 1969.

```
if mod(year,4) = 0 then elecyear = 1;
   else                  elecyear = 0;
if year <= 1969 then do;
   era = '60''s, 50''s';
   earlyw = workers * 1000 ;
   end;
   else do;
        era = '70''s, 80''s';
        laterw = workers * 1000;
        end;
```

This example shows nested DO-groups and a group containing no statements.

```
if year <= 1959 then do;
   era = 'pre 60';
   cost = daysidle * 1000 * avgwage1 ;
   if year = 1958 | year = 1959 then do;
      end;
   end;
   else if (1960 <= year <= 1969) then do;
        era = '60''s';
        cost = daysidle * 1000 * avgwage2 ;
        end;
   else if year >= 1970 then do;
        era = '1970 +';
        cost = daysidle * 1000 * avgwage2 ;
        end;
```

We might not know what to do within the nested loop, and so build the framework as a reminder that it will be filled in later. The statements in the nested loop are executed only for observations where YEAR is either 1958 or 1959. Notice that if this DO-group were misplaced and put in the 1960 group, it would never be executed.

This example simply assigns a value to ERA based on the value of YEAR.

```
if year <= 1959 then do;
    era = 'pre 60';
    end;
    else if (1960 <= year <= 1969) then do;
        era = '60''s';
        end;
    else if year >= 1970 then do;
        era = '1970 +';
        end;
```

Why go to the trouble of setting up DO-groups when a simple IF-THEN-ELSE sequence would suffice? If you anticipate adding other assignments and conditions later on, it may be easier to build the grouped framework at the start and let the program grow into it.

A portion of the SAS Log for DO-groups gone wrong is reproduced in Exhibit 8.3. The number of DO's (three) and END's (two) is unequal. Notice that SAS complains only mildly (a NOTE after the DATA step) about the absence of the END but is more strenuous (an ERROR) about the seemingly floating ELSE statement. The problem is solved by inserting an END statement between lines 8 and 9.

The more complicated the DATA step, the more likely are situations in which SAS does not know how to identify the real source of the problem: the missing END caused the trouble in Exhibit 8.3, but all SAS saw was an ELSE not attached to an IF.

EXHIBIT 8.3   Log reporting unclosed DO blocks

```
1      data stoppage;
2      infile '\data\raw\workstop';
3      input year nstop workers daysidle;
4      if year <= 1959 then do;
5          era = "pre 60";
6          end;
7          else if (1960 <= year <= 1969) then do;
8              era = "60's";                              No END statement to
9          else if year >= 1970 then do;                  close the DO group
ERROR: No matching IF-THEN clause.
10              era = "1970 +";
11              end;
12     run;
NOTE: There were 1 unclosed DO blocks.      ◄───    SAS cannot identify which
                                                     DO group was not closed.
NOTE: The SAS System stopped processing this step because of errors.
NOTE: The data set WORK.STOPPAGE has 0 observations and 5 variables.
NOTE: The DATA statement used 6.00 seconds.
```

Exercises

8.3. Consider the following prose description of conditions used to assign values of STYLE and OK:

IF TYPE is 'f,' set STYLE to 'g' and OK to 1. When TYPE is 'm,' set STYLE to 'b.' Otherwise, set OK to 0 and STYLE to ' '.

Write the statements to assign these values to OK and STYLE. Do it two ways. First write them without using DO-groups. Then write them again, this time using DO-groups.

8.4. It bears repeating that there is no single "right" way to write programs. Rewrite one of the examples from this section. Adjust the method of statement indentation until you feel comfortable with a particular style. If there is, in fact, any rule, it is probably that the style should emphasize the start and end of each DO-group.

8.5. Consider the following program segment:

```
if year > 90 then do;
    cohort = 'y';
    int    = year - 90;
    end;
    else if year < 90 then do;
        cohort = 'n';
        end;
```

If YEAR takes the following values in five observations, what values will COHORT and INT have in each observation?

88   89   90   92   .

8.6. What, if anything, is wrong with the following program segment?

```
if (of s1-s5) < 5 then do;
    end;
    else if min(of s1-s5) >= 5 then do;
    size_cls = 'L';
else if min(of s1-s5) = . then do;
    size_cls = '?';
    end;
```

8.3 Restricting Output: OUTPUT, DELETE, and RETURN

The example programs have been reading and writing the same number of observations. What if we do not want all the observations from the raw data represented in the output dataset? We might want only years that were above average, males, East Coast states, and so on.

The DELETE, RETURN, and OUTPUT statements can be used together or separately in the DATA step to control the flow of observations to the output SAS dataset. While reading their descriptions in this section, remember that the *input* dataset (the raw data) remains unaltered. Only the *output* dataset is affected.

POINTERS AND PITFALLS

Be aware of possible statement interactions and dependencies when using two or more of DELETE, RETURN, and OUTPUT in the same DATA step.

Syntax

The format of each of these statements is straightforward.

OUTPUT.   The OUTPUT statement syntax is shown below:

```
OUTPUT;
```

The statement lets you assume control of writing *all* observations to the output dataset. OUTPUT instructs SAS to send the observation to the output dataset. Once it writes the observation, SAS will try to continue the DATA step, looking for statements below the OUTPUT to execute. An OUTPUT statement is implied at the end of the DATA step if it is not actually written in the program.

The OUTPUT statement may appear *anywhere* in the dataset. The examples below, however, show that its placement has a significant impact on the dataset's contents. Multiple OUTPUT statements are permitted: each will force the writing of an observation with the *current* values of the variables defined in the DATA step.

DELETE.   The DELETE statement syntax is shown below:

```
DELETE;
```

This statement is nearly the opposite of OUTPUT: it instructs SAS to return to the beginning of the DATA step and *not* write the current observation to the output dataset.

RETURN.   The RETURN statement is shown below:

```
RETURN;
```

This statement instructs SAS to return to the DATA statement (i.e., go to the beginning of the step). Before doing so, SAS looks for an OUTPUT statement in the DATA step. If it does not find one, SAS writes an observation to the output dataset. Thus the only distinction between RETURN and DELETE is that an observation may be written to the output dataset: RETURN has an implied attempt to OUTPUT.

Exhibit 8.4 compares the impact of the OUTPUT, DELETE, and RETURN statements.

EXHIBIT 8.4   Impact of OUTPUT, DELETE, and RETURN on dataset contents

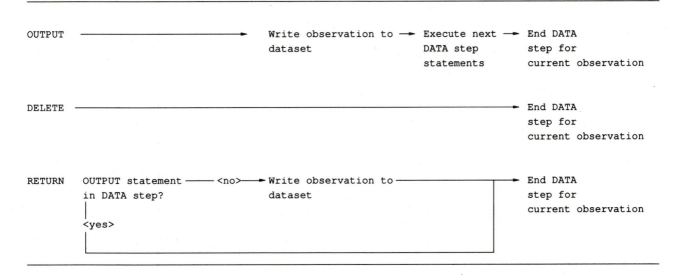

Examples

Any number and combination of DELETE, RETURN, and OUTPUT statements are acceptable. Once two or more occur in the same DATA step, however, the probability of unintended results increases. The following examples demonstrate simple, complicated, and suspect uses of the statements. In each, assume that only the example's statements can affect the contents of the output dataset.

You know that the mean of NSTOP is 267, and you want to analyze those observations that are above average. IF and OUTPUT statements are used to restrict the contents of the SAS dataset:

```
if nstop > 267 then output;
```

You want to output only those observations where NSTOP exceeds the average, 267. Before doing so, however, two calculations need to be performed:

```
if nstop > 267 then do;
   nworkers = nworkers * 1000;
   daysidle = daysidle * 1000;
   output;
   end;
```

This is an example of zag before you zig: if NSTOP exceeds 267, the observation will be output, *then* the calculations performed. NWORKERS and DAYS-IDLE will be multiplied by 1000, but only *after* the observation is written to the dataset! The output dataset will contain the original, unmultiplied values of the variables. The statements are identical to those in the preceding example; only their order differs.

```
if nstop > 267 then do;
   output;
   nworkers = nworkers * 1000;
   daysidle = daysidle * 1000;
   end;
output;
```

You want the DATA step to stop executing for the current observation if the value of YEAR is less than 1970. If the condition is true, SAS will return to the beginning of the current DATA step and attempt to read the next observation from the data source.

```
if year < 1970 then delete;
```

You want to stop executing the observation's pass through the DATA step if its value of YEAR is less than 1970. Before returning to the DATA statement, you want to output the observation. The value of COST will be missing for these pre-1970 observations.

```
if year < 1970 then return;
cost = daysidle * 1000 * wage3 ;
```

You want to stop executing the observation's pass through the DATA step if its value of YEAR is less than 1970. Since an OUTPUT statement is present in the DATA step, SAS assumes that observations should be written to the output dataset only at that point. RETURN in this case has exactly the same effect as DELETE, since observations with YEAR less than 1970 will never get to the bottom of the DATA step and execute the OUTPUT statement.

```
if year < 1970 then return;
cost = daysidle * 1000 * wage3 ;
output;
```

This final example will create some confusion. The observation will be output in the DO-group. Then it will be written again at the second OUTPUT. Observations where NSTOP exceeds 267 will be written twice, others only once. This may or may not be what you intended, but it *is* what you told SAS to do.

```
if nstop > 267 then do;
   nworkers = nworkers * 1000;
   daysidle = daysidle * 1000;
   output;
   end;
output;
```

Exercises

8.7. Examine the sample program below (the line numbers are included only for reference purposes; they are not part of the program).

```
(1) data test;
(2) infile 'a:\import';
(3) input keyval s1-s5;
(4) tot = sum(of s1-s5);
(5) avg = mean(of s1-s5);
(6) valid = n(of s1-s5);
```

Write statements to meet the following requirements. In each case, specify where you would place the statement(s) (e.g., "at line 4.5" or "between lines 4 and 5").

a. Insert an OUTPUT statement that duplicates the default action of DATA step processing.

b. You want observations with KEYVAL values of <= 100. Insert a DELETE statement to accomplish this. There are two logical locations for this. What are the advantages of each?

c. Accomplish the same result as in part b but use a single pair of RETURN and OUTPUT statements.

d. How would the values of TOT, AVG, and VALID be affected in the following scenarios? Assume that only the statement being discussed is used in the DATA step (e.g., if the scenario uses RETURN it assumes OUTPUT and DELETE are not used).

(1) A conditional RETURN is located at statement 3.5 rather than 6.5.

(2) An OUTPUT statement is located at statement 3.5 rather than 6.5.

(3) A conditional DELETE statement is located at statement 3.5 rather than 6.5.

8.8. Dataset SAMPLE has the following values:

STATUS	T1	T2
4	0	1
6	2	1
.	0	0

What does dataset RESULT look like in the following situations? (Assume other required DATA step statements that are not shown are correct and have no effect on your answer.)

a.
```
if status <= 5 then output;
index = (2*t1) + t2;
```

b.
```
if status <= 5 then do;
    index = (2*t1) + t2;
    output;
    end;
```

```
c. if status <= 5 then do;
     index = (2* t1) + t2;
     end;
     else delete;
d. if status > 5 then delete;
   index = (2*t1) + t2;
e. if status >= 5 then return;
   index = (2*t1) + t2;
   output;
```

Summary

Chapter 8 describes some of the tools used for controlling the flow of statement execution in the DATA step. Not all statements should be executed for every observation in a SAS dataset. The IF statement and its companion, ELSE, provide the means to bypass execution of a statement if a logical condition does not apply (is "false").

DO-groups, starting with a DO and ending with an END, are a convenient way to identify a group of related statements. They are often used together with IF and ELSE statements, thus allowing the user to specify a group of statements to execute when a condition is true and another group to execute when a condition is false. This simple construct is the foundation for even the most complex programs.

The last section of the chapter describes ways to restrict how and when an observation is written to the SAS dataset. The OUTPUT, DELETE, and RETURN statements control the timing of output to the dataset. Some of the subtleties of and interactions between these statements are described. Here, as with IF statements and DO-groups, a few simple concepts are the basis for large, logically complicated programs.

9 DATA Step IV: Dressing Up and Trimming Down

So far even though we have discussed maybe 10 percent of SAS's capabilities, we can perform a wide range of activities. The preceding chapters have demonstrated SAS's flexibility in performing calculations, controlling which observations will be kept, and selectively executing statements. This chapter illustrates tools for "dressing and trimming." These features allow you to control the appearance of output and select variables to store in the SAS dataset. Display formats, labels, and the DROP and KEEP statements are discussed.

9.1 Dressing Up: The Format Statement

Up to this point we have mentioned input formats only indirectly and not paid much attention to how variables are displayed by PROCs. We now consider why control over these input and output characteristics is desirable.

The data in the sample datasets have been well behaved: variables are always in the same place on each input record, and the numbers and characters are in a readily understandable format. A realistic wrinkle in this picture is data in more complicated formats such as scientific notation, packed decimal, or written with commas. The simple list and column formats described in Chapter 4 are not able to read this type of data.

On the display side, you may want to "dress up" the variables. Commas, leading zeros, and dollar signs not only make output more attractive but enhance the clarity and meaning of the data.

Using formats can increase the power of the INPUT statement and improve the output's appearance. This section discusses the concepts behind formats and illustrates the uses of the formats that come with SAS. Formats used to control display are emphasized. Advanced forms of the INPUT statement formats are discussed in detail in Chapter 11.

Concepts

A format is SAS's way to control how variables are read, displayed, and, in some cases, manipulated by DATA steps and procedures. If a format is not explicitly entered for a variable, SAS assigns one based on the variable's data type (numeric or character) and size. SAS uses the format's rules to copy raw data from the dataset, translate them into SAS variables, and store their values in memory.

When displaying variables the process is largely the same. SAS uses the output format's instructions to copy the SAS variable from memory and write it with the designated dollar signs, commas, and so on using the number of columns specified by the user. Input and output formatting are illustrated in Exhibit 9.1.

POINTERS AND PITFALLS

Input formats do not alter data stored in the raw dataset. Display formats do not alter the variables stored in the SAS dataset. Formats only affect how data are *moved*, that is, how they are translated from raw data to SAS variables and from SAS variables to procedure listings.

Before examining the syntax of SAS formats, we look at some features relevant to their use.

System Formats.   Many formats come as part of the SAS System. All formats described in this chapter are these *system formats*. To read or display data in a way not accommodated by system formats, users can write their own formats. This extremely powerful capability, known as *user-written formats*, is described in Chapters 14 and 15.

Format Types.   The two major format groups correspond to the two SAS data types, numeric and character. Just as character fields in INPUT statements begin with a dollar sign ($), so do character formats.

Note that the terms *numeric* and *character* refer to the way the variable is *stored*, not the way it *appears* on input or output. The number 1,000 may look like character data because it has an alphabetic character (the ","), but you can

---

EXHIBIT 9.1   Use of input and display formats

```
Input


Field of              Input               SAS variable
raw data  ─────────▶  format  ─────────▶  in memory


Output


SAS variable          Display             Variable displayed
in memory  ─────────▶  format  ─────────▶  on output listing
```

---

use a numeric format (COMMA) to strip off the comma and store the numeric portion (the value of 1000).

Data Values.   When a raw data field is read or a variable displayed, the data value itself remains unaltered. The format controls only the variable's *use*: transferring from raw data to memory or copying from memory to the output file. Using the COMMA format when printing a numeric variable will not change the data type of the variable, nor will it insert a comma in the stored variable.

Input and Output Relationship.   There is no correspondence between the format used to read a field of raw data and that used to display the variable. Thus a numeric field can be read with the COMMA format and displayed using COMMA, dollar, or scientific notation formats. Of course, in both the input and display processes, the format choice can be left to SAS.

Specifying Output Formats.   Output/display formats are usually specified in the FORMAT or ATTRIB statements. These can be either part of the DATA step or procedures.

If a format is assigned to a variable as part of the DATA step, it "travels with" the dataset. Any time the dataset is used in the current SAS session or in later ones, in PROCs or in other DATA steps, SAS will know how to display the variable. This is referred to as assigning *permanent formats*. Both system and user-written formats may be used this way.

If the FORMAT or ATTRIB statements are part of procedures, the instructions for display are in effect only for that PROC. The examples in Chapters 2, 4, 5, and 6 used this style of format specification. If a variable has a permanent format and a PROC specifies a different format, the permanent format is overridden for the execution of the PROC.

"Turning off" a Format.   Whether used in a PROC or a DATA step, the format assigned to a variable can be deleted, or "turned off," by not specifying a format in the FORMAT statement. Once a format is removed, SAS chooses the best way to display the variable. Thus if variables SALARY1 through SALARY5 were stored in the DATA step with a format of DOLLAR10.2 and you wanted to use PROC PRINT to display the data without the dollar signs and commas, you could use either of these FORMAT statements:

```
format dept $3. superv $15. salary1-salary5 ;

format salary1-salary5 ;
```

The syntax of the FORMAT statement requires that variables that have formats turned off must be the last ones specified. This avoids association with other, formatted variables.

Width and Alignment.   Most formats have widths and alignments. *Width* refers to the number of columns used by the variable in the raw data or when displayed on output. *Alignment* refers to how SAS behaves when the variable's formatted length is less than its specified width. Generally, numeric formats right-align and character formats left-align. This means that if there are leftover columns, numbers are shifted flush right in the columns and characters are shifted flush left in the column.

Mistyped Formats.   SAS issues an error message if you try to use a character format for a numeric variable or a numeric for a character. Exhibit 9.2 shows a SAS Log with both types of errors.

EXHIBIT 9.2 Log reporting incorrect match of format-variable types

---

```
------------------- SAS Log ---------------------------

3     data one;
4     infile '\data\raw\preciv';
5     input name $ 1 - 20 party $ 21-29 born 31-34;
6     format born $4.;
ERROR: Variable BORN is numeric but is being used as a character variable.
7     run;
NOTE: The SAS System stopped processing this step because of errors.
NOTE: The data set WORK.ONE has 0 observations and 3 variables.
NOTE: The DATA statement used 6.00 seconds.
8
9     proc print;
10    format party 2.;
ERROR: You are trying to use the numeric format F with the character variable
      PARTY.
11    run;
NOTE: The SAS System stopped processing this step because of errors.
NOTE: The PROCEDURE PRINT used 6.00 seconds.
```

> BORN read as *numeric* but displayed as *character*

> PARTY read as *character* but displayed as *numeric*

Problem Formats.   Other problems using formats include field overflow and decimal places exceeding field width. Field overflow is a common problem and is what its name implies: SAS tried to use the format on an observation's variable but could not fit the numbers, commas, and so on in the columns provided.

SAS's reaction depends on the format. For example, if the COMMA format is used and the value cannot fit in the columns, SAS will omit the comma(s). If there is still not enough room, decimal places are dropped. If there is still not enough space, the number will be printed in scientific notation. If even this does not work, SAS fills the field with asterisks (*). If *any* of these actions is taken, a note is printed in the SAS Log saying that one or more variables could not be printed using the format specifications. Unfortunately, the message does not indicate *which* variables caused the problem!

If decimal places exceed the number of columns (e.g., a format of 5.6), SAS will simply print an error message and not execute the DATA step or procedure. The message will highlight which variable is causing the error.

Syntax

Input Formats.   The general forms of input formats are shown below. Chapter 11 discusses the possible uses of input formats in much greater detail. The formats are presented here to give a feel for the similarity in form and interpretation between INPUT and output formats.

```
INPUT var fmtw[.d] ... ;
INPUT (varlist)(fmtw[.d] [fmtw[.d]]) ... ;
```

Some common applications of input formats follow:

```
input name $20. age 3. sex $1. race $1.;

input (score1-score10)(5.);

input model $5. (rate1-rate5)(7.2);
```

Input formats have the following elements:

1. The *var* is a variable name.

2. The *varlist* is the name of a variable list referring to any combination of character and numeric variables.

3. The *fmt* is the name of a system or user-written format.

4. The *w* is the format's width. This is the number of columns used when reading the variables referred to by *var* or *varlist*.

5. The *d*, a numeric format option, specifies the number of decimal places.

Output Formats. The FORMAT statement used in the DATA step is *exactly the the same* as that used with procedures. Refer to Chapter 5 for a description of its

---

EXHIBIT 9.3   Frequently used SAS system formats

Format name	Description	Input Alignment	Input	Disp	Min width	Max width	Min dec	Max dec
w.	Numeric, no decimals	r	x	x	1	32	na	na
w.d	Numeric with decimals	r	x	x	1	32	0	31
COMMA	Embedded commas	r	x	x	1	32	0	31
E	Scientific notation	r	x	x	7	32	na	na
HEX	Numeric hexadecimal	l	x	x	1	16	0	0
PD	Packed, with decimal places [1]	l	x	x	1	16	0	10
PK	Packed, no decimals [1]	l	x		1	16	0	10
RB	Real, binary	l	x	x	2	8	0	10
IB	Integer, binary	l	x		1	8	0	10
PIB	Positive integer, binary	l	x	x	1	8	0	10
BEST	SAS selects appropriate decimal places	r		x	1	32	na	na
DOLLAR	Dollars and commas added	r		x	2	32	0	2 [2]
Z	Insert leading zeros	r		x	1	32	0	30
SSN	Format Social Security numbers [3]	na		x	11	11	na	na
$w.	Character, leading blanks trimmed	l	x	x	1	200	na	na
$CHAR	Character, leading blanks preserved	l	x	x	1	200	na	na
$HEX	Character hexadecimal	l	x	x	1	200	na	na

---

[1] When PK is used, field's last 4 bits will contain a number.
    When PD is used, it contains a D or C to indicate positive
    or negative, respectively. Decimal points are implied.

[2] Must be either 0 or 2.

[3] Input values with fewer than nine digits are filled with
    zeroes in the leftmost positions, while those with more
    than nine digits are invalid and set to missing.

---

syntax. Any number of FORMAT statements can be used in a DATA step. The placement of the statement in the DATA step is usually not important. (The order of variables in the dataset will be affected, however, if the FORMAT statement is the first reference to a variable in the DATA step.) It is a good idea to develop some self-imposed rules for their placement. Many people put FORMAT, LABEL, and the other statements discussed later in this chapter at the bottom of the DATA step.

Frequently Used System Formats

Exhibit 9.3 (see p. 159) describes some of the more frequently used SAS system formats. Bracketed ([ ]) numbers refer to the qualifiers at the bottom of the exhibit. An "na" for an item means that the column's content is not relevant for the format.

Format Examples

Exhibit 9.4 demonstrates the use of some of the formats in Exhibit 9.3 using different variable values and different widths. The observations are different values of the variable ORIGINAL. The variables are duplicates of ORIGINAL formatted as indicated by the column headings. In each, notice how SAS reacts to the "problem format" conditions described earlier.

Exercises

9.1. Examine the following program (the line numbers are used for reference only, they are not part of the program):

```
(1) data testrun;
(2) infile '\t.dat';
(3) input group $ 1-5 salary rating;
(4) run;
(5) proc print data=testrun;
(6) run;
(7) proc freq data=testrun;
(8) tables rating;
(9) run;
```

a. Write a FORMAT statement to display RATING in three columns, with one decimal place. RATING should be displayed this way for both procedures, but you should use only one statment to assign the format. Indicate where you would locate the statement.

b. Write a FORMAT statement to display SALARY with a dollar sign ($) and commas (,). No decimal places are needed. The maximum salary is, alas, limited to six figures.

c. Since RATING has a maximum value of 999, permanently associate a format of 3. with RATING. Override it in the PRINT procedure to allow numbers with two decimal places (e.g., 9.12). Indicate where you would locate each statement.

9.2. The values of PROFITK in five observations are:

```
-21.3    0    5.2    19.0    120.1
```

What is displayed for each value when the following formats are used? Indicate blank spaces by using a character such as ^ or \.

a. comma4.0
b. comma4.2
c. z4.2
d. comma8.0
e. comma8.2
f. z8.2

9.3. Refer to Appendix E's description of the STATES dataset. Select five numeric variables and assign formats that enhance the display of the variables' values. For example, you might want to display an income figure with a DOLLAR format or large numbers with COMMA formats.

9.4. What, if anything, is wrong with the following program? An ellipsis (. . .) indicates all or part of a statement that is not germane to the exercise.

```
data ... ;
infile ... ;
input name $ 1-20 temp 22-27 3;
```

EXHIBIT 9.4   Display formats: different values formatted with different widths

---

```
------------------ Output listing --------------------

    ORIGINAL         9.       7.       5.    3.   2.   1.        10.2       10.3

-10000.000000     -10000    -10000    -1E4   ***  **   *     -10000.00   -10000.000
 -1000.000000      -1000     -1000    -1000   ***  **   *      -1000.00    -1000.000
    -5.000000         -5        -5       -5    -5   -5   *         -5.00       -5.000
    -1.234500         -1        -1       -1    -1   -1   *         -1.23       -1.234
    -0.500000         -1        -1       -1    -1   -1   *         -0.50       -0.500
     0.000000          0         0        0     0    0   0          0.00        0.000
     0.500000          1         1        1     1    1   1          0.50        0.500
     1.234500          1         1        1     1    1   1          1.23        1.234
     5.000000          5         5        5     5    5   5          5.00        5.000
  1000.000000       1000      1000     1000   1E3  **   *       1000.00     1000.000
 10000.000000      10000     10000    10000   1E4  **   *      10000.00    10000.000

    ORIGINAL        10.4              10.5         comma10.    comma10.2    comma8.

-10000.000000   -10000.000      -10000.000         -10,000   -10,000.00    -10,000
 -1000.000000    -1000.0000      -1000.0000          -1,000    -1,000.00     -1,000
    -5.000000       -5.0000         -5.00000              -5       -5.00         -5
    -1.234500       -1.2345         -1.23450              -1       -1.23         -1
    -0.500000       -0.5000         -0.50000              -1       -0.50         -1
     0.000000        0.0000          0.00000               0        0.00          0
     0.500000        0.5000          0.50000               1        0.50          1
     1.234500        1.2345          1.23450               1        1.23          1
     5.000000        5.0000          5.00000               5        5.00          5
  1000.000000     1000.0000       1000.00000           1,000     1,000.00      1,000
 10000.000000    10000.0000      10000.0000          10,000    10,000.00     10,000

    ORIGINAL   comma8.2        e6.        e5.        e4.       best10.    best7.

-10000.000000   -10000.0    -1.000E+04  -1.0E+04   -1.E+04      -10000    -10000
 -1000.000000   -1000.00    -1.000E+03  -1.0E+03   -1.E+03       -1000     -1000
    -5.000000      -5.00    -5.000E+00  -5.0E+00   -5.E+00          -5        -5
    -1.234500      -1.23    -1.234E+00  -1.2E+00   -1.E+00     -1.2345   -1.2345
    -0.500000      -0.50    -5.000E-01  -5.0E-01   -5.E-01        -0.5      -0.5
     0.000000       0.00     0.000E+00   0.0E+00    0.0E+00          0         0
     0.500000       0.50     5.000E-01   5.0E-01    5.0E-01        0.5       0.5
     1.234500       1.23     1.234E+00   1.2E+00    1.2E+00     1.2345    1.2345
     5.000000       5.00     5.000E+00   5.0E+00    5.0E+00          5         5
  1000.000000    1,000.00    1.000E+03   1.0E+03    1.0E+03       1000      1000
 10000.000000   10000.00    1.000E+04   1.0E+04    1.0E+04      10000     10000
```

EXHIBIT 9.4 *(continued)*

ORIGINAL	best5.	best3.	dollar10.	dollar10.2	dollar8.2	dollar8.
-10000.000000	-1E4	***	$-10,000	$-10000.00	-10000	$-10,000
-1000.000000	-1000	***	$-1,000	$-1,000.00	-1000.0	$-1,000
-5.000000	-5	-5	$-5	$-5.00	$-5.00	$-5
-1.234500	-1.23	-1	$-1	$-1.23	$-1.23	$-1
-0.500000	-0.5	-.5	$-1	$-0.50	$-0.50	$-1
0.000000	0	0	$0	$0.00	$0.00	$0
0.500000	0.5	0.5	$1	$0.50	$0.50	$1
1.234500	1.234	1.2	$1	$1.23	$1.23	$1
5.000000	5	5	$5	$5.00	$5.00	$5
1000.000000	1000	1E3	$1,000	$1,000.00	$1000.00	$1,000
10000.000000	10000	1E4	$10,000	$10,000.00	10000.0	$10,000

ORIGINAL	z10.2	z10.3	z10.4	z10.5
-10000.000000	-010000.00	-10000.000	-10000.000	-10000.000
-1000.000000	-001000.00	-01000.000	-1000.0000	-1000.0000
-5.000000	-000005.00	-00005.000	-0005.0000	-005.00000
-1.234500	-000001.23	-00001.234	-0001.2345	-001.23450
-0.500000	-000000.50	-00000.500	-0000.5000	-000.50000
0.000000	0000000.00	000000.000	00000.0000	0000.00000
0.500000	0000000.50	000000.500	00000.5000	0000.50000
1.234500	0000001.23	000001.234	00001.2345	0001.23450
5.000000	0000005.00	000005.000	00005.0000	0005.00000
1000.000000	0001000.00	001000.000	01000.0000	1000.00000
10000.000000	0010000.00	010000.000	10000.0000	10000.0000

```
format name 20. temp 6.3;
run;

proc print ... ;
format temp $6.3;
run;
```

9.5. Write a FORMAT statement to assign COMMA8. to SAL1 through SAL10, $CHAR20. to ID, and to turn off formats for DELTA1 and DELTA2. Then rewrite it using two statements, one to assign formats and one to turn them off.

9.2 Dressing Up: LABEL and ATTRIB Statements

LABELs Revisited

Most SAS procedures can display an extended description of a variable's meaning. This information is entered in one or more LABEL statements. The LABEL statement was discussed in Chapter 5's review of statements used with procedures. The LABEL statement used in the DATA step is exactly the same as the one used with procedures.

Like the FORMAT statement, when LABEL statements are entered in a DATA step they "travel" with the SAS dataset. If you defined a LABEL in the DATA

EXHIBIT 9.5 Comparison of ATTRIB and LABEL/FORMAT statements

```
attrib salary1 format=comma10. label='Salary, first year'
       salary2 format=comma10. label='Salary, second year'
       salary3 format=comma10.
       salary4 format=comma10.
       salary5 format=comma10.
       ;

label salary1='Salary, first year'
      salary2='Salary, second year'
      ;
format salary1-salary5 comma10. ;
```

step and then use the dataset in subsequent PROCs, it is not necessary to reenter LABEL: the labels are part of the dataset. Reentry of a LABEL specification is required only if you wish to change the text of the label for the duration of a procedure. Also like formats, if a variable is given a label more than once, only the last label found in the DATA step is kept.

The LABEL statement can be placed anywhere in the DATA step. Many users place the label at the bottom of the step, although some prefer to place it as close as possible to the first reference to the variable. Just as formats can be nullified, or "turned off," so can labels. This is done by specifying a blank or null (consecutive apostrophes, '') LABEL. The variable label for AREA is nullified in the following example:

```
label region = 'Census regional breakdown'
      area   = ''
      ;
```

Combining Labels and Formats: ATTRIB

The ATTRIB statement lets you identify the LABEL and FORMAT of a variable in one place. Its syntax is identical to the ATTRIB statement described in Chapter 5. Like its LABEL and FORMAT counterparts, it may be associated with a dataset or with a procedure.

When would the individual statements (LABEL and FORMAT) be chosen over the ATTRIB statement? ATTRIB's advantage is one of documentation: both the label and format are next to each other. The case for separate statements is stronger when many variables share the same format. In this case it is quicker to associate and change formats with variables than it would be using ATTRIB.

The ATTRIB, FORMAT, and LABEL statements in Exhibit 9.5 accomplish the same thing. Notice, though, how much less coding is involved when separate statements are used. Also consider how tedious writing the ATTRIB statement would be if there were 20 SALARY variables instead of just five, and how many extra keystrokes and commands would be required to change the COMMA10. format.

Exercises

9.6. Refer to Appendix E's description of the STATES dataset.
  a. Write an ATTRIB statement assigning formats and labels to variables STATE, NAME, POP, _5YRPLUS, _65PLUS, PCTIN, and PCTCROWD. Variables

representing counts should be formatted 12 columns wide with commas, while percentages should take six columns and three decimal places. Character variables can use as many columns as they do in the raw dataset.

   b. Repeat part a but use separate FORMAT and LABEL statements.

9.7. The label for variable O2UP is "Oxygen Uptake." O2UP is used in the following program (line numbers are for your reference only—they are not part of the program).

   (1) `proc print data=smokers;`
   (2) `var patient o2up;`
   (3) `run;`
   (4)
   (5) `proc freq data=smokers;`
   (6) `tables o2up;`
   (7) `run;`

   a. Write a LABEL statement to assign the label to O2UP. Indicate where you would place the statement in each PROC.
   b. If you wanted the PRINT column heading to split across two lines, one for each word, what would you add to the PROC statement and/or the label?

9.3   Trimming Down: DROP and KEEP Statements

Chapter 8 described ways to trim unwanted *observations* from the dataset. DELETE, RETURN, and OUTPUT force SAS to avoid saving in the SAS dataset every record used on input. In this section two statements that perform analogous functions for unwanted *variables* are discussed: DROP and KEEP are simple tools to help prevent dataset clutter.

Why DROP?

Why drop variables from a dataset if you bothered reading or creating them in the first place? A number of situations warrant excluding variables. An input variable may be converted to a more useful scale: for example, feet and inches might be converted to meters, and the original variables dropped. A variable may be needed only to determine whether the observation will be deleted: for example, do not keep GENDER if the DELETE statement was executed for GENDER ^='F'. This is a reasonable approach since we know that the dataset will contain only values of GENDER equal to 'F'. Finally, if the variable of interest is a summary measure, individual measures may not be needed: for average weekly temperature readings over a year, you may want only temperature means and variances, not the 52 individual variables.

Being selective about which variables are included in the output dataset helps control the amount of storage it uses. This control is especially important when you are handling datasets with many observations or variables.

POINTERS AND PITFALLS

DROP and KEEP encourage good database design by allowing you to specify what will and will not be needed for later analysis. These statements should not be used, however, until you are sure the DATA step is working correctly. Variables that will ultimately be unwanted may be useful in the "debugging" process and thus should be kept during the early stages of program development.

Syntax

The syntax of the DROP and KEEP statements is straightforward:

```
DROP varlist;

KEEP varlist;
```

Some common applications of DROP and KEEP statements follow:

```
drop ft in;
keep meter;

drop math1-math4 ver1-verb4;
```

In the DROP and KEEP statements, *varlist* is any legal variable or variable list. In a DROP statement, *varlist* indicates the variables that will *not* be included in the output dataset. In a KEEP statement, *varlist* indicates all the variables that *will* be included in the output dataset.

Usage Notes

DROP and KEEP have the following features:

- They can be placed anywhere in the DATA step. It is a good idea, especially in long DATA steps, always to put them in the same place so they can be easily located. The bottom of the DATA step is customary.

- DROPping or KEEPing a variable that does not exist causes an error. Part of the SAS Log from such a mistake is shown in Exhibit 9.6. Notice that even though an error message was printed, the dataset was, in fact, created and had the right number of variables.

- If a variable is specified in both a KEEP and a DROP statement, it will be kept.

- When writing complex DATA steps, avoid DROPs and KEEPs until you are sure that the step is running correctly. The seemingly DROPpable variables often hold the key to what went awry in the program.

EXHIBIT 9.6   Attempt to DROP a nonexistent variable

```
------------------- SAS Log ---------------------------

    1    data one;
    2    infile '\data\raw\preciv';
    3    input name $ 1 - 20 party $ 21-29 born 31-34;
    4    drop presname ;
    5    run;
ERROR: The variable PRESNAME in the DROP, KEEP, or RENAME list has never been
       referenced.
NOTE: The infile '\data\raw\preciv' is file C:\DATA\RAW\PRECIV.
NOTE: 16 records were read from the infile C:\DATA\RAW\PRECIV.
      The minimum record length was 57.
      The maximum record length was 57.
NOTE: The data set WORK.ONE has 16 observations and 3 variables.
```

Exercises

9.8. Examine the following DATA step. The ellipsis (. . .) indicates all or part of statements that are not germane to the exercise.

```
data runa;
infile ... ;
input run $ 1 def1-def3;
avg = mean(of def1-def3);
if run = 'a' | run = 'A' then output;
```

Write a DROP or KEEP statement that restricts the variables in dataset RUNA. Justify your answer. If no dropping or keeping is appropriate, explain.

9.9. What, if anything, is wrong with the following program? As usual, an ellipsis (. . .) represents portions of the program that have no bearing on the exercise.

```
data one;
infile ... ;
input id $ 1-8 subgrp $ 7-8 lev1 lev2;
total = lev1 + lev2;
drop lev1 lev2;
run;

proc freq data=one;
tables total lev1 lev2;
run;
```

9.10. Examine the following DATA step:

```
data x;
infile cards;
input batch $ 1-12 m1 d1 m2 d2 m3 d3;
if n(of m1--d3) > 4 then stat = 'ok';
   else                stat = '??';
run;
```

We want dataset X to contain only variables BATCH and STAT. Write a DROP statement to accomplish this, then write an equivalent KEEP statement.

Summary

Chapter 9 describes some techniques for enhancing the appearance of displayed variables and for reducing the size of SAS datasets. SAS format types, which are discussed in detail, parallel SAS data types: they are either character or numeric. The formats are used only to display data; they do not affect the variables stored in the SAS dataset. The format instructions stored in the SAS dataset may be overridden during the execution of a procedure. Format terminology, such as "alignment" of the formatted variable within a range of columns of a specified "width," is also discussed; and some commonly used SAS System formats are presented, along with examples of their use.

In addition, another means of enhancing output and legibility and both LABEL and ATTRIB statements are discussed. Finally, the rationale and syntax for dropping variables from a SAS dataset are presented. Frequently a DATA step has variables that are intermediate steps toward a desired calculated value. These excess variables do not need to be stored as part of the SAS dataset. The DROP and KEEP statements allow the user to be specific about the contents of the SAS dataset being written.

10 SAS Datasets

In the preceding chapters, SAS datasets have been used in many ways. They were created in DATA steps and sorted, printed, analyzed, and displayed with PROCs. This chapter looks at SAS datasets in greater detail. It shows that SAS datasets are a useful, economical way to store and process data. It also discusses how to create and use permanent SAS datasets (and why the adjective "permanent" can be a misnomer!). Finally, a few utility PROCs that help manage SAS datasets are presented.

Keep in mind that some of the chapter's material is not appropriate for all operating systems. The exceptions are noted in the text, and a more comprehensive treatment of system differences is found in Appendix D.

10.1 SAS Datasets in General

SAS datasets are specially formatted files containing all the information a PROC or a DATA step needs when processing a dataset. This section discusses the contents of the SAS dataset, which fall into two major categories: data and directory; dataset libraries, which are useful means of grouping collections of datasets; and the advantages of using SAS datasets.

Data

The data read with the INPUT statement and those created with assignment statements are stored as SAS dataset *variables*. The individual records, or rows, of the raw dataset correspond to the dataset's *observations*. The data are arranged by SAS in a special format that makes transferring the numbers and characters from disk to memory fast and efficient.

Directory

The other component of a SAS dataset, the directory, tells SAS where to find a variable in an observation and the variable's data type, label, and formats. Information about the dataset itself is also stored in the directory. These items include the date and time of dataset creation, read/write passwords, the dataset's size, and (in some operating systems) a copy of the program statements that created the dataset. The directory has four principal components: locations, labels, formats, and history.

Locations.   Once data are stored in a SAS dataset, their physical location within it is not a concern. The DATA step is no longer needed since datasets contain the name of a variable and where in each observation the variable can be found.

Labels.   Variable labels are stored in the dataset and will "travel" with it. This capability relieves the user of having to define variable labels each time they are used in a PROC. If a PROC can use labels and a variable has a nonblank label, the label will be used. A special *dataset* label is also kept in the directory. This label may be used to describe the nature of the entire dataset. Its creation and use are discussed on page 175.

Formats.   The names of display formats for variables are stored in the dataset. Like labels, these format references accompany the dataset whenever it is used. Also like labels, formats can be overridden for an individual PROC.

History.   On many systems, SAS stores several "generations" of programs used to create the dataset. These histories of the data's development can be an invaluable tool when trying to determine the exact nature and contents of the dataset. Though useful, the history feature of the directory has been discontinued and itself "become history" in version 6 of the SAS System.

POINTERS AND PITFALLS

SAS datasets are efficient and reliable ways to store data. They have numerous logistical advantages over raw datasets and should be stored permanently whenever possible.

When thinking of the differences between a raw dataset and a SAS dataset, keep the following analogy in mind. The raw dataset and the DATA step used to process it are similar to building a model airplane. The pieces (data) fit together according to a set of directions (INPUT statement). The painting and decals (LABELs and FORMATs) come later and enhance the appearance of the assembled product. The building process is also fairly time-consuming.

The SAS dataset is analogous to the finished model. It may sometimes take more space to store than the boxed, unassembled pieces, but it can be used over and over again without having to take time for assembly. Its ready-to-use form means quicker access: you do not have to build it from scratch each time you use it.

Dataset Libraries

Just as related variables can be grouped into datasets, related datasets can be grouped into **libraries**. The library's form varies by operating system: in IBM/MVS installations a library is contained in a single "OS dataset"; in microcomputer and minicomputer environments a library is contained within a single subdirectory.

> **POINTERS AND PITFALLS**
>
> All computer operating systems support the concept of libraries (though what they are called varies). Libraries can be used to organize data, programs, and documentation into logically separate entities. The benefits of such organization are particularly great in large projects.

Regardless of system type, libraries are a useful means of grouping collections of datasets. The concept may also be extended to raw datasets, user-written formats (discussed in Chapters 14 and 15), and SAS programs. Although there is nothing to stop you from scattering all your datasets in different libraries, in the long run it is easier to keep raw datasets in a library such as FISCAL\RAWDAT, SAS datasets in FISCAL\SASDAT, documentation and related reports in FISCAL\DOC, and programs in FISCAL\SASPGM. This organization is illustrated in Exhibit 10.1. It is specific to the IBM microcomputer environment, but its logic can easily be extended into other operating systems.

EXHIBIT 10.1   Directory organization of a hypothetical project

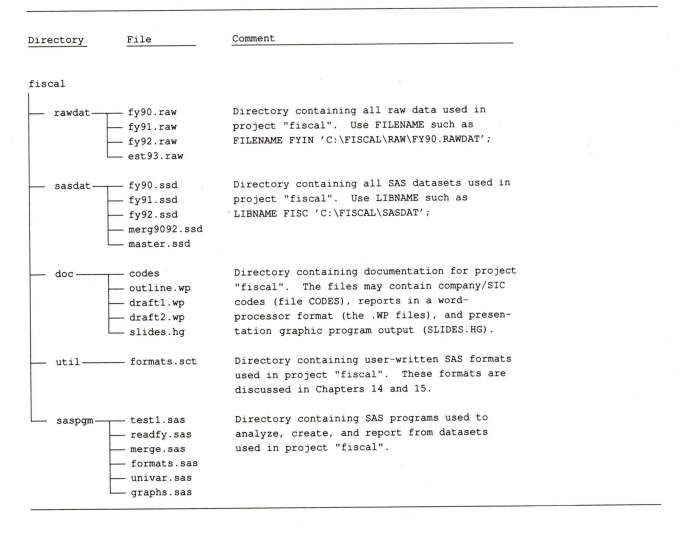

```
Directory        File            Comment


fiscal

├── rawdat ──┬── fy90.raw        Directory containing all raw data used in
│            ├── fy91.raw        project "fiscal".  Use FILENAME such as
│            ├── fy92.raw        FILENAME FYIN 'C:\FISCAL\RAW\FY90.RAWDAT';
│            └── est93.raw

├── sasdat ──┬── fy90.ssd        Directory containing all SAS datasets used in
│            ├── fy91.ssd        project "fiscal".  Use LIBNAME such as
│            ├── fy92.ssd        LIBNAME FISC 'C:\FISCAL\SASDAT';
│            ├── merg9092.ssd
│            └── master.ssd

├── doc ─────┬── codes           Directory containing documentation for project
│            ├── outline.wp      "fiscal".  The files may contain company/SIC
│            ├── draft1.wp       codes (file CODES), reports in a word-
│            ├── draft2.wp       processor format (the .WP files), and presen-
│            └── slides.hg       tation graphic program output (SLIDES.HG).

├── util ─────── formats.sct     Directory containing user-written SAS formats
│                                used in project "fiscal".  These formats are
│                                discussed in Chapters 14 and 15.

└── saspgm ──┬── test1.sas       Directory containing SAS programs used to
             ├── readfy.sas      analyze, create, and report from datasets
             ├── merge.sas       used in project "fiscal".
             ├── formats.sas
             ├── univar.sas
             └── graphs.sas
```

Refer to Appendix D for more detailed examples of how libraries are handled by the major operating systems.

SAS Dataset Advantages

Using SAS datasets provides several benefits, including their processing speed, self-contained nature, ease of handling, and multimedia storage capability.

Speed.   SAS datasets are specially designed to move as much data as quickly as possible. SAS will move x units of storage in SAS dataset format quicker and with less consumption of system resources than the same x units stored as raw data.

Self-Contained Nature.   The SAS dataset is a total, self-documenting package. Everything you need to know about the dataset is contained in the directory: you do not have to refer continually to hard copy or machine-readable documentation for details about data scale, location, and so on.

Ease of Handling.   As later chapters (especially Chapter 13) show, SAS datasets are easy to handle. They can be updated, combined with other SAS datasets, and so on with a few simple statements. SAS keeps track of "what's where" at all times during even the most complex operations.

Multimedia Storage Capability.   SAS datasets can be stored on any type of disk, tape, or mass-storage device. When they are created, no special commands are usually needed to indicate the type of device, and when they are used (even if disk and tape datasets are combined), device specification is not required.

10.2   Permanent SAS Datasets

The SAS datasets used so far in this book have had an important limitation imposed upon them. They were *temporary* datasets, deleted at the end of the SAS batch job or interactive session. They could be used any number of times *within* the batch job or interactive session. But once the job or session terminated, SAS would delete them as part of its routine housecleaning.

This section describes how to save SAS datasets so that they may be used in later jobs. Although the principles of this process are the same in any operating system, the examples in this chapter are system-specific. All the examples were run on IBM-compatible microcomputers running under DOS 3.3 or higher. Examples for other systems (IBM's MVS and CMS, DEC's VMS) can be found in Appendix D. The discussion is divided into three sections: creating permanent SAS datasets, using them, and dataset options.

Creating Permanent SAS Datasets

LIBNAME.   A new statement and extensions to two familiar ones are required. The new statement is LIBNAME. It serves much the same purpose with SAS datasets as FILENAME does with raw data: it links SAS dataset libraries to DATA steps. Its syntax is shown below:

```
LIBNAME ref 'pathname' ... ;
```

Some common applications of the LIBNAME statement follow:

```
libname input '[clin.sasfiles]';

libname master '[]';

libname datain 'd:\soils\sasdata';
```

The LIBNAME statement has two components:

1. The *ref* is a name assigned to link the location of the permanent SAS dataset to the rest of the program. The *ref* can be from one to eight characters, must begin with an alphabetic character (a through z), and may contain only alphabetic characters and numerals. The default *ref* is WORK. Some *refs* are reserved for use by SAS. Among these are SASUSER, MACAUTO, SASMAPS, and LIBRARY.

2. The *pathname* identifies the subdirectory or, in IBM mainframe environments, the OS dataset in which the dataset will be stored. See Appendix D for examples of LIBNAME statements in other operating systems.

You can specify more than one *ref-pathname* combination per LIBNAME statement. LIBNAME can be located anywhere in the program prior to *ref*'s first use. Again, the common practice is to put all LIBNAMEs near the beginning of the program, before any DATA step or PROC.

The DATA Statement. SAS datasets can be saved in permanent dataset libraries. This enables you to create the file in one SAS session, then use it both later in the same session and in subsequent sessions. The convenience and computer resource savings of permanent datasets are considerable. They require an extension of the DATA statement:

```
DATA [ref.]dsname[(options)] ... ;
```

An application of this new form of the DATA statement follows:

```
data input.master;
```

The DATA statement uses *dsname* and *options* and adds *ref*:

1. The *ref* has the same meaning and syntax conventions as the *ref* in the LIBNAME statement. This is the means used to link a DATA step or PROC to the operating system's physical datasets. Defining a LIBNAME of DATAHERE and using a *ref* of DATAHERE in the DATA statement tells SAS you want to store a SAS dataset named *dsname* in the path identified by DATAHERE.

2. The *dsname* and *options* are the same as in the initial description of the DATA statement. The *dsname* is the name of the SAS dataset being created in this DATA step. The *options*, discussed later (see page 174), are features that affect which observations are read from the dataset, which variables are kept and dropped, and so on.

"Two-Level" Datasets. Given the linkage of the program's dataset references to the outside "world" of subdirectories and paths, we can return to examples in the previous chapters with a better appreciation of what was happening. Many of the Log messages had notes like "Dataset WORK.PRES has *xx* observations and *yy* variables." WORK was never specified because we never wanted to save a dataset for later use. SAS supplied the WORK *ref* by default. However, SAS always refers to datasets by a so-called two-level name: a *ref* and a dataset name, separated by a period.

When SAS sees a two-level name and the first level is not WORK, it assumes you are trying to save the dataset permanently. It tries to match the *ref* name in the DATA and OUTPUT statements to a *ref* in a LIBNAME (remember that this varies somewhat by operating system: see Appendix D for system-specific details).

Changing Names.   The *ref* can change from job to job and session to session, but the SAS dataset name cannot. You can create a dataset called DATAOUT.FIS-CAL89, provided the LIBNAME refers to a path called DATAOUT. When you use the dataset in later jobs, you can change the DATA statement and LIBNAME to the more intuitive DATAIN. You could not, however, change FISCAL89. Exhibit 10.2 demonstrates the linkage between the name of the SAS dataset used by the operating system, the LIBNAME statement, and the DATA statement.

EXHIBIT 10.2   Linkage between file name, LIBNAME, and DATA statement

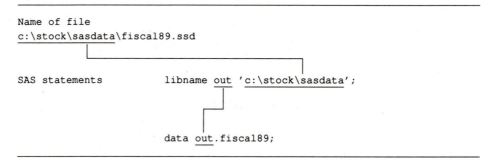

A simple example will illustrate the creation of permanent SAS datasets. More advanced examples are found later in this section.

We read the U.S. president vital statistics data introduced in Chapter 4, saving it with a SAS dataset name of PRES. PRINT is used to display the dataset (the rule of last use applies to both permanent and temporary SAS datasets). The input raw data come from file 'c:\data\raw\allpres.' The output SAS dataset is directed to fileref OUT, the path 'c:\data\sas\'. This subdirectory contains the SAS dataset, file name PRES.SSD (microcomputer SAS datasets in version 6.03 are typically stored with an extension of SSD). The SAS Log is shown in Exhibit 10.3.

Using Permanent SAS Datasets

The permanent SAS dataset can either be used in a DATA step or processed by a procedure. If the dataset does not require modification (calculation, subsetting, and so forth), simply enter the two-level dataset name in the PROC's DATA option. There is no need to copy the dataset with a DATA step and use the temporary copy.

If modification of the dataset is necessary before using it in a PROC, read it in a DATA step with a SET statement. SET performs functions similar to those of the INPUT statement. INPUT moves raw data into memory, then reads the raw data, translates it into a SAS variable, and moves it to another area in memory, possibly writing it to a dataset. The SET statement performs these data movement tasks for SAS datasets, both temporary and permanent. When a SET statement is executed, SAS moves the next observation from the SAS dataset into memory, ready for use by the DATA step. In addition to moving data, SET makes the dataset's formats and labels available to the DATA step. The SET statement

EXHIBIT 10.3   Create permanent SAS dataset PRES

```
NOTE: Copyright(c) 1985,86,87 SAS Institute Inc., Cary, NC 27512-8000, U. S. A.
NOTE: SAS (r) Proprietary Software Release 6.03
      Licensed to D**2 Systems, Site 123456789.

   1    libname out '\data\sas\';
   2
   3    data out.pres;
   4    infile '\data\raw\allpres';
   5    input name $ 1-20 party $ 21-29 born 31-34  bornst $ 37-38
   6          inaug 43-46   ageinaug 51-52   agedeath 56-57
   7          ;
   8    run;
NOTE: The infile '\data\raw\allpres' is file C:\DATA\RAW\ALLPRES.
NOTE: 40 records were read from the infile C:\DATA\RAW\ALLPRES.
      The minimum record length was 57.
      The maximum record length was 57.
NOTE: The data set OUT.PRES has 40 observations and 7 variables.
NOTE: The DATA statement used 11.00 seconds.
   9
  10    proc print data=out.pres;
  11    run;
NOTE: The PROCEDURE PRINT used 10.00 seconds.
```

These names must match if the dataset will be saved permanently.

can also be used to combine two or more datasets. This feature is discussed in Chapter 13.

Syntax.   The SET statement's syntax resembles that of the DATA statement:

```
SET [ref.]dsname[(options)] ... ;
```

The *ref*, *dsname*, and *options* are identical in meaning to their DATA statement counterparts.

Examples.   The following examples use the dataset created in Exhibit 10.3. More elaborate examples are found at the end of the "Dataset Options" section.

In the first example, select presidents inaugurated in the twentieth century. Notice that the *ref* in the LIBNAME and DATA statements has changed from OUT (used in earlier examples) to IN but that PRES remains the same:

```
libname in '\data\sas\';

data only20th;
set  in.pres;
if inaug >= 1900 then output;
run;
```

In this example a new permanent dataset containing twentieth-century presidents is created. The dataset in the DATA statement is changed to a two-level name:

```
libname in '\data\sas\';

data in.only20th;
set  in.pres;
```

```
if inaug >= 1900 then output;
run;
```

The \DATA\SAS subdirectory now has files PRES and ONLY20TH, both with extensions of SSD.

In this example the dataset created in the first example above is inadvertently overwritten. SAS correctly thought it was going to write a new copy of IN.PRES but only had a RUN statement in the body of the DATA step. ONLY20TH now has one observation and zero variables!

```
libname in '\data\sas';

data in.only20th;
run;

proc print data=in.only20th;
run;
```

This example uses the two permanent SAS datasets in some PROCs. Notice that the program does not require any DATA steps:

```
libname in '\data\sas';

proc univariate data=in.pres;
var ageinaug agedeath;
id name;
run;

proc print data=in.only20th;
id name;
run;
```

In this example a variable is added to each observation of the PRES dataset. Variable lists may now use YRDIED as the last variable in the dataset since it was appended to the list of variables already in PRES.

```
libname in '\data\sas';

data in.pres2;
set  in.pres;
yrdied = born + agedeath;
run;
```

Dataset Options

This section discusses some of the options available when reading and writing SAS datasets. The syntax of each option is covered first, and then a set of examples is reviewed. The options share an important characteristic: when they are used as an *input dataset* (in a SET statement or a PROC's DATA option), the option is in effect only for the duration of the DATA step or PROC: the dataset itself remains unchanged.

DROP and KEEP. The DROP and KEEP dataset options perform the same functions as their DATA step counterparts: they specify which variables will be dropped or kept in the SAS dataset being read (SET statement) or written (DATA statement). The syntax of these options is illustrated below:

```
DROP=varlist

KEEP=varlist
```

Some common applications of the DROP and KEEP options follow:

```
set in.fy83(keep=sic020q1--sic891q4);

proc print data=temp(keep=id level status score);
```

The *varlist* is any combination of variable lists and variable names. If a variable is specified in both the DROP and KEEP lists, it is dropped.

RENAME. RENAME is used to change the name of one or more variables in the dataset being read or written. RENAME's syntax is illustrated below:

```
RENAME=(old=new ...)
```

Some common applications of the RENAME option follow:

```
set in.geo(rename=(zipcode=zip st=state));

proc print data=test(rename=(grade=level));
label level = "Attainment level (SJHA index)";
run;
```

In the RENAME option, *old* is the current variable name and *new* is the new name. SAS performs DROPs and KEEPs before the RENAME, so you should DROP and KEEP the *old* names.

LABEL. Just as you could enhance the meaning of a variable's name by using a label, so can you attach a label to a dataset. The LABEL option enables you to clarify the meaning and content of the dataset. The label will be added only when the option is used in the DATA statement (when a dataset is being written). LABEL's syntax is illustrated below:

```
LABEL='text'
```

A common application of the LABEL option follows:

```
data out.fy9091(label='Earnings data, fisc yr 90-91');
```

The *text* is the extended description of the dataset. It follows the same rules as a variable label: 40 characters maximum, enclosed in quotation marks ('') or single quotes ('), single apostrophe represented by two consecutive single quotes ('') if *text* is enclosed by single quotes.

OBS and FIRSTOBS. The OBS and FIRSTOBS options specify which observations will be read from an existing dataset (one used in the SET statement or in the DATA option of a PROC). These options are especially useful when you are debugging a program and do not want to process all the observations in a dataset. Programmers typically experiment with a reasonable size subset, then remove OBS/FIRSTOBS once the program is running satisfactorily.

FIRSTOBS specifies where to begin reading, and OBS indicates where to finish. Think of OBS being equivalent to "lastobs" and not how many observations to process. The default *n* for FIRSTOBS is 1, the default *n* for OBS is the last observation in the dataset. Specifying FIRSTOBS = 50 without OBS reads all observations from 50 to the end of the dataset. Specifying OBS = 50 without FIRSTOBS reads the first 50 observations. The syntax of these options is shown below:

```
FIRSTOBS=n

OBS=n
```

Some common applications of the FIRSTOBS and OBS options follow:

```
set in.labdata(obs=40);
```

```
set invntory(firstobs=100 obs=200);
```

The *n* is an integer constant indicating the starting and ending observations to be read. If a FIRSTOBS *n* value exceeds the number of observations in the dataset or if an OBS *n* value is less than FIRSTOBS, SAS behaves as if the dataset had no observations. An OBS *n* value exceeding the number of observations in the dataset has no effect on execution.

Examples.   The following examples continue using the presidential dataset. The SAS Log is displayed when the option causes an error or does not cause an error when we would expect it to.

Key features and options used in this section are summarized in the following list:

Exhibit	Features/Options
10.4	Restrict (KEEP) only a subset of SET dataset's variables
10.5	Dataset label; KEEP option on output dataset
10.6	KEEP, OBS used with a dataset in a PROC
10.7	Use OBS, FIRSTOBS to print middle of a dataset; also use RENAME option
10.8	DROP and KEEP the same variable
10.9	Attempt to RENAME a nonexistent variable
10.10	Attempt to KEEP a nonexistent variable
10.11	Ineffective, misplaced use of OBS option
10.12	Conflicting use of FIRSTOBS, OBS options
10.13	Specify one instead of two-level name

Exhibit 10.4 creates a copy of PRES, keeping only a few of its variables.

EXHIBIT 10.4   Only a subset of a dataset's variables copied

```
libname out 'd:';
libname in  '\data\sas';

data out.subset;
set  in.pres(keep=name born ageinaug agedeath);
run;
```

Exhibit 10.5 expands on Exhibit 10.4, adding a dataset label (LABEL=) and a variable to the KEEP list (YRDIED). Notice that variables specified in the KEEP list do not have to be from the SET dataset: any variable known to the DATA step may be kept. Exhibit 10.6 uses the PRES data directly in a PROC, printing only the first five observations and only a few variables.

Exhibit 10.7 prints 10 PRES observations from the middle of the dataset. Some of the kept variables are renamed.

Exhibit 10.8 accidentally KEEPs and DROPs the same variable. No note about the conflict is printed, and the duplicated variable is dropped from the SET dataset.

EXHIBIT 10.5   Dataset label added; only some variables selected to keep in the output dataset

```
libname out 'd:';
libname in  '\data\sas';

data out.subset(keep=name ageinaug born agedeath
          yrdied
          label='Subset: critical dates');
set  in.pres;
yrdied = born + agedeath;
run;
```

EXHIBIT 10.6   Variables and observations used in a PROC restricted

```
libname in  '\data\sas';

proc print data=in.pres(keep=born inaug agedeath obs=5);
run;
```

EXHIBIT 10.7   Print middle observation

```
libname in  '\data\sas';

proc print data=in.pres(keep=born inaug agedeath firstobs=16
                   obs=25
                   rename=(born=yrborn inaug=yrinaug));
run;
```

EXHIBIT 10.8   DROP and KEEP the same variable

```
1     libname in '\data\sas\';
2     libname out 'd:';
3
4     data out.subset;
5     set  in.pres(keep=ageinaug  name  agedeath
6                  drop=name );
7     run;
NOTE:  The data set OUT.SUBSET has 40 observations
       and 2 variables.
NOTE:  The DATA statement used 8.00 seconds.
```

If in both KEEP and DROP lists, the variable will be dropped.

EXHIBIT 10.9   Log reporting a reference to a nonexistent variable

```
1     libname in '\data\sas\';
2     libname out 'd:';
3
4     data out.subset;
5     set  in.pres(drop=bornst rename=(bornst=whreborn));
ERROR: The variable BORNST in the DROP, KEEP, or RENAME list has never been
       referenced.
ERROR: Invalid DROP, KEEP, or RENAME option on file IN.PRES.
6     run;
NOTE: The SAS System stopped processing this step because of errors.
NOTE: The data set OUT.SUBSET has 0 observations and 0 variables.
NOTE: The DATA statement used 6.00 seconds.
```

Any versions of SUBSET in path 'D:' will be replaced by this new, incorrect dataset. See NOREPLACE option, Chapter 12.

Exhibit 10.9 tries to rename a variable that was dropped. Even though the variable exists in the dataset, it does not exist during this DATA step: SAS cannot rename something that is not there.

Exhibit 10.10 tries to keep a variable that does not exist in either the SET dataset or elsewhere in the DATA step.

Exhibit 10.11 uses the OBS option where it does not belong. OBS and FIRSTOBS are valid only when reading the dataset (in the SET statement and DATA option of a PROC). SAS ignores the option and processes all observations in the SET dataset.

EXHIBIT 10.10   Log reporting attempt to KEEP a nonexistent variable

```
1     libname in '\data\sas\';
2     libname out 'd:';
3
4     data out.subset(keep=name ageinaug agedeath termlnth);
5     set  in.pres;
6     run;
ERROR: The variable TERMLNTH in the DROP, KEEP, or RENAME list has never been
       referenced.
NOTE: The data set OUT.SUBSET has 40 observations and 3 variables.
NOTE: The DATA statement used 9.00 seconds.
```

The dataset is created with only present, valid variables.

EXHIBIT 10.11   OBS option used without effect on output dataset

```
1     libname in '\data\sas\';
2     libname out 'd:';
3
4     data out.subset(obs=5);
5     set  in.pres;
6     run;
NOTE: The data set OUT.SUBSET has 40 observations and 7 variables.
NOTE: The DATA statement used 9.00 seconds.
```

OBS has no impact when used with *output* datasets.

Exhibit 10.12 mistakenly thinks that OBS indicates *how many* observations to process. But you cannot start at 15 and end at 10, so no observations are written to the output dataset. If the intent was to print 10 observations, the OBS value must be changed to 24 (10 observations starting at 15 end at 24).

Exhibit 10.13 shows the most honest of honest mistakes: forgetting to refer to the input (SET) dataset with a two-level name. Similar problems but with a different error message would occur if a two-level name were entered and the LIBNAME statement omitted.

EXHIBIT 10.12   Conflicting FIRSTOBS and OBS specifications

```
1      libname in '\data\sas\';
2      libname out 'd:';
3
4      data out.subset;
5      set  in.pres(obs=10 firstobs=15);
6      run;
NOTE: The data set OUT.SUBSET has 0 observations and 7 variables.
NOTE: The DATA statement used 8.00 seconds.
```

EXHIBIT 10.13   Log reporting missing SET dataset (should have been two-level name)

```
1      libname in '\data\sas\';
2      libname out 'd:';
3
4      data out.subset;
5      set  pres(keep=name ageinaug agedeath
6                  drop=name);
ERROR: Data set WORK.PRES not found.
7      run;
NOTE: The SAS System stopped processing this step because of errors.
NOTE: The data set OUT.SUBSET has 0 observations and 0 variables.
ERROR: Data set OUT.SUBSET was not written because of NOREPLACE option.
NOTE: The DATA statement used 5.00 seconds.
```

Exercises

10.1. Dataset TEMPS will be stored in path C:\DISS\DISSDATA. Write the LIBNAME and DATA statements that will direct TEMPS to the correct path.

10.2. Use the dataset in Exercise 10.1 in a PRINT procedure. Use a first-level name (ref) of DATAIN.

10.3. Write the statements to create a subset of the TEMPS data in Exercise 10.1 and direct it to the same path (the path C:\DISS\DISSDATA will now have two SAS datasets). The new dataset, SUBSET, should have only values of REG equal to 'E.'

10.4. Follow the instructions in Exercise 10.2 but use the KEEP dataset option to print only variables STATE, HIGH, and LOW.

10.5. Follow the instructions in Exercise 10.4 but use the RENAME dataset option to rename HIGH to HIGHEST and LOW to LOWEST.

10.6. Rewrite Exercise 10.1 adding the label "Preliminary survey: instrument 1" to the dataset TEMPS.

10.7. Using TEMPS, write LIBNAME and PRINT procedure statements to print the following portions of the dataset:
   a. First 10 observations
   b. 20th through 30th observations
   c. 20th through last observations

10.8. What, if anything, is wrong with the following programs?
   a.
```
data two;
  set one;
  keep v1-v5;
run;

proc print data=two(drop=v1-v5);
run;
```
   b.
```
data two(obs=20);
  set one(label='Sample dataset');
run;
```
   c.
```
libname datain ''; * Default directory;

data datain.two;
  set dataout.one(keep=st area inc80 inc90);
  avginc = mean(inc80,inc90);
run;
```

10.3   After the Fact: PROC CONTENTS

Now that you have a SAS dataset, what do you do with it? You use it in PROCs, add variables, thin it out with DROP and KEEP statements, and so on. As datasets accumulate and time passes, you will also become familiar with other aspects of life with data: forgetting what is in the dataset and wanting to change its contents once you remember.

This section discusses CONTENTS, the first of two *utility* procedures. The CONTENTS procedure lists the names of SAS datasets in a library and directory information from the datasets. It is very handy to use when dataset size and variable names, labels, and order are of interest.

POINTERS AND PITFALLS

The CONTENTS procedure lists information from the directory of a SAS dataset. It is especially useful for determining variable order (essential information when writing variable lists).

Syntax

The CONTENTS procedure's syntax is described below:

```
PROC CONTENTS [DATA=dset|_ALL_] [SHORT] [POSITION]
              [DIRECTORY] [NODS];
```

Some common applications of PROC CONTENTS follow:

```
proc contents data=in._all_ short;

proc contents data=newmast position directory;
```

The CONTENTS procedure has the following components:

1. The *dset* is the one- or two-level name of the SAS dataset to process. If not specified, SAS selects the most recently created dataset (refer to the rule of last use in Chapter 5). If _ALL_ is specified, SAS selects all SAS datasets found in the path or OS dataset identified by the first level of the dataset name (WORK or a user-supplied fileref).

2. *SHORT* requests an abbreviated listing of the dataset's contents. Only an alphabetical listing of the dataset's variables will be displayed.

3. *POSITION* requests a listing of the variables, whether in the default or *SHORT* format, by their position in the dataset. The first variable name encountered when SAS built the dataset will be listed first, and so on through the last variable. This feature is very useful when you want to use variable lists (discussed in Chapter 3) but are not certain of variable order.

4. *DIRECTORY* asks to display a listing of the SAS dataset names in the path referred to by the first level of the dataset name. This listing is in addition to the dataset information normally printed by CONTENTS (unless *NODS* was specified—see 5). A specification of DATA = LAB8.MICE0391 would list all SAS dataset names in the path referred to by LAB8. *DIRECTORY* is the default if _ALL_ is used as the dataset name in the DATA option.

5. *NODS* suppresses the listing of individual dataset information. It is useful only when the *DIRECTORY* or *DATA = fileref._ALL_* options are selected.

Examples

The CONTENTS procedure's output is straightforward and easy to grasp. In Exhibit 10.14 the special dataset name _ALL_ and the NODS options are used to obtain a listing of all datasets found in the path pointed to by the first-level name SASDSETS. The examples in this section were run with release 6.03 of the SAS System. The appearance and content of the listings will be somewhat different in other releases.

Exhibit 10.15 lists complete information on dataset PRES. All defaults are taken (verbose listing, list variables in alphabetical order).

In Exhibit 10.16 an abbreviated listing of information about the same dataset (SHORT) is requested. A listing of the variables by position in the dataset (POSITION) is also requested.

Exercises

10.9. The SAS dataset TRIAL0 is stored in the path named [MASTER.TRIAL]. You want to use variable lists (discussed in Chapter 3) in a procedure but are not certain of the variable order. Write the LIBNAME and PROC CONTENTS statements required to determine the order of the variables in the dataset.

10.10. Compare the full and abbreviated alphabetical CONTENTS listings for dataset PRES in Exhibits 10.15 and 10.16. What variable information is lost when using the abbreviated listing? Describe a situation where the lost data may not be important. Describe another situation where the information in the abbreviated listing would not be adequate for a data analysis or management task.

EXHIBIT 10.14   Use CONTENTS to list SAS dataset names in a library

```
------------------- Program listing -------------------

libname sasdsets 'c:\data\sas\';

proc contents data=sasdsets._all_ nods;
```

Process *all* datasets in
LIBNAME SASDSETS.

```
------------------- Output listing -------------------

CONTENTS PROCEDURE

-----Directory for Library SASDSETS------

              #    Name       Memtype
              1    CCDB83     DATA
              2    COASTAL    DATA
              3    CSTSUMM    DATA
              4    FORMATS    CATALOG
              5    HALLFAME   DATA
              6    LIFESPAN   DATA
              7    MISCDATA   DATA
              8    NAMES      DATA
              9    NATLPARK   DATA
             10    PARKPRES   DATA
             11    PRES       DATA
             12    STATES     DATA
             13    SUMMSTAT   DATA
             14    TOP15      DATA
```

EXHIBIT 10.15   Complete CONTENTS information for a dataset (SAS release 6.03)

```
------------------- Program listing -------------------

libname sasdsets 'c:\data\sas\';
proc contents data=sasdsets.pres;

------------------- Output listing -------------------

CONTENTS PROCEDURE
```

Form is libname,
SAS dataset name.

```
  Data Set Name:   SASDSETS.PRES           Type:

  Observations:    40                      Record Len: 67
  Variables:       7
  Label:           US President vital stats
```

Position of variable
in each observation.

```
-----Alphabetic List of Variables and Attributes-----

  #   Variable   Type   Len   Pos   Format   Label
  7   AGEDEATH   Num     8    59
  6   AGEINAUG   Num     8    51
  3   BORN       Num     8    33    F4.      Year born (4 digits)
  4   BORNST     Char    2    41
  5   INAUG      Num     8    43    F4.      Year took office, NOT year elected
  1   NAME       Char   20     4
  2   PARTY      Char    9    24    $FREE1.
```

EXHIBIT 10.16   Compact alphabetical, positional CONTENTS of a dataset

---

```
------------------- Program listing -------------------

libname sasdsets 'c:\data\sas\';

proc contents data=sasdsets.pres short position;

------------------- Output listing -------------------

------------------- Begin new page of output ------------------

CONTENTS PROCEDURE

-----Alphabetic List of Variables for SASDSETS.PRES-----

AGEDEATH AGEINAUG BORN     BORNST   INAUG    NAME     PARTY

- - - - - - - - - - New page of output - - - - - - - - -

CONTENTS PROCEDURE

----Variables Ordered by Position----

 NAME     PARTY    BORN     BORNST   INAUG    AGEINAUG AGEDEATH
```

> Display is created by POSITION option. It is useful when you need variable lists.

10.4   After the Fact: PROC DATASETS

The CONTENTS procedure is a quick, convenient means for determining what is contained in the directory of a SAS dataset. It cannot, however, change directory information. Such modification is left to the DATASETS procedure. Understanding when DATASETS should be used can often avoid time-consuming, resource-intensive DATA steps.

PROC DATASETS

The DATASETS procedure has distinct library, dataset, and data management capabilities. These, and a special syntactical feature of its interactive use, are discussed in separate subsections.

Syntax I: General.   PROC DATASETS has two features that set it apart from the DATA step and most other PROCs. First, when DATASETS is run in full-screen or line modes, the RUN statement does not end the PROC. It simply tells SAS to process the statements entered since the beginning of the PROC or the last RUN. The QUIT statement actually ends the procedure. This subtlety is completely transparent when running batch programs but can be puzzling when running any truly interactive form of SAS. The second DATASETS feature is that, unlike most other PROCs, its output is usually written to the Log.

The DATASETS statement is described below:

```
PROC DATASETS [DDNAME|LIBRARY=ref] [NOLIST] [NOFS];
```

The DATASETS statement has three elements:

1. The *ref* identifies the SAS dataset library to manage. If not specified, WORK is used. Unless working with the default WORK library, *ref* should be associated with a LIBNAME statement.

2. *NOLIST* suppresses the listing of dataset names in the library. By default, DATASETS begins execution by displaying an alphabetical listing of dataset names.

3. *NOFS* instructs SAS not to use its full-screen capabilities when prompting for input and displaying output. This option is ignored if the program is being run in batch mode or command-line mode.

The SAS Log for a simple use of PROC DATASETS is shown in Exhibit 10.17.

EXHIBIT 10.17   DATASETS used to list dataset names in a library

```
-------------------- SAS Log ---------------------------

   1     libname presdata '\data\sas';
   2
   3     proc datasets library=presdata;
                   -----Directory for Library PRESDATA------

                       #   Name      Memtype
                       1   NAMES     DATA
                       2   PRES      DATA
                       3   SUBSET    DATA
   4     run;
NOTE: The PROCEDURE DATASETS used 5.00 seconds.
```

Syntax II: Library Maintenance.   Once you know which datasets are in the library, you may want to rename or delete one or more of them. The CHANGE, EXCHANGE, DELETE, and SAVE statements accomplish this. Notice that here, as with other statements, we could get exactly the same results by using one or more DATA steps or operating system commands. Using DATASETS is preferable to DATA steps since it changes *directory* information rather than data. This requires minimal computer resources. The DATASETS-DATA step trade-off becomes increasingly significant as dataset size increases.

The library maintenance statements' syntax is described below:

```
CHANGE old=new;
EXCHANGE name1=name2;
DELETE dataset ... ;
SAVE dataset ... ;
```

Some common applications of library maintenance statements follow:

```
change current=old;
exchange basis=assess;
```

```
delete trial: ;
save final: master;
```

The library maintenance statements have three components:

1. The *old* and *new* specify the old, or current, and new names for a dataset in the library.

2. The *name1* and *name2* identify datasets to swap. The contents of *name1* become known to SAS as *name2*, and vice versa.

3. The *dataset* is the name of a dataset to erase (if used in the *DELETE* statement) or keep (if used in the *SAVE* statement). The *dataset* may be any of an explicit dataset name, a list of the form data*x*–data*z* (e.g., YEAR5–YEAR8 to use YEAR5, YEAR6, YEAR7, and YEAR8), or a common prefix followed by a colon (e.g., REG: to use REG1, REGION1, REG, and REGULAR).

SAS Logs demonstrating use of the library maintenance statements are shown in Exhibits 10.18 and 10.19.

In Exhibit 10.18 dataset NAMES is changed to ROSTER, and then ROSTER and SUBSET are exchanged. Notice that the RUN; statement has to be used so that the CHANGE statement can take effect and the EXCHANGE statement will be able to locate ROSTER.

Exhibit 10.19 shows what happens if we decide the exchange shown in Exhibit 10.18 was not needed and that ROSTER can be deleted. The names are exchanged, and then ROSTER is deleted. We could have said SAVE SUBSET PRES and ended up with the same library contents. You can usually achieve the same results by using both the DELETE and SAVE statements: delete what is not needed or save what is needed. In Exhibit 10.19 we simply chose to use DELETE.

Syntax III: Dataset Maintenance. Recall that it is more efficient to alter directory information with DATASETS than with a DATA step. Two DATASETS statements, CONTENTS and MODIFY, display what is in the dataset and change directory information. The CONTENTS *statement* syntax is presented below. The

---

EXHIBIT 10.18   Log reporting successful dataset name change

```
1    libname presdata 'd:';
2
3    proc datasets library=presdata;
              -----Directory for Library PRESDATA------

                   #    Name      Memtype
                   1    NAMES     DATA
                   2    PRES      DATA
                   3    SUBSET    DATA
4    change names=roster;

5    run;                                                    RUN before using new
                                                             dataset name ROSTER
NOTE: Changing the name PRESDATA.NAMES to PRESDATA.ROSTER (memtype=DATA).
6    exchange roster=subset;
7    quit;
NOTE: Exchanging the names PRESDATA.ROSTER and PRESDATA.SUBSET (memtype=DATA).
NOTE: The PROCEDURE DATASETS used 7.00 seconds.
```

EXHIBIT 10.19   Log reporting two separate data management tasks

---

```
1      libname presdata 'd:';
2
3      proc datasets library=presdata;
                -----Directory for Library PRESDATA------

                     #    Name      Memtype
                     1    PRES      DATA
                     2    ROSTER    DATA
                     3    SUBSET    DATA
4      exchange subset=roster;
5      run;
NOTE: Exchanging the names PRESDATA.SUBSET and PRESDATA.ROSTER (memtype=DATA).
6      delete roster;
7      quit;
NOTE: Deleting PRESDATA.ROSTER (memtype=DATA).
NOTE: The PROCEDURE DATASETS used 6.00 seconds.
```

---

syntax and output closely resemble the CONTENTS *procedure* described earlier in this chapter.

```
CONTENTS [DATA=[dataset|_ALL_]] [POSITION] [SHORT];
```

The CONTENTS statement has three components:

1. The *dataset* specifies the dataset whose directory contents will be displayed. The special dataset name _ALL_ requests a contents listing for every dataset in the library. If neither *dataset* nor _ALL_ is entered, DATASETS will use the most recently *modified* dataset.

2. *POSITION* displays variable information in the order in which the variables were made known to SAS. This option is especially useful when you are trying to write variable lists. The normal, alphabetical listing is also produced.

3. *SHORT* produces an abbreviated listing. Only the variable names are displayed.

The MODIFY statement usually requires one or more additional statements that change the features of the dataset's directory. Only one dataset can appear in a MODIFY statement:

```
MODIFY dataset[(LABEL='ds_text');
[LABEL var=['var_text'] ... ;]
[FORMAT var=[fmt] ... ;]
[RENAME old=new ...;]
```

Some common applications of DATASETS' dataset directory operations follow:

```
modify crop(label='Horticulture output, 1980-89');
label kw = 'Kiwi fruit planting (hectares)';
format kw=hect6. minsize maxsize;
rename totacre=size district=region;
```

The MODIFY statement has four elements:

1. The *dataset* is the name of the dataset whose variable characteristics you want to change.

2. The *ds_text* and *var_text* are labels for *dataset* and *var*, respectively. They follow exactly the same rules as their dataset and variable label counterparts in the DATA, LABEL, and ATTRIB statements. A zero length or blank entry removes the label.

3. The *fmt* is the new format specification for *var*. Enter it as you would in a FORMAT statement. If *fmt* is not entered, the existing format is used. This may be an earlier user-specified format or the default picked by SAS.

4. The *old* and *new* identify pairs of variables to rename.

The SAS Logs and output files showing the dataset maintenance statements are shown in Exhibits 10.20, 10.21, and 10.22. Notice that unlike the library maintenance statements, which sent all output to the SAS Log, CONTENTS output is directed to the output file.

Exhibit 10.20 displays the contents of the directory for the PRES dataset.

EXHIBIT 10.20   CONTENTS statement output

```
------------------- SAS Log ---------------------------

  1     libname presdata '\data\sas';
  2
  3     proc datasets library=presdata;
                -----Directory for Library PRESDATA------

                    #    Name      Memtype
                    1    NAMES     DATA
                    2    PRES      DATA
                    3    SUBSET    DATA
  4     contents data=pres;
  5     quit;
NOTE: The PROCEDURE DATASETS used 8.00 seconds.

------------------- Output listing --------------------

                    DATASETS PROCEDURE

  Data Set Name:   PRESDATA.PRES        Type:
  Observations:    40                   Record Len: 67
  Variables:       7
  Label:

           -----Alphabetic List of Variables and Attributes-----

  #   Variable  Type  Len  Pos  Label
  7   AGEDEATH  Num    8   59
  6   AGEINAUG  Num    8   51
  3   BORN      Num    8   33
  4   BORNST    Char   2   41
  5   INAUG     Num    8   43
  1   NAME      Char  20    4
  2   PARTY     Char   9   24
```

> CONTENTS *statement* output closely resembles output from the CONTENTS *procedure*.

EXHIBIT 10.21   Abbreviated contents of all datasets in a library

---

```
------------------- SAS Log ---------------------------

    1     libname presdata '\data\sas';
    2
    3     proc datasets library=presdata nolist;
    4     contents data=_all_ position short;
    5     quit;
NOTE: The PROCEDURE DATASETS used 12.00 seconds.

------------------- Output listing --------------------

                       DATASETS PROCEDURE

          -----Directory for Library PRESDATA------

              #    Name       Memtype
              1    NAMES      DATA
              2    PRES       DATA
              3    SUBSET     DATA

- - - - - - - - - - New page of output - - - - - - - - -

                       DATASETS PROCEDURE

          -----Alphabetic List of Variables for PRESDATA.NAMES-----

INAUG     NAME      PARTY

                    ----Variables Ordered by Position----

NAME      PARTY     INAUG

- - - - - - - - - - New page of output - - - - - - - - -

                       DATASETS PROCEDURE

          -----Alphabetic List of Variables for PRESDATA.PRES-----

AGEDEATH AGEINAUG BORN     BORNST    INAUG     NAME      PARTY

                    ----Variables Ordered by Position----

NAME      PARTY     BORN     BORNST    INAUG     AGEINAUG AGEDEATH

- - - - - - - - - - New page of output - - - - - - - - -

                       DATASETS PROCEDURE

          -----Alphabetic List of Variables for PRESDATA.SUBSET-----

AGEDEATH AGEINAUG BORN     BORNST    INAUG     NAME      PARTY

                    ----Variables Ordered by Position----

NAME      PARTY     BORN     BORNST    INAUG     AGEINAUG AGEDEATH
```

EXHIBIT 10.22   Modify dataset contents, RUN, then run CONTENTS

```
------------------- SAS Log ----------------------------

  1     libname in 'd:';
  2
  3     proc datasets library=in;
                    -----Directory for Library IN------

                        #    Name      Memtype
                        1    NAMES     DATA
                        2    PRES      DATA
                        3    SUBSET    DATA
  4     modify pres(label='US President vital stats');
  5     rename born=yrborn
  6            inaug=yrinaug;
NOTE: Renaming variable BORN to YRBORN.
NOTE: Renaming variable INAUG to YRINAUG.
  7     run;
  8     modify pres;
  9     label yrborn='Year born (4 digits)'
 10           yrinaug='Year took office, NOT year elected'
 11           ;
 12     format party $1.;
 13     run;
 14     contents data=pres;
 15     quit;
NOTE: The PROCEDURE DATASETS used 11.00 seconds.

------------------- Output listing --------------------


                    DATASETS PROCEDURE


Data Set Name:   IN.PRES            Type:
Observations:    40                 Record Len: 67
Variables:       7

Label:          US President vital stats

          -----Alphabetic List of Variables and Attributes-----


 #  Variable  Type  Len  Pos  Format    Label
 7  AGEDEATH   Num    8   59
 6  AGEINAUG   Num    8   51
 4  BORNST     Char   2   41
 1  NAME       Char  20    4
 2  PARTY      Char   9   24  $FREE1.
 3  YRBORN     Num    8   33            Year born (4 digits)
 5  YRINAUG    Num    8   43            Year took office, NOT year elected
```

> Changes made by DATASETS are boxed.

Exhibit 10.21 displays the abbreviated contents of all datasets in the library identified by *ref* PRESDATA. The listing of dataset names at the beginning of the output is produced because the _ALL_ keyword was used in the DATA option.

In Exhibit 10.22 some changes are made to PRES and then CONTENTS is used to display the new, updated directory. Notice the use of RUN to make changes take effect immediately. Without this, references to the renamed vari-ables would have failed and *none* of the MODIFY commands would be imple-mented.

Syntax IV: Data Movement.   DATASETS has a group of statements that enable you to copy one or more SAS datasets. COPY, SELECT, and EXCLUDE enable you to copy from one medium to another (e.g., disk to tape), optionally selecting specific datasets from the input library.

Before discussing the statements, two points about data movement need to be mentioned. First, some operating systems allow these copy and move activities with operating system commands. You could, for example, copy a SAS dataset with the TSO, VMS, CMS, or MS/PC-DOS COPY command. In some systems you may rename a SAS dataset with a system command. *These approaches are not always reliable.* Their success depends on the version of SAS being used and the operating system. As a general rule, it is usually best to let SAS handle its datasets. The SAS site representative at your location will be able to advise you on using non-SAS commands to move SAS datasets.

Second, COPY, SELECT, and EXCLUDE operate only on complete datasets. That is, the KEEP, DROP, RENAME, and OBS and FIRSTOBS options are not available and have no effect if specified. If you want to copy only certain vari-ables and observations, use a DATA step.

Data movement usually requires the COPY statement and one or more ad-ditional statements that identify datasets to copy:

```
COPY OUT=outref [IN=inref] [MOVE] ;
[SELECT dset ... ;]
[EXCLUDE dset ... ;]
```

Some common applications of the data movement statements follow:

```
proc datasets library=in;
copy out=dataout;
select final: ;
```

The data movement statements have four components:

1. The *outref* identifies the ref of the output library. This library can be an existing dataset or can be created during the current job or interactive ses-sion. Even though DATASETS does not modify tape dataset directories, it can COPY to tapes.

2. The *inref* identifies the ref of the input library. The default is the library specified in the DATASETS statement.

3. The *MOVE* deletes each dataset in the input library once it has been success-fully moved to the output library. If *MOVE* is not entered, SAS leaves the input library intact. Although SAS does its best to ensure that the transfer of data is completed successfully, it is best to use *MOVE* sparingly. Instead, use COPY without *MOVE*, then run a CONTENTS on the new dataset and DELETE the original once you are satisfied the copied dataset is correct.

4. The *dset* identifies datasets to *SELECT* for or *EXCLUDE* from copying. By default, COPY will operate on all datasets found in the library in the input library. The *dset* can be any combination of an explicit listing of dataset names, a list of the form data*x*–data*z* (e.g., YEAR5–YEAR8 to copy YEAR5, YEAR6, YEAR7, and YEAR8), or a common prefix followed by a colon (e.g., REG: to copy REG1, REGION1, REG, and REGULAR).

Two examples in a PC-DOS environment illustrate the ease of use of DATA-SETS' data management statements.

This example uses the same information as the previous example but is less restrictive about output. All datasets are copied and the input library is left intact once the copying is completed:

```
libname out 'a:';
libname in  'c:\analysis\sas';

proc datasets library=in;
copy out=out;
quit;
```

In this example all datasets beginning with FINAL are copied to diskette (defined with a LIBNAME of OUT). Once copied, the FINAL: datasets are deleted from the input library (defined with a LIBNAME of IN):

```
libname out 'a:';
libname in  'c:\analysis\sas';

proc datasets library=in;
copy move out=out;
select final: ;
quit;
```

Exercises

10.11. The path defined by LIBNAME DATAHERE contains SAS datasets. Write the DATASETS statement(s) to list the names and sizes of the datasets.

10.12. Suppose the DATASETS run in Exercise 10.11 revealed dataset names of VER1, VER2, and VER3. Write DATASETS statements to do the following (assume that each question is independent of the others):
   a. Delete VER2 and VER3. Do this two ways, using different commands.
   b. Change VER1's name to CURRENT.
   c. Change VER3 to OLDEST, VER2 to VER1, and VER1 to NEWEST.
   d. Display information about VER1. First produce an abbreviated listing, then list all available information.
   e. Using only one statement, display variable lists of all datasets in the library.
   f. Make the following changes to dataset VER1:
   (1) Add a dataset label indicating that it is the most recently downloaded data for study group 5.
   (2) Add a FORMAT of COMMA5. to variable DEFICIT and remove the variable for SURPLUS (notice that you do not have to know the name of the format when you remove it).
   (3) Change the name of variable Q1 to Q191 and Q2 to Q291. Use a single statement for this renaming.
   g. Copy datasets VER1 and VER2 to the library with a LIBNAME of THERE. Use two different statements.
   h. Copy *all* datasets to LIBNAME THERE. Once they are successfully moved, delete them from LIBNAME DATAHERE.

10.13. What, if anything, is wrong with the following programs? Assume datasets STATES and NATIONS are the only ones in the library defined by LIBNAME HERE.
   a. ```
proc datasets ddname=here;
   modify states;
   format region $2.;
   rename region=area;
   label area='Part of country';
   run;
```
```

    b. `proc datasets ddname=here;`
       `delete nations;`
       `modify nations(label='UN roster');`
       `quit;`

    c. `proc datasets ddname=here;`
       `copy out=here;`
       `run;`
       `modify states;`
       `contents;`
       `quit;`

Summary

Chapter 10 examines the format, advantages, and use of permanent SAS datasets. SAS datasets contain data and related "directory" information such as variable and dataset labels, formats, and creation dates. Since they are in a format directly readable by procedures, SAS datasets can be processed much more rapidly than raw data. Creating the permanent dataset involves use of a two-level dataset name and, optionally, dataset options. These options affect which variables will be stored int he dataset, which variables will be renamed, and which observations will be processed.

The CONTENTS and DATASETS procedures can be used to examine and modify SAS datasets. CONTENTS simply provides a listing of the dataset's directory information. DATASETS may be used to rename, copy, move, and label datasets. It can also affect dataset variables with a variety of rename, label, and formatting options.

11 The DATA Step Revisited

This chapter introduces some new concepts and fine-tunes some old ones. We will see how to control the amount of storage used for a variable, do like-styled calculations en masse, create multiple SAS datasets in a single DATA step, halt DATA step execution prematurely, and read complicated raw data. The DATA step is the low-level, "down and dirty" way to attack even the most complicated datasets and computational problems. The tools covered in this chapter enable you to perform vast amounts of work without having to become obsessive about detail.

11.1 Multiple Output Datasets

So far, writing observations to an output dataset has been confined to the implicit or explicit use of the OUTPUT statement. A simple extension to the syntax of the DATA and OUTPUT statements allows more than one SAS dataset to be created in a single DATA step.

Why would we want to do this? We might want to split the datasets by category (e.g., male-female, North-South). Or we might want a "clean" dataset and one for observations that were somehow "defective" (e.g., too many missing values for critical variables). It is always more efficient to pass through data in a single DATA step and write out several output datasets than it is to make a separate pass for each dataset.

Syntax

The extended form of the DATA and OUTPUT statements is illustrated below.

```
DATA dataset1 dataset2 ... ;

OUTPUT datasetn ... ;
```

Some common applications of these forms of DATA and OUTPUT follow:

```
data out.master rejects;

output out.master;

data save.testmast save.grp1;
```

The DATA and OUTPUT statements contain two elements:

1. The *dataset1*, *dataset2*, and so on are datasets that will be written in the DATA step. Their syntax for both one- and two-level names is exactly the same as described in Chapters 4 and 10. Any number of datasets can be created in a single DATA step.

2. The *datasetn* refers to one or more datasets in the DATA statement's list.

Usage Notes

When creating multiple datasets, keep the following guidelines in mind:

- The rule of last use (discussed in Chapter 4) considers the rightmost entry in the DATA statement's dataset list the most recently created dataset. In the statement DATA TWO ONE; dataset ONE would be considered the most recently created.

- Any combination of permanent and temporary datasets is allowed in the DATA statement's list.

- An observation may be written to any dataset any number of times.

- The DROP and KEEP *statements* refer to every dataset specified in the DATA statement. The DROP and KEEP *dataset options* control variable selection for individual datasets.

- If an OUTPUT statement does not specify a dataset name, the observation is written to all datasets listed in the DATA statement.

Examples

The following program segments illustrate correct and inappropriate uses of the enhanced DATA and OUTPUT statements.

In the first example two datasets from the input SAS dataset REGIONS are written. REGION2 contains only observations where variable REGION is 2. MISSING contains observations where REGION is missing or the number of missing values for key numeric variables is unacceptable (more than five). MISSING could have been saved as a permanent SAS dataset but was simply printed instead. Notice that if the PRINT procedure did not use the DATA option, the rule of last use would select dataset REGION2, resulting in a misleading TITLE for the procedure's output.

```
data missing(keep=area gnp1-gnp20)
     out.region2;
set  in.region;
if area = ' '   |
   nmiss(of gnp1-gnp20) > 5 then output missing;
if area = '2' then output out.region2;
run;

proc print data=missing;
title1 "Obs w AREA missing and/or too many missing GNP's";
run;
```

In this next example the REGIONS dataset is split in two, one dataset for AREA values of 1, another for the 2's. Notice that the dataset name was not specified in the OUTPUT statements. This forces both the 1's and the 2's into *both* output datasets and makes identification of the area in each dataset impossible since AREA was DROPped.

```
data out.region1
     out.region2;
set  in.region;
if          area = '1' then output;
   else if area = '2' then output;
drop area;
run;
```

In this example the program segment presented in the preceding example is corrected by adding dataset names to the OUTPUT statements:

```
data out.region1
     out.region2;
set  in.region;
if          area = '1' then output out.region1;
   else if area = '2' then output out.region2;
drop area;
run;
```

This example uses the random-number generator (discussed in Chapter 7) to assign observations to two output datasets. The number of observations in each of these datasets should be roughly equal:

```
data part1 part2;
set in.total;
select = ranuni(0);
if select <= .5 then output part1;
   else              output part2;
drop select;
run;
```

Exercises

11.1. Dataset MASTER has five observations. Values of variable PERIOD are shown below (observation numbers are included for your reference—they are not part of the dataset).

(1) 1900
(2) 1900
(3) 1920
(4) 1930
(5) .

Which observations (identify them by number) are written to datasets PRE and POST in the following DATA steps?

a.
```
data pre post;
   set master;
   if period > 1925 then output post;
```
b.
```
data pre post;
   set master;
   if period <= 1920 then output pre;
   output post;
```
c.
```
if period <= 1920 then output pre;
   else        output post;
output;
```

11.2. What, if anything, is wrong with the following DATA steps?

a.
```
data x.p1 x.p2;
   set inmast;
   if v = 't' then output p1;
      else    output p2;
```
b.
```
data x.p1 x.p2;
   set inmast;
   if v = 't' then output;
```

11.2   Controlling Storage: The LENGTH Statement

Two important characteristics of variables are their length and location in the SAS dataset. **Length** refers to the number of units of disk storage, or *bytes*, used by the variable in each observation. By default SAS uses eight bytes for numeric variables. Depending on the context, character variables require eight bytes or the same number of bytes as number of columns used in the INPUT statement or character constant. **Location** refers not to column location in raw data but position of the variable stored in the SAS dataset. By default SAS stores variables in the order in which it finds them in the DATA step. Recall the use of PROC CONTENTS in Chapter 10 to determine a dataset's variable order.

Why are these features important and why would you want to take control over them? One reason is that once you know locations (possibly by using the CONTENTS command of the DATASETS procedure or PROC CONTENTS) you can safely write variable lists of the form start–end. Another reason is economy. The eight-byte numeric default allows a maximum value of about 2 to the 31st power. Unless you anticipate handling such numbers, the default eight-byte allocation wastes disk storage. Most integers can be safely stored in a fraction of the space. Such storage reduction is a nicety in some programs and absolutely critical in others. The LENGTH statement allows you to specify the order and amount of storage used by variables in a SAS dataset.

> **POINTERS AND PITFALLS**
>
> The LENGTH statement's placement in the DATA step affects its ability to affect variable storage and arrangement. It can greatly reduce the amount of storage used by SAS datasets and, like DROP and KEEP, encourages well-thought-out dataset design.

Syntax

The two forms of the LENGTH statement are presented below. They can either be combined in a single statement or used separately in the same DATA step.

```
LENGTH varlist [$] len ... ;

LENGTH DEFAULT=len ... ;
```

Some common applications of the LENGTH statement follow:

```
length name $20 style $1;

length default=3 debt 5 income 5;
```

The LENGTH statement has four elements:

1. The *varlist* is any combination of variables and variable lists. All variables must be of the same data type (i.e., all character or all numeric). They do not have to be known to SAS from a SET or INPUT statement!

2. The $ indicates that *varlist* consists of character variables.

3. The *len* specifies the length, in bytes, of the variable. Numeric variables range from two (three on microcomputers) to eight. Character variable lengths range from 1 to 200.

4. *DEFAULT=* is a quick specification valid only for numerics. It is sometimes helpful to specify all numerics this way and then follow the DEFAULT option with the variables that are exceptions to the rule (e.g., specify a DEFAULT value of 4, then indicate that variables X, Y, and Z have a length of 2).

Usage Notes

Keep in mind the following points when using the LENGTH statement:

- LENGTH's placement in the DATA step determines its effectiveness. If it is placed *before* the step's first reference to a variable, it will store it in the indicated number of bytes. If it is placed *after* the step's first reference to a variable, it will have no effect, nor will SAS produce an error message.

- There is absolutely no correspondence between the number of columns used for a numeric variable in the raw data and the number of bytes specified in the LENGTH statement. It is perfectly acceptable to read "wide" numbers (e.g., 13-digit national budget figures) from a raw dataset and store them in seven bytes: SAS's internal storage of numbers is efficient enough to permit the reduction in size.

- Exhibit 11.1 presents the largest integers that can be stored at different lengths in the principal operating environments supported by the SAS System.

EXHIBIT 11.1   Maximum integers allowed at varying lengths

Length	IBM Mainframes, Minicomputers	IBM-compatible Microcomputer Environments
2	255	.... not allowed ....
3	65,535	8,192
4	16,777,215	2,097,152
5	4,294,967,295	536,870,912
6	1,099,511,627,775	137,438,953,472
7	281,474,946,710,655	35,184,372,088,832
8	72,057,594,037,927,935	9,007,199,254,740,992

IBM mainframe environments include the CMS, OS, VSE, and VM/PC operating systems.

IBM-compatible microcomputer environments include the MS-DOS, PC-DOS, and OS/2 operating systems.

Minicomputer environments include DEC's VMS, Prime Computer's PRIMOS, and Data General's AOS/VS operating systems (maximum values for PRIMOS are somewhat lower; use the next highest length to be safe).

- To avoid worrying about rounding and other misrepresentation problems with real, fractional numbers, a good rule of thumb for numeric variables is to specify lengths only for integers. The full eight bytes may be needed for variables with fractional values.

- As discussed in Chapter 18, it is usually a good idea to specify lengths for *all* calculated or assigned character variables. The second example below (in which SIZE is specified) gives an example of the confusion that can be avoided by this practice.

- If a variable in a LENGTH statement is not assigned a value with an INPUT, SET, assignment, or other statement, SAS prints a Note in the Log saying that the variable is "uninitialized": set up but never used. The variable is kept in the output dataset and stored with a missing value. This is often a clue that you misspelled the variable name. For example, WIDGET may be set up in the LENGTH statement and referred to as WIGDET in the INPUT statement.

Examples

The examples in this section emphasize the use of LENGTH to define explicitly the order of variables. When reviewing them, also keep in mind that each time you specify a numeric length of less than eight, you economize on disk storage.

This example uses LENGTH to define variables A, B, C, and D in the order in which they are found in the raw dataset. Since the stored order agrees with the dataset order, the INPUT statement can use a variable list:

```
length b c a 3 d 4 ;
input b--d ;
```

In this example the length of the character variable SIZE is specified. SIZE is assigned values in an IF-THEN-ELSE sequence. Without the LENGTH statement SAS would assign a length of five, since the first value of SIZE encountered ("small") was five characters long. Subsequent, longer values (like "medium") would be truncated to five characters!

```
data newvar;
set  geog;
length size $6;
if        (0    <= sqmiles <= 1000) then size = 'Small';
   else if (1001 <= sqmiles <= 5000) then size = 'Medium';
   else                                   size = 'Large';
```

In this example the amount of storage taken by variables in an existing SAS dataset is reduced. Dataset OLD, created in a previous job or session, did not use a LENGTH statement and defined variables in the order REGION, GNP88, and GNP89. Placement of LENGTH *before* SET changes the amount of storage used by these variables as they are read from IN.OLD. It also affects the order of the LENGTH statement's variables within the observation. If the LENGTH statement *follows* the SET, it would have no effect on either the variables' length or order.

```
data new;
length gnp88 gnp89 5 region $3;
set  in.old;
```

Exercises

11.3. Write LENGTH statements to meet the following requirements.
  a. Establish a default length for all numeric variables in the dataset. No values exceed 1,000.

  b. Variables NAME_F, NAME_M, and NAME_L are each 20 characters long.

  c. Variable CITYNAME is a character variable 30 bytes long, and ZIP is a numeric requiring 4 bytes.

11.4. Dataset TRIALS consists of numeric variables LEV1-LEV20, SITE, and LABID. Write a DATA step to create dataset TRIALS2. It has variables in the order SITE, LABI, and LEV1-LEV20.

11.3 Defining Groups of Variables: The ARRAY Statement

A common requirement in data analyses is using sets of variables to perform calculations. You may, for example, have five variables in a series of verbal scores and five mathematical reasoning scores. If you wanted to compute the ratio of each verbal score to its corresponding math score, you might write a series of assignment statements:

```
ratio1 = verbal1 / math1;
ratio2 = verbal2 / math2;
ratio3 = verbal3 / math3;
ratio4 = verbal4 / math4;
ratio5 = verbal5 / math5;
```

This is fine for a few repetitions of the calculations, but consider the programming effort for 50 sets of scores. There is a pattern in these statements that is clear to our eyes but not, apparently, to SAS. Clearly, it would be nice if we had a way to say, "line up the verbal scores, line up the math scores, then divide them and put them in a set of ratios."

Another common problem involving repetition and patterns arises when you want to change a set of variables from one number to another. If raw data contained missing values coded as –9 or –8, you would have to write IF statements for every numeric variable in the dataset:

```
if date = -9 | date = -8 then date = .;
if center = -9 | center = -8 then center = .;
```

As with the score ratio example, this sort of coding is fine for a few variables but becomes cumbersome when many variables are involved. (The term *many*, of course, is relative.)

SAS lets you associate a group of related variables by giving them a collective name. Variables VERBAL1 through VERBAL5, for example, could be grouped with the name VERBAL. Likewise MATH1 through MATH5 and RATIO1 through RATIO5 could be given names MATH and RATIO. The three groups VERBAL, MATH, and RATIO are known to SAS as **arrays**. Calculations in the DATA step can operate on arrays just as they can on variables, but with the added advantage of being compact. Rather than repeat essentially the same assignment statement five times (once for each variable in the array), SAS syntax is flexible enough to express the calculation in a single statement.

POINTERS AND PITFALLS

Assignment statements that use parallel groups or other predictable arrangements of variables are candidates for ARRAY processing. If you find yourself writing essentially the same assignment statement many times or changing only variable names from statement to statement, arrays may be used to your advantage.

Using arrays requires two statements: ARRAY and an expanded form of the DO statement (described in Section 11.4). A special function, HBOUND (described in Section 11.3), helps simplify and generalize array processing. Array processing saves great amounts of coding effort and is appropriate any time there is a regular, predictable pattern in calculations.

Syntax

Arrays are set up in the ARRAY statement. It can appear anywhere in the DATA step prior to its first use. Arrays come in several varieties but all share a few features: the variables that make up the array (the verbal and math scores from the first example) are called its **elements**. You can use the array either as a whole or as a specific element in the DATA step. Individual elements are identified by **subscripts**, numbers that identify an element's position in the array. Array definitions are not stored with the DATA step, nor can they be used in the PROCs described in this book. The math scores used earlier in this section may be represented graphically in the array MATH in Exhibit 11.2.

The ARRAY Statement. The syntax of the ARRAY statement is presented below:

```
ARRAY name{dim} [$] [len] [[elements] [(st_values)]] ;
```

Some common applications of the ARRAY statement follow:

```
array rate{40};

array rate{*} inc1-inc15 rev1-rev15;

array dayname{7} $2 ('S', 'M', 'Tu', 'W', 'Th', 'F');
```

The ARRAY statement has six elements:

1. The *name* is the name of the array. It follows the usual SAS naming conventions: eight-character maximum, beginning with an underscore (_) or *a* through *z*, containing only alphabetic characters, numbers, and underscores. *Name* cannot already be a variable or an array in the dataset.

2. The $ indicates that the array elements of the array are character variables that have not yet been defined to SAS by any of the LENGTH, INPUT, SET, or assignment statements.

3. The *len* indicates the length of any variables that have not yet been defined to SAS by LENGTH, INPUT, SET, or assignment statements.

4. The *elements* are the names of the variables in the array. Any combination of variable lists and variable names is permitted. All elements in the array must be the same data type (all numeric or all character). A variable can be in more than one array in the same DATA step.

───────

EXHIBIT 11.2   Pairing of variables and element numbers

```
array math(5) math1-math5;
```

Variable	MATH1	MATH2	MATH3	MATH4	MATH5
Element number	1	2	3	4	5
Element reference	math(1)	math(2)	math(3)	math(4)	math(5)

5. The *dim* is the number of elements in the array. If the number is not known, an asterisk (*) is entered. The *dim* may also be enclosed in brackets (e.g., [12]) or parentheses [e.g., (12)]. The asterisk may be specified only in version 6 of the SAS System.

6. The *st_values* indicate initial values for array elements. These values will be assigned to each observation in the dataset being processed. Values are separated by a comma and/or one or more blanks. Character values should be enclosed in quotation marks or single quotes. If fewer values than there are elements in the array are specified, the remaining elements will have missing values. For example, if the array has five elements and is given only three starting values, the fourth and fifth elements are set to missing. The *st_values* do not replace variables already known to SAS: if a dataset containing variable X is SET, then an array using X is defined, the starting values of X are ignored. The *st_values* may be defined only in version 6 of the SAS System.

Enumeration of array elements is optional. If element names are omitted *and* a number for *dim* is identified, SAS creates variables NAME1, NAME2, up through NAME*dim*. If array name PERIOD is set up with the ARRAY statement ARRAY PERIOD20; SAS will use variables PERIOD1, PERIOD2, . . ., PERIOD20 in the array. If an element list is not entered, starting values can still be specified.

The HBOUND Function. Some array-handling tasks require the user to specify the size of the array (i.e., the number of elements). For arrays defined as SCORES(20) this is a simple matter: the number of elements is predetermined by the programmer. What happens, however, if the array were set up as follows:

```
array scores{*} name-numeric-avg;
```

The asterisk (*) indicates that SCORES should hold as many elements as there are numeric variables between NAME and AVG. Specifying the number of elements would be difficult, since the order and number of variables may change. The HBOUND function provides a clean way around this potentially awkward programming problem. Its syntax is shown below:

```
HBOUND(array)
```

Here *array* is the name of a numeric or character array in the DATA step. HBOUND returns a numeric value equal to the number of elements in the array. HBOUND is used in the READINGS array on page 204. HBOUND is available only in version 6 of the SAS System.

Using the Array

You can refer to the entire array or just one of its elements when performing logical comparisons or arithmetic calculations. To refer to an element, simply refer to it by using a subscript. The subscript can be either a numeric constant or variable or a numeric value that evaluates to an integer. In either case, the subscript should not be less than one and should not exceed the number of elements in the array.

An entire array may be used in a function by entering the array name with an asterisk (*) subscript. The following example shows some valid forms of array usage:

```
array widgets{10} wid1-wid10;
allwid = sum(of widgets{*});

array score{5};
if score{1} ^= . then do;
```

Examples

At this point we do not have all the tools to work effectively with arrays. The examples here are confined to array declarations and small program segments.

In this example arrays are set up for the verbal/math ratio example used at the beginning of this section. The two groups of ARRAY statements are equivalent:

```
array ratio{5} ratio1 - ratio5;
array verb{5}  verb1  - verb5;
array math{5}  math1  - math5;

array ratio{5} ;
array verb{5}  ;
array math{5}  ;
```

This next example declares an array to equate weekday numbers with day names. Seven character variables nine bytes long are set up and given initial values corresponding to the days of the week:

```
array wkday{7} $ 9 mo  tu we th fr sa su ('Monday' 'Tuesday'
        'Wednesday' 'Thursday' 'Friday' 'Saturday' 'Sunday') ;
```

In this last example, among the variables in dataset INVNTORY are COUNT_Q1 and FENCE_Q1. The starting values in array C1 will not overwrite the values of these two variables, since they were known to SAS prior to the execution of the ARRAY statement. Notice how the effective use of multiple lines for the statement and blank space separating groups of starting values clearly conveys how values are established:

```
set curr.invntory;
array c1{8} count_q1-count_4 fence_q1-fence_q4
        (0,0,0,0,       100,100,100,100) ;
```

Exercises

11.5. Set up arrays to meet the following requirements:
  a. Array name TEST has 20 numeric elements, TEST1 through TEST20. Write the ARRAY statement two ways.
  b. Array MTH has three elements, JAN, FEB, and MAR. Their initial values are "January," "February," and "March." Since the ARRAY statement will be the first reference to these variables, pay particular attention to their length.

11.6. What, if anything, is wrong with the following statements?
  a. `array test20 testbase test0-test18;`
  b. `array test20 testbase test0-test19;`
  c. `array init5 $1 (1,2,3,4,5);`
  d. `array values100;`
  e. `array q10;`
      `if nmiss(of q) > 10 then delete;`

11.4   Repetitive Execution: DO-Loops

Many situations require similar types of calculations on groups of variables. You want to perform essentially the same operation with different variables or using different constants. The only tool at your disposal so far is a simple assignment statement: if you need the same operation done on five pairs of variables, you would code roughly the same statement five times.

This chapter describes an extension of DO-groups called DO-loops. While the DO-group (discussed in Chapter 8) was simply a group of related DATA step statements that were executed or skipped as a block, the **DO-loop** defines a group of statements that may be executed more than once. If the assignment statements are written a bit differently and put in a DO-loop, statements can be repeated many times, making the program more compact and easier to maintain. In this section simple applications are discussed in detail and the syntax for more complex operations is presented.

Syntax

DO-loops have different levels of complexity. Using a DO-loop requires extending the DO statement and usually demands refinement of the normal program statements executed within the loop. The following are some advanced forms of the DO statement:

```
DO index = begin TO end ;

DO index = begin TO end BY increment ;

DO index = value [, value ... ] ;
```

Some common applications of this form of the DO statement follow:

```
do rep = 1 to 5;

do i = 1 to 50 by 2;

do type = 'A', 'B', 'F';
```

The DO statements above have five components:

1. The *index* is a numeric variable controlling the execution of the loop. This variable is often used within the loop to refer to array subscripts. It is legal, but usually not advisable, to change its value within the range of the loop.

2. The *begin* is a numeric variable, constant, or expression that defines the beginning value taken by *index*.

3. The *end* is a numeric variable, constant, or expression that defines the last value taken by *index*.

4. The *increment* is a numeric variable, constant, or expression that controls how the value of *index* changes. Its default value is 1. If the *start* value plus the *increment* exceeds the *end* value (e.g., 1 to 5 by 6), the loop will be executed once for the beginning value of the loop. Similarly, if *increment* moves the *start* value further from *end* instead of closer to it (e.g., 1 to 5 by −1), the loop is executed only for the beginning value. Generally, once *index* plus *increment* exceeds *end*, the loop terminates.

5. The *value* is a numeric or character variable, constant, or expression. This form of the index variable's specification is useful when its values are not orderly enough for the begin TO end BY increment format.

What Goes on in the Loop?

When SAS encounters an indexed DO-loop, it starts the "index" variable at the number indicated by "begin." The body of the loop (the statements up to the next END statement) is executed. References to the array may use either a numeric constant or the index variable as a subscript. This is a simple but extremely useful and powerful facility. When SAS reaches the END statement, it returns, or "loops" back, to the DO statement. The index variable is increased by *increment*: if it exceeds the *end* value, execution of the DATA step resumes at the next statement beyond the END statement.

Usage Notes

These forms of the DO statement are extremely powerful and flexible. Keep in mind the following points when using them:

- Loops execute until the *increment* (default or specified) exceeds the *end* value. This means that *end* must be reachable from *begin*. You could not have a *begin* at 100 and *end* at 50 unless you used a negative increment.
- The *index* becomes part of the SAS dataset being created unless it is included in a DROP statement.
- The *begin*, *end*, *increment*, and *value* must be nonmissing.
- You can combine the various forms of the indexed DO statements, using *begin*, *end*, and, optionally, *increment*, with one or more "value" specifications. An example of this is shown in Exhibit 11.4.
- Each form of these DO-loops can be nested within each other or a DO-group.

Examples

Now that you know about arrays and the expanded form of the DO statement, you are ready to look at some of the work you can do with DO-loops. Such work includes calculations referencing array elements, writing observations to SAS datasets, reading data with INPUT or SET statements, and any other repetitive DATA step activity. Because some of these activities produce subtleties that confound even the most experienced SAS user, the examples here are confined to simple cases involving calculations and writing observations to a SAS dataset.

This example completes the test scores example on page 202. Notice that three arrays (RATIO, VERB, MATH) are used in the loop, but only one of them (RATIO) is specified in the DO statement. Each implicitly subscripted array has the same number of elements. If the array sizes were unequal, problems could arise, as seen in Exhibit 11.3.

```
array ratio{5} ratio1 - ratio5;
array verb{5}  verb1  - verb5;
array math{5}  math1  - math5;
do i = 1 to 5;
    ratio{i} = verb{i} / math{i};
end;
drop i;
```

In this example we do not have the luxury of knowing the size of array READINGS. The HBOUND function is used to establish the upper bound of the DO-loop. Upon completion of the loop, variable MAXDIFF will contain the largest difference between successive elements in array READINGS.

```
array readings(*) y84q1--y92q2;
n_elem = hbound(readings);
maxdiff = 0;
do i = 2 to n_elem;
    diff = readings(i) - readings(i-1);
    if diff > maxdiff then maxdiff = diff;
end;
```

In this example a set of quarter-to-quarter differences from a series of 20 quarters of economic indicators are created: the first element in the new array is the difference between quarter 1 and quarter 2, the second element is the difference between quarter 3 and quarter 2, and so on, yielding 19 elements in the new array. The end value is set at 19 so the I + 1 subscript will equal but never exceed the INDICS array dimension of 20.

```
array indics{20} q1 - q20;
array diffs{19}  diff1-diff19;
do i = 1 to 19;
    diffs{i} = indics{i+1} - indics{i};
end;
```

This example is similar to the preceding one only with a subscripting problem. The end value in the DO statement is changed to 20. When the value of I is 20, SAS attempts to locate an element I + 1, or 21, in INDICS. Since INDICS has only 20 elements, SAS complains by saying there is an "Array subscript out of range." Part of the SAS Log is reproduced in Exhibit 11.3: part of the diagnostic dump includes the location of the error (line 6, column 36) and the value of the index variable I, showing its value at the time of the error to be 20. The programmer's task is to trace through the program's logic to see why 20 would produce an error.

In this example even-numbered elements of an array are added. The loop index, WK, begins with an even number (2) and has an even-number increment (2). Notice that EVEN_TOT is set to zero before the loop. This is not required since the SUM function ignores missing values and will, in effect, start at zero. It is, however, good practice to initialize running totals outside the loop.

```
array weeks{52} w1-w52;
even_tot = 0 ;
do wk = 2 to 52 by 2;
    even_tot = sum(weeks{wk},even_tot);
end;
```

EXHIBIT 11.3   Log report array subscript problem

```
------------------- SAS Log ----------------------------

    1    data econ;
    2    input id $4. q1-q20;
    3    array indics{20} q1 - q20;
    4    array diffs{19}  diff1-diff19;
    5    do i = 1 to 20;
    6        diffs{i} = indics{i+1} - indics{i};
    7    end;
    8    cards;
```

ERROR: Array subscript out of range at line 6 column 36. ◄——————— Subscript error message

```
RULE:----+----1----+----2----+----3----+----4----+----5----+----6----+----7---
  10      1.7 1.9 1.8 2.0 1.9
ID=ADV Q1=1.1 Q2=1.2 Q3=1.3 Q4=1.2 Q5=1.2 Q6=1.3 Q7=1.5 Q8=1.6 Q9=2 Q10=2.1
Q11=1.9 Q12=1.8 Q13=1.8 Q14=1.8 Q15=1.8 Q16=1.7 Q17=1.9 Q18=1.8 Q19=2 Q20=1.9
DIFF1=0.1 DIFF2=0.1 DIFF3=-0.1 DIFF4=0 DIFF5=0.1 DIFF6=0.2 DIFF7=0.1 DIFF8=0.4
DIFF9=0.1 DIFF10=-0.2 DIFF11=-0.1 DIFF12=0 DIFF13=0 DIFF14=0 DIFF15=-0.1
```

```
DIFF16=0.2 DIFF17=-0.1 DIFF18=0.2 DIFF19=-0.1  I=20  _ERROR_=1 _N_=1  ◄———
```
Value of index variable I at the time of the error

NOTE: SAS went to a new line when INPUT statement reached past the end of a
      line.
NOTE: The data set WORK.ECON has 0 observations and 41 variables.
NOTE: The DATA statement used 12.00 seconds.

In the next example a dataset for loan repayments is created. The MORT function is used to estimate repayments under a variety of interest rates. The proposed loan is $50,000, payable over 360 months. There are a few interesting features of this small program: first, it mixes index value styles in the DO statement. Also notice that the OUTPUT statement is in the loop. This means a separate observation is output for *each* pass through the loop. Notice also that the DATA step does not read any data. There are no INPUT or SET statements. All the numbers are generated by the DO-loop in one execution of the DATA step. This technique is very handy when you want to develop some test data prior to entering the "real" data for a project: you can easily control the number of observations and the values of their variables. The SAS Log and output files are presented in Exhibit 11.4.

In this example a test dataset is created using random numbers. The program, though short, contains many useful features of DO-loops in particular and SAS in general. No INPUT or SET statements read data: everything is generated by the DO-loops and function calls. The placement of OUTPUT in the *inner* loop produces an output dataset *n* larger than if the statement were placed just before the end of the *outer* loop (the statement is executed more times when it is in the inner loop). Finally, the output listing in Exhibit 11.5 indicates that this "synthetic" dataset is already sorted: by the outer loop index and (in descending order) the inner loop index.

This example, like the previous one, shows the sort of manipulation that is handy when you have a rough idea of what your data will look like and want to rough it out, or "prototype," ahead of time. The SAS Log and output files are presented in Exhibit 11.5 (see p. 208).

Exercises

11.7. Describe how you would fix the problem illustrated in Exhibit 11.3 (the "subscript out of range" problem).

11.8. Create a dataset prototype. Variable YEAR should range from 1990 to 1995. For each level of YEAR there should be MONTH values of 1, 2, . . ., 12. Variable FLUX can be a random number between 0 and 1. Write the DATA step so that the dataset is in BY-group order: YEAR, MONTH. Exhibit 11.5 illustrates these techniques.

11.9. Array ALL has 20 numeric elements. Write DO statements to refer to the following elements:
  a. All elements
  b. Even-numbered elements
  c. Every third element, beginning with 1 (i.e., 1, 4, 7, . . .)
  d. Elements 1 through 10, 15, and 20

11.10. Refer to the economic indicators on page 205. Rewrite it so that array DIFFS contains the average of period $n + 1$, n, and $n - 1$. Remember to adjust the ARRAY size and ending index! You might also want to rename DIFFS to something that more accurately reflects its content.

11.11. What, if anything, is wrong with the following DATA step program fragments?
```
a. array x20;
   total = 0;
   do i = 1 to 25 by 2;
      total = sum(x(i), total);
   end;
b. array x20;
   n = 0;
```

```
        do i = 1 to 25 by 2;
           n = n + 1;
           total = sum(xn, total);
        end;
  c. do i = .07 to .08;
```

11.12. How many observations are created by the following DO-loops?

```
  a. do i = 1 to 5;
        do k = 1 to 15;
          output;
        end;
     end;
  b. do i = 1,2;
        do k = 1 to 10 by 2;
          output;
        end;
     end;
```

EXHIBIT 11.4   DO-loop used to generate loan payments

```
------------------- SAS Log ----------------------------

   1      options nocenter nodate nonumber;
   2
   3      data mort;
   4      do rate = .07, .075, .0775 to .125 by .0050, .14, .15;
   5         pay = mort(30000,.,rate/12,360);
   6         output;
   7      end;
   8      run;
NOTE: The data set WORK.MORT has 14 observations and 2 variables.
NOTE: The DATA statement used 9.00 seconds.
   9
  10      proc print data=mort;
  11      title 'Monthly payments on loan of $30,000 with 30 year life';
  12      run;
NOTE: The PROCEDURE PRINT used 8.00 seconds.

------------------- Output listing --------------------

Monthly payments on loan of $30,000 with 30 year life

OBS     RATE       PAY

   1    0.0700    199.591
   2    0.0750    209.764
   3    0.0775    214.924
   4    0.0825    225.380
   5    0.0875    236.010
   6    0.0925    246.803
   7    0.0975    257.746
   8    0.1025    268.830
   9    0.1075    280.044
  10    0.1125    291.378
  11    0.1175    302.823
  12    0.1225    314.369
  13    0.1400    355.462
  14    0.1500    379.333
```

EXHIBIT 11.5   Nested loops and OUTPUT used to create test data

```
-------------------- SAS Log --------------------------

     1     options ps=66 nodate nonumber nocenter;
     2
     3     data proto;
     4     do length = .1 to .3 by .05;
     5        do width = 10 to 1 by -2;
     6           area = length * width;
     7           id   = 1000 * ranuni(7654321);
     8           output;
     9        end;
    10     end;
    11     run;
NOTE: The data set WORK.PROTO has 25 observations and 4 variables.
NOTE: The DATA statement used 10.00 seconds.
    12
    13     proc print;
    14     title 'Data prototyping';
    15     format id 3.;
    16     run;
NOTE: The PROCEDURE PRINT used 10.00 seconds.

-------------------- Output listing --------------------
```

Data prototyping

OBS	LENGTH	WIDTH	AREA	ID
1	0.10	10	1.0	882
2	0.10	8	0.8	428
3	0.10	6	0.6	582
4	0.10	4	0.4	499
5	0.10	2	0.2	965
6	0.15	10	1.5	973
7	0.15	8	1.2	888
8	0.15	6	0.9	120
9	0.15	4	0.6	460
10	0.15	2	0.3	754
11	0.20	10	2.0	340
12	0.20	8	1.6	134
13	0.20	6	1.2	995
14	0.20	4	0.8	317
15	0.20	2	0.4	811
16	0.25	10	2.5	1
17	0.25	8	2.0	142
18	0.25	6	1.5	863
19	0.25	4	1.0	798
20	0.25	2	0.5	256
21	0.30	10	3.0	480
22	0.30	8	2.4	130
23	0.30	6	1.8	475
24	0.30	4	1.2	24
25	0.30	2	0.6	69

Exercises
(continued)

```
c. do i = 1, 2, 5;
      do j = 1 to 3;
        do k = 1 to 5;
          output;
        end;
      end;
   end;
d. do i = 1 to 5;
      do k = 1 to 10;
        rand = ranuni(1234567);
      end;
      output;
   end;
```

11.5   Halting Execution: STOP and ABORT

SAS provides many statements that allow the normal execution sequence of the DATA step to be bypassed. IF-THEN-ELSE constructs, DO-groups, and OUTPUT allow selective execution of one or more statements. Another situation arises when you want to terminate the DATA step or the SAS program if a condition arises. This section describes the STOP and ABORT statements, two related but subtly different ways to stop DATA step execution.

Syntax

The STOP and ABORT statements are presented below:

```
STOP;
```

```
ABORT [ABEND];
```

STOP tells SAS to terminate DATA step execution. Execution of the program resumes at the next unit of work (DATA step or procedure). *ABORT* tells SAS to terminate DATA step execution. If *ABEND* is specified, the entire SAS program will stop, or abnormally end. Omission of the ABEND specification allows SAS to resume execution at the next DATA step or PROC.

Usage Notes

Although the STOP and ABORT statements appear to perform the same function, they have some important differences:

- STOP allows the dataset(s) in the DATA statement to be replaced. ABORT does not.
- STOP allows SAS to continue with units of work beyond the current DATA step. ABORT, when used with ABEND, does not, and terminates the entire program.

Both STOP and ABORT do not write the current observation to the output dataset.

Examples

The first example shows how to stop building a dataset if an observation has more than five missing values for a series of test scores:

```
data clean;
set  master;
if nmiss(of score01-score25) > 5 then stop;
```

The second example performs calculations "in place" on an existing dataset's variables. If a division by zero is attempted, execution of the DATA step is stopped and the input SAS dataset is not replaced:

```
data clean;
set  master;
if prev = 0 then abort;
   else change = (curr-prev) / prev ;
```

Exercises

11.13. A DATA step is building dataset MASTER. The variable NBAD is important in the step's logic. Write the STOP and ABORT statements to meet the following requirements:
  a. Do not replace MASTER if NBAD exceeds 0. As soon as NBAD exceeds 0, stop execution of the DATA step.
  b. Stop the execution of the entire program if NBAD exceeds 0.
  c. Stop the DATA step when NBAD exceeds 0, but allow execution to proceed with the next DATA step or PROC.

11.6   Complex Raw Data: INPUT Revisited

So far the examples have used raw data in fairly simple formats. This section shows how to simplify reading raw data and introduces some new tools to read more difficult raw dataset layouts. Extensions to the INPUT statement are discussed.

The INPUT statement is the most complex and most powerful of the DATA step statements covered in this book. Some of its more useful properties are explored in this section: specifically, how to read datasets that have more than one line of data per observation; how to express input formats compactly; how to read more than one observation per line of input; and how to read part of a line of data, then later decide how to read the rest of it.

Multiple Lines per Observation

Data for an observation frequently use more than one line in the raw dataset. This may be done for the sake of legibility or simply because of the large volume of data for an observation.

Syntax. SAS provides two INPUT statement features that enable you to skip from one line down to another. These **line pointers** are shown below:

```
#n
```

```
/
```

The #n tells SAS to skip to the nth line of the raw data for the observation. The n can be a numeric variable or, more commonly, a numeric constant. The / tells SAS to move to the next line of the raw dataset. Enter as many as needed to skip more than one line. To keep observations aligned, the INPUT statement should have one fewer / than there are lines for an observation.

Usage Notes. Both forms of line pointers obey the following rules:

- Unless an INFILE option (N, described in Chapter 19) is specified, movement through the observation's raw data can only go down. Once all the data from a line of input are read and a / or #*n* instruction received, the line can no longer by read by the INPUT statement. Read *all* the data from a line, then go to the next line.

- Lines of raw data may be skipped. Just because there are four lines for each observation does not mean that you must read data from all four.

- The INPUT statement must stop reading on the last line of the observation's raw data, even if there are no variables to read from it. If there are three lines per observation, the INPUT must move to the third line. If the INPUT statement used only the first two lines and did not read the third line, the third line of the *current* observation would be used as the first line for the next observation!

- The / and #*n* line pointers can be used in the same INPUT statement.

Examples. Exhibit 11.6 shows a dataset that uses two lines of data per observation. The data are consumer price indices for several years and categories by country.

In this next example two equivalent INPUT statements using different forms of the line pointer are presented. Data are read from both lines:

```
input country $ 1-15 cpi85 16-18 cpi84 20-22 cpi83 24-26 cpi80 28-30
      / food 16-18 clothing 20-22 housing 24-26 transit 28-30;
```

```
input country $ 1-15 cpi85 16-18 cpi84 20-22 cpi83 24-26 cpi80 28-30
      #2 food 16-18 clothing 20-22 housing 24-26 transit 28-30;
```

Another pair of equivalent INPUT statements are shown in this example. Each reads data from the first of the two raw data lines and skips the second:

```
input country $ 1-15 cpi85 16-18 cpi84 20-22 cpi83 24-26 cpi80 28-30
      / ;
```

```
input country $ 1-15 cpi85 16-18 cpi84 20-22 cpi83 24-26 cpi80 28-30
      #2 ;
```

EXHIBIT 11.6 Consumer price index sample data

```
Australia       440 412 396 295
                423 415 497   .
Japan           322 316 309 281
                329 346 263 280
United Kingdom 599 565 538 423
                615 332 642 612
United States   322 311 298 246
                309 191 349 319
Italy           767 702 634 403
                689 832 765 917
```

In this example we mistakenly think that since all the information we need is on the first data line, there is only one line per observation. SAS will do what it was instructed to do: read COUNTRY from the first 15 columns, CPI85 from the next three, and so on. The program has no way of knowing it should skip the second line of each country's data and so reads the second line of the first country as a separate observation! This explains the doubling of the dataset's number of observations. Notice that this is a error in logic rather than syntax. The Log and output files are shown in Exhibit 11.7.

Column Pointers

Every time SAS follows an instruction from an INPUT statement, it adjusts its column indicator, or *pointer*. By default, the pointer is poised ready to read the

EXHIBIT 11.7   Syntactically correct, logically incorrect INPUT directions

```
------------------- SAS Log ---------------------------

   1     options nodate nonumber nocenter ps=60;
   2     filename x 'c:\chap10\figs\10f270';
   3
   4     data cpi;
   5     infile x;
   6     input country $ 1-15 cpi85 16-18 cpi84 20-22 cpi83 24-26 cpi80 28-30 ;
   7     run;
NOTE: The infile X is file C:\CHAP10\FIGS\10F270.
NOTE: 10 records were read from the infile C:\CHAP10\FIGS\10F270.
      The minimum record length was 30.
      The maximum record length was 30.
NOTE: The data set WORK.CPI has 10 observations and 5 variables.
NOTE: The DATA statement used 9.00 seconds.
   8
   9     proc print;
  10     title 'Read one line instead of two per observation';
  11     run;
NOTE: The PROCEDURE PRINT used 8.00 seconds.

-------------------- Output listing --------------------

Read one line instead of two per observation
```

OBS	COUNTRY	CPI85	CPI84	CPI83	CPI80
1	Australia	440	412	396	295
2		423	415	497	.
3	Japan	322	316	309	281
4		329	346	263	280
5	United Kingdom	599	565	538	423
6		615	332	642	612
7	United States	322	311	298	246
8		309	191	349	319
9	Italy	767	702	634	403
10		689	832	765	917

next physical column beyond the field just read. Thus if TASK_NUM were read in columns 20 through 23, the column pointer would be pointing to column 24.

Sometimes it is helpful to deliberately move the column pointer rather than take SAS's default location. Only the syntax is presented here; an example is given in the "Common Input Formats" section below.

Syntax. The two forms of the column pointer are illustrated below:

```
@col
```

```
+[(-)movement[)]
```

Some common applications of column pointers follow:

```
input @30 model $10. @42 year $4.;
```

```
input @start fullname $15. +(-10) partname $10.;
```

In the first form of the column pointer, *col* is the column number to move to in the current line. The *col* must be a numeric variable or constant greater than zero. Use this form of the column pointer for absolute movement.

In the second form, *movement* is the number of columns to move. The *movement* can be a numeric variable or constant. If it is a constant with a negative value enclosed in parentheses, the pointer moves "backwards," to the left. Otherwise, it moves *movement* columns to the right on the current line. Use this form of the column pointer for relative movement. The (−movement) pointer directive may be used only in version 6 of the SAS System.

Usage Note. Both forms of column pointers can be used in the same INPUT statement.

Input Formats

The only methods of reading data presented so far are list (raw data separated by spaces) and column (variable name followed by column locations) styles. Another, more powerful form is described in this section. It reads data that column input cannot and can be abbreviated to read many variables with minimal coding effort. The input formats described here are similar in appearance and opposite in function to the output formats described in Chapter 9. An *output* format took a variable from a SAS dataset and displayed it with commas, dollar signs, and so on. An *input* format tells SAS to read the raw data, stripping out the commas, dollar signs, and so on when converting the raw data field to a SAS variable.

Syntax. The syntax of the input formats is exactly like the output format. The general form is shown below:

```
fmtw[.d]
```

The input formats have three elements:

1. The *fmt* is the name of the format.
2. The *w* is the variable's width, in columns, in the input dataset.
3. The *d* is the number of decimal places to insert into numeric fields. If a decimal is actually coded in the data, it overrides *d*.

Common Input Formats. Exhibit 11.8 describes frequently used input formats. Many of these are identical to the output formats described in Chapter 9. There

EXHIBIT 11.8   INPUT formats

Format	Description	Width range	Decimal range
w.	Numeric, no decimals	1-32	0-10
w.d	Numeric, with decimals	1-32	0-10
BZw.d	Treat blanks as zeros	1-32	0-10
COMMAw.d	Embedded commas	1-32	0-31
HEXw.	Numeric hexadecimal	1-1	
IBw.d	Integer binary	1-8	0-10
PIBw.d	Positive integer binary	1-8	0-10
RBw.d	Floating point binary	2-8	0-10
PDw.d	Packed, with decimal places	1-16	0-10
PKw.d	Packed, no decimal places	1-16	0-10
$w.	Character	1-200	
$CHARw.	Character, embedded blanks	1-200	
$EBCDICw.	Convert EBCDIC to ASCII [1]	1-200	
$HEXw.	Character hexadecimal	1-200	

[1] Available only in IBM microcomputer environments

are subtle differences in maximum and minimum w values, so do not assume that what works on output will behave identically on input.

The following example combines input formats and column pointers to rewrite the INPUT statement in the line pointer example on page 211. The statements are equivalent. The "best" is a matter of style and convenience:

```
input country $15. cpi85 3. @20 cpi84 3. @24 cpi83 3. @28 cpi80 3.
      / @16 food 3. @20 clothing 3. @24 housing 3. @28 transit 3.;

input country $15. cpi85 3. @19 cpi84 4. cpi83 4. cpi80 4.
      / @16 food 3. @19 clothing 4. housing 4. transit 4.;

input country $15. cpi85 3. @20 cpi84 3. +1 cpi83 3. +1 cpi80 3.
      / @16 food 3. @20 clothing 3. +1 housing 3. +1 transit 3.;
```

Format Lists

The seemingly bulky notation used above raises the question of why input formats and column pointers are such an asset. Instead of simply specifying the beginning and ending column, there is the movement of the pointer and the use of a format, hardly the compact expression touted earlier.

To take full advantage of pointers and input formats requires one more IN-PUT statement feature: **format lists**. These let you take advantage of repetition and patterns in the input data.

Syntax

The syntax of the format list is presented below. The variables lists and the formats are all items that have already been covered. What is new is their presentation:

```
(var_list)(format_list)
```

Some common applications of format lists follow:

```
(rate1-rate20)(3.);

(st--last)(4.2 +1);

(name add1 add2)($20. $24. $24.);
```

Format lists have two components:

1. The *var_list* is any combination of SAS variables and variable lists.
2. The *format_list* is a list of input formats used to read the corresponding *var_list* variables.

Usage Notes. Although format lists permit extremely compact notation, their use also requires several cautions:

- A simple form of the format list has as many formats as there are variables in *varlist*. If the format list has more formats than the variable list has variables, the excess formats are ignored. If the format list has fewer formats than the variable list has variables, SAS reuses, or "recycles," the list from the beginning. This is both a useful and potentially problematic feature of the list, since it can cause peculiar data values if you are not aware of the *format_list* recycling feature. The height-weight example illustrates this problem.
- The format list may have any combination of character and numeric variables and may include +*movement* style column pointers.
- There is no limit to the number of variable list–format list pairs allowed in an INPUT statement. Sometimes multiple lists can express an input format more succinctly than a single one.

POINTERS AND PITFALLS

When using format lists be sure you are aware of the correspondence between the variables in the variable list and the formats that will be used in the format list. This is especially important for recycled lists.

Examples. The first example rewrites the INPUT statements from the "Input Formats" section using format lists. The statements shown here are equivalent:

```
input country $15. @16 (cpi85 cpi84 cpi83 cpi80)(3. +1 3. +1 3. +1 3. +1)
      / @16 (food clothing housing transit)(3. +1 3. +1 3. +1 3. +1) ;

input country $15. @16 (cpi85 cpi84 cpi83 cpi80)(3. +1)
      / @16 (food clothing housing transit)(3. +1) ;

input country $15. @16 (cpi85 cpi84 cpi83 cpi80)(4.)
      / @16 (food clothing housing transit)(4.) ;
```

In this next example, assume a raw dataset's observations have five pairs of height-weight readings. The patient's name is in columns 1 through 10, followed by height in meters (three columns wide, two implied decimal places) and weight in kilograms (four columns wide, one implied decimal place). Four equivalent INPUT statements follow, along with explanatory comments.

First, write the statement using column input, indicating decimal places where necessary.

```
input name $ 1-10      height1 11-13 .2 weight1 14-17 .1
       height2 18-20 .2 weight2 21-24 .1 height3 25-27 .2
       weight3 28-31 .1 height4 32-34 .2 weight4 35-38 .1
       height5 39-41 .2 weight5 42-45 .1 ;
```

Using w.d formats in the next example saves some coding effort. Notice that the column locations are *implied*, not directly specified.

```
input name $10. height1 3.2 weight1 4.1 height2 3.2 weight2 4.1
                height3 3.2 weight3 4.1 height4 3.2 weight4 4.1
                height5 3.2 weight5 4.1 ;
```

In the next statement we take advantage of the repetition in the data. Once NAME is read, all remaining variables repeat formats of 3.2 and 4.1. We enclose the height-weight variable pairs in parentheses and cycle the formats.

```
input name $10. (height1 weight1 height2 weight2 height3 weight3
                height4 weight4 height5 weight5)(3.2 4.1) ;
```

Finally, we make the statement even more compact, although possibly at the cost of making it more difficult to read. First read NAME, then move to column 11 and read all heights (3.2, skipping the four columns used by weights). Then move to column 14 and repeat for weights. The advantage of this format is that specification of the variable list is more compact (the variable list can represent related height and weight variables compactly). Notice that the order of the variables is different than the earlier versions of the statement: NAME, HEIGHT1-HEIGHT5, WEIGHT1-WEIGHT5 instead of NAME, HEIGHT1, WEIGHT1, . . ., HEIGHT5, WEIGHT5.

```
input name $10. @11 (height1-height5)(3.2 +4)
                @14 (weight1-weight5)(4.1 +3) ;
```

In this last example a variable list–format list mismatch is created. A character and two numeric variables are specified in the variable list but only a single, numeric format is specified in the format list. The Log in Exhibit 11.9 shows the error messages created by the misspecification. The message is not what you would expect (why did SAS not complain about *character* variable CV1 being read with a *numeric* format?). This anomalous result may serve only to highlight the importance of having an accurate pairing of variable and format lists.

EXHIBIT 11.9   Improper recycling of a format list

```
    1     data sample;
    2     length nv1 nv2 3 cv1 $3 ;
    3     input (cv1 nv1 nv2)(2.) ;
ERROR: Variable NV1 is numeric but is being used as a character variable.
ERROR: Variable NV2 is numeric but is being used as a character variable.
    4     cards;
NOTE: The SAS System stopped processing this step because of errors.
NOTE: The data set WORK.SAMPLE has 0 observations and 3 variables.
NOTE: The DATA statement used 5.00 seconds.
```

Duplicate Formats: The Repetition Factor

Frequently data layouts will not lend themselves to neat, tidy format lists but do have some degree of repetition and order. You can exploit this order by using a format prefix called the **repetition factor**.

Syntax. The repetition factor syntax is shown below:

```
nrep*format
```

The *nrep* is the number of times to repeat the format. It can be a numeric variable or constant. The *format* is any numeric or character format.

Usage Notes. The repetition factor is not available in the DEC VMS operating system.

Correct reading of the data is unpredictable if the repetition factor forces SAS to read beyond the end of the current raw data line. Use of 40*3. on a standard 80-column dataset, for example, may cause data to be lost or misread.

Example. Assume two groups of variables are arranged side by side in the raw data. The first group is five two-column variables, the second group three five-column variables. The reduction in coding effort gained by using the repetition factor is shown below:

```
input (type1-type5 level1-level3)(2. 2. 2. 2. 2. 5. 5. 5.) ;

input (type1-type5 level1-level3)(5*2. 3*5.) ;
```

Multiple Observations per Line

In some situations where there are relatively few variables per observation, data can be entered very compactly using a special feature of the INPUT statement. This feature allows data to be entered for more than one observation in a single line of input. If, for example, you had three numeric variables, GENDER, RACE, and SCORE, you could enter a line of data such as

```
1 1 25 1 1 30 2 3 23 2 2 45 1 1 23 1 2 19 2 2 18 2 1 17 1 2 19
```

Rather than enter nine separate lines of data for the observations, only a single line is used. The INPUT statement must tell SAS not to automatically move to the next line of raw data once it completes the INPUT statement's instructions. Instead, the statement should tell SAS to position itself just beyond the last field read from the current record, ready to read data for the next observation. SAS must "hold" the current line of data.

Syntax. The line holder, also known as the *double trailing at*, is presented below:

```
INPUT vars @@ ;
```

The INPUT statement specified *vars* to read, as usual. The @@ tells SAS to hold this line of raw data and use it when processing the next observation. The @@ *must* be the last entry in the INPUT statement.

Usage Note. The use of the @@ does not require you to put more than one observation's data in a line. It simply tells SAS to try to use it for more than one observation. If no more data are found, it will go to the next data line.

Example. Exhibit 11.10 shows the SAS Log and data lines used to create a small test dataset. Notice that one of the pairs of variables splits across lines. Also notice the "SAS went to a new line . . . " note in the Log. This is not an error, just a reminder that a nonstandard feature was being used.

EXHIBIT 11.10   Use @@ to read multiple observations per line of raw data

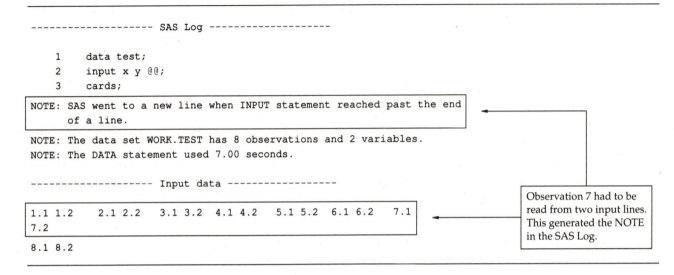

```
-------------------- SAS Log --------------------


  1    data test;
  2    input x y @@;
  3    cards;
```

NOTE: SAS went to a new line when INPUT statement reached past the end
      of a line.

NOTE: The data set WORK.TEST has 8 observations and 2 variables.
NOTE: The DATA statement used 7.00 seconds.

```
-------------------- Input data ------------------
```

```
1.1 1.2    2.1 2.2    3.1 3.2   4.1 4.2    5.1 5.2   6.1 6.2    7.1
7.2
```

```
8.1 8.2
```

Observation 7 had to be read from two input lines. This generated the NOTE in the SAS Log.

Multiple INPUTs per Observation

Whether by choice or necessity, there is occasional need to read variables from a raw data record, do some DATA step processing, then continue reading the same record. The @ INPUT statement option, commonly known as the *trailing at*, tells SAS that once the current INPUT statement completes execution it should not discard the record. The next INPUT statement in the DATA step will read from the same line of data. If the DATA step does not execute any other INPUT statements during the current observation, the record is discarded and a new one read for the next pass through the DATA step. The single @ allows multiple INPUTs for a *single* observation per line of raw data, while @@ allows *multiple* observations to be created from a single line of raw data.

Why go to this trouble? Why not just read all the variables at once? There are two reasons.

First, the file might have more than one type of record layout. Such *hierarchical* files are fairly common and are easy to read with the *trailing at*. Unlike the previous sample files, that always had the same variables in the same columns in each record, hierarchical files may have different sets of variables in different records, each record possibly having a different length and each using a different data layout. There could be, for example, a "house" record describing the location and structure of a survey respondent, a "family" record describing the characteristics of a family in the house, and "person" records with social, economic, and demographic data for each person in the family. This example is developed in Exhibit 11.11.

The second reason for partial reading of a line of data is computational efficiency when processing large numbers of variables. If not all records will be output to a SAS dataset, it is usually more efficient to read the selection variables, then read the rest of the record for observations that will be output. For example, if only large states are going to be kept in the dataset being built, first use an INPUT that would read the state-size variable, using a trailing @. If the variable meets the criteria for inclusion in the output dataset, read the rest of the data using another INPUT statement, this one without a trailing @. This technique can increase execution speed dramatically and is demonstrated in the efficient reading example on page 219.

Syntax. The syntax for the *trailing at* is presented below. Notice that a single @ is used, unlike the multiple observation per line's @@.

```
INPUT specs @ ;
```

The *specs* indicates input specifications: variable names, line and column pointers, and formats. The @ tells SAS to hold this line of data for possible use by INPUT statements later in the DATA step. The @ *must* be the last entry in the INPUT statement.

Usage Notes. Keep in mind the following guidelines when using more than one INPUT statement per DATA step:

- Failure to use the @ causes SAS to read the next line of raw data if another INPUT is executed later in the DATA step. The output SAS dataset's number of observations can drop dramatically in these cases. Rather than continue reading the *same* data line, the next INPUT uses a *new* line. This effectively cuts the observation count in half, since two lines rather than one were used to satisfy the INPUT statements' requirements.

- Use of the trailing @ almost always implies more than one INPUT statement in the DATA step. Any number of INPUTs can be used to read the same line of data, provided all but the last use the @.

- The raw data record is always discarded, or "released," at the end of the DATA step, even if the last INPUT statement had a trailing @.

- The column pointer is always positioned to the next available column. If you read columns 4 to 6 in the first INPUT, the pointer would be positioned to column 7. The next INPUT is not automatically pointing to column 1 but remains ready to read column 7. It is usually good practice to use explicit column pointers (@col) to start all INPUT statements.

Examples. The two hypothetical examples described in the introduction to this section are briefly developed here.

Exhibit 11.11 uses hierarchical data: house, family, and person-level datasets are created from a single hierarchical raw dataset. Chapter 19 revisits this example, showing how a single dataset with all levels of analysis can be generated.

This next example shows how to instruct SAS to read raw data efficiently. The two DATA steps shown below produce the same SAS dataset. The second uses fewer computer resources since the bulk of the work is done only if a record is used in the output dataset. Needless processing is avoided.

```
data large;
infile in;
* Read all variables regardless of whether we'll keep
  the observation in dataset LARGE. ;
input @20 area 6. @90 (gnp_t1-gnp_t240) (7.1);
if area >= 50000 then output;

data large;
infile in;
input @20 area 6. @ ;
if area >= 50000 then do;
   * Read the variables only if necessary. ;
   input @90 (gnp_t1-gnp_t240) (7.1);
   output;
   end;
```

EXHIBIT 11.11   Using the trailing @ to read a hierarchical dataset

```
data house(keep=state nrooms)
     family(keep=sal_inc div_inc int_inc n_in_fam)
     person(keep=sal_inc int_inc age race);
infile in;
input @1 id $4.   @5 type $1. @ ;  ◄───────────────
if type = 'H' then do;
   input @20 state $2. @55 nrooms 2. ;
   output house;
   end;
   else if type = 'F' then do;
        input @21 (sal_inc div_inc int_inc)(7.) @60 n_in_fam 2. ;
        output family;
        end;
   else if type = 'P' then do;
        input @21 (sal_inc int_inc)(7.) @50 age 3. race $1. ;
        output person;
        end;

------------------ Input Data ------------------
```

> Trailing '@' holds the current line of raw data for other INPUT statements.

```
0020H          CA                              03
0020F               52000      0    850                    2
0020P               32000    850      0      34W
0020P               20000      0      0      28W
0094H          NE                              07
0094F               85500   2500    900                    1
0094P               85500   2500    900      52H
0120H          MI                              05
0120F               20000      0      0                    1
0120P               20000      0      0      21W
0120F               46500    500    290                    3
0120P               30000    450    290      31B
0120P               13500     50      0      19B
0120P               13000      0      0      17B
```

> Raw data has information on three households, four families, and seven people.

Exercises

11.14. Refer to the description of the STATES dataset in Appendix E. Write an INPUT statement to read variables STATE, NAME, CROWD, and PCTCROWD. Write the statement several ways, and remember that there are five 'cards' per observation!

11.15. Variables ALPHA1 through ALPHA10 were read with the following INPUT statement:

```
input @10 (alpha1-alpha3)($1. +1)
      @5  (alpha4-alpha6)($1.)
      / alpha7 $1. +1 (alpha8-alpha10)(+1 $1.);
```

An observation's values of these variables were ALPHA1 = 'a,' ALPHA2 = 'b,' ..., ALPHA10 = 'j.' What did the raw data record look like for this observation? Clearly indicate blank columns.

11.16. A DATA step reads the following data lines:

```
1    1.1    2    2.1    3    3.1
4.     4.1
5.
5.1
```

What does the SAS dataset look like if it read the data with the following INPUT statements?

a. `input x y;`
b. `input x y @@;`
c. `input x;`

11.17. Simplify the following INPUT statement. (*Hint*: Notice the way the lines of the statement are arranged and look for repeated column widths.)

```
input v1    1- 3   v2    4- 5   v3    6- 9 .1
       v4   10-12   v5   13-14   v6   15-18 .1
       v7   19-21   v8   22-23   v9   24-27 .1
       v10  28-30   v11  31-32   v12  33-36 .1 ;
```

11.18. Examine the following DATA step:

```
data all;
infile 'ubaacs.survey.tariff(un87)';
input type $1. @;
if type = 'm' then do;
   input nreps 2. id $2.;
   do i = 1 to nreps;
      input year rate;
      output;
   end;
   end;
run;
```

a. Give a prose description of what the program does.
b. What variables will be in dataset ALL? Which ones will be repeated NREPS times (i.e., duplicated across observations)?
c. Suppose the raw dataset identified by the INFILE statement looked like this:

```
m 3US
87 5.4
88 5.15
89 5.0
i 1US
9.7 11.2
m 2it
88 6.0
89 5.8
```

What will dataset ALL look like?

11.7 INFILE Options

Until now, the only use of the INFILE statement has been to tell the DATA step about the location of raw datasets. This section reviews some of the more useful INFILE options. These include options that let you specify variable delimiters in the data, where to start and end the reading of the raw data, and how to react to an insufficient number of data values in a line of raw data.

Variable Delimiters

Blanks are the typical separators, or delimiters, for variables when list-style input is used (see Chapter 4 for a discussion of list input). The DLM option of the INFILE statement allows more than one delimiter to be used.

Syntax. The DLM option's syntax is shown below:

```
DLM='delimiters'
```

The *delimiters* is a list of characters used to separate variables in each observation. The default is a blank (dlm=' '). At least one of the delimiters should be used between data fields in the raw data. Consecutive occurrences of the delimiters are counted as a single occurrence. This has an important implication: consecutive occurrences do *not* imply a missing numeric value for a variable. DLM is available only in version 6 of the SAS System.

Example. The DLM string used in Exhibit 11.12 allows reading of raw data with variables separated by blanks and/or commas. This format is especially popular with many microcomputer packages such as dBASE IV and Lotus 1-2-3.

Reading Only Part of the Data

During the early stages of program development, it is usually sufficient to use only part of the raw dataset as a test file. The FIRSTOBS and OBS options tell SAS to read select portions of the raw dataset.

Syntax. The syntax of the FIRSTOBS and OBS INFILE options is illustrated below:

```
FIRSTOBS=recnum
```

```
OBS=recnum
```

The *recnum* indicates a record number in the raw dataset. When used with FIRSTOBS, it specifies the first record read from the raw data. When used with OBS, it specifies the last record read.

Usage Notes. Keep in mind the following points when using FIRSTOBS and OBS:

- Either or both of FIRSTOBS and OBS may be used in the same INFILE statement. They may appear in any order in the statement.
- When both FIRSTOBS and OBS are entered, FIRSTOBS should be smaller than OBS. Otherwise, no records will be available to the DATA step.
- If the FIRSTOBS record number exceeds the size of the raw dataset, no records are processed. An excessively large value for OBS has no effect; SAS simply stops processing when it runs out of raw data records.
- Remember that these options refer to the number of raw data records, not SAS observations. There can be confusion if there is not a one-to-one correspondence between raw data lines and SAS dataset observations. For

---

EXHIBIT 11.12　DLM used to read data separated by blanks and commas

---

```
data neonatal;
infile cards ;
input id lb oz len apgar1 apgar5 ;
cards;
90001,8,7,21.5,.,9
90002,6,6,20.5,9,9
90003, 7, 7, ., 10, 10
```

---

example, if there were two raw data records per observation (as in the examples in Section 11.6) and FIRSTOBS = 7 was specified in INFILE, SAS would begin processing at the beginning of the fourth SAS observation, not the seventh.

Examples.   This first example reads up to 20 records from the raw dataset:

```
data test;
infile cards obs=20;
input id 3. name $10. age 2. sex $1.;
```

This next example also reads up to 20 records from the raw dataset. Rather than read one line of data per observation as in the previous example, we read two lines (#2 pointer). This yields a dataset with 100 observations:

```
data test;
infile cards obs=20;
input id 3. name $10. age 2. sex $1.
      #2 (inctot pubasst socsec)(7.) ;
```

The last example begins reading at the 11th line of raw data, stopping at the 20th line. Since there are two lines per observation, the output dataset has five observations:

```
data test;
infile cards obs=20 firstobs=11;
input id 3. name $10. age 2. sex $1.
      #2 (inctot pubasst socsec)(7.) ;
```

Handling "Short" Observations

The final group of INFILE options discussed in this chapter deal with handling "short" observations. These are lines of data where you expected to read, say, 20 variables but found fewer. The FLOWOVER, MISSOVER, and STOPOVER options permit different ways of handling the discrepancy.

Syntax.   Use these options by specifying MISSOVER, FLOWOVER, or STOP-OVER in the INFILE statement. MISSOVER tells SAS that it should simply assign missing values to variables for which it cannot find data on the current raw data line. FLOWOVER allows SAS to continue reading data for the current observation from the next raw data line. STOPOVER terminates the creation of the SAS dataset if a variable list–raw data discrepancy is found. Only one of the "missoptions" may be specified.

Example.   Exhibit 11.13 displays a raw dataset, a short DATA step, and PROC PRINT output when using each of the three missing data options. Using no options or FLOWOVER allows the third data line to be read as part of the second observation. This explains why the datasets have only two observations.

Exercises

11.19. The raw dataset read by a DATA step looks like this:

```
1,2,,3
4,5,,
,,6
```

Answer parts a–c reading the data with this statement:

```
input v1-v4;
```

EXHIBIT 11.13   Comparison of FLOWOVER, MISSOVER, and STOPOVER

```
------------------- Program listing -------------------

data test;
infile 'sample.dat' [option] ;
input a b c;

------------------- Output listing -------------------

raw data                        default (no options)

1 2 3                           OBS      A     B     C
4 5
6 7 8 9                          1        1     2     3
                                 2        4     5     6


flowover                        missover

OBS    A     B     C            OBS      A     B     C

1      1     2     3            1        1     2     3
2      4     5     6            2        4     5     .
                                3        6     7     8


stopover

OBS    A     B     C

1      1     2     3
```

a. What will be the values of V1 through V4 if the INFILE option DLM = ' ,' is used?

b. What are the values if DLM = ','?

c. What if DLM = ' ' is used?

11.20. Write INFILE statements to meet the following requirements. Assume in each case that the statement links to a FILENAME of RAWDATA.

a. One raw data line per observation. Use the 10th through 20th lines.

b. Each observation requires three lines of raw data. Construct the dataset using what will be the 10th through 15th observations in the SAS (output) dataset.

11.21. What, if anything, is wrong with the following statements?

a. infile datahere obs=20 firstobs=20;

b. infile datahere obs=20 firstobs=21;

11.22. Suppose a raw dataset looked like this:

```
1 2 3
4
5 6
```

If list INPUT is used to read variables V1 through V3, what will the dataset look like when the following INFILE statements are used?

a. infile cards missover;

b. infile cards flowover;

c. infile cards stopover;

Summary

The DATA step is a rich and powerful programming environment. Chapter 11 discusses some of its more advanced options. A single DATA step may create more than one output SAS dataset by specifying multiple dataset names on the DATA statement and, optionally, using an OUTPUT statement. The LENGTH statement is used to explicitly specify the order and storage requirements of dataset variables. This feature gives the programmer more control over the dataset's contents and often improves processing efficiency.

Logically related groups of variables may be processed with arrays. These are declared in the ARRAY statement and are usually processed in DO-loops. Execution of the DATA step and program can be stopped by using the STOP and ABORT statements, respectively.

The INPUT statement is the most complex in the SAS language. The chapter looks at some additional options and uses of the INPUT statement—namely, the handling of raw data with multiple lines per observation, use of column "pointers," special input format descriptors, and other advanced features for unusual data conditions. Also discussed are some of the more advanced features of the INFILE statement: using raw data file delimiters, restricting data lines to be read, and handling incomplete data lines.

PART FOUR

Special Tools for Special Needs: Data Management and Reporting Topics

12 Controlling the Environment: System Options

The DATA step does not operate in a vacuum. Whether running in a batch or interactive environment, SAS has many "environmental" features that set dozens of switches to control data flow, output presentation, and routing of Log and output files. Normally the user does not notice these features because, like other parts of the SAS System, they have sensible defaults. Awareness of these system options and how to exploit them can be an important asset in SAS programming.

Dataset and INFILE options such as OBS, KEEP, DROP, and RENAME have already been covered. System options follow the same principles only on a broader, programwide scale. This chapter introduces SAS System options: it discusses what they are, how and why to use them, and how to identify their current settings.

12.1 What Are Options?

SAS makes a host of assumptions whenever running even the simplest program. It assumes, for example, that execution should stop if it encounters a syntax error. It assumes that a dataset named in a DATA statement should replace one of the same name. It assumes that the output should be centered in the output file and that the date and page numbers will be printed.

SAS makes dozens of other assumptions, too, some subtle, others obscure and specialized. They are all "intelligent defaults": the option's assumed value is the one that users normally chose. For example, most users want to stop processing if a syntax error is detected and prefer centered output with dates and page numbers. The default setting for an option varies from system to system and between batch and interactive environments: a default setting in one environment may not be appropriate or suitable in another. These points are addressed throughout this chapter.

SAS provides several ways to specify and list option values. The OPTIONS statement sets one or more SAS System options. Options may also be set when the SAS command is executed in an interactive environment. The OPTIONS procedure is a quick way to display the current settings of all the options on the system.

POINTERS AND PITFALLS

Understanding system options and their capabilities is important for users doing all but the simplest programming. Options address a variety of aesthetic and logistical issues and can often save large amounts of tedious DATA step coding.

12.2 Specifying OPTIONS

Both the OPTIONS statement and SAS command line options use the same logic and roughly the same syntax. Both forms may be used in the same SAS session.

OPTIONS Statement

The OPTIONS statement specifies option settings and may appear *anywhere* in the SAS program, between and within DATA steps and some procedures. Some of the procedures discussed in Chapters 21, 23, and 24 (PLOT, REG, and ANOVA) are *interactive*. Among their advantages are the ability to alter display-oriented options *during* the procedure. For example, you could run a scattergram with dimensions suitable for your terminal, then alter the display size options and rerun the plot. The OPTIONS statement's syntax is described below:

```
OPTIONS opt ... ;
```

The *opt* is a SAS System option. Any number of options, in any order, may be specified in a single OPTIONS statement. Some are indicated simply by their name (e.g., NUMBER) and can be reversed, or toggled, by entering a prefix of NO (e.g., NONUMBER). Others require a numeric or character value, indicated by an equals sign (=) and the appropriate value (e.g., PAGESIZE=66). If an option is specified more than once in the statement, only the latest value is used.

Command Line Specification

The specification of options on the command line in interactive SAS varies from system to system. Each system has its own method of separating SAS program file names, if any, from options. Appendix D gives complete examples and rules for option specification. The following is a sample of specifying NODATE in the four major environments supported by SAS:

```
TSO:       SAS OPTIONS(NODATE)
CMS:       SAS (NODATE)
VMS:       SAS /NODATE
MS-DOS:    SAS -NODATE
```

Where to Specify? Command line specification is handy for running a program with an option, then rerunning without it to evaluate its impact: the program itself does not have to be modified, just the command that ran the program. Options specified in the program itself (via the OPTIONS statement) may be continually reset and are better able to respond to analysis and program needs. Generally, however, there are no clear advantages to where options are specified. Most of the options described in this chapter may be set in the command line in interactive SAS and then reset within the session.

Exercises **12.1.** Assume an interactive SAS session was started with the following command:

```
sas -nodate -nonumber
```

If the following program statements are executed, what will be the settings for the date, numbering, and centering options?

```
options center date;
proc print;
run;
```

12.2. Assume an interactive SAS session was started with the following command:

```
sas /nocenter
```

Based on the program below, what user-specified options are in effect when PRINT is executed? when FREQ is executed?

```
options date ps=22;

options nonumber;
proc print data=examp;
run;

proc freq data=examp;
run;

options nodate nonumber nocenter;
```

12.3 Frequently Used Options

The options in this section fall into four categories: display, data management, session control, and others. Each SAS site's system administrator sets defaults to an appropriate value. Keep in mind the following points while reading this section:

- Typical defaults are listed for each option. Your site may default differently. Use the OPTIONS procedure to be sure of your installation's settings.
- If an option is not available in an operating system, the restriction is noted.
- All display, data management, and other options may be specified on the SAS command line or in the OPTIONS statement. Session control options are available only in the SAS command line of an interactive session.

Display Options   SAS has several options that control the display of LOG and procedure output, as described here:

Option	Purpose
DATE	Prints the current date at the top right of each page of output (the default). The date is a handy way to archive output. To suppress date printing, enter NODATE.
NUMBER	Prints the page number of the output file at the top right of each page (the default). Numbering, like dating, permits easy reference to and identification of runs. To suppress page numbering, enter NONUMBER.
CENTER	Centers the procedure's output in each page (the default). To print starting flush left on the page, enter NOCENTER. The appearance of some procedures changes markedly when the centering is turned off. The information conveyed by the output, however, is unaffected.
LINESIZE=width	Specifies the number of columns available for procedure output and custom report printing (covered in Chapter 20). The *width* can range from 64 to 256 characters. The default for batch jobs is often set to the maximum available for the site's printers (usually 133). Interactive sessions usually default to the screen width (78 or 80).
PAGES=[count¦MAX]	Specifies the maximum number of pages a SAS job or session can print. If a MAX is entered for *count*, SAS will not place any restrictions on output. Remember that the output is still subject to any restrictions imposed by the operating system! *This option is not available in microcomputer environments running SAS version 6.*
PAGESIZE=count	Specifies the number of lines per page of printed SAS output. The *count* can range from 20 to 200 lines. In interactive sessions it is adjusted to the screen size. In most batch jobs it is set to give a reasonable margin at the top and bottom of the page. Some computing environments (notably microcomputers) use the same *count* value for both interactive and batch processing.
PAGENO=number	Specifies the page number for the next page of output. The *number* can be any positive value. PAGENO is useful when you have several procedures and want to reset the page counter for each one's output. *This option is available only in microcomputer environments running SAS version 6.*
MISSING='char'	Specifies a single character (default is '.') printed for missing values of all numeric variables. This option is useful when you need to dress up output and must draw attention to or away from missing values. The *char* may be blank. MISSING only affects the

	display of the data. It has no impact on the way the number is stored.
PROBSIG=level	Controls the format of *p* values in statistical procedures. The default (0) prints to four decimal places. The *level* of 1 prints at least one significant digit, and 2 prints at least two significant digits.
CAPS	Translates all program statements and data to uppercase. This can be especially useful if you want to read character data in uppercase but do not want to modify the data itself. CAPS performs the conversion automatically. No manual intervention with a text editor is necessary, and the data remain in upper- and lowercase. The typical default, NOCAPS, preserves case in in-stream data, labels, character constants, and PROC FORMAT values. *This option is not available in microcomputer environments running SAS version 6.*
LABEL	Makes variable labels accessible to procedures. Specify NOLABEL to deny access. *This option is not available in microcomputer environments running SAS version 6.*

Example. For the sake of comparison and not with an eye toward aesthetics, let us compare two PRINT listings. The first listing, shown in Exhibit 12.1 (see p. 234), is "standard" output, taking all defaults [date and page number printed on output listing, lines in output page equal to the number of rows in the screen (22), output listing centered in page, missing values printed with a period (.)].

The second listing, shown in Exhibit 12.2 (see p. 235), is "fully loaded," using the same PRINT procedure with the DATE, NUMBER, CENTER, LINESIZE, PAGESIZE, and MISSING options set. Wholesale use of the options like this is not usually required. It is done here simply to give a feel for how different the same data can appear when you take control of the system options.

Data Management Options

SAS has system options that control dataset access and replacement, as described here:

Option	Function
OBS=[n¦MAX]	Specifies the last observation number to process from a SAS dataset. This option is similar in principle to the OBS = dataset option but applies to *all* datasets being read and written. The default, MAX, is all observations. An *n* of 0 (together with NOREPLACE, described below) is useful for checking syntax and not actually processing data. An *n* less than the maximum is useful for testing programs: once the output is acceptable, remove the OBS restriction. This usually speeds program development time and saves computer resources.
FIRSTOBS=n	Specifies the number (default is 1) of the first observation to process from any dataset. This option is similar in principle to the FIRSTOBS = dataset option but applies to *all* datasets being read or written.

EXHIBIT 12.1 "Before" output from PRINT: no options used

```
------------------- Program listing -------------------

proc print data=test;
by type;
run;

------------------- Output listing -------------------

                    Take All Defaults                          1
                              21:27 Thursday, December 15, 1988

-------------------------------- TYPE=a ----------------------------------

           OBS    M1    M2    M3

            1      .     1     2
            2      9     0     .
            3      .     .     .


-------------------------------- TYPE=b ----------------------------------

           OBS    M1    M2    M3

            4      2     3     2
            5      4     3     .
            6      3     4     3
            7      2     3     4

- - - - - - - - - - New page of output - - - - - - - - -

                    Take All Defaults                          2
                              21:27 Thursday, December 15, 1988

-------------------------------- TYPE=c ----------------------------------

           OBS    M1    M2    M3

            8      .     .     .
```

REPLACE	Allows permanent SAS datasets to be replaced; that is, if IN.ANNUAL existed prior to a SAS job or session and you tried to write a new version of IN.ANNUAL, SAS would allow it. Specify NOREPLACE with OBS = 0 to check program syntax and prevent overwriting.
DSNFERR	Allows SAS to generate an error message and stop normal execution when a SAS dataset used by the job or session cannot be located. Specifies NODSNFERR to allow execution to continue. Because the NODSNFERR option can produce unexpected results, you should use it sparingly. *This option is not available in microcomputer environments running SAS version 6.*

EXHIBIT 12.2 PRINT output appearance altered by SAS System options

---

```
------------------- Program listing --------------------

options nodate nonumber nocenter linesize=64 missing='?' pagesize=66;

proc print data=test;
by type;
run;

------------------- Output listing ---------------------

------------------------- TYPE=a ---------------------------

OBS    M1    M2    M3

 1     ?     1     2
 2     9     0     ?
 3     ?     ?     ?


------------------------- TYPE=b ---------------------------

OBS    M1    M2    M3

 4     2     3     2
 5     4     3     ?
 6     3     4     3
 7     2     3     4


------------------------- TYPE=c ---------------------------

OBS    M1    M2    M3

 8     ?     ?     ?
```

Example.   As with the display-oriented option example, the example in Exhibit 12.3 does not purport to be a typical or even useful program. The intent is simply to illustrate how the combination of data management options can alter results. Refer to the comments and the NOTEs in Exhibit 12.3 to see how the options work.

Session Control Options

These options are available only when SAS is being run interactively. They control the appearance of the screen, allow statements entered during the session to be saved as a disk file, and override the usual routing of Log and procedure output files.

Option	Function
TPS=nlines	Requests that *nlines* lines be used on a terminal screen. The default is the number of lines usually associated with the terminal being used. The *nlines* may range from 20 to 500. *This option is available only in MVS/TSO and VM/CMS systems.*

────────

EXHIBIT 12.3　Dataset and system options used to control dataset size

────────

```
NOTE: Copyright(c) 1985,86,87 SAS Institute Inc., Cary, NC 27512-8000, U.S.A.
NOTE: SAS (r) Proprietary Software Release 6.03
      Licensed to D**2 Systems, Site 12345678.

    1     options obs=50;
    2     libname out 'c:\';
    3
    4     * A dataset is being written, not read, so OBS system option
    5       has no effect. ;
    6     data out.all;
    7     do i = 1 to 100;
    8        output;
    9     end;
   10     run;
NOTE: The data set OUT.ALL has 100 observations and 1 variables.
NOTE: The DATA statement used 6.00 seconds.
   11
   12     * SUBSET1 is a copy of observations 30 through 50 in OUT.ALL.
   13       FIRSTOBS begins reading at observation 30 of OUT.ALL and OBS
   14       tells SAS to end reading at observation 50, hence 21
   15       observations (30 through 50 inclusive = 21).
   16     options firstobs=30;
   17     data subset1;
   18     set  out.all;
   19     run;
NOTE: The data set WORK.SUBSET1 has 50 observations and 1 variables.
NOTE: The DATA statement used 4.00 seconds.
   20
   21     * Reset FIRSTOBS back to 1. If this was not done, SAS would have
   22       tried to start with the 30th observation of a 21-observation
   23       dataset. The result would have been a 0 observation dataset.;
   24     options firstobs=1;
   25     data subset2;
   26     set  subset1;
   27     run;
NOTE: The data set WORK.SUBSET2 has 50 observations and 1 variables.
NOTE: The DATA statement used 5.00 seconds.
   28
   29     * Reset OBS= and use OBS= dataset option as well. ;
   30     options obs=15;
   31     data subset3;
   32     set  subset2(obs=10); * Dataset OBS overrides OBS in OPTION
   33                                  statement. ;
   34     run;
NOTE: The data set WORK.SUBSET3 has 10 observations and 1 variables.
NOTE: The DATA statement used 5.00 seconds.
   35
   36     * Finally, attempt to replace OUT.ALL when NOREPLACE option is
   37       in effect. ;
   38     options noreplace;
   39     data out.all;
   40     set  subset3;
   41     run;
```

EXHIBIT 12.3 *(continued)*

```
NOTE: The data set OUT.ALL has 10 observations and 1 variables.
ERROR: Data set OUT.ALL was not written because of NOREPLACE option.
NOTE: The DATA statement used 5.00 seconds.
NOTE: SAS Institute Inc., SAS Circle, PO Box 8000, Cary, NC 27512-8000
```

TLS=ncols	Requests that *ncols* columns be used on a terminal screen. The default is usually 78 or 80. The *ncols* may range from 64 to 132. *This option is available only in VAX and VM/CMS systems.*
TLOG	Requests a disk dataset named SAS.SAS be created in the directory current when SAS was run. The file contains all SAS statements entered from a line mode SAS session. Specify NOTLOG to suppress the option or to turn off an earlier *TLOG*. *This option is available only in VAX and VM/CMS systems.*
DMS	Requests that the interactive SAS session utilize the SAS Display Manager, a powerful full-screen program editing, submission, and retrieval system (for more information on the Display Manager, see Appendix D). Specifying NODMS when running interactive SAS begins a line-mode session. NODMS can significantly reduce SAS's use of system resources and should be used if memory limitations become problematic. Be aware that even if NODMS is specified, some procedures (e.g., DATASETS, in version 5) may try to utilize your terminal's full-screen capabilities. These PROCs' defaults may be suppressed with the NOFS option, discussed below. In some systems, using DMS requires answering a prompt from SAS about "device type." This is the type of terminal you are using: VT220, IBMASYNC, and so on. NODMS and DMS may be specified only when calling SAS: they may not be part of an OPTIONS statement.
FS	Allows SAS procedures to use full-screen features. If NOFS is specified, even if DMS is selected, procedures utilizing the terminal's full-screen capabilities will revert to a line-oriented environment. NOFS and FS may be specified only when calling SAS; that is, they may not be coded in an OPTIONS statement.
LT, LD, LP PT, PD, PP	This family of options controls the disposition of Log and procedure output files. The first letter of each specifies whether the option affects the *Log* or *Procedure* output. The second letter requests *typing* (i.e., listing at the terminal), writing to *disk*, or *printing*. A prefix of NO with any of these suppresses the activity. Specifying an option such as LD does not interrupt the display of the Log at the terminal. It simply echoes it: the Log is displayed at the terminal *and* written to a disk file. Any number of these options may be specified at any time in the SAS session. LD datasets are usually named SASLOG.LOG. PD datasets are usually named SASLIS.LIS. *This option is available only in VAX and VM/CMS systems.*

Example. Exhibit 12.4 illustrates the use of session control options in a VAX/VMS environment. SAS is called with the NODMS and NOFS options,

EXHIBIT 12.4   Routing Log, output in a DEC VAX interactive session

```
$ sas /nodms/nofs

  <DATA step>

  /* Now run the PROC on a few observations.
     Procedure output, like the Log, is sent to
     the terminal. */
  options obs=20;
  proc print ;
  run;

  /* Remove size restriction and send PRINT
     output to a disk file and the default printer.
     Suppress PRINT's display on the terminal. */
  options obs=max pd pp nopt;
  proc print;
  run;

  /* Resume sending procedure and Log output to the
     terminal.  Begin recording of program statements
     to a disk file. */
  options tlog pt nopd nopp;

  <DATA step><PROC><PROC>

  /* Now that the next PROC is "tuned," resume routing
     procedure output to disk.  The next PROC's output
     is appended to the PD file written earlier. */
  options pd;
  proc chart;
  vbar region / sumvar=volume type=mean discrete;
  run;

  /* End the session */
  endsas;
```

sending the session into line mode. A DATA step is run first, then a PRINT on the first 20 observations (OBS=20). The observation restriction is lifted (OBS=MAX), and the procedure output to the terminal is halted (NOPT) and redirected to a disk file (PD) and printer (PP). Once the printing is complete, the program statements (TLOG) begin recording and the procedure output is sent to its original destinations (PT, NOPD, and NOPP).

Later, once another procedure is ready to be saved, output is redirected to the disk file (PD). Procedure output is appended to the file begun earlier. Once the session ends (the ENDSAS statement), the files SAS.SAS, SASLOG.LOG, and SASLIS.LIS are available in the default directory and the first PRINT (PP option) output is sent to the default printer.

Had these statements been run in batch rather than interactive mode, the files would be named fn.SAS, fn.LOG, and fn.LIS, where "fn" is the file name of the program of SAS statements. For example, if you stored the program in file

TEST.SAS, then ran SAS by entering the command "SAS TEST," the files would be TEST.LOG and TEST.LIS.

Other Options

Several other options may be used to control SAS's reaction to errors, display of NOTEs and source statements, and other features.

Option	Function
FMTERR	Allows SAS to print an error message and stop execution when the format specified for a variable cannot be located. Specifying NOFMTERR causes SAS to use the appropriate default numeric or character format. *This option is not available in microcomputer environments running SAS version 6.*
INVALIDDATA='char'	Specifies that all numeric variables be assigned the special missing value *char* when an INPUT statement or function encountered invalid data. Valid *char* values are period (.), underscore (_), and alphabetic characters (a through z). See Chapter 19 for more information on special missing values. *This option is not available in microcomputer environments running SAS version 6.*
NOTES	Allows printing of information NOTEs in the SAS Log. Specify NONOTES to suppress these messages. Because these are usually useful diagnostic messages, this option should be used sparingly.
SOURCE	Displays the SAS program statements in the SAS Log (the default). To suppress printing, enter NOSOURCE. This option is usually used only when you want to save paper and when your program is "clean": nothing is more frustrating than using NOSOURCE, having a syntax error, and not knowing where the error occurred. Use this option sparingly.
DQUOTE	Allows quotation marks (") to delimit character literals in assignment statements, TITLEs, FOOTNOTEs, LABELs, and other situations requiring quoted text. This option allows a single quote (') to be entered as is in a literal (e.g., "Didn't answer"). If DQUOTE is used, represent a quote (") in a literal by two consecutive quotes (e.g., """Don't Know"""). *This option is not available in microcomputer environments* since it and the single quote are a fixed part of the SAS language and not an option.
ERRORS=n	Specifies the number of observations (usual default is 20) for which error information will be printed. The *n* can be any nonnegative number. This option reduces the volume messages caused by out-of-bound array indices, invalid input data, division by zero, and so on (refer to Chapters 7 and 11 for details). Frequently, examining the results of a small number of errors is sufficient to get an idea of what might be going wrong with the DATA step. NOTEs about the

error condition are printed in the Log regardless of ERRORS' value.

ERRORABEND   Aborts the program as soon as SAS encounters an error situation (bad syntax, ABORT ABEND statement, and so on). The usual default, NOERRORABEND, sets OBS = 0 and continues scanning the rest of the program's statements when an error occurs. *This option is not available in microcomputer environments running SAS version 6.*

Exercises

12.3. Write OPTIONS statements for the following scenarios. You may also indicate how the options would be specified if you were running interactive SAS in your operating system (refer to Appendix D for system-specific details).

a. Procedure output will be sent to a page 60 lines long, not centered, without page numbers, and 130 columns wide.

b. Missing values will be displayed as the character *x*. Page length and line width are set to be suitable for display at your terminal (22 rows by 78 columns).

c. Process only the first 20 observations from SAS datasets.

d. Process only observations 20 through 100 from SAS datasets. Do not allow SAS to overwrite datasets; allow diagnostics to be printed for 50 observations if error conditions are raised.

12.4. SAS dataset ALL has 50 observations. How many observations will there be at the successful end of each DATA step in the following sequence?

```
data all2;
set  all(obs=20);
run;

options firstobs=5;
data all3;
set  all2;
run;

options obs=20;
data all4;
set  all3;
run;

options firstobs=10;
data all5;
set  all4(firstobs=10);
run;
```

12.4 The OPTIONS Procedure

It is easy to forget default option settings or the names of the options. A simple way to list option names and current settings is to use the OPTIONS procedure. Its syntax is shown below:

```
PROC OPTIONS [SHORT];
```

The *short* requests a condensed listing. This listing will simply list the current setting of all system options, listing multiple options per line. The default listing is more verbose and explanatory. *This option is not available in microcomputer environments.*

Usage Note

Users executing SAS interactively with the SAS Display Manager may view and alter option settings by opening the OPTIONS window. Enter OPTIONS on the command line of any window, then press the ENTER key. A two-column listing of option names and settings will be displayed. Reset option values as needed, then move to the command line, enter END, and press the ENTER key. The new options will now be in effect.

Examples

The output from OPTIONS in several environments is presented in Exhibits 12.5 through 12.7. Notice that the option listing is written to the SAS Log, not the output file. Exhibit 12.5 shows options in a microcomputer environment running SAS version 6.

EXHIBIT 12.5   OPTIONS output in a microcomputer environment

```
------------------- SAS Log ----------------------------

1     proc options;
2     run;
  OPTION           VALUE
  ------           -----
  CENTER           ON
  CHARCODE         OFF
  DATE             ON
  DEVICE
  ERRORS           20
  FIRSTOBS         1
  FORMCHAR         |----|+|---+=|-/\<>*
  KANJI            OFF
  LINESIZE         78
  MISSING          .
  MPRINT           OFF
  MTRACE           OFF
  NEWS
  NOTES            ON
  NUMBER           ON
  OBS              MAX
  OVP              OFF
  PAGENO           1
  PAGESIZE         21
  PROBSIG          0
  PROFILE          OFF
  REMOTE
  REPLACE          ON
  SOURCE           ON
  SOURCE2          OFF
  STIMER           ON
  SYMBOLGEN        OFF
  _LAST_           _NULL_
NOTE: The PROCEDURE OPTIONS used 4.00 seconds.
```

Exhibit 12.6 shows a condensed (SHORT) listing in an IBM TSO environment.

EXHIBIT 12.6   Condensed OPTIONS output in an IBM TSO environment

---

```
-------------------- SAS Log ----------------------------

  1? proc options short ;
  2? run;

SYSTEM PARAMETERS AND OPTIONS

BAUD=1200 BLDLTABLE BLKSIZE=0 BUFNO=2 BYERR NOCAPS CARDS=MAX C60 CENTER
NOCHARCODE NOCHKPT NOCLIST NOCMDMAC NODATE DB2SSID=DB2 NODBCS DEFAULT=TSO
DEVADDR= DEVICE=LSR48 DISK=DISK NODMS DMSXPGS=2 DQUOTE DSNFERR DSRESV NODUMP
NODUMPSYS NODYNALLOC NOERRORABEND ERRORS=20 FILLMEM='FEFF'X FILSZ FIRSTOBS=1
FMTERR FS FSMODE=IBM FSP GEN=2 GRAPHICS NOIMPLMAC NOIMS INCLUDE INITSTMT=''
INTERACTIVE INVALIDDATA=. NOKANJI LABEL _LAST_=_NULL_ LEAVE=0 LINESIZE=132
LOG=FT11F001 MACRO NOMACROGEN NOMAUTOSOURCE NOMCOMPILE NOMEMERR NOMEMFILL
MEMRPT MERROR NOMEXTMOD MISSING='.' MLEAVE=6144 NOMLOGIC MODECHARS='?>*'
NOMPRINT NOMRECALL MSIZE=12288 MSYMSIZE=1024 MWORK=2048 NDSVOLS= NEWS NOTES
NONUMBER OBS=MAX OFFLINE=0.0000 ONLINE=0.0000 NOOPLIST NOOVP PAGES=MAX
PAGESIZE=60 PARM='' PARMCARDS=FT15F001 PRINTDEVICE=STANDARD PRINTINIT
PROBSIG=0 PROCSIZE=16776192 PROMPTCHARS='000A010D05000000'X REPLACE S=0 S2=S
S370 SASAUTOS=SASAUTOS SASHELP=SASHELP SASMSGS=SASMSGS SASNEWS=SASNEWST
SASUTL=SASUTL SASUSER=SASUSER SEQ=8 SERROR SKIP=0 NOSNP NOSNPPROG SORT=1
SORTDEV=SYSDA SORTLIB='SYS2.SORT.SORTLIB' NOSORTLIST NOSORTMSG SORTMSG=SYSOUT
SORTPGM='SORT' SORTSIZE=MAX SORTWKDD='SASS' SORTWKNO=3 NOSOURCE NOSOURCE2
SPOOL NOSTIMER NOSYMBOLGEN NOSYNCSORT SYSIN SYSIN=SYSIN SYSPARM='' TAPE=TAPE
TAPECLOSE=DISP NOTEXT82 TIME=MAX TLS=0 TPS=0 TRANTAB=GTRANTAB TSO UNITS=11 12
13 14 15 16 17 18 19 20 USER=WORK USERPARM='' VNFERR VSAMLOAD VSAMREAD
VSAMUPDATE WORK=WORK NOWORKINIT
```

Exhibit 12.7 shows a full listing from a VAX/VMS environment.

EXHIBIT 12.7   OPTIONS output in a DEC/VAX environment

```
INTERACTIVE          SET INTERACTIVE OR BATCH DEFAULTS FOR OPTIONS
BUFSIZE=4096         SIZE OF INTERNAL I/O BUFFER
C48=96              EITHER C48, C60 OR C96 FOR 48, 60 OR 96 CHARACTER SET
NOCAPS              TRANSLATE QUOTED STRINGS AND TITLES TO UPPER CASE
CC=FORTRAN          CARRIAGE CONTROL FORMAT FOR PRINT FILES
CENTER              CENTER PRINTED OUTPUT
NOCHARCODE          USE CHARACTERS ?-, ?/, ?=, ?(, ?) FOR _, |, ^, [, AND ]
                    RESPECTIVELY
DATE                PRINT CURRENT DATE ON EACH PAGE OF SAS LOG
NODBCS              SUPPORT FOR DOUBLE BYTE CHARACTER SETS, INCLUDING KANJI
NODMS              INVOKE DISPLAY MANAGER
DQUOTE              ALLOW DOUBLE QUOTES (") AROUND LITERALS
DSNFERR             TREAT DATA SET NOT FOUND AS AN ERROR. NO = SET TO _NULL_
ERASE              DEFAULT WORK FILES AND LIBRARY DELETED AT END OF SAS SESSION
NOERRORABEND        ABEND ON ERROR CONDITION
ERRORS=20           MAXIMUM NUMBER OF OBSERVATIONS WITH PRINTED ERROR MESSAGES
FIRSTOBS=1          NUMBER OF FIRST OBSERVATION TO BE PROCESSED
```

EXHIBIT 12.7 *(continued)*

FMTERR	TREAT MISSING FORMAT OR INFORMAT AS AN ERROR
FS	PROCEDURES TO OPERATE IN FULL SCREEN MODE
FSDEVICE=	NAME OF FULL-SCREEN DEVICE USED TO INVOKE SAS DISPLAY MANAGER
NOFULLSTIMER	PRINT LONG VERSION OF STIMER MESSAGES
INCLUDE	PROCESS %INCLUDE STATEMENTS
INVALIDDATA=.	MISSING VALUE ASSIGNED FOR INVALID DATA
NOKANJI	SUPPORT FOR DOUBLE BYTE CHARACTER SETS, INCLUDING KANJI
LABEL	MAKE VARIABLE LABELS AVAILABLE TO SAS PROCEDURES
LCC=FORTRAN	CARRIAGE CONTROL FORMAT FOR LOG/LIST
LDISK	ROUTE SAS LOG TO DISK
LINESIZE=80	PRINTER LINE WIDTH FOR OUTPUT FILES
NOLPRINT	ROUTE SAS LOG TO DEFAULT SYSTEM PRINTER
LTYPE	ROUTE SAS LOG TO TERMINAL
MISSING=.	CHARACTER TO BE PRINTED FOR MISSING VALUES
MODECHARS=?>*	PROMPT CHARACTERS FOR INTERACTIVE SESSION
NONEWS	PRINTS NEWS FROM SAS HELP LIBRARY TO LOG AT START-UP TIME
NOTES	PRINT INFORMATORY SAS MESSAGES TO LOG
NUMBER	PRINT PAGE NUMBER ON EACH OUTPUT PAGE
OBS=2147483647	NUMBER OF LAST OBSERVATION TO BE PROCESSED
NOOVP	ALLOW OVERPRINTING OF OUTPUT LINES
PAGES=2147483647	MAXIMUM NUMBER OF PAGES OF PRINTED OUTPUT
PAGESIZE=24	NUMBER OF LINES PRINTED PER PAGE OF SAS OUTPUT
NOPDISK	ROUTE SAS PROCEDURE OUTPUT TO DISK FILE
NOPPRINT	ROUTE SAS PROCEDURE OUTPUT TO DEFAULT PRINTER
NOPRINTHOVP	DARKEN LINES BY OVERPRINTING IN PRINTED OUTPUT
PROBSIG=0	NUMBER OF SIGNIFICANT FIGURES WHEN PRINTING STATISTICAL P-VALUES
PTYPE	ROUTE SAS PROCEDURE OUTPUT TO TERMINAL
REPLACE	ALLOW REPLACE OF PERMANENTLY ALLOCATED SAS DATA SETS
RETURN=RETURN	RETURN KEY FUNCTIONALITY UNDER DISPLAY MANAGER
SKIP=0	NUMBER OF LINES TO SKIP BEFORE PRINTING TITLE
SORTPGM=BEST	SORT PROGRAM NAME SPECIFICATION
NOSOURCE	LIST SAS SOURCE STATEMENTS ON LOG
SOURCE2	LIST %INCLUDED SAS SOURCE STATEMENTS ON LOG
SPARM1=	TRACING AND DEBUGGING INFORMATION FOR USER-WRITTEN PROCEDURES
STIMER	PRINT CPU AND MEMORY UTILIZATION NOTES AFTER EACH SAS STEP
SYNCHECK	ENTER SYNTAX CHECK MODE IF ERRORS ARE DETECTED
TAB=TAB	TAB KEY FUNCTIONALITY UNDER DISPLAY MANAGER
TAPECLOSE=LEAVE	VOLUME POSITIONING AT CLOSE OF SAS DATA LIBRARY ON TAPE
TLINESIZE=80	TERMINAL LINE WIDTH FOR PRINTED OUTPUT
NOTLOG	COPY SAS SOURCE STATEMENTS TO DISK FILE
UNLOAD=	MODULES TYPES UNLOADED
USER=WORK	LOGICAL FILE NAME FOR USER DATA SET
WORK=WORK	LOGICAL FILE NAME FOR WORK DATA SET
WRKINIT	WORK LIBRARY INITIALIZED WHEN CREATED.
XWAIT	WAIT FOR USER TO END X COMMAND DISPLAY
LAST=_NULL_	NAME OF LAST DATA SET CREATED

Exercise

12.5. Run the OPTIONS procedure on your system and print the results. Note the default settings for the options described in this chapter and consider if they differ from how you would normally use them (e.g., is the default CENTER when you would prefer NOCENTER?) If any do, in fact, differ from your preference, be ready to change them routinely in your jobs. You may also want to think about start-up files (see Appendix D for system-specific details). Keep this output for future reference and rerun the procedure periodically (since system administrators may change settings).

Summary

SAS offers many ways to control the way it displays output, reacts to error or unusual conditions, and handles other "environmental" features. These SAS System options may be specified on the command line when you invoke interactive SAS and possibly altered during the SAS session. They may also be set when submitting a batch SAS program. Display options control such features as output centering, line width, page number and date display, and use of special characters for missing values.

Data management options instruct SAS to process part of SAS datasets and control whether SAS replaces datasets if a syntax error is detected. Session control options are used with interactive SAS and enable Display Manager, screen size, and other, sometimes esoteric, features to be set. A variety of other options controls items such as displaying invalid numeric data, flagging misspelled formats as errors, displaying NOTEs and SAS statements that should be on the Log, and printing a specified number of error diagnostics.

13 Combining SAS Datasets

The SAS environment is not limited to simple single-dataset manipulation. Building the "final" dataset for a project is seldom a one-time event. Variables and observations must be added to the dataset, and existing values in the dataset may have to change. The need for these actions is not a shortcoming of the analyst but a natural part of the business and research process. Survey data may come in batches, and preliminary analysis of a dataset may produce unexpected results that require still more data to explore effectively. This chapter describes how to combine datasets holding different pieces of the analysis puzzle. The SET statement's capabilities are extended, and MERGE and UPDATE are introduced.

13.1 Data-Combining Basics

Virtually any combination of two or more datasets is possible in SAS. This chapter presents five methods that are usually sufficient for all but the most exotic analyses. This section looks at what each technique accomplishes and then outlines data-combining terminology.

Method 1: Concatenation

When SAS datasets are stacked on top of each other they are said to be concatenated. All of the first dataset's observations are written to the output dataset, followed by all of the second's, and so on. The datasets may have exactly the same variable names (the usual case) or some or even none in common.

Concatenation is often used in situations where study data (surveys, lab measurements) become available at different times. The dataset can be sorted later if physical order is important. Thus it would be feasible to have an observation for study subject "Smith" in March of 1989 at observation 25 and have another "Smith" observation, this one for December 1989 at observation 840.

Method 2: Interleaving

SAS datasets may be interleaved. This method produces an output dataset that includes all the observations of the input datasets. Observations with identical values of key variables appear consecutively in the output dataset. These values may be either sorted or grouped. SAS looks for the smallest (or largest, if the DESCENDING option is used in the BY statement) value of the sort key variables and places all the observations with that value into the output dataset. The process is repeated for all the sort key groups until no observations are left in the input datasets.

Interleaving is used when you want to keep observations with the same value of the sort variables physically adjacent in the output dataset. This requires that one or more variables be common to all datasets being interleaved. Interleaving often makes listings of the data more readable and allows BY-group processing for analytic and graphic procedures. If the study dataset described in the previous section were interleaved rather than concatenated, subject Smith's March 1989 and December 1989 observations would follow each other in the output dataset.

Method 3: One-to-One Merge

Rather than placing datasets on top of each other or interleaving them, a one-to-one merge places datasets side by side. The first observation in the first dataset specified is paired with the first observation in the second dataset specified. The process is repeated until all observations in all datasets are read. For this type of merge to work correctly, all datasets should be ordered identically. That is, if the 12th observation in dataset JUNE contains data for Lab 25, the 12th observation in dataset JULY should also be data for Lab 25.

Because it is so easy for the pairing of observations to go awry, one-to-one merges are rarely used. The only situation in which they are appropriate is when you are certain of the order of all the datasets *and* there are no sort key variables common to each dataset.

Method 4: Matched Merge

The most widely used method for combining datasets, the matched merge, relaxes the one-to-one merge's constraint of identically ordered observations. Instead, SAS forms new observations based on similar values of the specified sort (BY statement) variables. If the merged datasets have values equal to the sort variable, the observation will have data from both observations. If not, the new observation will have missing values for the variables found in the unmatched, "missing" observation.

Matched merges are used when new variables and observations must be added to an existing SAS dataset. As long as the datasets have an identification variable in common, SAS can link their observations correctly.

Method 5: Updating

Updating is a restricted form of the matched merge. The restriction is placed on missing values. In a matched merge a missing value for like-named variables will overwrite a nonmissing value. This is usually an undesirable result and is not allowed in an update.

Only two datasets may be combined, and the sort (BY statement) variable's occurrence must be unique in the first, or "master," dataset. Nonmissing values in the second, or "transaction" dataset, are used to replace, or update, those in the master. Updates are usually performed when only nonmissing transaction dataset values should replace values for matched observations in the master dataset.

Terminology and Concepts

The five methods outlined above allow you to combine the same input SAS datasets with drastically different results. Though different in result, the methods share some concepts and terminology. This section lays the foundation for the presentation of the syntax, usage notes, and examples of each of the five data-combining methods later in the chapter.

Dataset Order and Value Changes.   The SET, MERGE, and UPDATE statements identify the SAS datasets to be combined. The order of the datasets in the list affects not only the sequencing of observations for the output dataset but their values as well. Such value changes are possible with the MERGE statement: the value from a dataset to the right of another in the dataset list will overwrite a like-named variable. Here, as with all forms of dataset manipulation, the data type of a variable must be identical in all datasets in that it is referenced.

To illustrate this replacement, if datasets ONE and TWO are merged, both with numeric variable EDUC, new observations would first contain ONE's EDUC value. Then TWO's value of EDUC would replace it. Notice that the original value, that in dataset ONE, is not changed: only the value in the observation being built is affected. Notice too that the order of the datasets in the MERGE list is important: TWO's values replaced ONE's in this case because of the order they were specified in the MERGE statement.

The process of adding variables and replacing values proceeds from left to right. When a new variable name is encountered (i.e., it is not in a dataset already read), it is added to the variables in the dataset being built.

Datasets Eligible for Combining.   Any combination of permanent, "two-level," and temporary SAS datasets may be combined. "Raw" datasets cannot be combined without first creating SAS datasets!

Sort Order.   When interleaving, match merging, or updating, the datasets must be sorted by one or more common variables. These may be in any combination of numeric and character data types. These could be items such as state and county FIPS codes and subject ID numbers. The sort order permits SAS to read sequentially through the datasets and include their observations in the correct order. If necessary, the SORT procedure is used to arrange the observations in the correct order.

Adding Variables to a Dataset.   All five combination methods may be used to add variables to a dataset. In practice, however, only the one-to-one merge, matched merge, and update add the variable in a way that is useful for analytic purposes. In each of these methods SAS reads variable names from the leftmost dataset in the dataset list, then reads the variable names from the next dataset in the list. If this dataset has names not already in the name list being built, SAS adds the names to the list.

Replacing Variable Values.   The one-to-one merge, matched merge, and update can be used to replace variable values. If two datasets have similarly named variables, the dataset on the right will overwrite, or replace, the value of the variable in the observation being readied for output. UPDATE will replace a value *only* if the right-hand dataset has a nonmissing value. (A "special missing" value of "_" will set master dataset values to missing. See Chapter 19 for details.)

Note that like-named variables must *all* have the same data type: RACE cannot be a character variable in one dataset and a numeric in another!

Adding Observations to a Dataset.   All of the methods outlined above allow you to add observations to a dataset. Adding an observation during a one-to-one merge, however, is often an indication that something was amiss, that the datasets did not have the requisite one-to-one correspondence required for successful matching. Observations are added during matched merging or updating if an observation has a unique combination of the BY variable(s) used to link the datasets.

Observations "Belong" to a Dataset.   When SAS is building the new, combined observation using any of the five methods, it knows which datasets in the SET, MERGE, or UPDATE dataset list are being used. The IN dataset option may be used to identify which input dataset contributed to the output dataset for a particular observation. IN is described in the SET statement discussion in Section 13.2.

Using IN allows, for example, the output dataset to contain only those observations in the matched merge that were in both datasets being merged (i.e., the intersection of the datasets). It could also be used to identify error situations such as being unable to match a transaction observation to an observation in the master dataset.

Datasets Sensitive to OBS and FIRSTOBS.   The data-combining operations discussed here are sensitive to the OBS and FIRSTOBS systems and dataset options (discussed in Chapters 10 and 12). A dataset with 100 observations can, for example, contribute a maximum of 20 observations to the dataset being created if the OBS=20 system option is in effect.

Some key points of this section are summarized in Exhibit 13.1.

POINTERS AND PITFALLS

Understanding the conceptual and practical differences among the different data-combining methods will save frustration and confusion during analyses.

Exercises

13.1. Which data-combining method is appropriate for the following situations?
  a. Place dataset ONE and TWO "on top" of each other. ONE's observations will come first, then TWO's.
  b. Join ONE and TWO so that TWO's values will replace ONE's for like-named variables. The datasets are both sorted by variable SOC_SEC.
  c. Follow the instructions in part b but do not allow TWO's missing values to overwrite nonmissing values in ONE.
  d. Join the datasets so that ONE and TWO have similar values of a common variable adjacent in the output dataset.
  e. Place observations in ONE and TWO side by side in the output dataset, regardless of common variables' values.

13.2. Answer the following questions about combining datasets:
  a. Which method(s) rely on BY variables being in sorted order?

EXHIBIT 13.1   Comparison of data-combining methods

Feature	Concatenation	Interleaving	Merge	Matched Merge	Updating
Datasets must be arranged (sorted) identically	no	yes	yes [1]	yes	yes
Limit on number of datasets that may be combined	none	none	none	none	[2]
Can be used to add variables to the output dataset	yes [2]	yes [2]	yes	yes	yes
Can be used to add observations to the output dataset	yes	yes	yes [3]	yes	yes [4]
Order of dataset specification is significant	no	no	no	no	yes
Need to specify common (BY) variables	no	yes	no	yes	yes
Missing values may overwrite nonmissing ones	no	no	yes	yes	no [5]

[1] Program will execute if datasets are not in identical order, but results will probably be suspect.

[2] May produce undesirable "holes" in the output dataset.

[3] Often produces undesirable results (see examples in Section 13.4)

[4] New observations often suggest problems with second (transaction) dataset. See examples in Section 13.6.

[5] A special exception is discussed in Chapter 19.

---

b. Which methods are sensitive to the order in which the datasets are specified?

c. Which allow new variables to be added to a dataset?

d. Which allow overwriting of nonmissing values of a variable?

e. Which require a BY statement?

13.3. Without looking ahead to the next sections of this chapter, describe the contents (i.e., observation-variable values) of dataset COMBINE in each program. Datasets A and B look like this:

Dataset A		Dataset B	
ID	X	ID	X
10	1	10	1
15	2	15	.
25	3	20	3
		25	4

```
a. data combine;
   set a b;
b. data combine;
   set a b;
   by id;
c. data combine;
   merge a b;
d. data combine;
   merge a b;
   by id;
e. data combine;
   update a b;
   by id;
```

13.2 Concatenation

Concatenate datasets when you simply want the output dataset to contain all observations from the leftmost dataset, followed by the observations from the next dataset in the list, and so on for each dataset.

Syntax

Concatenation requires extending the SET statement:

```
SET [fileref.]dset[(IN=var other_opts)]
    [fileref.]dset[(IN=var other_opts)] ... ;
```

Some common applications of this new form of the SET statement follow:

```
set month1 month2 month3;

set mast.master addition;

set new(keep=store rev1-rev20)
    base(keep=store rev1-rev20);
```

The SET statement used for concatenation has four components:

1. The *fileref* is an optional reference to a permanent SAS dataset library.

2. The *dset* is the name of a SAS dataset.

3. *IN=var* identifies the name of a variable that will be equal to 1 if the current observation being built for the output dataset comes from *dset*, 0 otherwise. The *var* is not kept in the output dataset.

4. The *other_opts* identifies other dataset options.

Usage Notes

When concatenating datasets, keep in mind that the number of observations in the output dataset equals the sum of the number of observations in the datasets listed in the SET statement. This, of course, assumes no observations were explicitly deleted in the DATA step, thus altering the output dataset's size.

Examples   Key features and options used in this section are summarized in the following list:

Exhibit	Features/Options
13.2	Datasets with identical variable names
13.3	Not all variable names match in each dataset
13.4	RENAME option used to make variable names identical
13.5	IN option used to create a new variable in the new, combined dataset

In Exhibit 13.2 each dataset has identical variable names (REG and TEMP).

EXHIBIT 13.2   SET datasets with identical variable names

```
------------------- Input datasets -------------------

January                        June

OBS    REG    TEMP             OBS    REG    TEMP
 1     ne      34               1     ne      65
 2     sw      60               2     sw      88
 3     w       55               3     w       72


------------------- Program and output ----------------

data new;
set january june;

OBS    REG    TEMP
 1     ne      34
 2     sw      60
 3     w       55

 4     ne      65
 5     sw      88
 6     w       72
```

Observations from January

Observations from June

In Exhibit 13.3 only some variable names are common to each dataset (REG and NST). Variable TEMP_JAN appears only in dataset JANUARY, while TEMP_JUN is only in JUNE. Notice the pattern of missing values in dataset NEW: missing values appear in the dataset even though the input datasets had no missing data!

Exhibit 13.4 uses the same input datasets as Exhibit 13.3. This time, the differing variables are renamed to the same name (TEMP).

Exhibit 13.5 uses the same input datasets as Exhibits 13.3 and 13.4. Not only are the variables renamed to a common name, but the IN= dataset option is used to create the variable MONTH. This indicates which month (i.e., which dataset) the observation came from. Analyses and plots can use this variable to separate TEMPs. The INJAN notation in the IF statement is shorthand for testing INJAN=1. Since SAS evaluates 1 as a "true" value and others as "false," this abbreviated comparison is consistent with the program's logic.

EXHIBIT 13.3   Not all variable names match in each dataset being SET

------------------- Input datasets --------------------

January   June

```
OBS    REG    NST    TEMP_JAN        OBS    REG    NST    TEMP_JUN
1      ne     6        34            1      ne     6        65
2      sw     7        60            2      ma     4        67
3      w      5        55            3      sw     7        88
                                     4      nw     4        72
```

------------------- Program and output ----------------

```
data new;
set january june;
```

	Variable unique to January
	Variable unique to June

```
OBS    REG    NST    | TEMP_JAN |  | TEMP_JUN |

1      ne     6      |    34    |  |    .     |
2      sw     7      |    60    |  |    .     |
3      w      5      |    55    |  |    .     |
4      ne     6      |    .     |  |    65    |
5      ma     4      |    .     |  |    67    |
6      sw     7      |    .     |  |    88    |
7      nw     4      |    .     |  |    72    |
```

EXHIBIT 13.4   RENAME option used to make variable names identical

-------------------- Input datasets --------------------

January   June

```
OBS    REG    NST    TEMP_JAN        OBS    REG    NST    TEMP_JUN
1      ne     6        34            1      ne     6        65
2      sw     7        60            2      ma     4        67
3      w      5        55            3      sw     7        88
                                     4      nw     4        72
```

-------------------- Program and output ----------------

```
data new;
set january(rename=(temp_jan=temp))
    june    (rename=(temp_jun=temp));

OBS    REG    NST    TEMP
1      ne     6      34
2      sw     7      60
3      w      5      55
4      ne     6      65
5      ma     4      67
6      sw     7      88
7      nw     4      72
```

EXHIBIT 13.5   IN option used to create a new variable

```
-------------------- Input datasets --------------------

January                              June

OBS    REG    NST    TEMP_JAN       OBS    REG    NST    TEMP_JUN
 1     ne      6        34           1     ne      6        65
 2     sw      7        60           2     ma      4        67
 3     w       5        55           3     sw      7        88
                                     4     nw      4        72

-------------------- Program and output ----------------

data new;
set january(in=injan rename=(temp_jan=temp))
    june   (           rename=(temp_jun=temp));
if injan then month = 'Jan';
   else         month = 'Jun';

OBS    REG    NST    TEMP    MONTH
 1     ne      6      34     Jan
 2     sw      7      60     Jan
 3     w       5      55     Jan
 4     ne      6      65     Jun
 5     ma      4      67     Jun
 6     sw      7      88     Jun
 7     nw      4      72     Jun
```

Exercises

13.4. Datasets ONE and TWO look like this:

Dataset ONE		Dataset TWO	
AREA	SALEPROF	AREA	SALE
n	230	n	220
n	210	e	110
ne	340	ne	210
s	170	ma	180
		s	120

What does dataset BOTH look like if the following SET statement and program statements are used (assume the DATA statement is simply DATA BOTH;)?

a. `set one two;`

b. `set one`
`    two(rename=(sale=saleprof));`

c. `set two one;`

d. `set one one two;`

e. `set one (rename=(area=region))`
`        two;`

f. `set one(in=p1)`
`        two(in=p2);`
`    if p1 = 1 then which = '1st';`
`        else       which = '2nd';`

13.3   Interleaving

Interleave two or more datasets when you want all observations with identical sort variable values grouped in the output dataset. The datasets being interleaved should be in the same order (i.e., a similar BY statement if the SORT procedure was used to arrange the datasets).

Syntax

Interleaving does not require any new statements. Use the SET statement as described in Section 13.2, and use a BY statement to indicate the common variable(s) by which the datasets will be combined.

Usage Notes

Remember the following points when interleaving datasets:

- More than one BY variable may be specified. BY-group variables may be sorted in high-to-low order (the BY statement's DESCENDING option) or in the default, low-to-high order.
- If the datasets are not sorted in the BY statement order, the DATA step will stop with an error and the output dataset will not be created or replaced.
- SAS begins building the new dataset by looking for the smallest value(s) of the BY variable(s) in the datasets specified in the dataset list. It takes *all* observations with this value from the first dataset, then examines datasets to its right in the dataset list for similar BY value(s). SAS then looks for the next-smallest value of the BY variable in the datasets and repeats the process. This cycle is continued until all observations are read from all the SET statement's datasets.

Examples

Key features and options used in this section are summarized in the following list:

Exhibit	Features/Options
13.6	Interleave by a single variable
13.7	Interleave and reverse order of datasets being SET
13.8	Different variables names in the datasets; a variable indicating observation's input dataset source is created.

In Exhibit 13.6 datasets GROUP1 and GROUP2 are interleaved by common/joining variable ID.

Exhibit 13.7 (see p. 256) uses the same dataset as Exhibit 13.6 but reverses the order of datasets in the SET statement. The number of observations in dataset NEW is the same as in Exhibit 13.6, as is the sequence of the BY variable, ID. What changes is the *order* in which the observations are taken from the two datasets: GROUP2's observations are written to dataset NEW, then GROUP1's.

In Exhibit 13.8 (see p. 257) two datasets sharing only the BY variable (ID) are interleaved. A new variable (WHEREFRM) is created to indicate which dataset the observation comes from. This variable may be used as a classification/stratifying variable in such procedures as FREQ, TTEST, and CHART (as discussed in Chapters 6 and 22).

------------------- **EXHIBIT 13.6 Interleave by a single variable**

```
-------------------- Input datasets --------------------

group1                          group2

OBS    ID    SCORE         OBS    ID    SCORE

 1     20     1.4           1     19     2.3
 2     20     1.3           2     20     2.5
 3     23     1.5           3     21     2.3
 4     24     1.5           4     23     2.6
 5     30     1.2           5     30     2.2
                           6     31     2.4


------------------- Program and output ----------------

data new;
set group1 group2;
by id;

OBS    ID    SCORE

  1    19     2.3
  2    20     1.4
  3    20     1.3
  4    20     2.5
  5    21     2.3
  6    23     1.5
  7    23     2.6
  8    24     1.5
  9    30     1.2
 10    30     2.2
 11    31     2.4
```

Exercises

13.5. Datasets ONE and TWO look like this:

Dataset ONE			Dataset TWO	
LOT	NPASS		LOT	PCTFAIL
10	80		10	.01
10	81		11	.01
11	72		11	.05
15	70		15	.10
			20	.08
			20	.09

What does dataset BOTH look like when the following SET, BY, and other DATA step statements are used?

a. data both;
   set one two;
   by lot;

b. data both;
   set two one;
   by lot;

```
c. proc sort data=one;
   by descending lot;
   run;

   proc sort data=two;
   by descending lot;
   run;

   data both;
   set one two;
   by descending lot;
```

EXHIBIT 13.7   Order of datasets reversed in the SET statement

```
------------------- Input datasets --------------------

group1                          group2

OBS    ID    SCORE              OBS    ID    SCORE

 1     20     1.4                1     19     2.3
 2     20     1.3                2     20     2.5
 3     23     1.5                3     21     2.3
 4     24     1.5                4     23     2.6
 5     30     1.2                5     30     2.2
                                 6     31     2.4


------------------- Program and output ---------------

data new;
set group2 group1;
by id;

OBS    ID    SCORE

 1     19     2.3
 2     20     2.5
 3     20     1.4
 4     20     1.3
 5     21     2.3
 6     23     2.6
 7     23     1.5
 8     24     1.5
 9     30     2.2
10     30     1.2
11     31     2.4
```

13.4   One-to-One Merge

Use a one-to-one merge to join two or more datasets that have equal numbers of observations *and* identically ordered observations. For reasons that will become obvious in the examples at the end of this section, this method of combining datasets should be used sparingly.

EXHIBIT 13.8   Different variable names in the datasets; "source" variable created

```
-------------------- Input datasets ----------------

group1                        group2

OBS    ID    SCORE1           OBS     ID    SCORE2


1      20     1.4             1       19     2.3
2      20      .              2       20     2.5
3      23     1.5             3       21     2.3
4      24     1.5             4       23     2.6
5      30      .              5       30      .
                             6       31     2.4


-------------------- Program and output--------------------

data new;
set group1(in=grp1)
    group2(in=grp2);
if         grp1 then wherefrm = 'grp1';
    else if grp2 then wherefrm = 'grp2';
by id;

OBS    ID    SCORE1   SCORE2   WHEREFRM

 1     19      .       2.3      grp2
 2     20     1.4       .       grp1
 3     20      .        .       grp1
 4     20      .       2.5      grp2
 5     21      .       2.3      grp2
 6     23     1.5       .       grp1
 7     23      .       2.6      grp2
 8     24     1.5       .       grp1
 9     30      .        .       grp1
10     30      .        .       grp2
11     31      .       2.4      grp2
```

Variable unique to GROUP1

Variable unique to GROUP2

Source dataset identifier, created using IN dataset option

POINTERS AND PITFALLS

The one-to-one merge should be used only when you are confident that observation *n* in each of the datasets represents the same unit of observation (the 10th observation in each dataset contains data for study subject "J. Smith"). If a common, matching variable is present in all datasets, use the matched merge rather than the one-to-one merge.

Syntax

The one-to-one merge requires a new statement, MERGE:

```
MERGE [fileref.]dset[(IN=var other_opts)]
      [fileref.]dset[(IN=var other_opts)] ... ;
```

Some common applications of the MERGE statement follow:

```
merge m1 m2;
```

```
merge in.social in.demog;

merge in.fin(in=in_fin) in.mkt(in=in_mkt);
```

In MERGE, *fileref*, *dset*, *IN=var*, and *other_opts* have exactly the same meaning as they did in the discussion of the SET statement in Section 13.2.

Usage Notes

The following features of the one-to-one merge should be kept in mind:

- The output dataset is the product of a *side-by-side* pairing of the datasets in the MERGE list. Concatenating and interleaving made the output dataset "taller" by adding observations and possibly "wider" by adding variables. The one-to-one merge usually makes the dataset "wider" and seldom makes it "taller."

- SAS reads the first observation from each dataset in the MERGE list, starting from the left. The next observation in the output dataset contains data from the second observation from each dataset. The process is repeated until there is no more data to be read. *SAS is not sensitive to whether this combination of data makes any sense.* It simply uses the *n*th observation from each dataset in the MERGE list.

- If a variable is present in more than one dataset, the output dataset will contain the value of that variable from the rightmost dataset in the MERGE list. This means that a missing "right" dataset value can overwrite a nonmissing "left" dataset value for a like-named variable. All such like-named variables must be of the same data type (all numeric or all character) in each dataset.

- The position of variables in the output dataset is determined by the order of the datasets in the MERGE statement's dataset list. See the examples below for illustrations.

- Since unequal numbers of observations in the MERGE datasets can easily render the output dataset meaningless, the one-to-one merge should be used only when there are no common variables to link the datasets.

Examples

Key features and options used in this section are summarized in the following list:

Exhibit	Features/Options
13.9	Two datasets merged correctly (observation order identical in each)
13.10	Second (right) dataset replaces values in the first (left) dataset; missing data overwrites nonmissing
13.11	RENAME used to preserve common variables' values
13.12	Observation order not logically correct
13.13	Input datasets with unequal numbers of observations

In Exhibit 13.9 a state-level dataset and a corresponding dataset for the states' capital cities are merged. The datasets are ordered identically: the first observation in STATE and the first observation in CAPITAL contain data about the same entity, the state of Georgia. The pairings of STATE and CAPITAL observations continue correctly through the four observations.

In Exhibit 13.10 a new variable, RANK, is added to each dataset. It is nonmissing in every observation in STATE but is missing twice in CAPITAL. Since the variables are identically named, the right-hand dataset (CAPITAL) will overwrite the RANK values in the left-hand dataset (STATE). Valid, nonmissing rank values are lost.

EXHIBIT 13.9   Two datasets merged correctly

```
------------------- Input datasets -------------------

state                                    capital

OBS      STATE      INCOME          OBS      CAPITAL      AVGWIND
 1        ga         9583            1       Atlanta        9.1
 2        nc         9044            2       Raleigh        7.8
 3        sc         8502            3       Columbia       6.9
 4        va        11095            4       Richmond       7.5
------------------- Program and output ----------------

data allinfo;
merge state capital;

OBS      STATE      INCOME      CAPITAL      AVGWIND
 1        ga         9583      Atlanta        9.1
 2        nc         9044      Raleigh        7.8
 3        sc         8502      Columbia       6.9
 4        va        11095      Richmond       7.5
```

EXHIBIT 13.10   Second (right) dataset replaces values in first (left) dataset

```
------------------- Input datasets -------------------

state                              capital

OBS   STATE   INCOME   RANK      OBS   CAPITAL   AVGWIND   RANK
 1     ga      9583     21        1    Atlanta     9.1      .
 2     nc      9044     22        2    Raleigh     7.8      30
 3     sc      8502     35        3    Columbia    6.9      21
 4     va     11095     10        4    Richmond    7.5      .

------------------- Program and output ----------------

data allinfo;
merge state capital;

OBS   STATE   INCOME   RANK    CAPITAL    AVGWIND
 1     ga      9583      .      Atlanta     9.1
 2     nc      9044     30      Raleigh     7.8
 3     sc      8502     21      Columbia    6.9
 4     va     11095      .      Richmond    7.5
```

> RANK value is from CAPITAL, the rightmost dataset in the MERGE statement.

Exhibit 13.11 (see p. 260) works around the overwriting problem in Exhibit 13.10 by using the RENAME dataset option. The pairs of rank values are preserved in two separate variables. Remember that the variables are not *permanently* renamed in datasets STATE and CAPITAL. They are only renamed for the duration of the DATA step creating ALLINFO. The input data are identical to those used in Exhibit 13.10.

EXHIBIT 13.11   RENAME used to preserve common variables' values

------------------- Input datasets -------------------

state   capital

OBS	STATE	INCOME	RANK	OBS	CAPITAL	AVGWIND	RANK
1	ga	9583	21	1	Atlanta	9.1	.
2	nc	9044	22	2	Raleigh	7.8	30
3	sc	8502	35	3	Columbia	6.9	21
4	va	11095	10	4	Richmond	7.5	.

------------------- Program and output ----------------

```
data allinfo;
merge state(rename=(rank=incrank));
      capital(rename=(rank=windrank));
```

OBS	STATE	INCOME	INCRANK	CAPITAL	AVGWIND	WINDRANK
1	ga	9583	21	Atlanta	9.1	.
2	nc	9044	22	Raleigh	7.8	30
3	sc	8502	35	Columbia	6.9	21
4	va	11095	10	Richmond	7.5	.

Exhibit 13.12 performs some geographical alchemy by getting the last two CAPITAL observations out of order. Columbia becomes the capital of Virginia, and Richmond moves south to become South Carolina's capital. Remember, SAS lets you make this type of mistake. It is your responsibility to ensure the data are in the correct order!

EXHIBIT 13.12   Observation order not logically correct

------------------- Input datasets -------------------

state   capital

OBS	STATE	INCOME	OBS	CAPITAL	AVGWIND
1	ga	9583	1	Atlanta	9.1
2	nc	9044	2	Raleigh	7.8
3	va	11095	3	Columbia	6.9
4	sc	8502	4	Richmond	7.5

------------------- Program and output -------------------

```
data allinfo;
merge state capital;
```

OBS	STATE	INCOME	CAPITAL	AVGWIND
1	ga	9583	Atlanta	9.1
2	nc	9044	Raleigh	7.8
3	va	11095	Columbia	6.9
4	sc	8502	Richmond	7.5

Exhibit 13.13 uses datasets with unequal numbers of observations. Knoxville is added to the CAPITAL dataset, but Tennessee is not added to the STATE dataset. The one-to-one merge exhausts STATE after four observations and marks its variables as missing in the excess CAPITAL observation. Notice that the mismatching of Exhibit 13.12 was also performed here.

EXHIBIT 13.13   Unequal numbers of observations

```
------------------- Input datasets -------------------

state                              capital

OBS     STATE    INCOME            OBS      CAPITAL    AVGWIND
 1       ga       9583              1       Atlanta      9.1
 2       nc       9044              2       Raleigh      7.8
 3       sc       8502              3       Columbia     6.9
 4       va      11095              4       Knoxville    6.4
                                    5       Richmond     7.5


------------------- Program and output ---------------

data allinfo;
merge state capital;

OBS     STATE    INCOME    CAPITAL      AVGWIND
 1       ga       9583     Atlanta        9.1
 2       nc       9044     Raleigh        7.8
 3       sc       8502     Columbia       6.9
 4       va      11095     Knoxville      6.4
 5     [      ]   [   . ]  Richmond       7.5
```

Smaller dataset is given missing values in unpaired observations.

Exercises

13.6. Datasets ONE and TWO look like this:

Dataset ONE		Dataset TWO	
ID	AVG	ID	MEAN
200	93	200	93
180	82	170	92
170	81	180	94
210	88	210	98

What does dataset BOTH look like when the following programs are run?

a. `data both;`
   `merge one two;`

b. `data both;`
   `merge two one;`

```
c. proc sort data=two;
   by id;
   run;

   proc sort data=one;
   by id;
   run;

   data both;
   merge one two;
   run;
d. data both;
   merge one(in=in_one rename=(avg=mean))
      two(in=in_two);
   if in_one = 1 then where = 'one';
      else              where = 'two';
   run;
```

13.5   Matched Merge

Use a matched merge when you want to pair two or more datasets by one or more common variables. Unlike the one-to-one merge's sequential pairing of observations, the datasets must be sorted by the common variables and will be paired only when they match. The matched merge is the most common and powerful form of dataset combining.

Syntax

The matched merge does not require any new statements. Use the MERGE statement as described in Section 13.4 and use a BY statement to indicate the common variables by which the merge will take place.

Usage Notes

The following features of the matched merge should be kept in mind:

- One-to-one merge features such as overwriting of like-named variables, variable order in the output dataset, and similar data types for identically named variables also hold for the matched merge. Additionally, the matched merge requires that the datasets in the MERGE list be sorted by the variables in the BY list.

- The MERGE list datasets can have unequal numbers of observations. The BY statement controls how the datasets' observations will be paired. Suppose a BY variable of SUBJECT was specified. SAS would read the dataset list from left to right. The dataset containing the lowest value of SUBJECT would be used first. SAS then searches datasets to its right with an identical SUBJECT value. If a match is found, the values of the variables in the two datasets would be combined to form the output dataset's observation.

- The BY-groups as well as the entire datasets may have unequal numbers of observations. Within each BY-group, SAS will pair as many observations as possible from the MERGE datasets. When one dataset runs out of observations at a particular BY level, SAS keeps its last known values in memory and continues reading from other datasets until the BY-group is exhausted.

This is a very useful, important, and subtle feature of the matched merge. It is illustrated in Exhibit 13.16.

- If a BY variable level is present in dataset "A" but not in "B," variables unique to "B" are set to missing. This is illustrated in Exhibit 13.15.

Examples

Some of the examples in this section are presented in a somewhat different format than usual. Datasets are presented side by side rather than top to bottom in the exhibits. This emphasizes the logic SAS uses when merging the input datasets: "try to take an observation from the leftmost dataset, then work through the merge dataset list, eventually outputting to one or more datasets."

Key features and options used in this section are summarized in the following list:

Exhibit	Features/Options
13.14	All variable names identical; values overwritten in the same BY-group
13.15	IN option; some variable names unique
13.16	Unequal number of observations in the BY-groups
13.17	Master dataset updated based on IN option values
13.18	Dummy dataset used to avoid "missing observations"
13.19	IN option used to create merge pattern variable

Exhibit 13.14 merges two datasets with unequal numbers of observations and identically named variables. The right-hand dataset's values of variable VALUE overwrite those of VALUE in the left-hand dataset within the same BY-group.

EXHIBIT 13.14   All variable names match; common variables overwritten

```
                                      data merged;
                                      merge one two;
     dataset ONE     dataset TWO      by id;

     ID    VALUE     ID    VALUE      ID     VALUE

     a     1.1       a      .         a       .
     b     1.2       b     2.1        b      2.1
                     c     2.2        c      2.2
     d      .        d     2.3        d      2.3
     e     1.3                        e      1.3
                     f     2.4        f      2.4
     g     1.4       g     2.5        g      2.5
```

Exhibit 13.15 (see p. 264) adds unique variables to each dataset in Exhibit 13.14. Their values are preserved in the merged dataset. Note the IN= values: 1 in both ONE and TWO indicates a successful pairing by the BY variable, ID. A 0 for either dataset indicates a mismatch—an observation in MERGED was built with data that came from ONE but not from TWO, or vice versa.

Exhibit 13.16 (see p. 264) is a more complicated and realistic example showing unequal numbers of observations in BY-groups. BY variable ID values 1 and 2 pair up observations. Values 3 through 5 have unequal numbers in the two datasets. Notice how the short dataset's values are retained until the end of the

EXHIBIT 13.15   IN option demonstrated; not all variables names in common

```
data merged;
merge one(in=inone)
      two(in=intwo rename=(type=id));
by id;
```

dataset ONE

ID	VALUE	V_ONE
a	1.1	10.1
b	1.2	10.2
d	.	10.3
e	1.3	10.4
g	1.4	10.5

dataset TWO

TYPE	VALUE	V_TWO
a	.	20.1
b	2.1	20.2
c	2.2	20.3
d	2.3	20.4
f	2.4	20.5
g	2.5	10.6

ID	VALUE	V_ONE	V_TWO	In ONE?	In TWO?
a	.	10.1	20.1	1	1
b	2.1	10.2	20.2	1	1
c	2.2	.	20.3	0	1
d	2.3	10.3	20.4	1	1
e	1.3	10.4	.	1	0
f	2.4	.	20.5	0	1
g	2.5	10.5	10.6	1	1

BY-group: dataset TWO has only one observation at ID level 3, while dataset ONE has two such observations. The first observations from each BY-group are merged. When the second observation with ID=3 is taken from ONE, it uses the most recent value from TWO, a value of 2.1 for V2, rather than a missing value. Similar logic is applied to BY-groups 4 and 5. When there is a complete mismatch (ID values 6 and 7), the unique variable in the absent dataset (TWO when ID=6, ONE when ID=7) is missing in the output dataset.

EXHIBIT 13.16   Unequal number of observations in the BY-groups

```
data merge;
merge one two;
by id;
```

one

ID	COMM	V1
1	one	1.1
2	one	1.1
2	one	1.2
3	one	1.1
3	one	1.2
4	one	1.1
5	one	1.1
5	one	1.2
5	one	1.3
6	one	1.1

two

ID	COMM	V2
1	two	2.1
2	two	2.1
2	two	2.2
3	two	2.1
4	two	2.1
4	two	2.2
5	two	2.1
5	two	2.2
7	two	2.1
7	two	2.2

ID	COMM	V1	V2
1	two	1.1	2.1
2	two	1.1	2.1
2	two	1.2	2.2
3	two	1.1	2.1
3	one	1.2	2.1
4	two	1.1	2.1
4	two	1.1	2.2
5	two	1.1	2.1
5	two	1.2	2.2
5	one	1.3	2.2
6	one	1.1	.
7	two	.	2.1
7	two	.	2.2

"Short" BY-groups have values retained until the end of the BY-group.

Exhibit 13.17 is a simplified accounting application. Dataset MASTER contains an account number (variable ACCT) and balance (BALANCE). TRANSAC contains an account number and a summary of debit (DR) and credit (CR) activity for the update period. There are three possibilities when merging by account number: a match in both datasets (the account had activity during the period), an observation only in the master dataset (no activity), and an observation only in the transaction dataset (activity on nonexistent accounts). The merged dataset NEWMAST contains updated balances, while MISMATCH identifies attempted updates to nonexistent accounts. Calculations and routing to output datasets are based on the values of the IN= variables.

EXHIBIT 13.17   Master dataset updated based on IN option values

```
------------------- Input datasets -------------------


Master dataset                      Transaction dataset


ACCT     BALANCE                    ACCT     DR       CR
                                    138       0      120
140       -2100                     140     100       80
141         850
150        1525                     150       0       15
190       -2000
200           0
210        1900                     210       0     2000
                                    211     250       20

220          80

------------------- Program listing -------------------

data newmast(keep=acct balance)
     mismatch(keep=acct)
     ;
merge master(in=in_mast)
      transac(in=in_tran)
     ;
by acct;
if in_tran=1 & in_mast=1 then balance = balance + cr - dr;
if in_tran=1 & in_mast=0 then output mismatch;
   else                      output newmast;
run;

------------------- Output datasets -------------------


Updated master dataset              In transaction, not in master


ACCT     BALANCE                    ACCT
140       -2120                     138
141         850                     211
150        1540
190       -2000
200           0
210        3900
220          80
```

In Exhibit 13.18 the unit of observation in dataset ACTUAL is calendar year–years survived: in each of 1981 through 1984 it is possible for a study participant to survive with a disease from 1 to 5 years. For reasons beyond the scope of this example, the dataset should contain all possible combinations of calendar year and years survived, that is, 20 observations (4 calendar years times 5 survival times). ACTUAL has only 16 observations, a case of "missing observations" rather than "missing values."

EXHIBIT 13.18   Dummy dataset used to avoid "missing observations"

------------------- Program listing --------------------

```
data allyears;
do year = 81 to 84;
   do surv = 1 to 5;
      count = -99;
      output;
   end;
end;
run;
```

------------------- Input and output datasets ----------

						data complete; merge allyears in.actual; by year surv;			data complete; merge in.actual allyears; by year surv;		
actual			allyears								
YEAR	SURV	COUNT	YEAR	SURV	COUNT	YEAR	SURV	COUNT	YEAR	SURV	COUNT
81	1	50	81	1	-99	81	1	50	81	1	-99
81	2	45	81	2	-99	81	2	45	81	2	-99
81	3	37	81	3	-99	81	3	37	81	3	-99
81	4	29	81	4	-99	81	4	29	81	4	-99
81	5	20	81	5	-99	81	5	20	81	5	-99
82	1	42	82	1	-99	82	1	42	82	1	-99
82	2	40	82	2	-99	82	2	40	82	2	-99
			82	3	-99	82	3	-99	82	3	-99
82	4	30	82	4	-99	82	4	30	82	4	-99
82	5	23	82	5	-99	82	5	23	82	5	-99
83	1	60	83	1	-99	83	1	60	83	1	-99
83	2	55	83	2	-99	83	2	55	83	2	-99
			83	3	-99	83	3	-99	83	3	-99
			83	4	-99	83	4	-99	83	4	-99
83	5	42	83	5	-99	83	5	42	83	5	-99
84	1	43	84	1	-99	84	1	43	84	1	-99
84	2	40	84	2	-99	84	2	40	84	2	-99
84	3	36	84	3	-99	84	3	36	84	3	-99
84	4	32	84	4	-99	84	4	32	84	4	-99
			84	5	-99	84	5	-99	84	5	-99

Exhibit 13.18 shows one approach to filling in the dataset's gaps. A dummy dataset, ALLYEARS, is created, and DO-loops that account for all possible combinations of calendar year (variable YEAR) and survival time (SURV) are constructed. Variable COUNT is set to −99 to distinguish it from values in ACTUAL. Once ALLYEARS is created, the two datasets are merged. Pay particular attention to their order in the MERGE statement's dataset list (compare the values of COUNT in the results of the MERGE). Dataset COMPLETE has 20 observations and represents COUNT values initially missing with a −99.

Exhibit 13.19 merges three datasets (YEAR1, YEAR2, YEAR3), expecting LOC-ID combinations to match in each dataset. Exhibit 13.19 creates a variable PATTERN that describes the contribution of each dataset to the new dataset's observation. PATTERN reflects values of the IN= variables assigned in the MERGE statement. A value of 1 indicates the observation was present in YEAR3 but not in YEAR1 or YEAR2. A 10 indicates data only from YEAR2, and so on.

EXHIBIT 13.19   IN option used to create merge pattern variable

------------------- SAS Log ----------------------------

```
   1      options nodate nonumber nocenter;
   2      title '';
   3
   4      data all;
   5      merge input.year1(in=y1)
   6            input.year2(in=y2)
   7            input.year3(in=y3);
   8      by loc id;
   9      pattern = 100*y1 + 10*y2 + y3;
  10      run;
NOTE: The data set WORK.ALL has 989 observations and 14 variables.
NOTE: The DATA statement used 14.00 seconds.
  11
  12      proc freq data=all;
  13      tables pattern / nocum ;
  14      format pattern z3.;
  15      run;
NOTE: The PROCEDURE FREQ used 3.00 seconds.
```

------------------- Output listing --------------------

```
PATTERN    Frequency    Percent
--------------------------------
  001            1        0.1
  010           20        2.0
  011            2        0.2
  100           57        5.8
  101           12        1.2
  110            7        0.7
  111          890       90.0
```

A frequency distribution of PATTERN shows that most observations had three datasets contributing (89% had PATTERN values of 111). Since we wanted *all* observations to fall in this category, it may be worthwhile to print observations with PATTERNS other than 111. The listing may reveal LOC or ID coding problems.

Exercises

13.7. Datasets ONE and TWO look like this:

Dataset ONE			Dataset TWO	
ID	X	Y	ID	X
10	8	1	10	.
15	4	2	15	.
15	.	3	19	5
19	1	3	19	8
			21	2

What does dataset BOTH look like in the following programs?

a.
```
data both;
  merge one two;
  by id
```

b.
```
data both;
  merge two one;
  by id;
```

c.
```
data both;
  merge one(in=inone)
      two(in=intwo);
  by id;
  if inone = 1 & intwo = 1 then output;
```

d.
```
data both;
  merge two one(drop=y);
  by id;
```

e.
```
data both;
  merge one(rename=(x=z))
      two;
  by id;
```

13.8. A DATA step begins with the following statements:

```
data subset;
merge a(in=in_a)
    b(in=in_b)
    c(in=in_c);
by sector;
```

Write IF statements to output observations to SUBSET under the following conditions:

a. The observation's value of SECTOR is in each dataset being merged.

b. The SECTOR value is found in dataset A but not in datasets B or C.

c. The SECTOR value is found in datasets A and B but not in dataset C.

d. The SECTOR value is found in exactly two of the three datasets being merged.

e. The SECTOR value is found in one and only one dataset being merged.

13.9. What, if anything, is wrong with the following programs? An ellipsis (...) indicates portions of a statement that are not germane to the exercise.

a.
```
data p1;
  set mast;
  if sector = 'a4' then output;
```

```
data p2;
set m2;
if sector = 29 then output;

data sects;
merge p1 p2;
by sector;
```
b.
```
data one;
infile ... ;
length type 2;
input lab grp m1-m50;
run;

data two;
infile ... ;
input lab grp type mx1-mx10;
run;

data merge;
merge two one;
by lab grp;
run;
```
c.
```
data intersec;
merge q1(in=q1) q2(in=q2);
by cusip;
if q2 = 1 & q1 = 0 then do;
   rate = ((q2-q1) / q1) * 100;
   output;
   end;
run;
```

13.10. Refer to Exhibit 13.19. Extend the example to include four years of data, and output only observations that were *not* present in each year (you might also want to rename the dataset, since ALL is not an accurate name).

13.11. Refer to the description of the PARKPRES and PRES datasets in Appendix E. We want to add PARKPRES variables _PRESPRK and _PRESACR to PRES, and do *not* want PARKPRES's value of PARTY to be present in the output dataset. Write the program to merge the datasets. Consider the variable common to both datasets—a president's name. Since this is the only link available in the datasets, you have no choice but to use it when merging. What potential problem(s) exist when using character data as a BY variable? If you had the chance to redesign the datasets, what would you do to alleviate the problem?

13.6   Updating

Update a dataset when you want to avoid the left-to-right overwriting of like-named variable values performed by the one-to-one and matched merges. Updating replaces only nonmissing values in the left-hand, or master, dataset.

Syntax

Updating requires the UPDATE statement:

```
UPDATE [fileref.]master[(IN=var other_opts)]
       [fileref.]transac[(IN=var other_opts)] ;
BY var_list ;
```

Some common applications of the UPDATE statement follow:

```
update input.master transac;
by sic2 sic5;

update old(in=inold) changes(in=inchg);
by ssn;
```

The UPDATE statement has three elements:

1. The *fileref*, *IN=var*, and *other_opts* have the same meaning as in the SET and MERGE statements.

2. The *master* indicates the dataset that will be updated by values in *transac*.

3. The *var_list* specifies the variables by which the two datasets will be joined.

Usage Notes

Update is simply a specialized merge. Keep in mind the following differences between updating and merging:

- Matched merge features and rules, such as variable order in the output dataset, similar data types for identically named variables, and identical sort order, also hold for UPDATE.

- Although virtually any number of datasets may be MERGED, only two may be used in the UPDATE dataset list.

- Duplicate BY-group values are allowed in the MERGE datasets. In UPDATE, only the second, or transaction, dataset may have duplicates.

- If a variable is present in more than one MERGE dataset, the value used in the observation being built is the one found in the rightmost dataset. This blanket rule allows missing values to overwrite nonmissing ones when merging datasets. By contrast, updating allows overwriting of the master by the transaction only if the transaction is not missing.

- The transaction dataset may have several observations with identical BY-group level(s). In this case, nonmissing values of a variable are continually replaced as the BY-group is being read. The actual updating of the master takes place at the last observation of the transaction dataset's BY-group level. The effect is that values are updated "down" through the transaction dataset's BY-group before they move "across" to update the master dataset. See Exhibit 13.21 (p. 272) for details.

Examples

Key features and options used in this section are summarized in the following list:

Exhibit	Features/Options
13.20	Update with no duplicate BY values in transaction dataset
13.21	Update with multiple transactions per BY-group

Exhibit 13.20 shows account updating with no duplicate BY-variable values in the transaction dataset. UPDATES replaces values of BALANCE and CR_AVAIL and adds a new variable, AC_TYPE. UPDATES also adds a new observation (ACCOUNT=230). Thus the update replaces values in the original data matrix, adds a column, and adds a row.

Exhibit 13.21 extends Exhibit 13.20 by supplying multiple transactions for two accounts (150 and 190). The first transaction for account 150 replaces BAL-

---

EXHIBIT 13.20   Update with no duplicate BY values in transaction dataset

---

```
------------------- Input datasets -------------------

Master dataset                          Updates

ACCOUNT   BALANCE   CR_AVAIL     ACCOUNT   BALANCE   CR_AVAIL   AC_TYPE
  121       850        y           121        .                  pers
  140      -2120       y           140        .         n        corp
  150       1540       y           150       2500
  190      -2000       n           190        .         y        corp
  210       3900       n
  220        80        y           230       3200       y        pers


------------------- Program listing -------------------

Data MASTER;
update master updates;
by account;
run;

------------------- Output listing -------------------

ACCOUNT   BALANCE   CR_AVAIL   AC_TYPE
  121       850        y         pers
  140      -2120       n         corp
  150       2500       y
  190      -2000       y         corp
  210       3900       n
  220        80        y
  230       3200       y         pers
```

ANCE, the second AC_TYPE. The values in the new MASTER reflect the cumulative changes. Account 190 undergoes a similar treatment: BALANCE, CR_AVAIL, and AC_TYPE are updated in the first transaction, while the second replaces the first's value of BALANCE. The final values in the new dataset contain the new (effectively twice-updated) BALANCE and other values from UPDATES.

Exercises

13.12. Datasets ONE and TWO look like this:

Dataset ONE			Dataset TWO	
ID	X	Y	ID	X
10	.	a4	10	.
20	.	b4	10	8
25	21	lg	25	15
30	8	lg	35	21
			35	.

What does dataset BOTH look like in the following programs?

a. ```
data both;
   update one two;
   by id;
```

b. ```
data both;
   update one(in=one)
      two(in=two);
   by id;
   if two = 1 & one = 0 then delete;
```

13.13. What, if anything, is wrong with the following programs?

a. ```
data new;
   update one two three;
   by id;
```

b. ```
data new;
   update transac master;
   by id;
```

EXHIBIT 13.21   Update with multiple transactions per BY-group

---

```
-------------------- Input datasets --------------------
```

Master dataset   Updates

ACCOUNT	BALANCE	CR_AVAIL		ACCOUNT	BALANCE	CR_AVAIL	AC_TYPE
121	850	y		121	.		pers
140	-2120	y		140	.	n	corp
150	1540	y		150	2500		
				150	.		corp
190	-2000	n		190	-3500	y	corp
				190	-3850		
210	3900	n					
220	80	y					
				230	3200	y	pers

Multiple observations with
BY-group variable ACCOUNT

```
-------------------- Program listing --------------------
```

```
Data MASTER;
merge master updates;
by account;
run;
```

```
-------------------- Output listing --------------------
```

ACCOUNT	BALANCE	CR_AVAIL	AC_TYPE
121	850	y	pers
140	-2120	n	corp
150	2500	y	corp
190	-3850	y	corp
210	3900	n	
220	80	y	
230	3200	y	pers

Last nonmissing values of
multiple transaction groups
replace MASTER values.

13.7   Common Errors

Dataset combining does not always proceed according to plan. This section illustrates typical errors encountered during combining, particularly when merging and updating. In each case the datasets being combined, the Log's error messages, and the output dataset, if produced, are presented.

Dataset Not Sorted

In Exhibit 13.22 the datasets are not sorted by the common (BY) variable, ID. The Log message identifies the offending dataset. Notice that the creation of the MERGED dataset stops at the last successful observation: a dataset is produced despite the error.

EXHIBIT 13.22   Datasets not in required sort order

```
------------------- Input datasets -------------------

ONE                   TWO

ID      V1            ID      V2
 1      1.1            1      2.1
 2      1.2            3      2.3
 3      1.3            2      2.2


------------------- SAS Log -------------------------

   25    data merged;
   26    merge one two;
   27    by id;
   28    run;
ERROR: BY variables are not properly sorted on data set WORK.TWO.
NOTE: The data set WORK.MERGED has 2 observations and 3 variables.

------------------- Output dataset -------------------

MERGED

OBS   ID    V1     V2
 1     1    1.1    2.1
 2     2    1.2    .
```

Missing BY variable

In Exhibit 13.23 dataset TWO's observations can be matched to ONE's by variable IDENT, but the BY statement indicates a common variable named ID. The merge could proceed successfully if the RENAME dataset option were used for dataset TWO in the MERGE statement.

Conflicting Data Types

In Exhibit 13.24 STATE was read as numeric in ONE and as character (U.S. postal code) in TWO. Even though they represent the same entity (a U.S. state or territory), they cannot be used to join the datasets since they are different data types.

EXHIBIT 13.23   Dataset missing a BY variable

```
------------------- Input datasets --------------------

ONE                        TWO

ID     V1                  IDENT     V2
 1     1.1                    1      2.1
 2     1.2                    3      2.3
 3     1.3                    2      2.2


------------------- SAS Log ----------------------------

   25     data merged;
   26     merge one two;
   27     by id;
   28     run;
ERROR: BY variable ID is not on input data set WORK.TWO.
NOTE: The SAS System stopped processing this step because of errors.
NOTE: The data set WORK.MERGED has 0 observations and 4 variables.
```

EXHIBIT 13.24   Variable with conflicting data types: numeric in ONE, character in TWO

```
------------------- Input datasets --------------------

ONE                        TWO

STATE     V1               STATE     V2
  1       1.1                ak      2.1
  2       1.2                al      2.3
  3       1.3                ar      2.2


------------------- SAS Log ----------------------------

   25     data merged;
   26     merge one two;
ERROR: Variable STATE has been defined as both character and numeric.
   27     by state;
   28     run;
NOTE: The SAS System stopped processing this step because of errors.
NOTE: The data set WORK.MERGED has 0 observations and 3 variables.
```

Repeating BY Groups

In Exhibit 13.25 the datasets have an instance (ID=1) where *both* datasets have two or more observations with duplicate BY-variable values. A message is printed in the Log noting the duplication. The message is simply a helpful reminder, not a warning or an error.

Duplicates in an UPDATE Master

UPDATE has a special requirement of uniquely identified, nonduplicate master dataset observations. The master in Exhibit 13.26 (see p. 276) has two pairs of duplicate ACCOUNT numbers (150 and 210). An error message and a listing of

EXHIBIT 13.25   MERGE with multiple observations in a BY group

```
------------------- Input datasets --------------------

ONE                     TWO

ID     V1               ID     V2
 1     1.11              1     2.11
 1     1.12              1     2.12
 2     1.20              2     2.20
 3     1.30              3     2.30


------------------- SAS Log -----------------------------

  27     data merged;
  28     merge one two;
  29     by id;
  30     run;
NOTE: MERGE statement has more than one data set with repeats of BY values.
NOTE: The data set WORK.MERGED has 4 observations and 3 variables.

------------------- Output listing --------------------

MERGED

ID     V1     V2
 1     1.11   2.11
 1     1.12   2.12
 2     1.20   2.20
 3     1.30   2.30
```

variable values in the offending observations is printed. The new, replaced MASTER dataset reflects the changed value from UPDATE only in the first ID=150. Results are difficult to interpret and unreliable for multiple BY-group variables in the transaction dataset in these situations. It is always advisable to correct the BY-variable duplication in the master dataset.

Exercises

13.14. What, if anything, is wrong with the following programs?

a.
```
data both;
   merge one(rename=(st=stcode))
         two(rename=(state=st));
   by stcode;
```

b.
```
data both;
   merge a b;
   by c notsorted;
```

c.
```
data newmast;
   update master master;
   by id reclevel;
```

d.
```
proc sort data=a;
   by descending id;
```

```
                            proc sort data=b;
                            by id;

                            data both;
                            merge a b;
                        e.  data one;
                            infile ... ;
                            input batch $5. ... ;

                            proc print data=two;
                            format batch comma5.;

                            data both;
                            merge one two;
                            by batch;
```

▬▬▬▬▬▬

EXHIBIT 13.26   UPDATE with duplicate BY values in master dataset

```
------------------- Input datasets --------------------

Master dataset                 Updates

ACCOUNT   BALANCE   CR_AVAIL    ACCOUNT   BALANCE   CR_AVAIL   AC_TYPE
  121       850       y           121        .                 pers
  140      -2120      y           140        .         n        corp
  150       1540      y           150       2500
  150       2590      y
  190      -2000      n           190        .         y        corp
  210       3900      n
  210       3000      n
  220        80       y
                                  230       3200       y        pers

-------------------- SAS Log ----------------------------

   37    data master;
   38    update master updates;
   39    by account;
   40    run;
ERROR: The MASTER data set contains more than one observation for a
BY group.
ACCOUNT=150 BALANCE=2500 CR_AVAIL=y AC_TYPE=  FIRST.ACCOUNT=1
LAST.ACCOUNT=0
_ERROR_=1 _N_=3
ERROR: The MASTER data set contains more than one observation for a
BY group.
ACCOUNT=210 BALANCE=3900 CR_AVAIL=n AC_TYPE=  FIRST.ACCOUNT=1
LAST.ACCOUNT=1
_ERROR_=1 _N_=6
NOTE: The data set WORK.MASTER has 9 observations and 4 variables.
```

EXHIBIT 13.26 *(continued)*

```
-------------------- Output dataset --------------------

Updated MASTER

ACCOUNT     BALANCE     CR_AVAIL     AC_TYPE
  121          850         y          pers
  140        -2120         n          corp
  150         2500         y
  150         2590         y
  190        -2000         y          corp
  210         3900         n
  210         3000         n
  220           80         y
  230         3200         y          pers
```

Summary

Two or more SAS datasets may be combined in virtually any manner. Chapter 13 reviews five basic methods: concatenation, interleaving, one-to-one merge, matched merge, and updating. Among the factors to consider when selecting a method are the sort order of the datasets, whether there are single or multiple occurrences of key variables, the order of specification of the datasets, the impact of having like-named variables in the datasets, and the desired "shape" (number of observations and variables) of the output dataset. DATA step programming and dataset options may also be used to enhance the power of the dataset combining commands: dataset variables may be dropped, kept, or renamed and special dataset options may be used to identify the source dataset of an observation in the output dataset.

14 User-Written Formats I:
Display

Until now SAS formats have simply been a means to get raw data into SAS datasets (input formats) or from SAS datasets into a more presentable form for output (display or output formats). The formats discussed in Chapters 7 and 11 were those that come with SAS software, the so-called system formats. Although these formats are useful, it is not hard to think of situations in which they might fall short. The ability to define custom, or user-written, formats opens up a vast array of data display and manipulation possibilities. This chapter discusses PROC FORMAT, the mechanism for defining custom formats. The focus is on using custom formats to enhance display. Storing user-written formats in permanent format libraries is explained as well.

14.1 PROC FORMAT and Custom Formats

Why go to the trouble of customizing formats? There are several reasons. First, custom formats take the display of the numbers a step further than most system formats allow. Although a system format could display an income value with commas and a dollar sign, a user-written format could not only display the number that way but also add text such as "above average" or "below poverty level". This capability goes beyond simply "dressing up" the numbers; it can dramatically increase the information value of the display.

Another reason for defining user-written formats is to assist in recoding or regrouping data values. Formats can be used to assign branch offices to districts or test score ranges to achievement levels, and in any other situation that requires collapsing many values into relatively few categories. Using custom formats for recoding and regrouping variables is discussed in Chapter 15.

A user-written format could also perform a one-to-one transformation of the data. For example, in a dataset containing test results, levels of the variable TESTTYPE could be recoded from a value such as "12" to the more expressive text string, "English III". Formats may be used in a DATA step to create a new variable containing a recoded value. They can also be used in a procedure, collapsing variable categories only during the procedure.

Format Types

The two types of user-written formats discussed in this chapter are value and picture. **Value formats** convert one or more user-specified values into a single character string. In most cases the original values of the variable are not preserved when the format is used with a variable. For example, if score values from 0 to 3 were converted to the value "Low" in a PRINT listing, you would not know whether an observation's value was 0, 1, 2, or 3. Likewise, in a FREQ table all values of the score variable from 0 to 3 would contribute to a single frequency category, "Low".

Picture formats will usually preserve variable values. Like value formats, they accept the specification of a value or a range of values. Unlike values, picture formats simply reproduce the value of the variable in a given range rather than convert it to a string. The picture format is, in effect, simply a template, or instructions for how to print the number.

The template specifications are very powerful. They allow you to insert special characters before, within, and following the number. For example, if the value being formatted is a measure of revenue, one range might specify negative numbers. The picture template associated with this range could place the number in parentheses. This is common practice in financial statements. Picture formats are not as useful for regrouping as value formats and are used primarily for enhancing the display of individual observations.

POINTERS AND PITFALLS

User-written formats may be used to collapse multiple levels of a variable into a single category or simply to alter the appearance of a variable when displayed.

Syntax

The FORMAT procedure allows two styles of formats to be defined. This section outlines the syntax of VALUE and PICTURE display formats. The syntax of the FORMAT procedure is described below:

```
PROC FORMAT [LIBRARY=ref] [FMTLIB];
VALUE fmtname[([DEFAULT=dfltc] [MAX=maxc] [MIN=minc])]
      range1 = 'value1'
      [range2 = 'value2' ... ]
      ;
PICTURE fmtname[([DEFAULT=dfltc] [MAX=maxc]
                  [MIN=minc])]
      range1 = 'picture1' [([FILL='fillchar']
                            [PREFIX='prechar']
                            [MULT=scale]
                            [NOEDIT] )]
      [range2 = 'picture2' [(rangeoptions)] ... ]
      ;
```

Some common applications of the FORMAT procedure follow:

```
proc format;
value hilow  1-20 = 'Low' 21-high = 'High';
value grouping  12-17, 20-30, 34, 32 = 'Valid';
                other                = '?????';
picture inc  low -< 0 = '999999)' (prefix='(')
             other    = '999999' ;
value $st  'pa','nj','de','ny',
           'PA','NJ','DE','NY' = 'Mid-Atl'
           other              = 'Other';
```

VALUE and PICTURE formats have several features in common:

- *FMTLIB* requests a listing of the contents of the format library, and *LIBRARY* identifies a *LIBNAME* statement pointing to a permanent format library. Using these options is discussed in Section 14.2. FMTLIB is available only in version 6 of the SAS System.

- The *fmtname* assigns a name to the format. If the format will be associated with a character variable, *fmtname* must begin with a dollar sign ($). The name can be a maximum of eight characters; can consist of numerals (0 to 9) and alphabetic characters (a to z); cannot end with a number; and should not duplicate a SAS System format, procedure, or function name.

- *DEFAULT, MAX,* and *MIN* control the number of columns used when displaying data. All are optional.

- *DEFAULT* tells SAS to use *dfltc* columns if no width is specified when the format is used. The value must be in the range 1 to 40 columns. If not entered, SAS uses the maximum width of *valuex* or *picturex*.

- *MAX* restricts the format to a maximum of *maxc* columns. The *maxc* cannot exceed 40. If not entered, SAS uses the value assigned to *DEFAULT*.

- *MIN* restricts the format to a minimum of *minc* columns. The *minc* cannot be less than 1. If not entered, SAS uses the value assigned to *DEFAULT*.

- The *rangex* specifies one or more values that will be displayed as either *valuex* or *picturex*. The values in the range can be explicit or implied. Examples of implied ranges include 1–10 (values 1 through 10), 1<–10 (greater than 1 up through 10), and 1-<10 (1 up to but not including 10). Explicit and implied ranges may be combined in a single range specification. If the VALUE statement is formatting character variables, *rangex* must be enclosed by a pair of single quotes (') or quotation marks ("). Values in the different ranges must not overlap: a value can appear only once in all the ranges. If a value does not appear in the ranges, it will be displayed or used in a group as is, unformatted. The special values LOW, HIGH, and OTHER indicate the lowest nonmissing value, highest value, and values not elsewhere specified (including missing values). The special values do not need to be enclosed in single quotes or quotes if they are being used with a character format. These rules are illustrated in the examples later in this section.

A feature unique to VALUE formats is that *valuen* is the character string that will be used to display the values in the corresponding *rangen*. Single quotes (') in the text should be represented by a pair of single quotes (' ') if *valuen* itself is enclosed in single quotes. The *valuen* may be a maximum of 40 characters.

PICTURE formats have the following unique features:

- The *picturex* defines the template that will be laid over the numbers defined in *rangex*. The picture cannot exceed 24 characters and must begin with numbers indicating the placement of the formatted number. Using 1 through 9 tells SAS to print leading zeros, while using 0 prints leading zeros as blanks. Text may follow the numbers to enhance the explanatory value of the picture. Single quotes (') in the text must be represented by a pair of single quotes ('') if *picturex* is enclosed in single quotes.

- The *FILL, PREFIX, MULT,* and *NOEDIT* options may change from range to range. These give further control to the shape and form of the formatted values.

- *FILL* instructs SAS to insert *fillchar* to the left of the formatted value if it is less than the default width (indicated by the DEFAULT option). The default *fillchar* is a blank.

- *PREFIX* tells SAS to display *prechar* to the left of the formatted value, space permitting. The *prechar* can be one or two characters long. The default *prechar* is blanks.

- *MULT* multiplies the value in *range* by *scale*. The result is formatted. Default is copying the number as is (no multiplying). This option is useful when you do not want to alter large numbers in the dataset but do want to print them as, say, millions of units. The *mult* performs the required multiplication without having to create new variables in a DATA step.

- *NOEDIT* tells SAS to print *picturex* without translating numbers into formatted values.

Usage Notes

User-written formats are a powerful and unique feature of the SAS System. Keep in mind the following points about the types and definitions of user-written formats:

- Subject to the limitations of the computer system, there is no limit to the number of VALUE and PICTURE formats that may be defined in a single PROC FORMAT.

- User-written formats may be used in DATA step FORMAT statements. Remember that if the dataset is stored permanently you must have the format available any time the dataset is used, since the format name is stored as part of the permanent SAS dataset. Section 14.2 discusses how to store and reuse formats permanently.

- Once defined with FORMAT, the user-written format is used like any SAS System format. The only difference between user-written and system formats is that the width specification for user formats need not be specified if the default width is acceptable.

- The PICTURE format's range options and ability to display both text and numbers give it a display advantage over VALUE formats. Consider the merits of both forms on a case by case basis: do not get into the habit of automatically using one or the other.

- Although SAS has a set of rules for dealing with overlapping ranges (e.g., 0–2 in one range and 2–5 in the next), it is best to avoid their use and to be clear about which values belong in which range (e.g., 0–<2 and 2–5 or 0–2 and 2<–5).

- The catchall *other* range is usually placed at the end of the series of range specifications. Although SAS will interpret the ranges correctly if *other* is placed elsewhere, its location as the last in the series is a good habit to develop (think of it as the unconditional ELSE ending the series of IF-THEN-ELSE statements specifying the ranges).

- SAS does not balk at entering several range and text specifications per line. Even so, try to use one line per range. This makes the procedure easier to read and less susceptible to coding and proofreading errors.

Exercises

14.1. Write VALUE and PICTURE statements to meet the following requirements (you can specify the name of the format):

a. Negative values are displayed as (xx,xxx.). Positive values look like xx,xxx.

b. Values "A" through "J" are "Part one." Other letters are "Part two." Both upper- and lowercase letters are grouped.

c. Numeric values for the format are scaled by a factor of .01. Values less than 100 and greater than 0 are grouped as "Low," others are "Other."

d. All values are preceded by the string '->'. Spaces not used by the numbers are filled with the character '*'. The maximum value is 1 million. Insert commas when displaying.

e. Negative values display as "deficit," positive as "surplus," and zero as "balanced."

f. Values range from 0 to 30. Define three equal-sized ranges, each one represented by a bar of asterisks (*) indicating the relative size of the bar and the actual number of the variable. Thus 25 might be displayed as:

```
25 ******
```

while 15 might be shown as:

```
15 ******
```

14.2. The following formats are defined by the FORMAT procedure.

```
value samplea -1 -< 0 = 'negative'
               0 <-  1 = 'up to 1'
               1 - high = 'high'     ;
value sampleb -1 -< 0 = 999 (neg.)'
               0 <-  1 = 'up to 1'
               1 - high = '9999 (high)'   ;
```

a. What values are printed when you use the formats with variable SAMPLE and it has the following values?

```
-2  .  0  1  .5  2
```

b. Write a FORMAT statement that associates format SAMPLEA with variable SAMPLE and format SAMPLEB with SPLITSMP.

14.3. What, if anything, is wrong with the following VALUE and PICTURE statements? (Look for logical as well as syntactical problems.)

a.
```
value $grp. 'a-z' = 'lower case'
            'A-Z' = 'upper case'
            'other = 'numeric' ;
```

b.
```
picture other = '999,999 (upper income, 1990)'
        low -< 0 = '99999 (deficit)' ;
```

c.
```
picture $group 'a' - 'k' = 'low'
               'l' - 'z' = 'high' ;
```

d.
```
value $grp = ' ';
```

Examples

This section presents a complete example using a small test score dataset. Although the data are hypothetical, the format development process is realistic: we write the formats once, see how they might be deficient, then revise them. The examples at the end of the section illustrate some of the other capabilities of both VALUE and PICTURE formats.

Test Score Example. The test score dataset is presented in Exhibit 14.1. The user-written formats are presented in Exhibit 14.2. Format GRADE converts

EXHIBIT 14.1   Test score dataset

Ident	Grade	Gender	Special Ed. Class	Family Income	Verb1	Verb2	Math1	Math2
1102	6	F	N	29000	21	.	20	.
1932	8	F	n	32500	18	18	22	23
1108	7	M	Y	22000	19	18	20	21
1241	9	F	Y	21000	22	23	20	20
1200	10	M	N	52000	20	.	18	.
1498	11	F	N	48700	18	18	17	20
980	11	F	y	28700	19	.	21	.
1105	10	M	N	34500	21	.	20	.
1201	9	M		26200	18	19	19	19

numeric values 7 through 9 to "Jr High", and values 10 through 12 to "Sr High". If values outside this combined range of 7 through 12 were found in the dataset, they would be displayed as is, unformatted.

$GENDER formats a character variable, converting values of F and M to "Female" and "Male". $SPECIAL formats Y and N values to character strings indicating not only the inherent yes/no content but also what the yes and no pertain to (special education classes).

INC is a picture format displaying values through 30000 in a format such as "$15.7 (below avg)". Other values are displayed in a format such as "45.0 (at, above avg)". The advantage of a picture format for display purposes is evident here: the individual value is discernable, and the value is annotated by its status relative to the rest of the population ("below" or "at, above" average).

Finally, the SCORE format defines a range of 15 up to but not including 20 as "Pass" and 21 through high as "Honors". The lack of logical closure in the IF-THEN-ELSE discussion in Chapter 8 is revisited here. It is discussed and remedied in later examples.

EXHIBIT 14.2   User-written numeric and character formats

```
proc format;
value grade      7 -  9 = 'Jr High'
                10 - 12 = 'Sr High' ;
value $gender   'F' = 'Female'
                'M' = 'Male' ;
value $special  'N' = 'No special ed.'
                'Y' = '1 or more classes' ;
picture inc     low - 30000 = '99.9k (below avg)'
                                (prefix='$' mult=.01)
                30000 <- high = '99.9k (at, above avg)'
                                (prefix='$' mult=.01) ;
value score     15 -< 20   = 'Pass'
                21 - high   = 'Honors' ;
run;
```

The formats are now used in a PRINT procedure. The PROC statements and output are shown in Exhibit 14.3. A few items in Exhibit 14.3 need correcting:

- A GRADE value of 6 was displayed as is. It was not in any of the range specifications of the GRADE format.
- SPEC_ED values "n" and "y" were not caught by the "$special" format, which formatted only uppercase values.
- Score values of 20 and missing (.) were displayed as is. The 20 was not formatted because there was a "boundary problem": the first range ("Pass") went up to but did not include 20, and the next range ("Honors") started at 21.

EXHIBIT 14.3   First use of user formats with test score data

```
------------------- Program listing --------------------

proc print;
id ident;
format grade grade. gender $gender. spec_ed $special. faminc inc.
       verb1--math2 score. ;
title 'First display with user-written formats';
run;

------------------- Output listing ---------------------

First display with user-written formats

IDENT    GRADE      GENDER      SPEC_ED                     FAMINC
1102         6      Female    No special ed.       $29.0k (below avg)

1932     Jr High    Female     n                    32.5k (at, above avg)
1108     Jr High    Male      1 or more classes     $22.0k (below avg)
1241     Jr High    Female    1 or more classes     $21.0k (below avg)
1200     Sr High    Male      No special ed.        52.0k (at, above avg)
1498     Sr High    Female    No special ed.        48.7k (at, above avg)
 980     Sr High    Female     y                     $28.7k (below avg)

1105     Sr High    Male      No special ed.        34.5k (at, above avg)
1201     Jr High    Male                             $28.2k (below avg)

IDENT    VERB1     VERB2     MATH1     MATH2
1102     Honors      .        20        .
1932     Pass      Pass      Honors    Honors
1108     Pass      Pass       20       Honors

1241     Honors    Honors     20        20

1200       20        .       Pass       .

1498     Pass      Pass      Pass       20

 980     Pass        .       Honors     .

1105     Honors      .        20        .

1201     Pass      Pass      Pass      Pass
```

Values in boxes were not caught by the user-written formats. They are displayed unformatted.

EXHIBIT 14.4   Revised user formats for test score data

```
proc format;
/* Change from VALUE to PICTURE */
picture grade    7 -  9 = 'Jr High'
                 10 - 12 = 'Sr High'
                 other   = '99 <-- Invalid?';  /* extra category */
/* add lowercase */
value $gender    'F','f' = 'Female'
                 'M','m' = 'Male' ;
/* add lowercase */
value $special   'N','n' = 'No special ed.'
                 'Y','y' = '1 or more classes'
                 /* add catchall category */
                 other   = '** unknown **' ;
picture inc      low - 30000 = '99.9k (below avg)'
                               (prefix='$' mult=.01)
                 30000 <- high = '99.9k (at, above avg)'
                                 (prefix='$' mult=.01) ;
value score      15 -< 20  = 'Pass'
                 /* change boundary condition from 21 to 20 */
                 20 - high  = 'Honors'
                 /* add catchall category */
                 .          = '<didn''t take>' ;
run;
```

After some reflection on what the out-of-range and missing values mean, we revise the formats as shown in Exhibit 14.4. Notice the following changes:

- GRADE is changed to a PICTURE format. This allows acceptable values to be printed as simple strings and others to have a message *and* the value to be displayed. This data and message capability was not possible with the VALUE format.

- Lowercase values were added to the ranges in both the $SPECIAL and $GENDER formats.

- Unacceptable, including missing, values in the $SPECIAL format are now included in an OTHER range.

- SCORE format's second range is changed, and missing values are printed as "<didn't take>".

The PRINT statements are essentially unchanged. The output is shown in Exhibit 14.5.

Miscellaneous Examples.   This section presents VALUE and PICTURE statements that give an idea of the variety of user-written format applications.

Exhibit 14.6 is in simple debit-credit format. Leading zeros will not be printed since 0's are used for digit indicators. Commas will be inserted.

This next example illustrates character ranges. SAS knows that a value of "B" is "less than" one of "C", "D", and so on. (Recall the discussion of character variable/constant magnitude in Chapter 7. Also see Appendix A for a list of how different systems compare character values.) Even though it does not make sense arithmetically, you can specify ranges for characters as for numbers:

EXHIBIT 14.5   Print using modified, corrected formats

```
------------------- Program listing -------------------

proc print;
id ident;
format grade grade. gender $gender. spec_ed $special. faminc inc.
       verb1--math2 score. ;
title 'First display with user-written formats';
run;

------------------- Output listing -------------------
```

Revised display

IDENT	GRADE	GENDER	SPEC_ED	FAMINC
1102	06 <-- Invalid?	Female	No special ed.	$29.0k (below avg)
1932	Jr High	Female	No special ed.	32.5k (at, above avg)
1108	Jr High	Male	1 or more classes	$22.0k (below avg)
1241	Jr High	Female	1 or more classes	$21.0k (below avg)
1200	Sr High	Male	No special ed.	52.0k (at, above avg)
1498	Sr High	Female	No special ed.	48.7k (at, above avg)
980	Sr High	Female	1 or more classes	$28.7k (below avg)
1105	Sr High	Male	No special ed.	34.5k (at, above avg)
1201	Jr High	Male	** unknown **	$26.2k (below avg)

IDENT	VERB1	VERB2	MATH1	MATH2
1102	Honors	<didn't take>	Honors	<didn't take>
1932	Pass	Pass	Honors	Honors
1108	Pass	Pass	Honors	Honors
1241	Honors	Honors	Honors	Honors
1200	Honors	<didn't take>	Pass	<didn't take>
1498	Pass	Pass	Pass	Honors
980	Pass	<didn't take>	Honors	<didn't take>
1105	Honors	<didn't take>	Honors	<didn't take>
1201	Pass	Pass	Pass	Pass

> Values in boxes were formatted incorrectly in Exhibit 14.3. The changes to the formats in Exhibit 14.4 produce this new, correct display.

```
value $group 'a'-'m', 'A'-'M' = '1st'
             'n'-'z', 'N'-'Z' = '2nd' ;
```

The next example displays a rate between 0 and 1, inclusive, as a percentage:

```
picture ratepct other = '009.99%' (mult=10000) ;
```

Since it is the only category, OTHER is all-inclusive, executed for every value. The value of the MULT option depends on the number of decimal places used in the picture. The multiplier is large enough so that when the product is overlaid on the picture's template, the decimal places will be aligned correctly. For example, a value of .0123 would be multiplied by 10,000 so that the result

EXHIBIT 14.6   PICTURE format to display CR/DR suffix

```
------------------- Program listing --------------------

proc format;
picture drcr low-< 0 = '00,009.99DR'
             0-high  = '00,009.99CR'
             ;
run;

------------------- Results ------------------------

Unformatted         Formatted
  Value               Value
-30001.00          30,001.00DR
 -5876.00           5,876.00DR
  -800.00             800.00DR
   -44.00              44.00DR
    -4.00               4.00DR
    -0.54               0.54DR
     0.00               0.00CR
     0.65               0.65CR
     5.00               5.00CR
    25.00              25.00CR
   280.00             280.00CR
  2800.00           2,800.00CR
 45678.00          45,678.00CR
```

(123) would be placed over the 009.99 template correctly (1.23). Extending this logic, if we wanted integer percents, MULT's value would be 100.

In the last example, the 10-digit variable PHONENUM contains a phone number. The first three digits are the area code, the last seven are the local number (beginning with a three-digit exchange). The picture format PHONE places the area code in parentheses and inserts a hyphen after the exchange.

```
picture phone other='999)999-9999' prefix='(';
```

14.2   Permanent Format Libraries

Recall Chapter 10's discussion of the merits of using permanent SAS datasets rather than constantly rereading raw data. When using a SAS dataset, you can move data into procedures faster, can easily share the data among different programs, and can execute programs faster since you bypass the inherently slow process of moving and translating raw data.

Some of these same advantages can be realized by creating permanent format libraries. Rather than going through the tedious process of defining and storing the format for the duration of a SAS session, you can define them once, store them in a format library, and use them in later runs by making only a single addition to the program.

> **POINTERS AND PITFALLS**
>
> Permanent format libraries may be used in later SAS sessions without having to rerun the FORMAT procedure. The FORMAT procedure is placed in a separate program file and altered and rerun as needed.

Although you can obtain the same results by defining the user formats from scratch in every job, it is worth the minimal extra effort required to use libraries: programs are shorter since they do not have to have all the lines for PROC FORMAT. More important, the same format definition can be reliably shared between different programs. This means, for example, that the REGION display in one program will be identical in meaning to the REGION in another. There is no possibility that a value of "Midwest" formatted by REGION could include different states in different programs. If the REGION format's definition changes in the central format library, it is automatically picked up by all programs using it.

This section discusses how to develop and use formats in a microcomputer environment. The process is slightly different for each major environment that SAS supports. The development and use of formats for other environments is presented in Appendix D. Even if you are not running on a microcomputer, skim this section for the underlying concepts of permanent libraries.

Creating the Format Library

So far the examples in this book have used temporary formats; that is, they were available only for the duration of the current SAS session. The FORMAT procedure wrote the user formats to a special library automatically created when SAS was started. This library was deleted when the session terminated.

To make the library permanent, FORMAT must be told where to keep it. This is done via the LIBRARY option in the PROC statement (see the syntax description in Section 14.1 for details). LIBRARY identifies the location of the format library. The path can be used exclusively for format libraries, or it can be any mixture of SAS programs, datasets, and raw data.

> **POINTERS AND PITFALLS**
>
> The special LIBNAME of LIBRARY identifies a path containing the SAS format library. This is one of the few special, reserved LIBNAMES used by the SAS System.

If the SAS job is being run only to create the library, *ref* can be any valid name not reserved by SAS (see Chapter 10 for reserved names). If you are creating the permanent library and using it later in the session, use the special name LIBRARY. No changes to the VALUE and PICTURE statements are required.

A permanent format library can be created using the formats in the examples at the end of Section 14.1. The Log is shown in Exhibit 14.7.

Using the Format Library

To use the format library created in Exhibit 14.7, add a LIBNAME statement using the special *ref* of LIBRARY. The LIBNAME statement should precede any DATA step or PROC using the user formats. Its path must point to the same directory used when the formats were created. When SAS detects a format reference in a session, it first looks for the format in the path indicated by LIBRARY. If it cannot

EXHIBIT 14.7   Create a permanent library of user-written formats

```
-------------------- SAS Log ---------------------------

    1       libname library 'c:\scores';
    2
    3       proc format library=library;
    4       picture grade     7 -  9 = 'Jr High'
    5                        10 - 12 = 'Sr High'
    6                        other   = '99 <-- Invalid?';
NOTE: Format GRADE has been output.
    7       value $gender   'F','f' = 'Female'
    8                       'M','m' = 'Male' ;
NOTE: Format $GENDER has been output.
    9       value $special  'N','n' = 'No special ed.'
   10                       'Y','y' = '1 or more classes'
   11                        other  = '** unknown **' ;
NOTE: Format $SPECIAL has been output.
   12       picture inc     low - 30000 = '99.9k (below avg)'
   13                                     (prefix='$' mult=.01)
   14                       30000 <- high = '99.9k (at, above avg)'
   15                                     (prefix='$' mult=.01) ;
NOTE: Format INC has been output.
   16       value score     15 -< 20   = 'Pass'
   17                       20 - high   = 'Honors'
   18                        .          = '<didn''t take>' ;
NOTE: Format SCORE has been output.
   19       run;
NOTE: The PROCEDURE FORMAT used 19.00 seconds.
```

find the format in the LIBRARY path, SAS looks in its default, system library paths. If it is not made aware of a path explicitly named LIBRARY, SAS will not be able to locate the user-written formats! Remember that this format search logic description applies to SAS on IBM-compatible microcomputers. The default search path is different when running on an IBM mainframe. In all environments, however, the effect is the same: if you do not point to the user-written format library, it cannot be used by SAS.

Exhibit 14.8 uses some of the formats created in Exhibit 14.7. The FORMATS procedure is replaced by a reference (the LIBNAME) to a format library.

EXHIBIT 14.8   Use a permanent format library

```
libname library 'c:\scores';

<<< DATA step omitted >>>

proc print data=scores;
id ident;
format grade grade. gender $gender. spec_ed $special. faminc inc.
      verb1--math2 score. ;
title 'Revised display';
run;
```

Displaying Library Contents: FMTLIB

Just as the CONTENTS procedure and PROC DATASETS aid your recall of a SAS dataset's contents, the FMTLIB option in the PROC FORMAT statement outlines the names, ranges, and options of format libraries. Exhibit 14.9 illustrates FMTLIB's output when run on the library created by Exhibit 14.7. The FMTLIB option is available only in version 6 of the SAS System.

EXHIBIT 14.9 Output produced by FMTLIB option in PROC FORMAT

```
------------------ Output listing --------------------


---------------------------------------------------------------
|       FORMAT NAME: GRADE     LENGTH:    15    NUMBER OF VALUES:    3      |
|  MIN LENGTH:   1  MAX LENGTH:  40  DEFAULT LENGTH  15  FUZZ: STD         |
|-------------------------------------------------------------------------|
|START           |END              |LABEL  (VER. 6.03    04JAN90:20:30:14) |
|----------------+-----------------+--------------------------------------|
|              7|               9|Jr High           P   F   M1           |
|             10|              12|Sr High           P   F   M1           |
|OTHER           |OTHER            |99 <-- Invalid?    P   F   M1           |
---------------------------------------------------------------


---------------------------------------------------------------
|       FORMAT NAME: INC      LENGTH:    21    NUMBER OF VALUES:    2      |
|  MIN LENGTH:   1  MAX LENGTH:  40  DEFAULT LENGTH  21  FUZZ: STD         |
|-------------------------------------------------------------------------|
|START           |END              |LABEL  (VER. 6.03    04JAN90:20:30:18) |
|----------------+-----------------+--------------------------------------|
|LOW             |           30000|99.9k (below avg    P$  F   M0.01       |
|----------------+-----------------+--------------------------------------|
|      30000<HIGH              |99.9k (at, above    P$  F   M0.01       |
---------------------------------------------------------------


---------------------------------------------------------------
|       FORMAT NAME: SCORE    LENGTH:    13    NUMBER OF VALUES:    3      |
|  MIN LENGTH:   1  MAX LENGTH:  13  DEFAULT LENGTH  13  FUZZ: STD         |
|-------------------------------------------------------------------------|
|START           |END              |LABEL  (VER. 6.03    04JAN90:20:30:19) |
|----------------+-----------------+--------------------------------------|
|              .|                .|<didn't take>                         |
|             15|             20<Pass                                     |
|             20|HIGH             |Honors                                |
---------------------------------------------------------------


---------------------------------------------------------------
|       FORMAT NAME: $GENDER LENGTH:     6   NUMBER OF VALUES:     4      |
|  MIN LENGTH:   1  MAX LENGTH:   6  DEFAULT LENGTH   6  FUZZ: STD         |
|-------------------------------------------------------------------------|
|START           |END              |LABEL  (VER. 6.03    04JAN90:20:30:16) |
|----------------+-----------------+--------------------------------------|
|F               |F                |Female                                |
|M               |M                |Male                                  |
|f               |f                |Female                                |
|n               |n                |Male                                  |
---------------------------------------------------------------
```

EXHIBIT 14.9   *(continued)*

```
-----------------------------------------------------------------
|       FORMAT NAME: $SPECIAL LENGTH:   17   NUMBER OF VALUES:     5     |
|   MIN LENGTH:   1  MAX LENGTH:  17  DEFAULT LENGTH  17  FUZZ: STD      |
|---------------------------------------------------------------|
|START           |END            |LABEL  (VER. 6.03     04JAN90:20:30:17)  |
|----------------+---------------+--------------------------------------|
|N               |N             |No special ed.                        |
|Y               |Y             |1 or more classes                     |
|n               |n             |No special ed.                        |
|y               |y             |1 or more classes                     |
|OTHER           |OTHER         |** unknown **                         |
-----------------------------------------------------------------
```

Exercises

14.4. Select one of the formats created in Exhibit 14.4. If you have formats in mind for your own work, so much the better—write the required VALUE and/or PICTURE statements.

  a. Store the format in a permanent format library on your system or, at a minimum, write the statements to do this.

  b. Run FORMAT using the FMTLIB option to display the ranges and values for your format. Does the output give you a clear understanding of the work the format is doing?

  c. Finally, use the format in a procedure to prove to yourself that it really does work!

Summary

User-written formats are a powerful extension of SAS System formats. They enable the end-user to exercise control over the display of both character and numeric data, greatly extending the power and flexibility of the default, System formats. Creation of the format usually requires specification of one or more values and the single value they will be displayed as. Editing functions such as multiplication, blank filling, and single-character prefixes are available.

Formats may be stored in permanent format libraries. Once in a library, the user-written format may be treated exactly the same as a SAS System format. The contents of the library may be examined with the FMTLIB option in the FORMAT procedure.

15 User-Written Formats II: Recoding

Chapter 14 discussed how to create and use user-written formats. It also illustrated one of their principal uses: to enhance the display of data. This chapter takes their use a step further. It show how formats affect the classification and grouping of data values in some PROCs and the DATA step. The range feature of the VALUE and PICTURE statements can be exploited to collapse many values of a variable into a single category, thus relieving the programmer of large volumes of IF-THEN sequences. Recoding, the other principal use of formats, is also examined. This chapter also discusses why regrouping data values is often desirable, how approaches that do not use user-written formats are inefficient, and how to dramatically reduce the amount of effort required to regroup by using a simple but often overlooked use of formats.

15.1 Whats and Whys of Recoding

The mechanics of recoding the variable are straightforward. One or more levels of the variable are transformed into an equal or, more commonly, smaller number of levels. These new levels can replace the original value of the variable or be used to create a new variable. They may also exist temporarily, for the duration of an IF statement's logical comparison of values or during a PROC's grouping of individual values into groups.

The motivation for performing this transformation is best illustrated with a few examples: you may want to group states into regions, branches into districts, years into eras, subject identification numbers into random groups, fine-grained categories into coarser ones, or one coding scheme into another.

Recoding the Hard Way

For the user armed with the DATA step tools presented so far in this book, the response to the demands of recoding is an IF-THEN-ELSE sequence. Suppose a variable representing store branches needs to be grouped into districts, a district being a collection of stores. The following program segment performs the grouping.

```
if store = 100 | 110 <= store <= 125   then district = '1';
    else if 140 <= store <= 170        then district = '2';
    else if store = 170 | store = 180  then district = '3';
    else if 200 <= store <= 250        then district = '4';
    else                                    district = '5';
```

For cases with comparatively few levels of a grouped variable (DISTRICT) and with simple grouping criteria in the IF statement's conditions, this method is fine. The structure becomes unwieldy when a greater number of district levels and/or many stores must be assigned. There will be many lines of code to enter, and updates and other changes become difficult.

Using IF-THENs to recode is also susceptible to undetected coding errors. Notice that store number 170 is assigned a district number in *two* statements (those assigning values to DISTRICTs 2 and 3). SAS will not complain or detect such an inconsistency even though it will likely cause confusion and errors in later analyses.

Finally, the IF-THEN approach to recoding inhibits sharing the districting scheme with other DATA steps in the same program or other programs processing the same data. As the number of programs requiring the grouping statements increases, it is more likely that changes to the statements may not be applied to all programs using them. Consistency of group definitions and the uniform application of changes to all occurrences of the grouping scheme become unreliable.

POINTERS AND PITFALLS

In most cases, using user-written formats for recoding is more efficient than IF statements. The formats also ensure no duplicate assignment of categories and allow sharing of the grouping scheme among programs.

15.2   DATA Step Recoding

User formats can be used in the DATA step to create new variables or recode existing ones. They can also be used as "transient variables," not actually part of the observation, and calculated as part of a comparison or decision (usually in an IF statement).

The PUT Function

A tool common to both creating new variables and recoding existing ones is the PUT function. PUT complements INPUT: where the INPUT statement *reads* from a raw data file, PUT reverses the process and *writes*, in this case to a character string. Instructions for how to write the variable to the character string are contained in a format. The PUT function's syntax is presented below.

```
PUT(var,fmtw.);
```

The PUT statement has two components:

1. The *var* is a SAS variable, constant, or expression.

2. The *fmt* is a system or user-written format *w* characters long. The *fmt* must agree in data type (character or numeric) with *var*. If *w* is not specified, it will be set to *fmt*'s default length. The *fmt* may be a VALUE, PICTURE, or SAS System format.

It is important to remember that the result of the function, the string written by the PUT function, is *always* a character variable, regardless of *var*'s data type. Any reference to it later in the program must treat it like any other character variable.

POINTERS AND PITFALLS

The PUT function always returns a character value. Use the INPUT function if necessary to convert character values to numeric.

PUT will be used many times in this chapter. Just to get started, though, the grouping of store branches into districts performed "the hard way" in Section 15.1 is rewritten and corrected (Exhibit 15.1). PROC FORMAT defines the format, and then the PUT function carries out the recoding in the DATA step. DISTRICT is a character variable of length one (since one character was the default for format DIST). Had DIST allocated stores into 50 areas instead of just five, the DATA step would remain exactly the same: the work, the allocation to districts, is being done in the format. All the PUT does is communicate, via the DISTRICT variable, the outcome of the formatting.

EXHIBIT 15.1   PUT function used to perform grouping of a numeric variable

---

```
proc format;
value dist 100,110-125 = '1'   140-169 = '2'   170,180 = '3'
           200-250     = '4'   other   = '5'  ;
run;

data redist;
input store;
district = put(store,dist.);
run;
```

Assigning the Recode to a Variable

Exhibit 15.1 demonstrated that the result of the PUT function may be assigned to a variable. A new variable can be created with the PUT function when the variable must be part of the dataset. All that is required is using the PUT function in an assignment statement.

Remember that the new variable will be character. What if a numeric value is required? How can it be converted from character to numeric? The conversion can be done using the INPUT function:

```
INPUT(var,fmtw.);
```

The *var, fmt,* and *w* have the same meaning as in the PUT function. INPUT, however, reverses the process: rather than writing characters from *var* using *fmtw,* the INPUT reads from *var* using *fmtw.* The *var* mimics a field of raw data *w* columns wide but rather than reading from a raw data file, SAS reads from a string of characters in the computer's memory. Thus *w* should be as wide as *var* (e.g., if AREA has length four, read it with a format such as 4). The result, unlike that of PUT, can be either a numeric or character variable since there is no restriction on agreement of variable and format data types. Remember that numeric *fmt* values expect *var* to consist entirely of characters representing numeric data.

Exhibit 15.2 returns to the store districting example in Exhibit 15.1. First the store number is converted into its district (TEMPDIST). Then this character string is read with a numeric format of 1., storing the result DISTRICT. Notice that if a format of $1. were used instead of 1, we would be back where we started: DISTRICT would be a character variable.

EXHIBIT 15.2   Result of recode is numeric variable DISTRICT

```
proc format;
value dist 100,110-125 = '1'   140-169 = '2'   170,180 = '3'
           200-250      = '4'   other   = '5'  ;
run;

data redist;
input store;
tempdist = put(store,dist.);
district = input(chardist,1.);
run;
```

Because INPUT's first argument may itself be an expression, the assignment statements in Exhibit 15.2 could be written like this:

```
district = input(put(store,dist.),1.);
```

The PUT function is evaluated, and the result used as the first argument to INPUT. This form of the calculation has the advantage of being compact, but it can suffer from a lack of clarity.

Assigning to a Variable: Other Examples.   Instead of collapsing many categories of one variable into a few of another, a one-to-one correspondence can be established. The format CEN2FIPS in Exhibit 15.3 converts U.S. Census Bureau codes for New England states into the more common FIPS (Federal Information Processing Standards) codes.

Exhibit 15.4 demonstrates how character strings can be converted into numerics. State postal codes are converted into FIPS codes with a PUT function, then the character variable created by the PUT is converted into a numeric variable using the INPUT function.

Exhibit 15.5 passes a function call (MEAN) rather than a single variable as PUT's first argument. The average of variables YR1 through YR5 is computed first. This value is then passed to the RANK format, which assigns one of the format's four groups ("l", "m", "h", and " ") to character variable LEVEL, whose length is one.

EXHIBIT 15.3   One-to-one recoding using the PUT function

```
proc format;
value cen2fips 11 = '23'    12 = '33'    13 = '50'
               14 = '25'    15 = '44'    16 = '09'
      ;
run;

data states;
input censt;
fipst = put(censt,cen2fips.);
run;
```

EXHIBIT 15.4   Character to numeric conversion

```
proc format;
value $postfip 'me','ME' = '11'    'nh','NH' = '12'
               'ma','MA' = '14'    'vt','VT' = '13'
               'ct','CT' = '16'    'ri','RI' = '15'
      ;
run;

data states;
input st_post $;
char_st = put(st_post,$postfip.);
state   = input(char_st,2.);
run;
```

EXHIBIT 15.5   Result of a function formatted

```
proc format;
value rank  low-.25 = 'l'       .25 <- .75 = 'm'
            .75 <- high = 'h'    other      = ' ';
run;

data grouped;
input yr1-yr5;
level = put(mean(of yr1-yr5),rank1.);
run;
```

What if more than one variable is involved in the grouping? The preceding examples showed the grouping done on the basis of a single user-written format. Some situations call for blending the brute-force IF-THEN technique and the more compact and elegant PUT function. Exhibit 15.6 extends the mean rank example in Exhibit 15.5. Variable LEVEL is determined by the mean of YR1 through YR5 *and* the absolute value of variable CHG.

EXHIBIT 15.6   Combination of IF-THEN statements and user-written formats used to recode

```
proc format;
value rank  low-.25 = 'l'           .25 <- .75 = 'm'
            .75 <- high = 'h'        other      = ' ';
run;


data grouped;
input yr1-yr5;
length level $2;
lev1 = put(mean(of yr1-yr5),rank1.);
if         lev1 = 'l' & abs(chg) <= 100 then level='s1';
   else if lev1 = 'm' & abs(chg) <= 100 then level='s2';
   else if lev1 = 'h' & abs(chg) >  100 then level='m';
   else                                      level=' ';
run;
```

POINTERS AND PITFALLS

Format-PUT recoding should not be a struggle. Remember that sometimes it is necessary to use a combination of PUTs and IF statements.

Using the Recode to Create Transient Variables

If the recode is needed simply to make decisions in IF statements, there is no need to create a new variable. The PUT function can be used to create a transient variable, one whose value exists only during the execution of the IF statement. Chapter 7 demonstrated a similar use of arithmetic functions. All that is required is to use the PUT function where a variable would normally be placed in a comparison. Exhibit 15.7 shows several uses of the PUT function in this context. Notice that the result of the function is being compared to a *character* constant. The comparison must be exact. The lowercase values "out" and "in " in the IF statement must be the same as the "out" and "in " defined in the format. This subtle but important point can cause a great deal of head-scratching if forgotten.

EXHIBIT 15.7   PUT function used to create transient variable

```
proc format;
value grp   low-250 = '1'    250-299 = '2'    other = '3' ;
value $fld  '1b','2b','3b','ss','c','p' = 'inf'
            other                       = 'out';
run;


data infield outfield;
length pos $2;
input avg pos $;
if put(pos,$fld.) = 'out' & avg >= .250 then output outfield;
   else if put(pos,$fld.) = 'inf' then output infield;
run;
```

Exercises

15.1. Character variable BATCH will be used to create new variables and to select observations for output to a new dataset. To assist this process, a format is successfully created with the following VALUE statement:

```
value $bgrp '01A' - '50Z' = '1'
             '510' - '99Z' = '2'
             other         = '?' ;
```

   a. Without creating a new variable, use $BGRP to select formatted values of variable BATCH for output to dataset GRP1. This dataset should contain only formatted values equal to "1."

   b. Create a new variable, BTCH_GRP. It is the formatted value of BATCH (using format $BGRP). Write the necessary assignment statement (it uses the PUT function), and state BTCH_GRP's data type and length.

   c. Suppose you wanted to create a new numeric variable, BTCH_NUM, based on the values of BATCH. Valid values would be 1, 2, and missing (.). There are two ways to do this. You could change the VALUE statement shown above and use the PUT and INPUT functions or use IF-THEN statements in the DATA step. Use both approaches. Show where to change the VALUE statement, then write two IF-THEN sequences to create BTCH_NUM.

15.2. Refer to Exhibit 15.2. What would be the impact if the OTHER clause were omitted? What would DISTRICT's values be? Assume that values of STORE range from 100 through 350.

15.3. Are the following program segments equivalent? Justify your answer. Assume STATE is character and format $GRP actually exists.

```
if put(state, $GRP.) = 'NE' then output;


        grp = put(state, $GRP.);
if grp = 'NE' then do;
   output;
   end;
```

15.4. Refer to Exhibit 15.7. Suppose the following statement were added to the definition of $FLD:

```
'mgr', '3bc', '1bc' = 'adm'
```

If an observation had a POS value of "MGR," what dataset would it be sent to? Where would "1bc" go?

15.5. Refer to Appendix E's description of the BRIDGES dataset. Write a format that will convert values of TYPE to their full description: "s" is formatted as "Suspension," "c" as "Cantilever," and "a" as "Arch." Then use the format to create character variable BR_STYLE. Use a LENGTH statement to ensure the variable is long enough (10 bytes).

15.3 Recoding Directly in PROCs

Chapter 14 demonstrated the use of custom formats to enhance data display. Recall that all SAS did was use the custom formats to control how the display of each observation and its variables was to take place.

Recoding directly in PROCs takes the same principle and applies it to the group assignments required by some procedures. If left unformatted, a variable such as employee salary would have as many levels in PROC FREQ as there are unique salaries. FREQ sees each salary as a distinct level to report.

If FREQ is instructed to use a format when building the frequency table, however, the process is changed. Each value of salary is filtered through a format, and the formatted values will be the ones used to form groups. Rather than group by distinct salary levels, FREQ will group by the coarser-grained user-defined groups. The same category-collapsing process takes place in CHART's VBAR and HBAR variables, MEANS' CLASS variables, and most other PROCs described in later chapters.

POINTERS AND PITFALLS

Most PROCs are sensitive to formats when assigning variable values to categories. You can recode directly in a PROC, without altering the input dataset, by using system or user-written formats.

Examples

The dataset in Exhibit 15.8 is used to illustrate direct recoding with some PROCs introduced in Chapters 5 and 6. The observations are the 15 most populous countries in 1987. The variables are a country identification number, country name, population in millions, percent urban, energy consumption per capita, and the percent of land in cultivation, pasture, and forests.

EXHIBIT 15.8   Populous country demographic and resource data

OBS	ID	COUNTRY	POP	PCTURB	ENERGY	CULT	PASTURE	FOREST
1	101	China	1062	32	19	11	30	14
2	107	Japan	122	76	111	13	2	68
3	113	Philippines	62	40	9	38	4	40
4	102	India	800	25	7	51	4	21
5	112	Vietnam	62	19	4	20	1	40
6	109	Bangladesh	107	.	.	.	.	.
7	105	Indonesia	175	.	.	.	.	.
8	110	Pakistan	105	28	6	25	6	4
9	304	United States	243	74	280	20	26	28
10	311	Mexico	82	70	50	13	38	23
11	406	Brazil	141	71	19	9	19	66
12	508	Nigeria	109	28	7	34	23	16
13	603	USSR	284	65	176	10	17	42
14	614	West Germany	61	85	163	30	19	29
15	615	Italy	57	72	90	41	16	21

After giving some thought to how and if recoding should take place, we write the format library shown in Exhibit 15.9. The principle underlying each format is identical: identify groups of related values. These groups (e.g., the values 100 through 199) will be treated as a single value ("Asia") when used in procedures that are able to collapse categories. For this recoding to take place, the user-written format must be made known to the PROC. This is done just as you would for a SAS System format: via a FORMAT statement or by assigning the format in the DATA step creating the dataset.

Now the data are displayed using the formats (Exhibit 15.10). Notice that the land-use variables are truncated to one digit. Rather than display "Low," "Medium," and "High" for the grouped values, we decide that the single, first digit is adequate, thus the *landuse1.* specification.

EXHIBIT 15.9   FORMAT statements to be used with populous country data

```
libname out 'c:\fmtlib';

proc format libname=out;
value cont        100-199 = 'Asia'        200-299 = 'Australasia'
                  300-399 = 'N. America' 400-499 = 'S. America'
                  500-599 = 'Africa'      600-699 = 'Europe'   ;
value pop         low -< 100 = 'Under 100m'
                  100 - high = '100m +'   ;
value urban       low -< 30  = 'Low'
                  30   -< 70 = 'Medium'
                  70   -high = 'High'    ;
value landuse  low - 10  = 'Low'
                  10 <- 30  = 'Moderate'
                  30 <- high = 'High'     ;
run;
```

EXHIBIT 15.10   User-written formats used with the PRINT procedure

```
------------------- Program listing -------------------

proc print label data=lib.top15;
format id cont. pop pop. pcturb urban.
cult pasture forest landuse1.;
label id = 'Continent' ;
run;

------------------- Output listing -------------------
```

Continent	COUNTRY	POP	PCTURB	ENERGY	CULT	PASTURE	FOREST
Asia	China	100m +	Medium	19	M	M	M
Asia	Japan	100m +	High	111	M	L	H
Asia	Philippines	Under 100m	Medium	9	H	L	H
Asia	India	100m +	Low	7	H	L	M
Asia	Vietnam	Under 100m	Low	4	M	L	H
Asia	Bangladesh	100m +	.	.	.	.	.
Asia	Indonesia	100m +	.	.	.	.	.
Asia	Pakistan	100m +	Low	6	M	L	L
N. America	United States	100m +	High	280	M	M	M
N. America	Mexico	Under 100m	High	50	M	H	M
S. America	Brazil	100m +	High	19	L	M	H
Africa	Nigeria	100m +	Low	7	H	M	M
Europe	USSR	100m +	Medium	176	L	M	H
Europe	West Germany	Under 100m	High	163	M	M	M
Europe	Italy	Under 100m	High	90	H	M	M

Recoding with FREQ.   In Exhibit 15.11 the CONT format groups country identification numbers into continent names.   FREQ is run on ID using CONT to
obtain a distribution of countries by continent.

In Exhibit 15.11 the format CONT performs the recoding: what would have
been a table of 15 country ID values without CONT is reduced to a more mean-

---

EXHIBIT 15.11   Format used to group variables levels in FREQ procedure

---

```
------------------- Program listing -------------------

proc freq data=demog;
tables id;
format id cont.;
run;


------------------- Output listing -------------------

Continent
```

	ID	Frequency	Percent	Cumulative Frequency	Cumulative Percent
Asia		8	53.3	8	53.3
N. America		2	13.3	10	66.7
S. America		1	6.7	11	73.3
Africa		1	6.7	12	80.0
Europe		3	20.0	15	100.0

---

ingful listing of five continents. The continents are listed in their internal, unformatted order. Asia was coded in the 100s, North America in the 300s, and so on. To display the names in their formatted order, the ORDER=FORMATTED option in the PROC statement is used. Exhibit 15.12 shows the new, alphabetical order.

Recoding with MEANS.   The CLASS statement of PROC MEANS identifies the variables used as stratifiers when computing statistics. Formats allow MEANS

EXHIBIT 15.12   Format used to group, ORDER option to rearrange categories

---

```
------------------- Program listing -------------------

proc freq data=lib.top15 order=formatted;
tables id;
format id cont.; a
run;


------------------- Output listing -------------------

Continent
```

	ID	Frequency	Percent	Cumulative Frequency	Cumulative Percent
Africa		1	6.7	1	6.7
Asia		8	53.3	9	60.0
Europe		3	20.0	12	80.0
N. America		2	13.3	14	93.3
S. America		1	6.7	15	100.0

---

to group these variables. Statistics will be printed for each level of the recoded CLASS variables. If energy and land-use statistics are needed for each level of urbanization values, the URBAN format would group the appropriate CLASS variable. The program and output are presented in Exhibit 15.13.

EXHIBIT 15.13   Format used to group CLASS levels in the MEANS procedure

------------------- Program listing -------------------

```
proc means data=library.top15 maxdec=2 min max mean;
class pcturb;
var energy cult pasture forest;
format pcturb urban.;
run;
```

------------------- Output listing -------------------

PCTURB	N Obs	Variable	Minimum	Maximum	Mean
Low	4	ENERGY	4.00	7.00	6.00
		CULT	20.00	51.00	32.50
		PASTURE	1.00	23.00	8.50
		FOREST	4.00	40.00	20.25
Medium	3	ENERGY	9.00	176.00	68.00
		CULT	10.00	38.00	19.67
		PASTURE	4.00	30.00	17.00
		FOREST	14.00	42.00	32.00
High	6	ENERGY	19.00	280.00	118.83
		CULT	9.00	41.00	21.00
		PASTURE	2.00	38.00	20.00
		FOREST	21.00	68.00	39.17

Exercises

15.6. Calculate average span length by TYPE of the bridges in the BRIDGES dataset (see Appendix E). Use the format created in Exercise 15.5 (see p. 299) to perform the recoding. Write the MEANS procedure statements. Why would you want to use the formatted value, rather than the original value of TYPE?

15.7. Refer to Exhibit 15.9. Rewrite the URBAN format so it will display both the value for percent urban (two digits) *and* the category label ("Low", "Medium", "High"). Use a PICTURE format. How will this affect the PRINT procedure output appearance? Could you achieve the correct grouping by using this new format with PCTURB in a PROC FREQ? Justify your answer.

15.8. Refer to the data in Exhibit 15.8 (also described in the COUNTRYS dataset in Appendix E).
  a. Write CHART procedure statements to produce a vertical bar chart of average energy consumption by COUNTRY grouped into continent. Use the CONT format shown in Exhibit 15.11 to recode country number into continent name. What would the chart look like if the format were not used (i.e., no recoding)?
  b. How would you rewrite the CONT format to arrange the bars in alphabetical order? You could accomplish this by using a CHART statement (it is also the easier way to do it): write the statement.

Summary

User-written formats may be used not only for display but also for recoding and grouping character and numeric variables. This sometimes overlooked feature of SAS formats often saves the user from coding many tedious and error-prone IF-THEN statements. The format may be used with the PUT function to create a new or transient variable in the DATA step. The format may also be used directly in a grouping procedure such as CHART or FREQ or in any procedure where collapsing categories is desired, such as PRINT. Assign the user-written format to the dataset or use a FORMAT statement with the procedure; the individual categories of the formatted variables will be grouped into the new categories specified by the format. The original dataset remains unchanged.

16 Date Values

Data often represent the nature, measurement, and timing of an event. Dates of birth, death, tests, employment history, and dividend payments are a few examples of date-oriented data. Such date-oriented variables may be handled arithmetically: determining time between events, targeting a date in the future or past, and making other calculations are common forms of date calculations. The temptation in these cases is to use three different variables for each date: one for day, month, and year. If the date will be simply read from raw data and used directly in a procedure, it could even be read as a character string.

This chapter shows that such approaches are as limited as they are simple. For example, only the simplest date calculations can be performed with the DATA step tools presented so far. A character variable representing a date often suffers from a lack of flexibility when displayed. For example, it takes considerable effort to change an input character string in the format of month name, day, and year to an output display of day, month number, and year.

This chapter discusses SAS functions and formats for handling date-oriented data. These functions and formats are some of the most powerful and useful features of the SAS language, and they open up a wide variety of applications even for the beginning SAS user. Without them, you would have to write your own date-handling calculations and formats. This forbidding region of the computing landscape is fraught with tedium and potential for error. It is far easier to learn how to use the features in this chapter than to attempt to write your own programs to do what they do, especially for novice computer users.

Section 16.1 outlines date value concepts and terminology. SAS provides special input formats to allow convenient, readable entry of dates as single variables. These formats are discussed in Section 16.2. Just as you sometimes need character or numeric constants in DATA step calculations or user-written formats, so may you need them for date constants. These constants are discussed in Section 16.3.

Performing all but the simplest calculations on date values is tedious without some built-in help. Date-handling functions are illustrated in Section 16.4. Finally, a number of output formats that enhance and facilitate the date once it is ready for display are presented in Section 16.5.

16.1   Date Value Concepts and Terminology

The underlying concept behind the effective use of SAS date values is based on simple counting. All dates are represented as the number of days before or since January 1, 1960. Thus August 11, 1945, is stored as −5256 (5,256 days before January 1, 1960), and April 11, 1987, is stored as 9962 (9,962 days after January 1, 1960). Crossing decade and century boundaries is handled automatically. SAS always checks for legitimate values by considering periodic events such as leap years and leap centuries. A value of 29 is legitimate for the number of days in February only when the year number is evenly divisible by four (a leap year) and century number is evenly divisible by 400 (a leap century). For all its intrinsic special attributes, the date value is just another numeric variable.

POINTERS AND PITFALLS

SAS stores dates as the number of days relative to January 1, 1960. There is no special "date" data type, so it is up to the user to know which numeric variables represent dates.

SAS supports several types of date value calculations. *Duration* is the simplest. It addresses questions such as, How long is it between event x and event y? The unit of measurement in these questions may be days, months, quarters, or years. *Estimation* allows the user to pose questions such as, How many weeks or months has a person born on a given date lived, up to today? *Extraction* of information from a date value suggests questions such as, What day of the week will the first birthday fall on? and What year was the person born? The functions described in Section 16.4 perform these calculations with a minimum of programming effort.

Exercises

16.1. Think about the kinds of data you usually work with.
  a. Are any of the data fields (variables) suitable as date values?
  b. How would you manipulate these fields arithmetically? as durations? projecting into the past or future?
  c. A numeric variable has a value of 5423. It represents a date. Do not compute the date, but give a brief explanation of what the number means.

16.2   Reading Date Values

Although raw data may contain values that already represent SAS date values (i.e., their values contain days relative to 1/1/1960), it is far more common to take advantage of the special date input formats and let SAS convert the fields into dates. This section describes some useful date formats and presents examples of their use. These formats allow date-oriented values to be entered in their "normal" format and ensure that these values are correctly translated into SAS date values.

Syntax

The format name, form of date represented, and column restrictions of date formats are presented in Exhibit 16.1.

EXHIBIT 16.1   Date input formats

Format	Date Representation	Example	Minimum Columns	Maximum Columns
DATEw.	day/month name/year	13Feb1988	7	32
DDMMYYw.	day/month number/year	12-2-88	6	32
JULIANw.	year/day number in year	88044	5	32
MMDDYYw.	month number/day/year	2/13/88	6	32
MONYYw.	month name/year	FEB88	5	32
YYMMDDw.	year/month number/day	88-2 -12	6	32
YYQw.	year/quarter	88/2	4	32

The date values in Exhibit 16.1 have six components:

1. The *day* is the day number of the event. It ranges from 1 to the maximum allowed for the month.

2. The *month name* is the name of the month, indicated by its first three letters. The name may be any combination of upper- and lowercase letters.

3. The *year* is the year number, represented by either two or four digits. Two-digit *year* values are assumed to begin with 19.

4. The *day number* is the number of the day in the year. It ranges from 1 to 366. Day numbers are used in Julian dates: 56141, for example, represents the 141st day of 1956.

5. The *month number* is the number of the month in the year. A value of 1 indicates January, while 12 indicates December.

6. The *quarter* is the quarter of the calendar year. A value of 1 indicates the quarter of January through March. The largest value is 4. Variables read with the YYQ format specify the first day of the quarter (1 April 1990, 1 October 1992, and so on).

Usage Notes

Keep in mind the following points when using date input formats:

- An illegal value for any of the subfields (day, month, year, quarter) sets the date variable to missing.

- The subfields of the date may be separated by any combination of blanks, hyphens (-), or slashes (/). Blanks may be inserted to improve legibility. All these subfield delimiters are optional. If the appropriate format is used, a value such a 05JAN90 will be handled correctly.

- As long as the subfields of the date are entered in the correct order, the format of the field may vary from one observation to the next: one observation may have "25jun11," and another may have "02-FEB-1908." Both observations may be read with DATEw.

- Fields are usually written the same way across observations, and usually without separators. This style is almost always used in business data processing, commercial databases, and government and other public domain data.

Examples

Exhibit 16.2 shows legal entry of July 12, 1988, using different formats and widths. It is important to understand that *all* these fields (except YYQ) are stored as 10420, SAS's calculation of the number of days since January 1, 1960.

EXHIBIT 16.2   July 12, 1988, read with different formats and widths

Format	w=7	w=9	w=10	w=12
DATE	12jul88	12jul1988	12-jul-88	12/jul/1988
DDMMYY	120788	12/07/88	12 07 88	12-07- 1988
MMDDYY	071288	07-12-88	07 12 88	07 /12/ 88
MONYY	071988	07 1988	07 / 88	07 / 1988
YYMMDD	880712	88 07 12	88 /07/12	1988/07/12
YYQ	883	88 03	1988-3	1988 / 3

Exercises

16.2. Which formats would be required to read the following date values? Do not worry about the width of the field; just specify the format name: DATE, MONYY, and so on.
   a. 1/1/90
   b. 1/feb/90
   c. 90 1
   d. 901
   e. 210290
   f. 12-18-80
   g. 18-12-80

16.3. Why can the format DATE7. not be used to read the following data lines even though they both describe the same date, July 4, 1976?

```
04jul76
6029
```

16.3 Date Constants

When processing date-oriented data, it is common to base decisions on one or more date values. You may want to select employees who enrolled in a benefits program before a certain date, identify students who took a course within a range of dates, or select companies that paid dividends in the fourth quarter of the year.

Although it is possible to enter the SAS date value (10420, −2180, and so on) directly into the appropriate statements, it is easier to use date constants. **Date constants** are special constants that SAS converts into date values. They are used just like their numeric counterparts. The beauty of their use is that the date is readily apparent to anyone reading the program and it is correctly interpreted by SAS (an instance of having it both ways).

The syntax of the date constant is straightforward:

```
'ddmmmyy'd
```

The *ddmmmyy* is the constant's date specification. It is entered exactly the same way as a DATE7. format ('07feb90'd, '30aug32'd). The *d* at the end of the constant ensures that SAS does not confuse the string with a character constant. The value is used as a number, just as if it had been read from a raw data file with a DATE7. format or entered directly as the number of days since 1 January, 1960.

Examples

The following five examples give an idea of the variety of date constant applications. Keep in mind that in most of the examples, date variables could have been used just as readily.

The first example begins executing a DO-group if date of employment is prior to the 1980s.

```
if empdate < '01jan80'd then do;
```

The second example selects people whose birthdate is in the second half of the year:

```
if '01jul90'd <= bday <= '31dec90'd then output;
```

The third example writes a format to group dates in U.S. history into broadly defined eras:

```
proc format;
value eras
      '11nov18'd -< '29oct29'd = 'Easy credit'
      '29oct29'd -< '16jul45'd = 'Depression, war'
      '16jul45'd -< '04oct57'd = 'Cold War'
      '04oct57'd -< '23nov63'd = 'Space race' ;
```

The fourth example generates a listing of paydays for a year. Becuase employees are paid biweekly, the increment in the loop is 14 days:

```
data payday;
do date = '04jan91'd to '31dec91'd by 14;
   output;
end;
run;
```

Rather than compare dates as in the previous examples, the fifth example creates date values. Batch dates are assigned based on part identification numbers:

```
if        partid = 'FF89' then batchdt = '05jan83'd;
    else if partid = 'RS89' then batchdt = '12jan83'd;
```

Exercises

16.4. Use date constants when writing the following statements:
  a. Output if BDAY is earlier than 1 January 1950.
  b. Begin a DO-group if EMPSTART is in the range 1 January 1970 through 31 December 1979.
  c. A range in a user-created format groups dates from 1/1/92 to 3/31/92 as '92, qtr=1.'
  d. INCREMNT, a DO-loop index variable, begins at 1/1/90 and loops to 12/31/95 in increments of seven days.

16.5. What, if anything, is wrong with the following assignment statements?
  a. xt = '02/feb/80'd;
  b. baseline = d'22apr90';
  c. st = '12mar58';

16.4   Date Calculations

Some date calculations can be carried out using simple arithmetic operators. A simple measure of duration, for example, can be computed by subtracting two dates. But most tasks can be performed more easily by using SAS date functions. This section briefly reviews simple calculations, then describes the families of date functions.

Simple Calculations

Some manipulations of date values are simply calculations of the duration of an event, usually measured in days, months, or years. The number of weeks since an employee was last reviewed, a person's age, and the time between critical dates are examples of such calculations. Using SAS date variables makes these calculations simple to perform: simply subtract the dates to obtain the number of elapsed days, then, if necessary, divide this number to scale it to weeks, months, years, or other units. If either date is a missing value, the result of the entire calculation will be set to missing, just like any other arithmetic operation. These techniques are illustrated below:

```
days    = '31dec90'd - bdate ;
months  = ('31dec90'd - bdate) / 30.4 ;
years   = ('31dec90'd - bdate) / 365.25 ;
```

Although simple to program, the calculations are limited in power. They are also a bit awkward. What if you wanted to always use the current date in the first three statements? The only option is to reenter the date constant each day the program is run. What if you needed to know the day of the week of BDATE's value? These and other calculations are handled easily by using date functions.

POINTERS AND PITFALLS

All but the simplest date calculations should be handled with date functions. If your variables are not stored as date-oriented data (e.g., separate variables for day, month, and year), you can use functions to convert these variables to a date value, then use date calculation functions.

Functions I: Create Date Values

Sometimes dates are supplied in a format not amenable to any of the SAS input formats described earlier in this book. Other situations require calculating the current date. These date creation activities are handled by the following functions:

Function	Description
TODAY()	*Purpose*: Extract the date from the computer system's clock. Store the result as a SAS date value.
	Comment: No arguments are required for this function. The parentheses are required since TODAY would be confused with a variable name without them.
DATEJUL(var)	*Purpose*: Convert a Julian date into a SAS date value.
	Argument: The Julian date, stored as a single numeric variable. If the year number and day cannot be read as a single variable, read them as separate variables, then convert them to a single date by multiplying the year by 1,000 and adding the day number.
	Comment: Julian dates are often used in commercially available datasets.
MDY(month,day,year)	*Purpose*: Create a SAS date value from separate month, day, and year variables.
	Arguments: SAS numeric variables or constants representing month, day, and year, respectively. A missing or out-of-range argument creates a missing value.
	Comment: Enter a constant to obtain a fixed date for a given period: MDY(mth_emp,1,yr_emp) produces a date value for the first of the month in the year the employee began work.
YYQ(year,quarter)	*Purpose*: Create a SAS date value from separate year and quarter variables.
	Arguments: The *year*, a two- or four-digit numeric variable or constant, and *quarter*, a numeric variable, constant, or expression in the range 1 to 4. Missing or invalid arguments set the value to missing.
	Comment: The SAS date variable will be set to the first day of the quarter: arguments of 1952 and 3 will create a date value of July 1, 1952.

Examples

Notice in the examples below how meaningful variable names enhance the clarity of the program. Month variables begin with "m" or "mth," day variables with "d" or "day," and so on.

The first example shows how to calculate the number of elapsed days from a date value representing a birthday to the current day. Notice that the TODAY function is used so that the date constant does not have to be reentered each day the program is run.

```
elapsed = today() - birthdte ;
```

The next example illustrates how to create the numeric variable STARTDTE from three separate variables.

```
startdte = mdy(m_st, d_st, y_st);
```

In this next example the visit follow-up date is a year from the current visit date. Calculate the date value for the follow-up. Assume the visit date was stored in separate variables.

```
followup = mdy(mth_vis, day_vis, yr_vis + 1) ;
```

In this example the dataset has exact dates for when an employee began work. Benefits coverage is begun on the first of the month.

```
benef_st = mdy(m_emp, 1, y_emp);
```

In this final example a Julian date is replaced with a SAS date. Notice that rather than creating a new variable, the value of DIVDATE is overwritten.

```
input @40 divdate 5. ;        * Julian:  88130 ;
divdate = datejul(divdate);   * Now SAS: 9980  ;
```

Functions II: Converting to Non-SAS Dates

The first group of functions created SAS date values. It is just as simple to create nondate numeric variables. In the functions described below, the only argument is a SAS date value. If it is a missing value, the result of the function is missing.

Function	Description
JULDATE(sasdate)	*Purpose*: Convert a SAS date value into a Julian date.
DAY(sasdate) MONTH(sasdate) YEAR(sasdate) QTR(sasdate)	*Purpose*: Extract day, month, year, and quarter numbers from a SAS date value.
WEEKDAY(sasdate)	*Purpose*: Extract the day of the week from a SAS date value. *Comment*: The result of the function is a number from one to seven. A 1 indicates Sunday, 2 indicates Monday, and so on. A user-written format can display the weekday number in a clearer, day-name form.

Examples.   The examples below illustrate the functions used to convert to non-SAS dates.

In this first example a late payment fee is assessed if day of receipt is past the 10th day of the month:

```
if day(receipt) > 10 then amt_due = amt_due * 1.015 ;
```

In the next example dividend dates are collapsed into quarters of the calendar year:

```
div_q = qtr(div_dte);
```

If fiscal-year rather than calendar-year quarters were required, DIV_Q could be manipulated by a series of IF-THEN statements to perform the required adjustments. For example, if the fiscal year began on July 1 (calendar quarter 3), DIV_DTE could be regrouped as follows:

```
qtr = str(div_dte);
if        qtr >  3 then fisc_q = qtr - 2;
   else if qtr ^= . then fisc_q = qtr + 2;
```

Labor rates on weekends are more expensive than on weekdays. In this example an hourly rate based on the day of the week is set. Since the ELSE statement is not unconditional, a missing JOB_DATE creates a missing RATE.

```
if    2 <= weekday(job_date) <= 6 then rate = 17.00 ;
   else if weekday(job_date)  ^= . then rate = 20.00 ;
```

In the next example the follow-up date calculation introduced on page 312 is refined. Weekend values are reassigned to the nearest weekday:

```
followup = mdy(mth_vis, day_vis, yr_vis+1) ;
if         weekday(followup)=1 then followup = followup+1;
   else if weekday(followup)=7 then followup = followup-1;
```

In this last example the WEEKDAY function and a user-written format are used to describe attributes of some noted "personalities." The appropriateness of the attribute is anyone's guess. The program and output are shown in Exhibit 16.3.

Functions III: Counting and Advancing Intervals

SAS provides two functions that simplify counting the number of periods between dates and advancing a specified number of periods from a starting date.

Function	Description
INTCK(*'unit',fromdate,todate*)	*Purpose*: Count the number of *units* from *fromdate* to *todate*, both SAS date values ("How many weeks [*unit*] is it from September 15, 1950 [*fromdate*] until October 1, 1990 [*todate*]?"). The result for all *units* except quarters is an integer. Quarters can have fractional values, indicating months. If *todate* is earlier than *fromdate*, the result may be negative.
	Arguments: The *unit* is a character variable or constant equal to *day, month, year,* or *quarter*. The unit may be entered in lower- or uppercase, or mixed case. The *fromdate* and *todate* are SAS date values indicating the beginning and end of the interval. If any of the arguments are missing, the value returned by INTCK is also missing.

EXHIBIT 16.3   User-written formats used with date functions and formats

```
------------------- Program listing -------------------

proc format;
value birthday 1 = 'Fair of face'      2 = 'Full of grace'
               3 = 'Full of woe'       4 = 'Far to go'
               5 = 'Loving and giving'
               6 = 'Works hard for a living'
               7 = 'Bonnie, bright, and good and gay' ;
run;

data one;
input name $char15. birth date7.;
day_week = weekday(birth);
format birth date7. day_week birthday.;
run;

------------------- Output listing -------------------

WEEKDAY function combined with user format

NAME               BIRTH    DAY_WEEK

Bob Barker         12DEC23  Far to go
Chubby Checker     03OCT41  Works hard for a living
Robert DeNiro      17AUG45  Works hard for a living
Allen Funt         16SEP14  Far to go
Andy Griffith      01JUN26  Full of woe
Mick Jagger        26JUL43  Full of grace
Nastassia Kinski   24JAN60  Fair of face
Loretta Lynn       14APR35  Fair of face
Dudley Moore       19APR35  Works hard for a living
Gregory Peck       05APR16  Far to go
Anthony Quinn      21APR15  Far to go
Phil Silvers       11MAY12  Bonnie, bright, and good and gay
Twyla Tharp        01JUL41  Full of woe
John Travolta      18FEB54  Loving and giving
```

Comment: When it performs its calculations, INTCK looks only at the *unit* portion of *fromdate* and *todate*. The resulting unit counts are sometimes counterintuitive. Consider a *fromdate* of April 1, 1990, and a *todate* of April 30, 1990. Although it looks like a month has passed, INTCK will return a value of zero months, since the month number of the *todate* (4) minus the month number of the *fromdate* (4) is zero. The date has to cross a *unit* boundary for the interval count to increase!

INTNX(*'unit',fromdate,count*) *Purpose*: Calculate the date *count* number of *units* from SAS date value *fromdate* ("What is the date 5 [*count*] weeks [*unit*] from date of termination [*fromdate*]?"). The function returns a SAS date value.

Arguments: The *unit* is identical in syntax and meaning to its use in INTCK. The *fromdate* is a SAS date value indicating the beginning of an interval. The *count* is the numeric variable or constant *units* to advance from *fromdate*. If any of the arguments are missing, the value returned by INTNX is also missing.

Comments: A positive value of *count* moves the date value forward. A negative value moves it backward. The treatment of *unit* portions in calculations follows the same logic as the INTCK function. The examples below demonstrate the subtleties of this aspect of the function.

Examples. Exhibit 16.4 takes a pair of dates (variables DATE1 and DATE2) and manipulates them with the INTCK and INTNX functions.

EXHIBIT 16.4   Date value manipulation with the INTCK and INTNK functions

date1	date2	INTCK ('day', date1, date2)	INTCK ('month', date1, date2)	INTCK ('year', date1, date2)	INTCK ('qtr', date1, date2)
31DEC90	.	.	.	.	.
31MAR91	01APR91	1	1	0	0.33
31MAR91	30APR91	30	1	0	0.33
01APR91	30APR91	29	0	0	0.00
01APR91	30JUN91	90	2	0	0.67
01APR91	01JUL91	91	3	0	1.00
01APR91	30JUN92	456	14	1	4.67
01APR91	30JUN90	−275	−10	−1	−3.33

date1	INTNX ('day', date1, 1)	INTNX ('day', date1, −1)	INTNX ('month', date1, 1)	INTNX ('month', date1, −1)	INTNX ('year', date1, 1)	INTNX ('year', date1, −1)	INTNX ('qtr', date1, 1)	INTNX ('qtr', date1, −1)
31DEC90	01JAN91	30DEC90	01JAN91	01NOV90	01JAN91	01JAN89	01JAN91	01JUL90
31MAR91	01APR91	30MAR91	01APR91	01FEB91	01JAN92	01JAN90	01APR91	01OCT90
31MAR91	01APR91	30MAR91	01APR91	01FEB91	01JAN92	01JAN90	01APR91	01OCT90
01APR91	02APR91	31MAR91	01MAY91	01MAR91	01JAN92	01JAN90	01JUL91	01JAN91
01APR91	02APR91	31MAR91	01MAY91	01MAR91	01JAN92	01JAN90	01JUL91	01JAN91
01APR91	02APR91	31MAR91	01MAY91	01MAR91	01JAN92	01JAN90	01JUL91	01JAN91
01APR91	02APR91	31MAR91	01MAY91	01MAR91	01JAN92	01JAN90	01JUL91	01JAN91
01APR91	02APR91	31MAR91	01MAY91	01MAR91	01JAN92	01JAN90	01JUL91	01JAN91

Exercises

16.6. Describe the meaning of the variables created in the following assignment statements.
   a. `durw = (end_dt - st_dt) / 7 ;`
   b. `durq = ceil((end_dt - st_dt) / 91.25) ;`
   c. `durq = (end_dt - st_dt) / 91.25 ;`
   d. `dur  = round((end - st) / 7, .5) ;`
   e. `tot_t = sum((end1 - st1),(end2 - st2));`
   f. `curr  = (today() - stdate) / 365.25;`
   g. `approx = round((today() - stdate) / 365.25);`

16.7. The three variables M, D, and Y represent month, day, and year values, respectively. The event indicated by the variables happens in the future. You want to create variable ELAPSE, whose value is the number of days between today and the day indicated by M, D, and Y. Write the statement(s) to do this.

16.8. Use the M, D, and Y variables as in Exercise 16.7, but now you need to create two variables. One is a SAS date value and the other is a character variable holding the Julian date representation of M, D, and Y. Write the required statements. The SAS date value indicates the number of days since a bond was purchased.

16.9. Write statements to meet the following requirements:
   a. Delete an observation if variable WORKDAY indicates a Saturday or Sunday.
   b. Assign a value of "q1" to WHEN if DUE falls in the first quarter of the calendar year.
   c. A company's fiscal year begins on October 1. Variable Q represents the fiscal quarter number of variable DIVDUE. Write the statements necessary to adjust the quarter number extracted from DIVDUE.
   d. Create a dataset with 366 observations and one variable. Each observation represents a unique day in 1992, and the variable is the date value for the day. (*Hint*: Use a DO-loop containing an output statement.)
   e. Calculate a date value NEXT representing the same day, next month, based on variable APPT.

16.10. Variable GRAD has a value of 11097 (20 May 90). PUB1 is 1 Jul 91, and DEFEND equals 1 FEB 90. What is the value returned by each of the following function calls?
   a. `intck('month', grad, pub1)`
   b. `intck('year',  grad, defend)`
   c. `intck('qtr',   grad, pub1)`
   d. `intnx('month', grad, 0)`
   e. `intnx('month', grad, 2)`
   f. `intnx('year',  grad, 1)`

16.11. Using the values from Exercise 16.10, are the following comparisons true or false?
   a. `intnx('month', defend, 1) = defend + 30`
   b. `month(intnx('month',defend,1) = month(defend+30)`

16.5 Displaying Date Values

It is handy to read and manipulate dates represented in SAS's internal format. But what about displaying these dates? Even if 10492 is a correct and useful date value, it has no inherent meaning when printed. Fortunately, SAS provides a number of display formats that enhance date value readability. The formats fall into two groups. Some closely resemble their input format counterparts, while others go further in their description of the date.

> **POINTERS AND PITFALLS**
>
> Unless your display needs are extremely specialized or you need to categorize date values, user-written formats are not needed with date values.

Syntax, Group I

Display formats that have input format counterparts are presented in Exhibit 16.5. The format name, date representation, width restrictions, and suggested width are listed. The recommended widths are the smallest values that preserve both data content and legibility.

EXHIBIT 16.5 Date display formats

Format	Display Format	Example	Minimum Columns	Maximum Columns	Recommended Width
DATEw.	day/month name/year	13FEB88	5	9	7
DDMMYYw.	day/month number/year	13/02/88	2	8	8
JULIANw.	year/day number in year	1988034	5	7	
MMDDYYw.	month number/day/year	02/13/88	2	8	8
MONYYw.	month name/year	FEB88	5	7	5
YYMMDDw.	year/month number/day	88-02-13	2	8	8
YYQw.	year/quarter	88Q1	4	6	

The date display formats in Exhibit 16.5 have the same six components as the date input formats in Exhibit 16.1. (Their meaning, of course, refers to the *display* of date values rather than how they are *stored* or read.)

1. The *day* is the date's day number.
2. The *month name* is the first three letters of the month, in capital letters.
3. The *year* is the year number, either two or four digits long.
4. The *day number* is the number of the day in the year. It ranges from 1 to 366.
5. The *month number* is the number of the month in the year. A value of 1 indicates January, while 12 indicates December.
6. The *quarter* is the quarter of the calendar year. It ranges from 1 to 4. It is separated from the year number by the letter Q.

Usage Notes, Group I

Remember these caveats when working with date display formats:

- Most of these formats have minimum widths that do not allow all information about the date to be printed. Although SAS displays as much information as possible, some will be lost if the field width is insufficient.
- If these formats are not adequate for your display needs, they can be used in a PUT function, reread by an INPUT function, then formatted by a user-written PICTURE format. Although this is not necessarily a difficult process, it usually makes you look back on the default SAS formats more forgivingly.

EXHIBIT 16.6   March 31, 1992, displayed with different formats and widths

The date being formatted is March 31, 1992.

Format	w2	w=3	w=4	w=5	w=6	w=7	w=8	w=9
DATE				31MAR	31MAR	31MAR92	31MAR92	31MAR92
DDMMYY	31	31	3103	31/03	310392	310392	31/03/92	
MMDDYY	03	03	0331	03/31	033192	033192	03/31/92	
YYMMDD	92	92	9203	92-03	920331	920331	92/03/31	
MONYY				MAR92	MAR92	MAR1992		
YYQ			92Q1	92Q1	1992Q1			

Examples, Group I

Exhibit 16.6 shows the display of the same date using different output formats and widths.

The formats may also be used to perform temporary recoding. Exhibit 16.7 illustrates the use of the YYQ format to group project milestone dates. Without the format, the table entries would be displayed as their internal value (e.g., 11080) rather than their "presentable," formatted value.

EXHIBIT 16.7   Date format used to regroup in FREQ procedure

-------------------- Program listing --------------------

```
proc freq data=dateline;
tables miledate;
format miledate yyq4.;
run;
```

-------------------- Output listing --------------------

MILEDATE	Frequency	Percent	Cumulative Frequency	Cumulative Percent
89Q2	3	5.0	3	5.0
89Q3	6	10.0	9	15.0
89Q4	6	10.0	15	25.0
90Q1	12	20.0	27	45.0
90Q2	9	15.0	36	60.0
90Q2	12	20.0	48	80.0
90Q3	12	20.0	60	100.0

Syntax, Group II

The second group of date display formats has no parallel to input formats. They display the date value in extended, "chattier" forms. Their syntax is outlined in Exhibit 16.8.

The date formats in Exhibit 16.8 have five components:

1. The *weekday* is either the first three characters of the weekday name or the entire name, depending on available space. It is printed in uppercase.

EXHIBIT 16.8   Date formats unique to display

```
WEEKDATE weekday, mon day year
WEEKDATX weekday, day mon year
WORDDATE mon day year
WORDDATX day mon year
JULDATE year daynumber
```

2. The *mon* is either the first three characters of the month name or the entire name, depending on available space. It is printed in uppercase.

3. The *day* is the day number of the date.

4. The *year* is the year number of the date. Either the entire four-digit number is displayed or just the last two numbers, depending on available space.

5. The *daynumber* is the number of the day within the year. It ranges from 1 to 366 and is always printed using three positions.

Usage Notes, Group II   When using the WEEKxxxx and WORDxxxx formats, keep in mind that they have minimum widths that do not allow all information about the date to be printed. Although SAS displays as much information as possible, some will be lost if the field width is insufficient. The space required varies from date to date since the length of month and weekday names varies.

Example, Group II   Exhibit 16.9 shows the display of the same date using different output formats and widths. The column widths chosen for this display were those where the wording and content of the formatted value actually changed. Other widths simply adjusted the positioning of the string within the columns.

EXHIBIT 16.9   Demonstrations of different combinations of date display formats and widths

w.	WEEKDATE	WEEKDATX
3	Tue	Tue
9	Tuesday	Tuesday
15	Tue, Mar 31, 92	Tue, 31 Mar 92
17	Tue, Mar 31, 1992	Tue, 31 Mar 1992
23	Tuesday, Mar 31, 1992	Tuesday, 31 Mar 1992
29	Tuesday, March 31, 1992	Tuesday, 31 March 1992

w.	WORDDATE	WORDDATX
3	Mar	Mar
9	March	March
12	Mar 31, 1992	31 Mar 1992
18	March 31, 1992	31 March 1992

Exercises

16.12. What formats are required to display the date values shown below? Do not worry about column widths; just indicate which formats should be used (e.g., DATE, WORDDATE).
  a. `JUL90`
  b. `04JUL80`
  c. `04JUL1981`
  d. `Friday, February 23, 1990`
  e. `901021`
  f. `88Q3`
  g. `July 12, 1991`

16.13. A dataset has observations for events in March and April of 1991. There is at least one observation with nonmissing values of variable EVENT for each day. The following formats were used to group EVENT in a FREQ table. How many groups were formed with each of the following formats?
  a. YYQ4.
  b. DATE5.
  c. DATE7.
  d. MONYY5.
  e. YYMMDD5.
  f. YYMMDD2.

Summary

The SAS System has a very powerful and elegant method for handling date-oriented data. All dates are represented internally as the number of days since January 1, 1960. A variety of input date formats are supplied with SAS: a field such as 7-12-90 can be entered in the raw data and automatically converted into the correct date value. Date constants may also be written, thus avoiding the need to express a date of interest directly in SAS's internal format.

Date values may be manipulated directly in simple expressions or via functions that convert dates into other formats; extract day, month, and year; or compute duration between events. The values may be displayed with several default, System display formats.

17 Creating Datasets with PROCs

Calculating descriptive statistics was discussed in Chapter 5. The MEANS and UNIVARIATE procedures calculate a wide variety of descriptive, single variable statistics. Other measures not explicitly covered by these procedures are standard scores and variable ranks. Variables may need to be standardized to a given mean and standard deviation. Observations may require ranking by the values of one or more variables.

These concepts often have direct, practical uses in data management and reporting. For example, you may be interested in percentile rankings of a variable. You might also want to calculate summary statistics for a dataset, then merge these figures with the individual observations.

This chapter discusses three procedures—MEANS, STANDARD, and RANK—that not only print these useful statistics but also create SAS datasets. These datasets often contain more information than their default output listings. Once you understand the structure and content of the datasets, you will have a group of very powerful new tools at your disposal.

17.1 MEANS Revisited

Many of the features of the MEANS procedure were covered in Chapter 5. This section discusses MEANS' ability to create SAS datasets containing the data in the printed report and other, more detailed data. Users of version 5 of the SAS System should remember that the CLASS statement is not implemented. The SUMMARY procedure in version 5 has nearly all the MEANS features presented here.

The MEANS Output Dataset

The structure and content of the MEANS output dataset is rich in information but can be difficult to understand. The best way to comprehend it is to examine a

EXHIBIT 17.1   Creating and displaying a MEANS output dataset

```
------------------- Program listing -------------------

proc means data=survey;
class region sample;
var yrsed faminc;
output out=summdata mean=avg_ed avg_inc;
run;

proc print data=summdata;
id region;
run;

------------------- Output listing -------------------
```

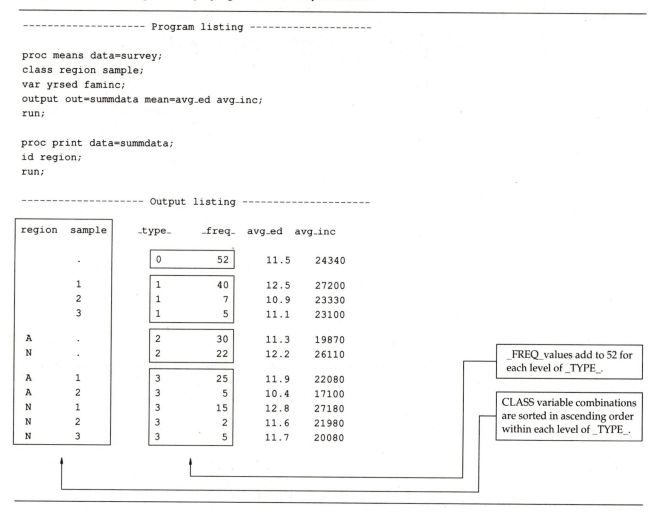

region	sample	_type_	_freq_	avg_ed	avg_inc
	.	0	52	11.5	24340
	1	1	40	12.5	27200
	2	1	7	10.9	23330
	3	1	5	11.1	23100
A	.	2	30	11.3	19870
N	.	2	22	12.2	26110
A	1	3	25	11.9	22080
A	2	3	5	10.4	17100
N	1	3	15	12.8	27180
N	2	3	2	11.6	21980
N	3	3	5	11.7	20080

FREQ values add to 52 for each level of _TYPE_.

CLASS variable combinations are sorted in ascending order within each level of _TYPE_.

fairly simple application. Exhibit 17.1 presents the statements analyzing a dataset and the PRINT procedure listing of the output dataset.

Let us look at the procedure statements first. The CLASS statement instructs MEANS to analyze dataset SURVEY by each level of REGION and SAMPLE. REGION is a character variable with values of either "A" or "N", and SAMPLE is a numeric ranging from 1 to 3. VAR functions exactly as discussed in Chapter 5: it requests that YRSED and FAMINC be the analysis variables. A new statement, OUTPUT, gives the name SUMMDATA to the output dataset and requests that it contain the MEAN of the analysis variables. These variables will be called AVG_ED and AVG_INC.

The output dataset, SUMMDATA, has unique and useful properties. Look first at two variables created by SAS, _TYPE_ and _FREQ_. _TYPE_ identifies the level of summary. A value of 0 indicates the summary record for the entire dataset; that is, it computes the statistics across all levels of REGION and SAMPLE. The _TYPE_ = 1 observations identify summary data for each level of SAMPLE across all REGIONs: REGION is, in effect, ignored. _TYPE_ = 2 observations contain the analogous data for REGION. They analyze across all SAMPLEs

(REGION is ignored). Finally, _TYPE_ = 3 observations contain statistics for each level of SAMPLE within each level of REGION.

In the more general case, there will be as many levels of _TYPE_ as there are unique combinations of CLASS variables: one CLASS variable has two unique levels (_TYPE_ values 0 and 1), two CLASS variables have four (0 through 3), three CLASS variables have 7 (0 through 7), and so on. The formula for calculating the number of levels is:

```
levels = 2^n - 1
```

where n is the number of CLASS variables.

FREQ is the number of observations found at each level of summary. At _TYPE_ = 0, _FREQ_ 52 means there were 52 observations in the entire dataset. At _TYPE_ = 1, SAMPLE = 1 there were 40 observations, and at SAMPLE = 2 there were 7. _FREQ_ values at each value of _TYPE_ always add to its value at _TYPE_ = 0 (_TYPE_ = 1 observation's _FREQ_ values of 40, 7, and 5 add to 52, as do _TYPE_ = 2's 30 and 22).

When the CLASS variables appear in the dataset, they do so in ascending order. Notice that both SAMPLE and REGION are in ascending order (1, 2, 3 and A, N). When several nonmissing CLASS variables appear in a level (as they do in _TYPE_ = 3), they are arranged as if specified in a SORT procedure's BY statement. In Exhibit 17.1 they are sorted by REGION and SAMPLE. If a particular CLASS variable combination does not exist, MEANS will not include it in the output dataset. There is no SAMPLE value of 3 in REGION A, so no summary statistics are generated and no records written to SUMMDATA.

The AVG_ED and AVG_INC values represent the mean years of education and family income *at each level of summary*. Thus the MEANS output dataset's information exceeds that of the printed output. In effect, only the _TYPE_ = 3 levels appear on the output listing.

POINTERS AND PITFALLS

MEANS output datasets contain summary statistics for the dataset as a whole as well as different combinations of CLASS variables. All or some of these levels of summary can be merged in later DATA steps with the original, unsummarized data. This allows you to, say, compare an individual observation's value of family income to the mean for its region and the mean for all regions (this comparison technique is shown in Exhibit 17.7).

Syntax

The complete syntax of the MEANS procedure is presented below. Items not in boldface have already been discussed in Chapter 6.

```
PROC MEANS [DATA=in_sasdataset] [NOPRINT] [FW=width]
           [MAXDEC=places] [MISSING] [NWAY]
           [IDMIN] [DESCENDING]
           [ORDER=FREQ|DATA|INTERNAL|FORMATTED]
           [printed_statistics] ;
    [VAR analysis_varlist ;]
    [CLASS class_varlist ;]
    [BY by_varlist ;]
    [FREQ freq_var ;]
```

```
[ID id_varlist ;]
OUTPUT [OUT=out_sasdataset] stat_request
        [MINID(an_var(dset_vars) ...) = new_names]
        [MAXID(an_var(dset_vars) ...) = new_names] ;
```

POINTERS AND PITFALLS

As is true with any complex procedure, first try MEANS with as many default options as possible. Once you are comfortable with the content of the output dataset, add options, print the dataset, and make sure you understand the options' impact. It usually takes a while to become comfortable with creating and using these datasets.

Some common applications of the new capabilities of the MEANS procedure follow:

```
proc means data=in.labs;
class lab study;
var m1-m30;
output out=summ mean=;

proc means data=fert nway;
var no_cont outcome;
id village;
output out=summfert n(no_cont outcome)=n_no n_out
        sum(no_cont outcome)=t_no t_out;
```

Eight new elements of MEANS are introduced:

1. *NOPRINT* suppresses the display of MEANS output. It is usually used when MEANS is being run only to produce an output dataset. The default is to display the statistics.

2. *MISSING* allows missing values to be valid levels of CLASS variables. By default, an observation with a missing value for any CLASS variable is excluded from the analysis.

3. *NWAY* includes only observations with the highest _TYPE_ value in the output dataset. The default is to include all _TYPE_ values. In Exhibit 17.1 *NWAY* would restrict the dataset to a _TYPE_ of 3.

4. *IDMIN* requests that the variable(s) specified in the ID statement use their minimum value. The default is to use maximum values. See the description of the ID statement in (8) for more details.

5. *DESCENDING* requests that the output dataset be ordered by high-to-low values of _TYPE_. The default is to arrange in ascending order. The class variables retain their sort order. Just the *groups* of _TYPE_ values are rearranged.

6. *ORDER=[FREQ|DATA|INTERNAL|FORMATTED]* specifies the arrangement of nonmissing CLASS variables in a level of _TYPE_. *FREQ* orders most frequent to least frequent. *DATA* orders the data in the order values were encountered in the analysis dataset. *INTERNAL* (the default) orders by unformatted variable values. *FORMATTED* orders by the formatted value of the variable. This may or may not be different than the order specified by *INTERNAL*, depending on the nature of the format being used.

7. The *class_varlist* specifies the variable(s) that will be used as levels for analysis. The statement is optional and has no defaults. If not entered, MEANS

summarizes over the entire dataset (i.e., creates a one-observation dataset with _TYPE_ = 0). The variables may be either numeric or character and may be grouped with formats. For example, rather than summarize individual INCOME values in the output dataset, a format could group them into "high", "medium", and "low".

8. The *id_varlist* specifies other variables to be included in the output dataset. The statement is optional. The variables are the largest (or if IDMIN was specified in the PROC statement, the smallest) values for the particular output observation's level of CLASS variables. Thus if ID STARTDTE were entered and the CLASS variable was SUBJCTID, the value of STARTDTE in the output dataset would represent each SUBJCTID's largest value of STARTDTE. Compare this option to the MINID and MAXID options of the OUTPUT statement, below.

The OUTPUT statement specifies the name and contents of the output dataset. The statement is required to create a dataset. Multiple OUTPUT statements may be entered. The OUTPUT statement has three components:

1. *OUT=out_sasdataset* indicates the name of the output SAS dataset. If the option is not specified, SAS uses its default dataset naming scheme (see Chapter 4 for details). Permanent datasets may be created. Dataset options (DROP, KEEP, RENAME) may be used.

2. The *stat_request* describes the statistics in the *OUT=* dataset. It has no default and must take one or more of the forms described in the "Specifying Statistics Requests" section.

3. *MINID* and *MAXID* are options that create new variables associated with one or more analysis variables. They have no default and use the format described in the "Specifying MINID and MAXID" section. They are available only in version 6 of the SAS System.

Specifying Statistic Requests

The statistics request section of the OUTPUT statement may take one or more of the following formats. In each form, *stat* is a statistics request keyword identical in form and meaning to the statistics keywords used in the PROC statement.

Form 1: stat=namelist. The *namelist* specifies the variables containing *stat*. The variables in *namelist* have a one-to-one correspondence with the variables listed in the VAR statement: the first variable in *namelist* contains *stat* for the first variable in the VAR statement; the second variable contains *stat* for the VAR statement's second variable; and so on.

Form 2: stat(varlist)=namelist. The *namelist* identifies variables containing statistic *stat* for analysis variables *varlist*. There is a one-to-one correspondence with the list of analysis variables. The list, however, is specified in the parenthetical expression rather than the VAR statement. There must be as many variables in *namelist* as there are in *varlist*.

Form 3: stat=. This form requests an output dataset that contains the same variables as in the VAR statement's list. The variables contain statistic *stat* for each of these variables. This form and Form 4 may not be used in the same OUTPUT statement.

Form 4: stat(varlist)=. This form specifies that VAR statement variables in *varlist* represent statistic *stat* in the output dataset. This form and Form 3 may not be used in the same OUTPUT statement.

Examples. The following three examples illustrate valid statistics requests. Only the VAR and OUTPUT statements are shown.

In the first example the output dataset contains variables GNP78 through GNP92, each representing a mean for the level of summary determined by _TYPE_ and CLASS variables:

```
var gnp78-gnp92;
output out=gnpsumm mean=;
```

In the second example the output dataset contains variables MATH, VERBAL, STD_MATH, STD_VERB, N_MATH, and MIS_MATH. These contain the means of MATH and VERBAL, their standard deviations, and the number present and missing of MATH, respectively. Notice that STD= and MEAN= could not both be specified, since duplicate variable names would be created.

```
var math verbal;
output out=scorstat std(math verbal)=std_math std_verb
       mean= n(math)=n_math nmiss(math)=mis_math;
```

In the third example the output dataset contains the standard deviations of MATH and VERBAL (STD_MATH and STD_VERB) and the count of nonmissing observations for VERBAL (implicitly named VERBAL). Notice, however, that since SAS simply pairs up the variables in the *namelist* with those in the VAR statement's list, STD_VERB contains a math statistic and STD_MATH contains a verbal statistic! There is no remedy for this except to be careful when assigning variable names.

```
var math verbal;
output out=scorstat std=std_verb std_math n(verbal)=;
```

Specifying MINID and MAXID

Minimum and maximum values of a variable can be associated with one or more analysis variables by using the MINID and MAXID options of the OUTPUT statement. Unlike the ID statement, which captures the maximum value of one or more variables for a *class* level, these options identify the extreme values within a class at the *observation* level. This enables you to create variables to answer questions like, "What is the name of the salesperson with the largest sales-to-quota ratio in each district?" These options are used in Exhibit 17.8.

Syntax. The syntax of MINID and MAXID is illustrated below:

```
[MINID|MAXID](an_var(dset_var ...))=namelist;
```

Some common applications of MINID and MAXID follow:

```
minid(dist(sale_name dist_name))=min_name min_dist;
maxid(profit(branch)) = max_bran;
```

MINID and MAXID contain three elements:

1. The *an_var* is a variable in the VAR statement.
2. The *dset_var* is any variable in the input dataset other than *an_var*. Enter as many of these as needed.
3. The *namelist* identifies the names the maximum/minimum values will take in the output dataset. They have a one-to-one correspondence with the *dset_var* list.

Example. The following example illustrates the use of both MINID and MAXID to identify observations containing high and low values of analysis variables SQRATIO and SALARY. Variables MIN_NAME and MIN_DPT represent the name and department number of the employee with the lowest value of SQRA-

TIO for each CLASS and _TYPE_ level in the output dataset. MAX_NAME and MAX_DPT serve similar purposes for the maximum values of SQRATIO. Finally, MAX_SAL contains the number of years employed for the observation containing the highest SALARY.

```
minid(sqratio(name dept))=min_name min_dpt
maxid(sqratio(name dept) salary(yremp))=
        max_name max_dpt max_sal
```

Complete Examples

This section presents examples of the new, more powerful capabilities of the MEANS procedure. All examples use the coastal mileage dataset used in Chapter 6. Key features and options used in this section are summarized in the following list:

Exhibit	Features/Options
17.2	Single CLASS variable; one statistic (MEAN) saved in output dataset
17.3	No CLASS or VAR statements
17.4	CLASS variable grouped with a user-written format
17.5	Two CLASS variables, several statistics
17.6	Multiple output datasets created
17.7	MEANS output dataset merged with original data
17.8	MINID and MAXID used
17.9	User-written format used to group; ORDER and MISSING options used
17.10	NWAY option, BY-group processing

Exhibit 17.2 (see p. 328) creates dataset SUMM containing the mean (MEAN option) of the analysis variables. Subgroups are defined by levels of variable COAST, and the results are printed with two decimal places (MAXDEC option). Notice that the MEANS output dataset has all the information contained in the MEANS printed output plus a summary level (observation with _TYPE_ = 0) that was not printed.

In Exhibit 17.3 (see p. 328) both the CLASS and VAR statements are omitted. The output dataset SUMM consists of a single, _TYPE_ = 0, observation containing the means of all numeric variables in the dataset.

Exhibit 17.4 (see p. 329) defines a user-written format $STGROUP to group the states by geographic region. MEANS is then run, suppressing printed output (NOPRINT) and using the FORMAT statement to specify $STGROUP. Note that if the format were absent, individual states would make up the different classes! The result is saved as a permanent, two-level dataset. The _TYPE_ = 1 observations are ordered by their internal values ("ak" in group "Offshore" comes before "al" in group "Gulf", and so on).

Exhibit 17.5 (see p. 330) takes a variety of statistics and counts for two CLASS variables. Notice the presence or absence of the CLASS variables and their sort order at each level of _TYPE_. The sum of _FREQ_ at each level of _TYPE_ is 26, the number of observations in the dataset.

In Exhibit 17.6 (see p. 330) two datasets (two OUTPUT statements) are created. Since the second OUTPUT statement does not use the OUT option, SAS assigns a name of the form DATA1, DATA2, and so on (see the discussion of default dataset naming in Chapter 4).

Exhibit 17.7 (see p. 331) takes mean GENCST for each COAST, then merges the output dataset with the original, state-level dataset. A new variable identifying states that are at or below the average for the coast and those that are above average is created. _TYPE_ is excluded from the output dataset: since NWAY was specified, *all* observations in SUMMSTAT have the same value of _TYPE_. Also

EXHIBIT 17.2   Single CLASS variable, one statistic in output dataset

------------------- Program listing --------------------

```
proc means data=in.coastal maxdec=2 mean;
class coast ;
var gencst tidalcst ;
output out=summ mean= ;
run;
```

------------------- Output listing --------------------

COAST	N Obs	Variable	Label	Mean
ec	20	GENCST	General outline	205.56
		TIDALCST	Detailed shoreline	2290.70
wc	6	GENCST	General outline	1447.17
		TIDALCST	Detailed shoreline	7136.50

------------------- Output dataset --------------------

COAST	_TYPE_	_FREQ_	GENCST	TIDALCST
	0	26	515.96	3408.96
ec	1	20	205.56	2290.70
wc	1	6	1447.17	7136.50

This level of summary was not available in the preceding printed output.

EXHIBIT 17.3   CLASS and VAR statements omitted

------------------- Program listing --------------------

```
proc means data=in.coastal maxdec=2 mean ;
output out=summ mean= ;
run;
```

------------------- MEANS printed output ---------------

N Obs	Variable	Label	Mean
26	GENCST	General outline	515.96
	TIDALCST	Detailed shoreline	3408.96

------------------- MEANS output dataset --------------

TYPE	_FREQ_	GENCST	TIDALCST
0	26	515.958	3408.96

EXHIBIT 17.4   CLASS variable grouped with a user-written format

```
------------------- Program listing -------------------

proc format;
value $stgroup 'me','nh','ma','ri','ct' = 'New England'
               'ny','nj','pa','de','md','va' = 'Mid Atlantic'
               'nc','sc','ga','fl' = 'South Atlantic'
               'al','ms','la','tx' = 'Gulf'
               'ca','or','wa' = 'West'
               'hi','ak' = 'Offshore'

run;

proc means data=in.coastal noprint ;
class state ;
var gencst tidalcst ;
output out=in.cstsumm sum= ;
format state $stgroup. ;
run;

------------------- MEANS output dataset ---------------
```

STATE	_TYPE_	_FREQ_	GENCST	TIDALCST
	0	26	12383	88633
Offshore	1	3	7390	34956
Gulf	1	4	861	12046
West	1	3	1293	7863
New England	1	5	473	6130
Mid Atlantic	1	6	428	10617
South Atlantic	1	5	1938	17021

notice that dataset NEW, with the new variable GEN_LEN, contains _FREQ_ and AVGGEN. These could be discarded from the dataset with a DROP statement or DROP dataset option.

Exhibit 17.8 (see p. 332) identifies minimum and maximum state and ocean names with the MINID and MAXID options. Notice that the statistics requested in the PROC statement differ from those in the OUTPUT statement. The output dataset contains more information than the printed report. Not only does it contain statistics for GENCST, but it also tells which observations have the extreme values: it shows, for example, that on the east coast (COAST = ec) New Hampshire (MINST = nh) has the smallest value of GENCST and Florida (MAXST = fl) the largest.

Exhibit 17.9 (see p. 333) uses the $STGROUP. format introduced in Exhibit 17.4. This time the output dataset's observations are reordered and displayed in formatted (or alphabetical) order with the ORDER option. Had any values of STATE been missing, they would form a separate category since the MISSING option was taken.

Exhibit 17.10 (see p. 334) processes the input dataset BY variable COAST. NWAY is specified in the PROC to avoid the lower levels of _TYPE_. The result is exactly the same as specifying both COAST and OCEAN in the CLASS statement. Although the saving in computer resources is trivial in this example, such BY-group processing is handy when processing large datasets with many CLASS variable categories.

EXHIBIT 17.5   Two class variables, several statistics

------------------- Program listing -------------------

```
proc means data=in.coastal noprint ;
class coast ocean ;
var gencst tidalcst ;
output out=tempstat sum(gencst tidalcst)=totgen tottide
       n(gencst)=ngen    mean(gencst)=avggen ;
run;
```

------------------- MEANS output dataset -----------------

COAST	OCEAN	_TYPE_	_FREQ_	TOTGEN	TOTTIDE	NGEN	AVGGEN
		0	26	12383	88633	24	515.96
	ar	1	1	1060	2521	1	1060.00
	at	1	15	2069	28673	13	159.15
	gu	1	5	1631	17141	5	326.20
	pa	1	5	7623	40298	5	1524.60
ec		2	20	3700	45814	18	205.56
wc		2	6	8683	42819	6	1447.17
ec	at	3	15	2069	28673	13	159.15
ec	gu	3	5	1631	17141	5	326.20
wc	ar	3	1	1060	2521	1	1060.00
wc	pa	3	5	7623	40298	5	1524.60

EXHIBIT 17.6   Multiple output datasets created

------------------- Program listing -------------------

```
proc means data=in.coastal noprint ;
class coast ocean ;
var gencst tidalcst ;
output out=tempstat sum(gencst tidalcst)=totgen tottide
       n(gencst tidalcst)=ngen ntide mean(gencst)=avggen ;
output n(gencst)=ngen nmiss(gencst)=missgen;
run;
```

Exercises

17.1. Refer to Exhibit 17.1.
  a. Which SAMPLE has the highest value of M_INC?
  b. Which REGION has the highest value of M_INC?
  c. Which combination of SAMPLE and REGION has the highest value of M_INC?
  d. Write a BY statement that corresponds to SUMMDATA's order.
  e. Which combination of SAMPLE and REGION is missing (not found in the dataset)?
  f. Why are REGION and SAMPLE values sometimes missing? (*Hint:* The pattern corresponds to different _TYPE_ values.)

EXHIBIT 17.7   Merge MEANS output dataset with original data

```
------------------- Program listing --------------------

proc means data=in.coastal nway noprint ;
class coast ;
var gencst ;
output out=in.summstat(drop=_type_) mean=avggen ;
run;

data new ;
merge in.coastal summstat ;
by coast;
if gencst <= avggen then gen_len = "at/below  avg";
   else if gencst > avggen then gen_len = "above  avg";
run;
```

```
------------------- MEANS output dataset ---------------------
```

COAST	OCEAN	STATE	GENCST	TIDALCST	_FREQ_	AVGGEN	GEN_LEN
ec	at	me	228	3478	20	205.56	above avg
ec	at	nh	13	131	20	205.56	at/below avg
ec	at	ma	192	1519	20	205.56	at/below avg
ec	at	ri	40	384	20	205.56	at/below avg
ec	at	ct	.	618	20	205.56	at/below avg
ec	at	ny	127	1850	20	205.56	at/below avg
ec	at	nj	130	1792	20	205.56	at/below avg
ec	at	pa	.	89	20	205.56	at/below avg
ec	at	de	28	381	20	205.56	at/below avg
ec	at	md	31	3190	20	205.56	at/below avg
ec	at	va	112	3315	20	205.56	at/below avg
ec	at	nc	301	3375	20	205.56	above avg
ec	at	sc	187	2876	20	205.56	at/below avg
ec	at	ga	100	2344	20	205.56	at/below avg
ec	at	fl	580	3331	20	205.56	above avg
ec	gu	fl	770	5095	20	205.56	above avg
ec	gu	al	53	607	20	205.56	at/below avg
ec	gu	ms	44	359	20	205.56	at/below avg
ec	gu	la	397	7721	20	205.56	above avg
ec	gu	tx	367	3359	20	205.56	above avg
wc	pa	ca	840	3427	6	1447.17	at/below avg
wc	pa	or	296	1410	6	1447.17	at/below avg
wc	pa	wa	157	3026	6	1447.17	at/below avg
wc	pa	hi	750	1052	6	1447.17	at/below avg
wc	pa	ak	5580	31383	6	1447.17	above avg
wc	ar	ak	1060	2521	6	1447.17	at/below avg

State GENCST compared to average for its coast

Average GENCST for each level of COAST

All CLASS levels are nonmissing since NWAY was specified.

EXHIBIT 17.8   MINID and MAXID used to aid identification of observations

---

```
------------------- Program listing -------------------

proc means data=in.coastal maxdec=0 min max nway ;
class coast ;
var gencst ;
output mean(gencst)=avgcoast std(gencst)=stdcoast
       minid(gencst(state ocean))=minst minocean
       maxid(gencst(state ocean))=maxst maxocean ;
run;

------------------- MEANS printed output -----------------


Analysis Variable : GENCST General outline


COAST  N Obs      Minimum        Maximum
------------------------------------------
ec      20           13             770
wc       6          157            5580
------------------------------------------


------------------- MEANS output dataset -----------------
```

COAST	_TYPE_	_FREQ_	AVGCOAST	MINGEN	MAXGEN	MINST	MINOCEAN	MAXST	MAXOCEAN
ec	1	20	205.56	13	770	nh	at	fl	gu
wc	1	6	1447.17	157	5580	wa	pa	ak	pa

Exercises
(continued)

17.2. In each statement below indicate the meaning of the variables contained in the output dataset. The following VAR statement is used:

```
VAR R1 R2 R3;
```

a. `output out=temp sum(r1) = _r1;`
b. `output out=temp mean=;`
c. `output out=temp n(r1 r2 r3)=count1-count3`
```
           mean(r1) = mean1
           std=stdr1-stdr3 ;
```

17.3. Write MINID and MAXID statements to meet the following requirements. The dataset contains SHRTNAME and CITY in addition to analysis variables WINPCT and TSCORE. As an example, identify via SHRTNAME the observation with the largest average TSCORE:

```
maxid(tscore(shrtname)) = mscoren;
```

a. `CITY with the smallest WINPCT`
b. `SHRTNAME and CITY for the largest TSCORE`
c. `CITY with smallest TSCORE, CITY with smallest WINPCT (use a single statement)`

17.4. A MEANS procedure uses the following CLASS and VAR statements:

```
class year rating;
var mpg curbwgt;
```

EXHIBIT 17.9   User-written format used to group; ORDER and MISSING options used

```
------------------ Program listing -------------------

proc means data=in.coastal maxdec=2 missing
          order=formatted sum n ;
class state ;
var gencst ;
output out=in.cstsumm sum= ;
format state $stgroup. ;
run;


------------------ MEANS printed output -------------------

Analysis Variable : GENCST General outline

STATE           N Obs   N          Sum
-----------------------------------------------
Gulf              4     4        861.00
Mid Atlantic      6     5        428.00
New England       5     4        473.00
Offshore          3     3       7390.00
South Atlantic    5     5       1938.00
West              3     3       1293.00
-----------------------------------------------


------------------ MEANS output dataset -------------------

STATE           _TYPE_   _FREQ_   GENCST
                  0        26      12383
Gulf              1         4        861
Mid Atlantic      1         6        428
New England       1         5        473
Offshore          1         3       7390
South Atlantic    1         5       1938
West              1         3       1293
```

ORDERED-FORMATTED arranges in ascending, alphabetical order.

Exercises
(continued)

Create OUTPUT datasets that meet the following requirements. Choose your own dataset and variable names, adding statements other than OUTPUT when necessary.

a. Output only nonmissing combinations of YEAR and RATING. The dataset should contain the sum and nonmissing count of each variable.

b. The dataset contains the sum and average of each variable. It should also identify the minimum MPG's observation value of variable MODEL.

c. The dataset should be sent to a permanent SAS dataset identified by a LIBNAME of SUMMOUT. Drop the _FREQ_ variable automatically created by MEANS. The dataset should contain the mean of each analysis variable.

17.5. Describe the contents of dataset COMP.

```
proc means data=master nway;
class year;
```

EXHIBIT 17.10   NWAY option, BY-group processing

```
------------------ Program listing --------------------

proc means nway mean  maxdec=2 data=in.coastal ;
class ocean ;
var tidalcst ;
output out=summ mean= ;
by coast;
run;

------------------ MEANS printed output --------------------

Analysis Variable : TIDALCST Detailed shoreline

East or West coast=ec ----------------------------------------------------

OCEAN  N Obs        Mean
-------------------------
at        15      1911.53
gu         5      3428.20
-------------------------


East or West coast=wc ----------------------------------------------------

OCEAN  N Obs        Mean
-------------------------
ar         1      2521.00
pa         5      8059.60

------------------ MEANS output dataset --------------------

COAST    OCEAN   _TYPE_    _FREQ_    TIDALCST

ec       at         1       15      1911.53
ec       gu         1        5      3428.20
wc       ar         1        1      2521.00
wc       pa         1        5      8059.60
```

Exercises
(continued)

```
var mpg;
output out=avgmpg mean=meanmpg;
run;

data comp(keep=pct year mpg model);
merge master avgmpg;
by year;
pct = (mpg / meanmpg) * 100 ;
run;
```

17.6. Refer to Exhibit 17.5 to answer the following questions.
   a. Why do the values for _FREQ_ and NGEN disagree?
   b. Which coast has the longest "general" shoreline?
   c. What is the average "general" coastline for the entire Gulf Coast? The Gulf Coast in level "E" of variable COAST? The _TYPE_ 1 and 3 values for the analysis variables agree even though level 3's values are supposed to be more finely broken out. Why does this equivalence exist?

17.2 PROC STANDARD

A common requirement for data display and analysis is that data be standardized to a given mean and/or standard deviation. One method produces standardized variables with a mean of 0 and a standard deviation of 1, often referred to as z-scores. Another method requires adjusting values so they have a particular mean, a process often called *centering*. STANDARD is a simple procedure that creates a new SAS dataset containing variables standardized on a user-specified mean and/or standard deviation.

Syntax

The syntax of the STANDARD procedure is shown below:

```
PROC STANDARD [DATA=in_sasdataset]
    [OUT=new_sasdataset]
    [NOPRINT] [MEAN=newmean] [STD=newdev]
    [REPLACE] ;
[VAR std_vars ;]
[FREQ freq_var;]
[BY by_varlist;]
```

Some common applications of the STANDARD procedure follow:

```
proc standard data=measures out=stdmsure mean=0 std=1;
var init-numeric-post;

proc standard data=gnp replace noprint std=1;
```

The STANDARD procedure has eight elements:

1. The *in_sasdataset* identifies the analysis dataset. If none is specified, the most recently created SAS dataset is used.

2. The *new_sasdataset* identifies the output dataset. It contains both the standardized variables and any other numeric and character variables in the dataset. If the OUT option is not selected, SAS creates a dataset using its default naming scheme (DATA1, DATA2, and so on).

3. The *MEAN=newmean* specifies the new mean for the analysis variables. If omitted, the mean of the new analysis variables is identical to their mean in the input dataset.

4. The *STD=newdev* specifies the new standard deviation for the analysis variables. If omitted, the standard deviation of the new analysis variables is identical to their standard deviation in the input dataset.

5. The *REPLACE* specifies that missing values of the analysis variable be replaced with the variable's mean. If MEAN= is also entered, missing values will be set to *newmean*.

6. The *NOPRINT* suppresses the display of STANDARD output. It is usually used when STANDARD is being run only to produce an output dataset. The default action is to display the mean, standard deviation, and number of nonmissing observations for each variable in the analysis. NOPRINT is available only in version 6 of the SAS System.

7. The *std_vars* specifies the numeric variables to be analyzed. The default is all numeric variables not used in the FREQ or BY statements.

8. The *by_varlist* and *freq_var* are identical in syntax and meaning to the BY and FREQ statements in PROC MEANS. Refer to their description in Section 17.1.

Usage Notes

Keep the following points in mind when using STANDARD:

- In order for the output dataset to differ from the input dataset, at least one of MEAN, STD, or REPLACE must be chosen.

- A dataset may contain both the original, input variables and their standardized counterparts. This requires merging the OUT and DATA datasets. See Exhibit 17.12 for details.

- If BY-group processing is performed, the sample statistics (mean and standard deviation) are taken for each BY-group, not for the entire dataset.

Examples

Exhibit 17.11 illustrates the impact of the standardizing and replacement options on a sample dataset. Tobacco-producing states are compared for the year 1986. The output from STANDARD is presented first. It will be the same regardless of user-specified options (REPLACE, STD, and MEAN). The exhibit then lists the original value, the value centered on 0, then adjusted with a standard deviation of 1, and finally with the missing values replaced by the specified mean of 0.

EXHIBIT 17.11   Comparison of STANDARD options

```
---------------------- STANDARD output --------------------

NAME                  MEAN              STD            N
TONS               74891.5      127418.77044          16


---------------------- Options compared -------------------

                                            replace
                 Original            std=1   std=1
        State     Value    mean=0   mean=0  mean=0

         al          .         .        .    0.000
         ct       3,124   -71,768   -0.563  -0.600
         de          .         .        .    0.000
         fl      13,303   -61,589   -0.483  -0.515
         ga      67,815    -7,077   -0.056  -0.059
         in      12,600   -62,292   -0.489  -0.520
         ky     331,055   256,164    2.010   2.140
         ma         702   -74,190   -0.582  -0.620
         md      24,300   -50,592   -0.397  -0.423
         ms       4,884   -70,008   -0.549  -0.585
         nc     444,380   369,489    2.900   3.087
         oh      15,080   -59,812   -0.469  -0.500
         pa      21,830   -53,062   -0.416  -0.443
         sc      75,480       589    0.005   0.005
         tn      92,734    17,843    0.140   0.149
         va      73,649    -1,243   -0.010  -0.010
         wi      14,480   -60,412   -0.474  -0.505
         wv       2,848   -72,044   -0.565  -0.602
```

Exhibit 17.12 merges the standardized scores with the original data. The OUT= dataset is merged with the DATA= dataset using the RENAME dataset option with the OUT= dataset to avoid overwriting similarly named variables. A matched merge is feasible since both the order and number of the datasets' observations are identical.

EXHIBIT 17.12   Standard scores merged with original values

```
---------------------- Part of program -----------------
proc standard out=run1(keep=tons) noprint mean=0 ;
var tons;
run;

data all;
merge weed run1(rename=(tons=stdtons));
run;

---------------------- Dataset ALL --------------------
```

STATE	TONS	STDTONS
al	.	.
ct	3,124	-71,768
de	.	.
fl	13,303	-61,589
ga	67,815	-7,077
in	12,600	-62,292
ky	331,055	256,164
ma	702	-74,190
md	24,300	-50,592
ms	4,884	-70,008
nc	444,380	369,489
oh	15,080	-59,812
pa	21,830	-53,062
sc	75,480	589
tn	92,734	17,843
va	73,649	-1,243
wi	14,480	-60,412
wv	2,848	-72,044

Exercises

17.7. Write the PROC STANDARD statements to analyze dataset INVNTORY (dataset STDINVNT is output in all cases).
  a. Variables ONHAND and ONORDER have a mean of 1 and a standard deviation of 0.
  b. Follow the instructions in part a, but replace missing values in INVNTORY with the mean of nonmissing values.
  c. Follow the instructions in part b, but suppress the printed output from the procedure (the descriptive statistics for the analysis variables).
  d. Replace the mean for each missing value of ONHAND and ONORDER with the mean of nonmissing observations in each BY-group (BY variable is PARTNUM).

17.8. What, if anything, is wrong with the following statements?

    a. `proc standard data=in.v1 out=in.v2;`
    b. `proc standard data=in.v1 out=in.v2 std=0;`

17.9. Describe the contents of dataset COMPARE:

```
proc standard data=in.m1
               out=all(rename=(totscore=t_all))
    mean=0 standard=1 replace;
var totscore;
run;

proc standard data=in.m1
               out=bygrp(rename=(totscore=t_by))
    mean=0 standard=1 replace;
var totscore;
by fy;
run;

data in.compare;
merge in.m1 all bygrp;
if totscore > . then score = 'nonmiss';
   else               score = 'miss';
run;
```

17.3   PROC RANK

Ranking one or more variables is a common data management, reporting, and analysis task. You might be interested in only the top 20% scores on a test, want to print only those observations with incomes at or above median, or want to prepare continuous, interval data for analysis as ordinal data by assigning ranked groupings. The RANK procedure produces an output dataset containing one or more variables' ranking compared to all nonmissing observations. The output dataset is identical to the input dataset in the order and number of observations. No printed output is produced.

Syntax

The RANK procedure syntax is described below:

```
PROC RANK [DATA=in_sasdataset] [OUT=out_sasdataset]
          [TIES=MEAN|HIGH|LOW] [DESCENDING]
          [GROUPS=ngrp|FRACTION|NPLUS1|PERCENT] ;
[VAR analysis_vars;]
[RANKS rank_vars;]
[BY by_varlist;]
```

A common application of the RANK procedure follows:

```
proc rank data=mast out=rankmast groups=5;
var pre1 pre2 post1 post2;
ranks r_pre1 r_pre2 r_post1 r_post2;
```

The RANK procedure has eleven components:

1. The *in_sasdataset* identifies the analysis dataset. If not entered, the most recently created SAS dataset is used.

2. The *out_sasdataset* identifies the new, output SAS dataset. It contains the ranked variables and any other numeric and character variables in the dataset.

If OUT and a dataset name are not specified, SAS uses the next name in the sequence DATA1, DATA2, and so on.

3. *TIES* tells SAS how to handle tied ranks. If *MEAN* (the default) is specified, the rank will be the sum of the tied ranks divided by the number of observations tied at that value. A value of *HIGH* selects the highest of the tied ranks to be used, and *LOW* selects the lowest.

4. *DESCENDING* ranks from largest to smallest; that is, the largest value is given a rank of 1, the second largest a rank of 2, and so on. The default is ranking from smallest to largest.

5. *GROUPS=ngrp* requests that the observations be assigned to *ngrp* groups of as equal numbers of observations as possible. Group numbering begins with 0. This option can be used to create deciles (*ngrp = 10*), quintiles (*ngrp = 2*), quartiles (*ngrp = 4*), and percentiles (*ngrp = 100*).

6. *FRACTION* requests fractional rankings. The observation's rank is divided by the number of nonmissing values. Thus if 20 observations had nonmissing values and an observation ranked fourth, the rank value would be 4/20, or .20.

7. *NPLUS1* is similar to FRACTION except the denominator is the number of observations *plus one*.

8. The *PERCENT* is also similar to FRACTION. Instead of expressing the value as a fraction (e.g., .43), it multiplies the value by 100 to get a percentage (e.g., 43.00). Use *GROUPS = 100* rather than *PERCENT* to compute percentiles!

9. The *analysis_vars* specifies the numeric variables to be ranked. The default is all numeric variables not used in the FREQ or BY statements. If the RANKS statement is used, however, VAR must also be used.

10. The *rank_vars* is a list of variables to be added to the OUT dataset. There is a one-to-one correspondence between *analysis_vars* and *rank_vars*: the rank of the *n*th variable in the VAR statement's list will be represented by the *n*th variable in the RANKS statement.

11. The *by_varlist* is identical in syntax and meaning to the BY statement in PROC MEANS. Refer to its description in Section 17.1.

Usage Notes

Keep the following points in mind when using the RANKS procedure:

- If RANKS is not present, the original values of the analysis variables will be replaced by their rank values. RANKS is the means by which you can keep both the original value and its rank.

- RANK requires that you ask three questions about the form of the output dataset: how to handle tied ranks (TIES), whether to go from low to high or the reverse (DESCENDING option), and what form of ranking to use (default, GROUPS, FRACTION, NPLUS1, and PERCENT).

- If BY-group processing is specified, the ranks will be within each BY-group. Care must be taken here to avoid interpreting multiple observations with the same rank as tied ranks! True ties can take place only within the same BY-group level.

Example

Exhibit 17.13 illustrates the default options and the impact of various combinations of handling ties, ordering, and assigning group options. The data are male life expectancies in selected North and South American countries.

EXHIBIT 17.13   Comparison of RANK procedure options

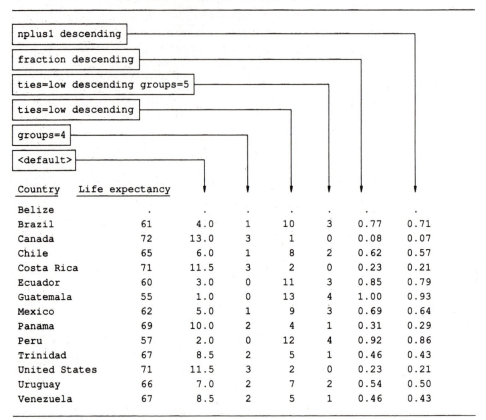

Country	Life expectancy						
Belize	.	.	.	.	.	.	.
Brazil	61	4.0	1	10	3	0.77	0.71
Canada	72	13.0	3	1	0	0.08	0.07
Chile	65	6.0	1	8	2	0.62	0.57
Costa Rica	71	11.5	3	2	0	0.23	0.21
Ecuador	60	3.0	0	11	3	0.85	0.79
Guatemala	55	1.0	0	13	4	1.00	0.93
Mexico	62	5.0	1	9	3	0.69	0.64
Panama	69	10.0	2	4	1	0.31	0.29
Peru	57	2.0	0	12	4	0.92	0.86
Trinidad	67	8.5	2	5	1	0.46	0.43
United States	71	11.5	3	2	0	0.23	0.21
Uruguay	66	7.0	2	7	2	0.54	0.50
Venezuela	67	8.5	2	5	1	0.46	0.43

Exercises

17.10. Dataset RECORDS contains numeric variables _88, _89, _90, and _91. Write RANK procedure statements to meet each of the following requirements. Use an output dataset name of RECRANK.

a. Assign decile (10% increments) rankings. The new variables should be named DEC88, DEC89, and so on.

b. Follow the same instructions as in part a, but replace the original values of _88 through _91 with their rankings.

c. Compute percentiles for the dataset. Replace the original values of the analysis variables with their percentile values.

d. Calculate upper and lower halves within each level of BY-group LEAGUE. If a tie exists, assign the higher group number. What is the value of this (higher) group number?

17.11. Input dataset LEAGUES looks like this:

LEAGUE	NONCONF
ACC	120
B10	110
B8	50
P10	22
EST	118

After running RANK, output dataset LEAGRANK looked like this:

LEAGUE	NONCONF	HALF
ACC	120	1
B10	110	1

B8	50	0
P10	22	0
EST	118	1

What PROC RANK statements were used to produce this dataset?

Summary

Chapter 17 discusses three procedures that create SAS datasets in addition to their printed output. These datasets may be used for graphical display and report writing, or they may be passed along to subsequent DATA steps for further processing, possibly comparing individual observations' values to higher-level summary data.

Extensions to the MEANS procedure allow you to capture summary detail for different combinations of the procedure's CLASS, or stratifying, variables. The same descriptive statistics that may be printed from MEANS may also be captured in one or more output datasets. The output dataset observations may be identified by nonanalysis (VAR statement) variables through the use of the MINID and MAXID statements.

The STANDARD and RANK procedures enable you to standardize and rank continuous and ordinal numeric data. These values may be used as input for report writing procedures or as a preparatory step in more sophisticated data analysis.

18 More About Character Data

This chapter turns away from "number crunching" and examines character data in more detail. The uses of character data covered so far have been fairly straightforward. We have read character values with $w. and $CHARw. formats, avoided embedded blanks when using list-style input, and restricted its use in the DATA step to character constants and decisions in IF-THEN statements.

This chapter introduces some DATA step tools that permit more complex and realistic character-data handling. These tools include an INPUT statement format that allows more flexible data entry, logical and comparison operators that increase the power of decision making and assignment statements, and functions for character-oriented tasks.

Two important principles of character-data handling are presented first. Then the new INPUT statement feature is introduced, followed by a description of the enhanced character-handling expressions and functions. The examples show that you can easily make order out of what appears to be textual chaos. Overall, the chapter demonstrates that even though SAS is not designed expressly for character data, it can nevertheless expedite its manipulation.

18.1 Character-Data Principles

When working with character data, two features are important: variable length and character-variable magnitude.

Variable Length

In Chapter 11 the LENGTH statement and its role in controlling variable size and order in the SAS dataset were examined. There and elsewhere different means were used to establish the length, or number of bytes of storage, of a character variable.

The character-handling facilities described in this chapter often make assumptions about variable lengths. Unfortunately, these are sometimes assumptions that new or even experienced SAS users would not make. Rather than discuss this seemingly anomalous behavior in this chapter, it is simpler to state a guideline: when creating a character variable, define its length in a LENGTH statement. This is not a required activity but simply a safe programming practice.

POINTERS AND PITFALLS

Avoid unpleasant surprises: always specify variable lengths for character variables being created in the DATA step.

Character-Variable Magnitude

It is easy to overlook an obscure but useful aspect of character data. Like their numerical counterparts, character variables can have a predictable order or rank: one character variable or constant can be "greater than" or "less than" another. You normally do not think of "a" being larger than "$" or "cat" being smaller than "dog". Nevertheless the *computer* does recognize character-data magnitude, and this aspect of character data may be exploited to help avoid cumbersome programming.

Magnitude is determined by the *character set* and *collating sequence* used by your computer. IBM and compatible mainframes use EBCDIC characters, while the rest of the industry uses ASCII. Although the meaning of the acronyms is not important, the way they order characters is: the comparison 'a' > '0' (the number zero), for example, will be true in ASCII but false in EBCDIC! This distinction affects not only comparisons but how the data are rearranged by PROC SORT and other order-sensitive procedures.

Appendix A compares the ASCII and EBCDIC sequences. Most new SAS users will use only one. Be aware that if you change computers (e.g., upload a PC SAS program for execution on an IBM mainframe), the behavior of comparison statements and ordering PROCs (SORT, MEANS, FREQ) may also change.

Exercises

18.1. What is the length of the character variables in dataset CHARVARS?

```
data charvars;
length name $25 add1-add4 $20 style $4;
input name $20. (add1-add4)($25.) style $4.
      year 4.;
if 1920 <= year <=1940 then era = 'Deco';
   else                     era = 'other';
run;
```

18.2. Which of the following comparisons are true? (Use the collating sequence used on the computer system you most frequently use—see Appendix A for details.)

a. 'Cat' > 'cat'
b. 'A00' > 'a00'
c. '000' < 'ZZZ'
d. '000' < 'aaa'
e. ' ' < '0'
f. ' A' > 'A '

18.2   Input Statement Extension

Character data may be read with formats ($ and $CHAR), column indicators (e.g., $2–20), or with list input. The first two forms simply pick the indicated number of columns off the raw data and move them into the SAS dataset variable, possibly removing the leading blanks. List input, however, cannot read variables with embedded blanks (those *within* a character string).

POINTERS AND PITFALLS

Peculiar results when reading character data with list input are often attributable to the omission of the & format modifier. If your character data consists of several "words" per variable and you plan to use list format, be sure you separate variables by at least two blanks.

The ampersand (&) format modifier instructs SAS to allow embedded blanks when reading character variables. It can either precede or follow the $. SAS stops reading the field when the number of columns read exceeds the variable LENGTH specified in the LENGTH statement or when SAS reads two or more consecutive blanks from the raw data field. These rules are illustrated in Exhibit 18.1.

EXHIBIT 18.1   Using the & format modifier

```
length with_amp  no_amp $ 10 ;
input with_amp $ &  no_amp $ ;
```

Entire raw data line	WITH_AMP	NO_AMP
one two three four	one two	three
one two three	one	two
one two three four five	one two th	four
one two three four five	one two th	

In the exhibit, notice that the variable WITH_AMP has multiple occurrences of single blanks. Also notice in the last line of data that SAS read a value for WITH_AM-P, and continued to scan the data beyond the length of WITH_AMP, looking for two or more blanks before starting to read NO_AMP. It did not find a pair of blanks before reaching the end of the raw data line and set NO_AMP to missing.

Exercise

18.3. The following statements are used to read five lines of raw data:

```
data chars;
infile in missover;
```

```
length list1 list2 fixed $5;
input list1 $
      list2 $ &
      @1 fixed $5.;
```

The dataset is shown below (an underscore—_—represents a blank). What values are assigned to variables LIST1, LIST2, and FIXED?

```
abc__def_g
a_b_c_d_e_fg
ab_cd__efg
abcdefghijk
abcdef_ghijk
```

18.3   Character Expressions

This section introduces several features of the DATA step that allow you to go beyond the simple character-variable assignment and logical comparisons that have been used so far. It presents two operators, for concatenation and comparison, and uses some familiar ones in a new context.

Concatenating Character Data

The concatenation operator (||) enables you to append character variables and/or constants to each other. Any number and combination of variables and constants may be used in the expression.

Trailing blanks are *not* trimmed from the operands: the value "East" stored in a variable with a length of 10 will contribute six blanks to the concatenated string. The TRIM function, described in Section 18.4, eliminates this unwanted "white space." Keep in mind that if the sum of the lengths of the operands exceeds that of the variable being created, some data will be lost.

POINTERS AND PITFALLS

Many character-handling operations require a series of functions and operators. For example, the || operator is often used with the TRIM function.

Exhibit 18.2 illustrates some applications of the concatenation operator.

Comparison Operators

Until now the only comparisons involving character data were tests of equality and inequality. All operators used for comparing numeric values may also be used with character data. This section illustrates their use and introduces an operator unique to character comparisons, the colon.

Comparing Magnitude.   The comparison operators $>$, $>=$, $<$, $<=$, $=$, and $^=$ may be used with character data. SAS follows two rules when performing comparisons. First, it compares the lengths of the operands and adds blanks to the right if one string is shorter than the other. Second, going from left to right it makes a character-by-character comparison of the two strings. SAS uses the

EXHIBIT 18.2   Using the concatenation operator ||

```
------------------ Program listing -------------------

length state $15 fullname $20;
input state & $;
fullname = state || ', USA';

------------------ Results -------------------------


STATE             FULLNAME

New  Hampshire  New  Hampshire  , USA
Alaska          Alaska          , USA
Wisconsin       Wisconsin       , USA
Washington,  DC Washington,  DC , USA


------------------ Program listing -------------------

length first last $10 degree $4 complete $25 ;
input last  $ first  $ degree $ ;
if degree = '' then complete = first || last || ', (unknown)' ;
else              complete = first || last || ', ' || degree ;

------------------ Results -------------------------


Line  of raw data          COMPLETE

Smith,   John              John      Smith,    , (un
Edwards,  Harry PhD        Harry     Edwards,  , PhD
Martinez,  Martin     LLD  Martin    Martinez, , LLD
```

computer's collating sequence for the comparisons, so it is sensitive to blanks, case, and special characters. These rules are applied in a few ASCII collating sequence comparisons in Exhibit 18.3. Spaces left blank in the columns indicate a "false" comparison.

Notice the importance of case and spacing in Exhibit 18.3. The first line's test for equality of "Jeff" and "jeff" failed because of mixed case. The third line compared the same text, but one variable had a leading blank and the other did not. Notice that in all cases the last two columns' tests against the constants 'J' and 'K' was equivalent to 'J   ' and 'K   '. The notational shortcuts used when writing numeric comparisons and lists may also be used with character data. These are illustrated below:

```
if 'A' <= group <= 'M'  |  'a' <= group <= 'm'
   then frsthalf = 'Y';

proc format;
value $assign 'a'-'m','A'-'M' = '1'
              'n'-'z','N'-'Z' = '2';
```

EXHIBIT 18.3   Using comparison operators with character data

---

```
V1 length = 5, V2 length = 10
```

v1	v2	v1 = v2	v1 > v2	v1 < v2	v1 > 'J'	v1 < 'K'
Jeff	jeff			True	True	True
Jeff	Jeff	True			True	True
Jeff	Jeff			True		True
Jeff	Jeff	True				True
Rick	Ricky			True	True	
Mark	Marc		True		True	

Comparing the First Characters of a String. What if you want to compare only the beginning characters of two character strings? Adding a colon (:) to the comparison operators restricts the comparison to the actual, not padded, characters in a character constant or to the length of the shorter character variable being compared. An ASCII-based example is presented in Exhibit 18.4.

EXHIBIT 18.4   Comparing the first character of a string

```
Spaces left blank in the columns indicate a "false" comparison.
```

District	District >: 'N'	District >: 'n'	District =: 'G'	District =: 'g'	District ^=: 'g'
Northland					True
Auckland					True
Egmont					True
East Cape					True
Bay of Plenty					True
Hawke's Bay					True
Wairapa	True				True
Wellington	True				True
Taupo	True				True
Golden Bay			True		True
Westland	True				True
Canterbury					True
Otago	True				True
Southland	True				True

Exercise

18.4. Dataset PRTFOLIO looks like this:

NAME	RATING
Gigatech	a1
Megatech	A1
Master Sys	A2
Digicorp	B
ExpSys	B1
HAL	B1

NAME's length is $10, and RATING's is $2.

a. Create variable HOLDING by concatenating NAME and RATING. RATING should be enclosed in parentheses.

b. Create variable JUDGMENT. Its value is based on RATING—those companies that are A or better have JUDGMENT values such as

```
BLUE CHIP: company name
```

Non-A RATINGs have values such as

```
SOME RISK: company name
```

c. Is the dataset already sorted by RATING?

d. How many observations will be in an output dataset if the following conditions are used in an OUTPUT statement? Indicate which collating sequence is being used.

(1) `rating > 'a'`
(2) `rating > 'A '`
(3) `rating >: 'A'`
(4) `rating >= 'B'`
(5) `rating <: 'A' | rating <: 'a'`
(6) `rating > ' '`

18.4 Character-Handling Functions

Some research and data management activities require sophisticated manipulations of character data. These manipulations include locating specific characters in a variable, replacing part of a character string, and altering the appearance of the string. This section describes a set of SAS functions designed to make these and other tasks as simple as possible. Recall that it is good practice to specify lengths for the variables being created by these functions.

In the explanation of the functions, the terms *string* and *character string* are used interchangeably and refer to any of character variables, constants, or expressions. Also keep in mind that numeric arguments to these functions must be positive integers. Missing or other invalid values will cause an error message in the Log and will set the value returned by the function to missing.

Getting Variable Information

The LENGTH function returns a numeric value indicating the length of a string. This is the position of the rightmost nonblank character in the string. If the argument is a missing value, its length is zero. The syntax of LENGTH follows:

```
LENGTH(string)
```

In the example below, if variable ANSWER is missing (length of zero) or too short to be useful (say, less than four characters), the observation is deleted from the SAS dataset being built.

```
if length(answer) < 4 then delete;
```

Other examples later in this chapter illustrate LENGTH's use with other functions. Very often you need to know the length of a character variable before using some of the other character-handling functions.

Extracting Part of a String

The SUBSTR and SCAN functions extract part or all of a character string. The syntax of SUBSTR follows:

```
SUBSTR(string,start[,length])
```

Extraction begins in the *start* column from the left and continues for *length* positions. If *length* is omitted or exceeds *string*'s length, the remainder of the string is extracted.

Exhibit 18.5 illustrates the impact of different *start* and *length* values. It also includes a portion of the Log indicating that an invalid argument was passed to SUBSTR.

EXHIBIT 18.5   Using the SUBSTR function

```
------------------- Program statement --------------------

SUB = substr('La Paz, BCS', start, length);


    START        LENGTH     SUB

      1          default    La Paz, BCS
      3            5        Paz,
      3           15        Paz, BCS
      7            1          ,
      7          default      , BCS
      1            0

------------------- SAS Log --------------------------
    2     length sub $11;
    3     sub = substr('La Paz, BCS',1,0);
NOTE: Invalid third argument to function SUBSTR at line 3 column 30.
```

The SCAN function locates and extracts a "word" within a string. Words are separated either by default or user-specified delimiters. Its syntax follows:

```
SCAN(string,word_num[,delimiters])
```

The *string* is a character string. The *word_num* is a positive integer variable, constant, or expression indicating the word, starting from the left of *string*, to extract. If *word_num* exceeds the number of words in *string*, SCAN returns a missing value. If *delimiters* is specified, SAS uses *any* of its characters as word separators. It may be either a character variable or constant. Consecutive occurrences of any combination of the separators count as one. If *delimiters* is not specified, SAS uses the string '.<>()&!$%;^-/\|'.

Exhibit 18.6 illustrates an application of the SCAN function.

Locating Strings

SAS offers several options for determining the presence or absence of a string in a character variable. The INDEX function returns the position of the first occurrence of a character variable or constant in a string. A zero is returned if the string was not located. The syntax of INDEX follows:

```
INDEX(string,loc_string)
```

EXHIBIT 18.6   Using the SCAN function

```
------------------ Program listing -------------------

* Program to extract first four words from a line of text.
  Words are delimited by commas, blanks, periods, and
  parentheses, and are a maximum of 10 characters. ;
data wordy;
input line $char40.;
array words{4} $10 word1-word4;
do i = 1 to 4;
   words{i} = scan(line, i, '(), .' );
end;
drop i;
run;

------------------ Output listing --------------------

LINE                                   WORD1       WORD2       WORD3   WORD4
  An   indented   line.                An          indented    line
Intentionally long word in this line   Intentiona  long        word    in
A word, in (paren's).                  A           word        in      paren's
```

The *string* is a character string and *loc_string* is the character variable, constant, or expression whose position is returned by the function.

INDEXC is a variation of INDEX. While INDEX was limited to a single *loc_string*, INDEXC allows more than one. The value returned by the function has exactly the same meaning as INDEX. The INDEXC syntax follows:

```
INDEXC(string,loc_string1[,loc_string2]...)
```

where *argument* and the *loc_strings* are identical in syntax and meaning to INDEX. However, if you are concerned with the location of *any* of a set of characters, you could specify them as separate *loc_strings*. If the characters were entered as a single *loc_string* using INDEX, SAS would have to find an exact match on *all* the letters to return a nonzero value. This is a subtle but important difference between the two functions.

VERIFY serves the opposite purpose of INDEX and INDEXC. While INDEX and INDEXC locate the position of a matched character, VERIFY returns the position of the first *unmatched* character. If all characters in the string-matching arguments are found in the first argument, VERIFY returns a zero. Its syntax follows:

```
VERIFY(string,loc_string1[,loc_string2]...)
```

The *string* is a character string and the *loc_strings* are lists of characters and/or constants to match in *string*. String-location functions are demonstrated in Exhibit 18.7.

Altering the Appearance of a String

SAS has several functions that change the appearance of a character variable.

COMPRESS.   The COMPRESS function removes characters from a string. The returned value does *not* preserve the locations of the eliminated characters: the string is rewritten as if the characters were never present. If the target variable's

EXHIBIT 18.7   Using the INDEX, INDEXC, and VERIFY functions

```
punc = ' ()"'',.' ;

indexc(line, punc); ─────────────────────────────────┐
index(line, punc); ───────────────────────────────┐  │
index(line, ',' ); ────────────────────────┐     │  │
index(line, ', '); ──────────────────┐      │     │  │
index(line, '.'); ────────────┐       │      │     │  │
                              ▼       ▼      ▼     ▼  ▼

Ray "Boom Boom" Mancini       0       0      0     0  5    5    0
Moses,Edwin                   0       0      6     0  6    6    0
Richard (Dick) Nixon          0       0      0     0  9    9    0
George H.W. Bush              9       0      0     0  9    9    0
Kareem Abdul Jabbar           0       0      0     0  0    0    0
Harry /Rabbit/ Walker         0       0      0     0  0    0    7
                                                          ▲    ▲
indexc(line, '(', ')', '"', '''', ',', '.'); ────────────┘    │
verify(line, 'abcdefghijklmnopqrstuvwxyz',                    │
             'ABCDEFGHIJKLMNOPQRSTUVWXYZ',                    │
             ' ()"'',.' ); ──────────────────────────────────┘
```

length was not already assigned by a LENGTH, INPUT, or assignment statement, it equals that of *string*. Its syntax follows:

```
COMPRESS(string[,list])
```

The *string* is a character string. The *list* is a character variable or constant containing characters to be "squeezed out" of *string*. If it is not entered, all blanks are eliminated. Some uses of COMPRESS are shown in Exhibit 18.8. The compressed variable could be the first step in standardizing an address (other steps might include changing strings such as "POB" and "BOX" to "PO Box").

EXHIBIT 18.8   Using the COMPRESS function

Original, input variable ADD	compress(ADD)	compress(ADD,'#:,.')
P.O. Box 4300	P.O.Box4300	PO Box 4300
Rt. 1, Box #34	Rt.1,Box#34	Rt 1 Box 34
Boston MA: 06146	BostonMA:06146	Boston MA 06146
PO Drawer 11203	PODrawer11203	PO Drawer 11203

SUBSTR.   The SUBSTR function was introduced earlier in this section as a means of extracting a portion of a character string. It may also be used in a different context. When used on the *left* of the equals sign in an assignment statement, it replaces the indicated substring with a string of equal or shorter length. This use of SUBSTR is sometimes called the *substring pseudovariable*. The syntax of SUBSTR in this new context is identical to the previous description.

If the *length* specification exceeds the length of the string, the remainder of the string is replaced.

An example of its use is presented in Exhibit 18.9. We want to eliminate the "Attn:" portion of an address line. To do so we must locate the string, convert it to a character we know is not used anywhere else in the line (in this case '>'), then eliminate that character from the variable. If we simply converted "Attn:" to blanks, there would be a gap of five spaces in the line. Using the method illustrated in Exhibit 18.9 requires a bit more effort but is more aesthetically pleasing.

EXHIBIT 18.9   Using SUBSTR to replace part of a string

```
------------------- Program listing --------------------

input name $char24.;
attn = index(upcase(name),'ATTN:');
if attn > 0 then do;
   substr(name,attn,5) = '>>>>>';
   name = compress(name, '>');
   end;

------------------- Results -------------------------
```

Name read from raw data	Name, after DATA step
attn: Marketing Dept.	Marketing Dept.
ATTN:Eastern Sales Mgr.	Eastern Sales Mgr.
JQ Adams, CEO	JQ Adams, CEO
Attn: Asst. to CEO	Asst. to CEO
AVCO, Attn: E. Abrams	AVCO, E. Abrams

Another example of this use of SUBSTR is shown in Exhibit 18.10. The array REPLACE has three elements, each representing a client's leasing of a company's product. An *x* indicates the product is currently leased. We are interested in determining the pattern of leased products.

A character variable PATTERN of length three is created (since three products are being leased), setting the appropriate column to a nonmissing value if it is leased. Then PATTERN is used in the FREQ procedure to observe its distribution. Notice that PATTERN's values tell which products were leased (the unique letters suggest product names): there were, for example, 42 occurrences of rentals of televisions and stereos (category "ts" had a frequency of 42).

UPCASE.   UPCASE converts all lowercase characters in its argument to uppercase. Its syntax follows:

    UPCASE(string)

The *string* is a character string. Nonalphabetic characters are unaffected by this function. There is no analogous function to convert to lowercase.

RIGHT and LEFT.   The RIGHT and LEFT functions align an argument. LEFT moves any leading blanks to the end of the argument, and right moves trailing

EXHIBIT 18.10   Using SUBSTR and FREQ to create PATTERN, a "distribution of distributions"

```
------------------- Program listing --------------------

data leasepat(keep=pattern);
set master.leases;
length replace1-replace3 $1 pattern $3;
array replace(3) ("f", "t", "s");
array lease(3) frniture tv stereo;
do i = 1 to 3;
   if lease(i) = 'x' then substr(pattern,i,1) = replace(i);
end;
run;

proc freq;
tables pattern;
run;
```

```
------------------- Output listing --------------------

                              Cumulative  Cumulative
PATTERN   Frequency   Percent  Frequency    Percent
-----------------------------------------------------

   s          80       47.1        80        47.1
   t           3        1.8        83        48.8
   ts         42       24.7       125        73.5
 f s          10        5.9       135        79.4
 ft           5        2.9       140        82.4
 fts         30       17.6       170       100.0
```

blanks to the beginning of the argument. Neither function changes the length of the argument. Their syntax follows:

    RIGHT(string)
    LEFT(string)

The *string* is a character string.

REVERSE.   The REVERSE function "flips" the argument, returning all of its characters, including blanks, in reverse order. The length of the argument is not changed. Its syntax follows:

    REVERSE(string)

The *string* is a character string.

The effect of the UPCASE, RIGHT, LEFT, and REVERSE functions is illustrated in Exhibit 18.11.

TRIM.   The TRIM function removes trailing blanks from a character string. TRIM is handy for making concatenated strings look more presentable. Its syntax follows:

    TRIM(string)

The *string* is a character string. The target variable in the expression has the same length as *string*, seemingly cancelling any benefit of using TRIM. TRIM is

EXHIBIT 18.11   Using the UPCASE, RIGHT, LEFT, and REVERSE functions

Original variable AUTH	upcase(AUTH)	right(AUTH)	left(AUTH)	reverse(AUTH)
Faulkner	FAULKNER	Faulkner	Faulkner	renkluaF
Joyce	JOYCE	Joyce	Joyce	ecyoJ
Wolfe	WOLFE	Wolfe	Wolfe	efloW
cummings	CUMMINGS	cummings	cummings	sgnimmuc
Russell	RUSSELL	Russell	Russell	llessuR

not typically used by itself in a DATA step, though. The other functions and operators that accompany it make its utility apparent. The TRIM function is demonstrated in Exhibit 18.12.

EXHIBIT 18.12   Using the TRIM function

```
-------------------- Program listing --------------------

length first last $10 degree $4 complete $25 ;
input last  $ first  $ degree $ ;
* Use TRIM to remove trailing blanks, then concatenate first name,
   a blank, last name, a blank, and degree status.  TRIM removes the
   excess blanks at the end of each name field, while the concatenated
   blanks ensure that the fields do not run together. ;
if degree = '' then complete = trim(first) || ' ' ||
                               trim(last) || ' (unknown)' ;
   else            complete = trim(first) || ' ' ||
                               trim(last) || ' ' ||
                               trim(degree) ;

-------------------- Result --------------------------------
```

Line of raw data	COMPLETE
Smith, John	John Smith, (unknown)
Edwards, Harry PhD	Harry Edwards, PhD
Martinez, Martin LSci	Martin Martinez, LSci

TRANSLATE.   The TRANSLATE function replaces characters in an expression. Unlike the SUBSTR pseudovariable, which required the starting position of a replacement string, TRANSLATE simply looks for all occurrences of "from" characters and replaces them with "to" characters. Its syntax follows:

TRANSLATE(string,to1,from1[,to2,from2]...)

The *string* is a character string. The *from* and *to* strings are character variables, constants, or expressions that control the replacement: all occurrences in *string* of the first character in *from* are replaced by the first character in the corresponding *to* string. TRANSLATE pairs other *from* and *to* characters and performs similar replacements. If a *to* string is shorter than a *from* string, TRANSLATE assumes a translation to blanks. Multiple *to* and *from* strings may be specified.

Exhibit 18.13 illustrates a use of the TRANSLATE function. Forward and backward slashes (/ and \) in a variable are converted to commas. Notice that if only one comma were specified in *to*, the remaining places would default to blanks.

EXHIBIT 18.13   Using the TRANSLATE function

```
------------------- Program listing --------------------

input line $char20.;
newline = translate(line, ',,', '/\');

------------------- Results ------------------------

LINE                    NEWLINE

14/  23 /./ 18.90       14,  23 ,., 18.90
12/21/17/19             12,21,17,19
8\9\12\20,8.9           8,9,12,20,8.9
21,22,17.3,18           21,22,17.3,18
```

Other Character-Handling Functions

Several character-handling functions perform tasks that may seem obscure at first but can be very handy in the right circumstances.

REPEAT.   REPEAT appends the first argument to itself a specified number of times. The target variable containing the repeated string should be long enough to hold it. If it is not, characters to the right will be truncated. The REPEAT syntax follows:

   REPEAT(string,nreps)

The *string* is a character string and *nreps* is a positive integer variable, constant, or expression indicating the number of repetitions. The returned value has *nreps* + 1 copies of *string* (the original value plus *nreps* copies). REPEAT is useful for handling some aspects of customized reports (this is discussed in Chapter 20).

An example using REPEAT is presented in Exhibit 18.14. Notice that if SCORE is less than 1, REPEAT cannot execute, thus setting BAR to a blank (character missing value). Also notice that the number of repetitions matches the score exactly since 1 was subtracted in the second argument. A user-written function could be written to create BAR but would require more work (20 assignments in a VALUE statement versus this single function call).

BYTE and RANK.   Two other functions return information about the computer's character set. BYTE returns a character in the collating sequence, and RANK returns a number indicating the rank of the character argument in the sequence. Their syntax follows:

EXHIBIT 18.14   Using the REPEAT function

```
------------------- Program listing ---------------------

length bar $20;
input id score @@ ;
bar = repeat('*', score-1);

------------------- Results -------------------------

ID      SCORE BAR

00890     19 ******************
00709      8 *******
00802     20 *******************
00764     12 ***********
00766      1 *
00877      .
```

```
        BYTE(seq_num)
        RANK(char)
```

The *seq_num* is a number in the machine's collating sequence. ASCII values usually range from 0 to 127. EBCDIC values may range from 0 to 255. The *char* is a one-character string whose position in the sequence is of interest. Exhibit 18.15 contains an example using the ASCII collating sequence. It identifies all printable nonalphabetic characters. For a complete enumeration of both ASCII and EBCDIC character sets, see Appendix A.

An Extended Example

Exhibit 18.16 (see p. 359) contains the SAS Log and output listing of a somewhat complicated application of functions handling character data. Each observation contains child-care data: the child's name, the date of attendance, and a variable containing up to five disposition characteristics ("happy" "curious," "fussy," "teary," "busy").

The characteristics may be entered in any order, in any mixture of upper- and lowercase, and are separated by any combination of blanks, commas, and slashes (/). The desired output is a set of indicators: 0 indicates the absence of a trait, and 1 indicates its presence. The strategy is to read the traits variable word by word and compare each word to a list of valid traits. If a trait matches, the indicator is set to 1. If not, the indicator remains 0. Multiple entries of a trait and entry of invalid traits are not flagged. Refer to the comments in the program for more details.

Exercises

18.5. Variable ZIPCODE is read with format $10. It reads ZIP codes in the form 07417 or 07417-1280. Create variable _9DIGIT. It equals 1 if ZIPCODE has a hyphen separating the fifth and seven digits. Otherwise it equals 0. Write the statements three ways, using the LENGTH, INDEX, and SUBSTR functions.

18.6. Write a statement that begins a DO-loop if variable MID_NAME ends in a period.

EXHIBIT 18.15   Using the BYTE and CHAR functions

```
-------------------- Program listing --------------------

data nonalpha;
* ASCII printable characters have values of 32 to at
  least 126, so use these as loop endpoints. ;
do i = 32 to 126;
   char = byte(i);
   rank = rank(char);
   if ^('a' <= char <= 'z' | 'A' <= char <= 'Z')
      then output;
end;
run;

proc print data=nonalpha label;
label char= 'char' rank='rank; ;
run;
-------------------- Modified output listing ------------
```

char	rank	char	rank	char	rank
	32	/	47	>	62
!	33	0	48	?	63
"	34	1	49	@	64
#	35	2	50	[	91
$	36	3	51	\	92
%	37	4	52	]	93
&	38	5	53	^	94
'	39	6	54	_	95
(	40	7	55	`	96
)	41	8	56	{	123
*	42	9	57	|	124
+	43	:	58	}	125
,	44	;	59	~	126
-	45	<	60		
.	46	=	61		

Exercises
(continued)

18.7. Variable name FULLNAME (length $30) has a respondent's first, middle, and last name. Create var SHORTNAM (length $20), which has only the first and last name, separated by a single blank.

18.8. Translate the first letter of variable LINE to uppercase.

18.9. Variable HEADER has length $40. Experiment with its appearance. Write the statements required to rewrite HEADER, placing its nonblank characters flush left, flush right, and centered in the string. The three versions of HEADER should appear similar to the following (the vertical bars — | — are shown simply to emphasize placement of the text within the 40-character limits):

```
|xxxx            |
|            xxxx|
|      xxxx      |
```

18.10. Count the number of times the string '?' occurs in variable LINE (length $72). One approach is to use the COMPRESS and LENGTH functions. Choose any method that works reliably. Regardless of method, assume that a line won't consist entirely of ?'s.

EXHIBIT 18.16   Character data and functions used to analyze child-care data

```
-------------------- SAS Log ----------------------------

1     options nocenter nodate nonumber;
2
3
4     data traits;
5     length trait1-trait5 $7          /* Disposition characteristics */
6            anytrait $7               /* An individual characteristic read
7                                         from TODAY string using SCAN */
8            today $40                 /* List of child's traits for the day */
9            happy curious fussy       /* Trait indicators (0 or 1) */
10               teary busy 3 ;
11
12    *-- Set up arrays: TRAIT's variables correspond to those in PRESENT
13        (PRESENT's variable order was defined in LENGTH statement).;
14
15    array trait{5} trait1-trait5 ('HAPPY','CURIOUS','FUSSY',
16                                   'TEARY','BUSY') ;
17    array present{5} happy--busy ;
18
19    *-- Read child's name, the date, and string of traits.  Then
20        convert the traits to uppercase to simplify matching to
21        variables in TRAIT array.;
22    input name $char10. @12 date date7. @20 today $char40.;
23    today = upcase(today);
24
25    *-- Clear indicators for the child. ;
26    do i = 1 to 5;
27       present{i} = 0;
28    end;
29
30    *-- Loop 5 times since there are up to 5 traits in var. TODAY. ;
31
32    do i = 1 to 5;
33       *-- Pick a trait out of the TODAY variable.  Traits do not have
34           to be entered in the same order that they are defined in
35           the TRAIT array.  ANYTRAIT and the TRAIT array variables
36           are the same length.;
37       anytrait = scan(today, i, ' ,/');
38       if anytrait ^= ' ' then
39          do j = 1 to 5;
40             *-- Compare each trait SCANned from TODAY to the list of
41                 valid traits in the TRAIT array.  If there is a
42                 match, set the corresponding indicator to 1. ;
43             *-- We >>do not<< flag observations with traits that are
44                 not "legal."  They are ignored.  Only valid traits
45                 affect the indicator variables.;
46             if anytrait = trait{j} then present{j} = 1;
47          end;
48    end;
49
50    drop i j anytrait trait1-trait5;
51    cards;
65    run;
```

EXHIBIT 18.16   *(continued)*

```
NOTE: The data set WORK.TRAITS has 13 observations and 8 variables.
NOTE: The DATA statement used 14.00 seconds.
    66
    67    proc print;
    68    id name date;
    69    format date date7.;
    70    title1 'Child characteristics, week of October 9-13, 1989';
    71    run;
NOTE: The PROCEDURE PRINT used 5.00 seconds.


------------------- Input data ---------------------------


Amelia        09oct89 busy
Amelia        10oct89 fussy/Teary
Amelia        12oct89 happy busy happy
Amelia        13oct89 chatty,happy,curious
Andrew        12oct89 Busy,Fussy
Andrew        13oct89 happy,busy,curious
Oscar         13oct89 FUSSY,TEARY
Tim           12oct89 curious/happy
Tim           09oct89 fussy, busy
Zak           09oct89 happy curious
Zak           10oct89 curious
Zak           11oct89
Zak           12oct89 happy,curious


------------------- Output dataset --------------------
```

Child characteristics, week of October 9-13, 1989

NAME	DATE	TODAY	HAPPY	CURIOUS	FUSSY	TEARY	BUSY
Amelia	09OCT89	BUSY	0	0	0	0	1
Amelia	10OCT89	FUSSY/TEARY	0	0	1	1	0
Amelia	12OCT89	HAPPY BUSY HAPPY	1	0	0	0	1
Amelia	13OCT89	CHATTY,HAPPY,CURIOUS	1	1	0	0	0
Andrew	12OCT89	BUSY,FUSSY	0	0	1	0	1
Andrew	13OCT89	HAPPY,BUSY,CURIOUS	1	1	0	0	1
Oscar	13OCT89	FUSSY,TEARY	0	0	1	1	0
Tim	12OCT89	CURIOUS/HAPPY	1	1	0	0	0
Tim	09OCT89	FUSSY, BUSY	0	0	1	0	1
Zak	09OCT89	HAPPY CURIOUS	1	1	0	0	0
Zak	10OCT89	CURIOUS	0	1	0	0	0
Zak	11OCT89		0	0	0	0	0
Zak	12OCT89	HAPPY,CURIOUS	1	1	0	0	0

Summary

The SAS System offers a good variety of operators and functions that facilitate the manipulation of character data. Using an ampersand (&) in a list INPUT statement enables SAS to allow for embedded blanks within a character variable. Adding the colon (:) operator to expressions restricts comparisons to the beginning portion of the character strings, thus avoiding the implicit blank padding normally assumed by the DATA step.

Many functions are available for handling character data. LENGTH identifies the number of nonblank characters in a string; SCAN and SUBSTR extract and possibly replace portions of the string. Functions that can alter the appearance of a string include COMPRESS, UPCASE, RIGHT, LEFT, REVERSE, TRIM, and TRANSLATE. The chapter's examples emphasize the functions' highly specialized nature: the more practical, real-world applications of character manipulation usually require use of several functions. Even simple manipulations demand attention to character variable lengths. Explicitly defining calculated variable sizes with a LENGTH statement is a good idea.

19 Advanced DATA Step Features

The DATA step features and PROCs discussed so far have given you a solid base on which to write simple to somewhat complex programs. You can recode data, summarize it, display results and reorder data with PROCs, combine datasets, and create new variables.

But there are still many gaps to fill, some that are fairly esoteric. This chapter looks at a collection of DATA step features used for data management and reporting. Calculations across observations, methods for identifying "breakpoints" in the data, special missing values, advanced data management options, automatic variables, and several new statements and options for reading raw data are discussed.

19.1 Calculations Across Observations: The RETAIN Statement

Earlier chapters used many examples of calculations involving both numeric and character data. All of these calculations have been *within* the observation; that is, the variables used in the calculation came from the same observation. Values of one observation, for example, were not added to values of another.

This cross-observation capability, however, is desirable in many situations. You may want to compute a running total for one or more variables, count the number of occurrences of a variable's value, set indicators within a BY-group, and so on. Because SAS automatically resets all values to missing each time the DATA statement is executed, we need a way to tell it *not* to reset.

The RETAIN statement performs this function. It allows values to be held from observation to observation. It also turns the responsibility of resetting the retained variables over to the user!

POINTERS AND PITFALLS

Retaining variables across observations is a useful and common DATA step technique. You should have a clear idea of when the retained variable will be initialized, have its value altered, and be reset. Many curious analytical results can be traced to retained variables that were not handled correctly.

Syntax

The syntax of the RETAIN statement is presented below:

```
RETAIN varlist [initial_values] ... ;
```

Some common applications of the RETAIN statement follow:

```
retain _numeric_;
retain flag1-flag5 ' ';
retain cum_sale 0  use_group 'N';
```

The RETAIN statement has two components:

1. The *varlist* lists variables whose values will not be reset to missing by the DATA statement. It may be any combination of numeric and character variables and variable lists.

2. The *initial_values* is the starting value of each variable in the preceding *varlist*. Initial values for retained character variables must be enclosed in single quotes (e.g., 'y').

Usage Notes

The ability to RETAIN variables is a powerful but sometimes misunderstood feature of the DATA step language. Keep the following points in mind when using it:

- Multiple RETAIN statements may be entered in the same DATA step.
- A single RETAIN may specify both numeric and character variables.
- If the first reference to a variable in a DATA step is in a RETAIN statement, SAS assumes it is numeric. To indicate a character variable, provide an initial value of the proper length (number of characters). An alternate, better strategy is to define the variable in a preceding LENGTH statement.
- Only variables created by assignment and INPUT statements may be retained. Retaining a variable brought into the DATA step with a SET, MERGE, or UPDATE statement is not an error, just an action with no effect.
- Some SAS programmers prefer to begin RETAINed variables' names with an underscore (_).

Implicit RETAINs

Consider the following syntactically correct DATA step statements:

```
n + 1;
totsal + sal;
if nmiss(of v1-v10) > 5 then badcount + 1;
totsum + sum(of pers--gram);
```

Statements of this form are known as *sum* statements. Although they appear to be missing their left-hand portion, they are, in fact, valid. SAS recognizes the first variable in this form of statement as a retained numeric variable: the retention is *implied*.

If the variable was not given an initial value in a RETAIN statement, SAS assumes a value of zero. Retained variables used in sum statements have to be specified in a RETAIN statement only when the order of the variables is important or a different, nonzero initial value is required.

This form of assignment is also unique in its insensitivity to missing values. Adding a missing value to the retained variable does *not* set it to missing. Missing values, in effect, are treated as zeros.

Examples

The first example reads a raw dataset, counting the number of times bad data values are found for a key variable. If the number of bad data occurrences exceeds 10, SAS stops building the dataset.

```
data out.result1;
infile '\test\exam1.dat';
input id 3.    totscore 3.    (part1-part4)(3.);
if totscore < 0 then nbad + 1;
if nbad > 10 then stop;
run;
```

Exhibit 19.1 reads data containing annual earnings data sorted in ascending order by year. Cumulative totals are calculated and an observation is output if two consecutive years show negative earnings.

EXHIBIT 19.1   OUTPUT on basis of RETAINed values

```
-------------------- Program listing --------------------

data conloss;
input yr earn @@;
retain _cumearn _consec 0;
if earn < 0 then _consec + 1;
   else          _consec = 0;
_cumearn = _cumearn + earn;
if _consec > 1 then do;
   output;
   end;
   else if _consec = 0 then _cumearn = 0;
cards;
13 60 -8 14 61   5 15 62   6 16 63 -2 17 64 -3
18 65   0 19 66 -1 20 67 -2 21 68 -5 22 69   .2
run;

-------------------- Output listing --------------------

YR    EARN    _CUMEARN    _CONSEC

64    -3       -5          2
67    -2       -3          2
68    -5       -8          3
```

This next example modifies Exhibit 19.1. An observation is output only if at least one year in a decade had negative earnings. The retained variables are reset at the end of the decade.

```
data one;
length era $9 ;
input yr earn @@;
retain _flag 'n' _cumearn 0;
if          60 <= yr <= 69 then era = 'Sixties';
   else if 70 <= yr <= 79 then era = 'Seventies';
   else if 80 <= yr <= 89 then era = 'Eighties';
if earn < 0 then _flag = 'y';
_cumearn + earn;
if yr = 69 | yr = 79 | yr = 89 then do;
   if _flag = 'y' then output;
   _cumearn = 0;
   _flag      = 'n';
   end;
cards;
```

Notice that the program relies on there being observations for 69, 79, and 89 to indicate the end of each decade. In Section 19.2, a simpler and more reliable method for detecting such boundaries is used.

Exercises

19.1. Write RETAIN statements to meet the following requirements:
   a. Retain numeric variables RATE1 to RATE10.
   b. Follow the instructions in part a, but assign initial values of −1 to RATE1 through RATE5 and 0 to RATE6 through RATE10.
   c. Retain character variable TYPE with an initial value of "Rental". In the same statement, retain numeric variable BASIS. No initial value is required.

19.2. Examine the following DATA step:

```
data testretn;
set in.debug;
init = -1;
length r1-r5 $4;
retain r1-r5 '<- ? ->' r0;
length r0 3;
```

   a. What are the lengths of r0 through r5?
   b. What are the initial values of retained variables r0 through r5?
   c. What is the order of the variables as they are stored in the dataset? That is, if you ran a CONTENTS procedure with the POSITION option, what would be the order in which the variables are displayed?

19.3. What, if anything, is wrong with the following program segments?
   a. ```
retain lowareas 0;
input area median_y;
lowareas = 0;
if median_y <= 12500 then lowareas + 1;
```
   **b.** ```
retain lowareas 0;
input area median_y;
if median_y <= 12500 then lowareas + 1;
   else        lowareas = 0;
```

19.2   Identifying BY-Group Boundaries: FIRST. and LAST. Prefixes

Many report-writing and data management tasks require identifying the beginning or end of a group of observations. A report with subtotals, for example, must reset one or more accumulators at the beginning of a group and write them at the end of the group. A data management application would be to identify duplicate observations in a group.

Two special variable prefixes, FIRST. and LAST. (spoken "first dot" and "last dot"), allow the beginning and end of BY-groups to be identified. They are used only in a DATA step using SET, MERGE, or UPDATE with a BY statement. SAS keeps track of the status of the levels of the BY variable(s): is the current observation the first of this value of BY variable "region"? If so, FIRST.-REGION equals 1. If not, it is 0. Similarly, at the end of the group of like-valued regions, LAST.REGION is 1. Otherwise it is 0. If the observation's REGION value is neither first nor last, both FIRST.REGION and LAST.REGION are 0.

POINTERS AND PITFALLS

Do not use FIRST. and LAST. until you are comfortable with what they represent. Exhibits 19.2 and 19.3 illustrate how SAS maintains the variables.

Usage Notes

Keep the following points in mind when using FIRST. and LAST. prefixes:

- They may be used only with a DATA step's BY variables. Either numeric or character BY variables may be prefixed. Regardless of data type, the FIRST. and LAST. values are always numeric, either 0 or 1, never missing.

- The BY-variable list used with FIRST. and LAST. may include the DESCENDING or NOTSORTED keywords. FIRST. and LAST. have exactly the same meaning when used with these keywords.

- The FIRST. and LAST. values are maintained automatically by SAS. They cannot be changed in an assignment statement, nor are they kept in the SAS dataset being built.

- When used in an IF statement, the condition FIRST.var=1 may be shortened to simply FIRST.var. A test for FIRST.var=0 may be expressed as ^FIRST.var. Similar shorthand is valid for LAST.

- An observation can be true for *both* FIRST. and LAST. for one or all of the BY variables.

- With multiple BY variables, when a BY variable's value changes, the FIRST. and LAST. values are reset for this variable and all BY variables following it in the BY statement's variable list. This process is illustrated in Exhibit 19.3.

FIRST. and LAST. Illustrated

This section illustrates the values of FIRST. and LAST. in two datasets.

Single BY Variable. Exhibit 19.2 illustrates the values of FIRST. and LAST. with a single BY variable.

Multiple BY Variables. Exhibit 19.3 illustrates the maintenance of FIRST. and LAST. when two BY variables are used. Notice that despite the appearance of

---

EXHIBIT 19.2   Values of FIRST. and LAST. automatic variables

---

code	first.code	last.code
a	1	0
a	0	1
b	1	1
c	1	0
c	0	1
d	1	0
d	0	0
d	0	0
d	0	1
e	1	1
f	1	1
g	1	0
g	0	1

EXHIBIT 19.3   FIRST. and LAST. values with multiple BY variables

---

REGION: First BY var	BRANCH: Second BY var	first. region	last. region	first. branch	last. branch
n	100	1	0	1	0
n	100	0	0	0	0
n	100	0	0	0	1
n	105	0	0	1	1
n	110	0	1	1	1
s	110	1	0	1	0
s	110	0	0	0	1
s	120	0	0	1	0
s	120	0	0	0	1
s	125	0	1	1	1
w	125	1	0	1	1
w	130	0	0	1	1
w	140	0	1	1	1

BRANCH numbers 110 and 125 cutting across REGION levels, the values of FIRST.BRANCH and LAST.BRANCH are maintained strictly within the particular level of REGION.

Examples

This first example reworks Exhibit 19.1. Notice how FIRST. and LAST. make the logic of the program easier to grasp: it is obvious that certain accumulators and flags are being reset at the beginning of a decade. In the original program, the placement and purpose of these statements were more obscure.

```
data negdec;
set  allyrs;
by era;
retain_flag 'n' _cumearn 0;
if first.era then do;
   _cumearn = 0;
   _flag    = 'n';
   end;
if earn < 0 then _flag = 'y';
if last.era & _flag = 'y' then output;
run;
```

In Exhibit 19.4 (see p. 370) a dataset is reshaped. Dataset "initial" has 11 observations containing a respondent ID and a test value. There may be multiple observations per respondent. The exhibit shows how to create a **dataset** with a single observation per ID, holding up to four test values in separate variables. Notice the clearing of counter (N) and score variables in the DO loop for FIRST.ID. The scores in the output dataset must be RETAINed for their values not to be reset each time the DATA statement is executed. Also note that output to the dataset takes place only when LAST.ID is true (=1).

In Exhibit 19.5 (see p. 371) two datasets based on values of several BY variables are created. A sales quota dataset is read, identifying quotas for salesperson–sales area combinations. Then a quarter's worth of sales data for the salespeople is read. The quota and sales datasets are merged, and two datasets are created. The first is at the area level and contains total sales and the number of salespeople who were above quota. The second is at the salesperson level and contains his or her area, quota, sales volume, and an indicator of sales above or below quota. As in Exhibit 19.4, notice the use of FIRST. variables to reset RETAINed variables used to accumulate sales volumes.

In Exhibit 19.6 (see p. 373) FIRST. and LAST. are used to control output to a dataset. The dataset INTERIM should contain unique combinations of ID and DATE. If it does, an observation is written to dataset FINAL. Otherwise, the first occurrence of a duplicate group is written to FINAL and the remaining observations in the group are written to DUPS. Variable OBSNUM contains the observation number in the original dataset.

Exercises

19.4. Refer to Exhibit 19.2. How many observations will be in an output dataset if the following conditions are used in this statement:

```
IF condition THEN OUTPUT;
```

a. `last.code`
b. `first.code`
c. `first.code & last.code`
d. `first.code = 0 & last.code = 0`
e. `first.code = 0 | last.code = 0`
f. `first.code = 1`
g. `first.code = 1 & last.code = 0`
h. `first.code = 1 & last.code = 1`

19.5. Which of the conditions in Exercise 19.4 are equivalent?

EXHIBIT 19.4   Reshaping datasets with FIRST. and LAST.

```
------------------- Program listing --------------------

data scores;
set  initial;
by id notsorted;
array score{4} score1-score4;
retain score1-score4;
if first.id then do;
   n = 0;
   do i = 1 to 4;
      score{i} = .;
   end;
   end;
n + 1;
if n <= 4 then score{n} = testval;
if last.id then output;
keep id score1-score4;
run;

------------------- Output listing --------------------
```

Input Dataset

```
 ID    TESTVAL
100       20
100       18
100       20
100       14
100       18
104       19
101       17
101       18
103       20
103       20
103       18
```

```
- - - - - - - - - - New page of output - - - - - - - - -
```

One Observation per ID

ID	SCORE1	SCORE2	SCORE3	SCORE4
100	20	18	20	14
104	19	.	.	.
101	17	18	.	.
103	20	20	18	.

Missing values occur when there are fewer than four (4) observations per ID level.

EXHIBIT 19.5   FIRST., LAST., and RETAIN used to create multiple output datasets

```
------------------- Program listing --------------------

data area(keep=area _areatot _aboveq)
     id(keep=area saleid quota _idtot _idabove) ;
merge sales quota;
by area saleid ;
if first.area then do;
   _areatot = 0;
   _aboveq  = 0;
   end;
if first.saleid then _idtot = 0;
_areatot + volume;
_idtot + volume;
if last.saleid then do;
   if _idtot >= quota then do;
     _aboveq + 1;
     _idabove = 'y';
     end;
     else _idabove = 'n';
   output id;
   end;
if last.area then output area;
run;

------------------- Output listing --------------------
```

Dataset QUOTA: Area-Salesperson Quotas

AREA	SALEID	QUOTA
1	122	9,000
1	124	10,500
2	122	12,000
2	130	5,700
2	135	2,000
		=======
		39,200

- - - - - - - - - - - New page of output - - - - - - - - -

Dataset SALES: Sales Volume Dataset

AREA	MONTH	SALEID	VOLUME
1	1	122	3,400
1	2	122	.
1	3	122	3,100
1	1	124	5,800
1	2	124	6,500
1	3	135	4,040
2	1	122	8,900
2	2	122	8,700
2	1	130	2,010
2	2	130	1,890
2	3	130	1,905
2	1	135	700
2	2	135	980
2	3	135	280
			=======
			48,205

EXHIBIT 19.5 *(continued)*

- - - - - - - - - - New page of output - - - - - - - - -

Dataset AREA: Area Totals

AREA	_AREATOT	_ABOVEQ
1	22,840	2
2	25,365	2
	========	
	48,205	

- - - - - - - - - - New page of output - - - - - - - - -

Dataset ID: Area-Salesperson Totals

AREA	SALEID	QUOTA	_IDTOT	_IDABOVE
1	122	9,000	6,500	n
1	124	10,500	12,300	y
1	135	.	4,040	y
2	122	12,000	17,600	y
2	130	5,700	5,805	y
2	135	2,000	1,960	n
			=======	
			48,205	

Exercises
(continued)

19.6. In Exercise 19.4, which statement(s) satisfy the following conditions?
  a. Output if the level of CODE is unique.
  b. Output if the observation is at the beginning of a group of CODEs and there is more than one observation in the BY-group.
  c. Output if the observation is neither at the beginning nor end of the BY-group.

19.7. What, if anything, is wrong with the following DATA step?

```
data two;
set one;
array s{5};
retain _s1 - _s5;
if first.id then do;
   do i = 1 to 5;
      s{i} = 0;
   end;
end;
```

19.3 Special Missing Values: The MISSING Statement

In previous chapters the absence of valid numeric data was represented by a missing value (.). Although several numbers indicating missing values may be converted to missing, the single missing data value can be restrictive in some circumstances. The use of special missing values specified by the MISSING statement allows the user to keep missing data out of analyses and to be able to distinguish different categories of missing values.

EXHIBIT 19.6   FIRST. and LAST. used to identify duplicate BY-group values

---

```
------------------- Program listing --------------------

data final dups;
set  interim;
by   id date;
if sum(first.id, last.id, first.date, last.date) = 4
   then output final;
   else if first.date then output final;
   else if first.date = 0 then output dups;
run;

------------------- Output listing ---------------------

interim
  ID    DATE    OBSNUM
  1      2        1
  1      3        2
  1      4        3
  2      8        4
  2      8        5
  2      9        6
  2      9        7
  2      9        8
  2     11        9
  2     12       10
```

Duplicate ID, DATE values

- - - - - - - - - - - New page of output - - - - - - - - -

```
final
  ID    DATE    OBSNUM

  1      2        1
  1      3        2
  1      4        3
  2      8        4
  2      9        6
  2     11        9
  2     12       10
```

Only first of duplicate observations is output.

- - - - - - - - - - - New page of output - - - - - - - - -

```
dups
  ID    DATE    OBSNUM
  2      8        5
  2      9        7
  2      9        8
```

There are two reasons for using special missing values. The first is to keep track of why a value is missing. Special missing values allow accounting for up to 28 reasons for the absence of data. Rather than simply coding a period (.) to indicate a person's missing weight, you can refine the value by specifying "respondent refused," "respondent didn't know," or "value subject to an unusual amount of measurement error." The degree of detail in recording the different types of missing values is limited only by the questionnaire or data collection strategy. If one of the special missing values is later judged valid or nonmissing, it can be assigned a value using an IF-THEN statement.

The second reason for using special missings is more esoteric. The underscore (_) special missing value is the only one that can replace a master dataset's nonmissing value in an UPDATE operation.

POINTERS AND PITFALLS

Special missing values subdivide the overall "missing" category: they specify the reason for missing data and, like the "." missing value, avoid treating the value as a legitimate value.

Syntax

Special missing values are used when reading raw data. The MISSING statement associates the values with one or more variables. Its syntax is presented below:

```
MISSING special_missings ... ;
```

The *special_missings* identifies one or more special missing values that may be used by all numeric variables. These values are the underscore (_) and uppercase alphabetic characters (A through Z).

The special missings are referred to by their value in either upper- or lowercase, preceded by a period (.). If Q10's special missings were A and B, they could be used in a statement such as

```
IF Q10 = .A | Q10 = .B THEN OUTPUT;
```

provided the DATA step had a MISSING statement resembling

```
MISSING A B;
```

Usage Notes

Keep in mind the following points when using special missing values:

- The MISSING statement is required only when reading variables with the INPUT statement. Other variables may be assigned a special missing anywhere in the DATA step without using a MISSING statement.

- Because the MISSING statement is handled before SAS begins executing the DATA step, its placement is significant only if variable order is important to the program. Most SAS programmers place it near the top of the DATA step, before the first INPUT statement.

- Special missing values in the raw data must be entered in uppercase and only once in the data field (i.e., do not fill a data field read with a 3. format with AAA; a single A will suffice). The data field should not begin with a period (.).

- Special missing values have ordinal values and may be used in comparisons. The smallest special missing value is the underscore (_), followed by the period (.), A, B, and so on through Z.

- Special missings values can be used in the value ranges of user-written formats and in any other situations where the traditional missing value (.) is allowed.

- When a variable is set to missing by an UPDATE statement, it is stored in the dataset as a period (.) even though an underscore (_) was used to replace it.

Examples

Exhibit 19.7 (see p. 376) accounts for three types of nonresponse in a questionnaire by using special missing values. A frequency distribution of the data is created.

In Exhibit 19.8 (see p. 377) "no opinion" is considered a valid response for question 3. "No opinion" is recoded to 3 and a frequency table is created.

Exhibit 19.9 (see p. 378) recodes a range of special missings to an underscore (_). The dataset is used as an UPDATE transaction file. Missing values in the master dataset may be overwritten to missing.

Exhibit 19.10 (see p. 378) takes a different approach to the update problem in Exhibit 19.9. The categories of the missing values are to be preserved in the updated master dataset. The special missing values are changed to unique nonmissing values that are not normally used in the dataset (negative numbers work well in this case since the data in question is a person's weight). The master is updated, and the values are then converted back to the original special missing values.

Exercises

19.8. A SAS dataset has the following values for its 35 numeric variables (Q1--Q28Y):

```
99   Do not know
98   Refused to answer
97   Skipped question
```

Write the DATA step statements to convert these values to the following:
a. The standard SAS missing value (.)
b. Unique missing values (i.e., preserve the distinctions among the three levels but treat all of them as missing)

19.9. A program contains these statements:

```
data testmiss;
infile cards;
missing d r u;
input (v1-v3)(1.);
datalines;
ddd
DDd
...
191
182
adu
RDD
A8C
112
run;

proc freq data=testmiss;
tables v1-v3 / missing;
run;
```

What will FREQ's output look like? Just calculate individual category counts.

19.10. Which of the following conditions are true?
a. ._ < .   **c.** . < .a   **e.** .A = .a   **g.** ._ < -999999
b. . < .A   **d.** .a < .Z   **f.** ".a" = ".A"   **h.** ._ = '._'

19.11. Write IF statements to calculate MISSRNGE. Values of SCREEN between .a and .m have MISSRNGE values of −1, and SCREEN values between .n and .z set MISSRNGE to −2.

EXHIBIT 19.7   Using special missing values

```
------------------- Program listing --------------------
proc format;
value ans .d = 'Don''t know'
          .n = 'No opinion'
          .r = 'Refused'
       ;

data survey;
missing d r n ;
input respid q1-q3;
cards;
run;

proc freq;
tables q1-q3 / missing;
format q1-q3 ans.;
title 'Survey Data with Three Special Missing Values';
run;

------------------- Input data ------------------------
```

```
RESPID     Q1    Q2    Q3

  14        4     3     .
  11        D     D     2
  12        R     3     N
  13        .     4     4
  15        5     5     N
  10        R     2     R

------------------- Output listing ---------------------
```

Survey Data with Three Special Missing Values

Q1	Frequency	Percent	Cumulative Frequency	Cumulative Percent
.	1	16.7	1	16.7
Don't know	1	16.7	2	33.3
Refused	2	33.3	4	66.7
4	1	16.7	5	83.3
5	1	16.7	6	100.0

Q2	Frequency	Percent	Cumulative Frequency	Cumulative Percent
Don't know	1	16.7	1	16.7
2	1	16.7	2	33.3
3	2	33.3	4	66.7
4	1	16.7	5	83.3
5	1	16.7	6	100.0

EXHIBIT 19.7 *(continued)*

Q3	Frequency	Percent	Cumulative Frequency	Cumulative Percent
.	1	16.7	1	16.7
No opinion	2	33.3	3	50.0
Refused	1	16.7	4	66.7
2	1	16.7	5	83.3
4	1	16.7	6	100.0

EXHIBIT 19.8   Converting special missing values to nonmissing

```
------------------- Program listing -------------------

proc format;
value ans .d = 'Don''t know'
          .n = 'No opinion'
          .r = 'Refused'
      ;
run;

data survey;
missing D R N ;
input respid q1-q
cards;
run;

data survey2;
set survey;
if q3 = .n then q3 = 3;
run;

proc freq;
tables q3 / missing;
format q3 ans.;
title 'Survey Data with Three Special Missing Values';
run;

------------------- Output listing -------------------
```

Survey Data with Three Special Missing Values

Q3	Frequency	Percent	Cumulative Frequency	Cumulative Percent
.	1	16.7	1	16.7
Refused	1	16.7	2	33.3
2	1	16.7	3	50.0
3	2	33.3	5	83.3
4	1	16.7	6	100.0

EXHIBIT 19.9   Special missing values recoded for use in UPDATE

------------------- Program listing -------------------

```
data transac;
infile '\trial\data\raw\group020';
missing a b c;
input id height weight;
if .a <= weight <= .c then weight = ._;
run;

data newmast;
update master transac;
by id;
run;
```

EXHIBIT 19.10   Special missing values preserved in UPDATEd dataset

------------------- Program listing -------------------

```
data transac;
infile '\trial\data\raw\group020';
missing a b c;
input id height weight;
if          weight = .a then weight = -98;
   else if weight = .b then weight = -97;
   else if weight = .c then weight = -96;
run;

data newmast;
update master transac;
by id;
if          weight = -98 then weight = .a;
   else if weight = -97 then weight = .b;
   else if weight = -96 then weight = .c;
run;
```

19.4   More Automatic Variables: _N_ and _ERROR_

Some uses of the FIRST. and LAST. variables were described in Section 19.2. SAS automatically creates and maintains these variables whenever a DATA step uses a BY statement. Among the other "automatic variables" are the observation counter _N_ and the error indicator _ERROR_. _N_'s value represents the number of times the DATA statement has begun executing. It is often identical to the number of observations processed so far by the DATA step. _ERROR_ is 0 if no statements have caused warnings or errors to be printed, 1 otherwise. Its use is demonstrated in Section 19.7. This section presents some caveats about and examples of _N_'s use.

Usage Notes

Keep in mind the following points when using _N_:

a. It is not kept in the SAS output dataset(s). To keep it, use an assignment statement (e.g., count = _n_;).

b. Although _N_'s value can be changed, it is hard to think of a valid reason to do so. Most SAS programmers simply leave it untouched.

c. When reading raw data with multiple lines per observation, _N_ still represents the number of times the DATA step has begun executing, not the number of raw data lines read.

Examples

Three of the more common uses of _N_ are computing at the start of the DATA step that values to RETAIN, sampling observations, and using the counter in an end-of-execution condition. The three examples below illustrate these uses.

Rather than repeat the TODAY function for every observation, its value can be taken the first time through the DATA step and RETAINed. This saves some computing time.

```
data terms;
set  in.tnotes;
retain current;
if _n_ = 1 then current = today();
daysleft = term - current;
run;
```

Say you want to take a 20% nonrandom sample from a dataset. _N_ divides evenly by 5 only every fifth (i.e., .2 or 20%) observation. When _N_ does divide evenly, the observations are output to dataset OUT.SUBSET.

```
data out.subset;
set  in .master;
if mod(_n_,5) = 0 then output;
run;
```

During the development of a DATA step, you may want a subset with 50 observations or 25 occurrences of a condition, whichever comes first. _N_ is more flexible than using the OBS= option. Once the program is satisfactory, the IF statement can be deleted or commented out.

```
data trial;
merge perm.master perm.xref;
retain excess 0;
if tranrate > .250 then excess + 1;
if _n_ = 50 | excess > 25 then stop;
run;
```

Exercises

19.12. The method for calculating CURRENT in the "data terms;" example above is the most efficient way to calculate a variable once and retain its value across observations. Rewrite this DATA step without using the _N_ = 1 technique. Why does this revised program consume more computer resources?

19.13. Examine the following DATA step:

```
data partial;
set complete;
cal1 = (rate * dist) / 100 ;
if cal1 > 50 then output;
```

Suppose you want a variable indicating which observation numbers from COMPLETE were written to PARTIAL. Write the statement(s) to do this. Call the new variable COMPLOBS.

19.5   Data Management Extensions: END, NOBS, and POINT

A few more options are available in the SET, MERGE, and UPDATE statements. END identifies when the last observation was read; NOBS identifies how many observations are in the dataset; and POINT enables direct access of SAS dataset observations. These options are especially useful when performing calculations across observations with RETAINed variables and when writing custom reports (discussed at length in Chapter 20).

The END Option

The END option is used with a SET, MERGE, or UPDATE statement. It identifies a numeric variable whose value is 1 when the last observation in the dataset has been read, 0 otherwise.

Syntax.   The syntax of the END option is presented below:

```
END=end_file_var
```

The *end_file_var* is the name of the numeric variable that indicates end-of-file status.

Usage Notes.   Keep in mind the following points about END's use:

- An IF statement testing for *end_file_var* being true can be shortened in the same way as FIRST. and LAST. tests. If a SET statement used END=END-FILE, the statement IF ENDFILE THEN OUTPUT; is acceptable.

- If the OBS system or dataset options are used, the END variable may never be set to 1 since the last observation may not be read. End-file conditions in these situations are more reliably simulated by using _N_ and STOP.

- The *end_file_var* is not added to the dataset(s) being created.

- When the END option is used to output an observation only at the end of the data, it usually means that the output dataset contains totals accumulated across the dataset. In these cases some or all of the input dataset's variables for individual observations do not make sense in the context of the output dataset (see Exhibit 19.11).

- The END option identifies the end of *all* the data; that is, it is set to 0 at the end of the last dataset processed by the SET, MERGE, or UPDATE statement's dataset list. In DATA steps using more than one dataset, it is not easy to identify when an *individual* dataset runs out of observations.

Example.   Exhibit 19.11 merges two datasets, counting the number of times some conditions occur in each. It outputs only when the last observation has been read (the only OUTPUT statement in the program is located within a DO-group that is executed only at the end of the dataset). Notice that the variable TOT will never be missing: since _EXCESS1 and _EXCESS2 were both implicitly retained, they were assigned an initial value of 0. Even if the IF statement conditions were never met, TOT could still be computed and would have a value of 0.

EXHIBIT 19.11   END used to output when last observation is read by SET

```
-------------------- Program listing --------------------

data summary;
set period1(in=in1)
    period2(in=in2) end=eof;
retain _excess1 _excess2;
if        in1 & reject > .0020 then _excess1 + 1;
   else if in2 & reject > .0015 then _excess2 + 1;
if eof then do;
   tot = _excess1 + _excess2;
   output;
   end;
keep _excess1 _excess2 tot;
run;

---------------------- Datasets ----------------------
```

period1			period2	
APPL_ID	REJECT		APPL_ID	REJECT
24	.0021		87	.0034
28	.0020		88	.0024
32	.0017		91	.0019
50	.0010		95	.0016

summary

_EXCESS1	_EXCESS2	TOT
1	4	5

The NOBS Option

The NOBS option is valid only in a SET statement. It identifies a numeric variable whose value is the number of observations in all the datasets in the SET's dataset list. Since the dataset directory contains information that enables SAS to compute the number of observations, the NOBS variable is assigned and may be used before any data are actually read. Like the END option, NOBS is useful when accumulating across observations and when writing custom reports.

Syntax.   The NOBS option syntax is presented below:

```
NOBS=n_of_obs
```

The *n_of_obs* is the variable name whose value represents the number of observations in the dataset(s) being SET.

Usage Notes.   Remember the following points when using NOBS:

a. The *n_of_obs* is not added to the dataset(s) being created.

b. Although *n_of_obs* can be reset in an assignment statement, it is inadvisable and probably counterproductive to do so.

Example.   In this example two datasets are concatenated. When the last observation has been read, an observation is written to the output SAS dataset. The observation contains the percent of observations contributed by each dataset in the SET statement.

```
data percent(keep=pctf pcts);
set  first(in=f)
     second(in=s) end=eof nobs=count;
if      f then _f + 1;
   else          _s + 1;
if eof then do;
   pctf = 100 * (_f / count);
   pctf = 100 * (_s / count);
   output;
   end;
run;
```

The POINT Option

Like NOBS, the POINT option may be used only in a SET statement. It identifies a numeric variable whose value represents the number of an observation to read from the SET dataset(s). POINT is extremely useful for sampling datasets and for performing complex data management tasks such as searching and indexing. Examples of its use for sampling are presented on page 391.

Syntax.   POINT's syntax is presented below:

```
POINT=obs_num
```

The *obs_num* is a variable name whose value is an observation number in the SET dataset(s).

Usage Notes.   Remember the following "rules" when using POINT:

- If *obs_num* is missing, has an integer portion less than 1, or exceeds the number of observations in the dataset, SAS will not perform reliably and will possibly cause an error. The example on page 391 gives some suggestions for preventing this situation.

- The *obs_num* must not be the name of a variable in any of the SAS datasets being SET.

- Incorrect use of the POINT option sometimes causes the DATA step to execute without any instructions about when to stop building the dataset. A STOP statement is used to avoid this situation (an instance of what is sometimes called an *infinite loop*). The use of STOP and POINT in a DATA step are demonstrated in the example on page 391.

Exercises

19.14. Refer to Exhibit 19.11. Where else could you place the calculation of TOT and get identical results? Why do these alternate locations consume more computer resources than the program in the exhibit?

19.15. Describe the activities performed by the following program:

```
data summ;
set x.inv nobs=count;
if sum(of alg--geol) > 500 then _nlarge + 1;
if _n_ = count then output;
```

19.16. What, if anything, is wrong with the following DATA step?

```
data tot;
set master end=nomore;
if nmiss(of v1--v90) > 10 then stop;
[assignment statements affecting implicitly
RETAINed variables]
if nomore then output;
run;
```

19.6 Reading Raw Data: Extending INFILE and INPUT

A number of INFILE and INPUT features are available for reading more complex raw datasets. This section describes these features.

INFILE Extensions

The INFILE statement has several options that give the user a great deal of control over the movement of raw data. Recall that whenever SAS reads raw data, it first makes a copy of the data record. Then it moves this copy into the computer's memory. Once in memory, the data are ready to be handled by the INPUT statement. The options described below control how the data are moved into memory and how the record is presented to the INPUT statement for processing. An end-of-file option is also discussed.

Data Movement Options. Options to control data movement include LRECL, RECFM, and PAD. LRECL is short for *logical record length*, the "width" of the input file's data lines. RECFM is an abbreviation for *record format*, the type of file layout. PAD controls the *pad*ding, or filling out, of short records with blanks. None of these options are needed for routine processing. They are used to simplify reading variable-length data lines and to handle other, more complex situations.

POINTERS AND PITFALLS

LRECL, RECFM, and PAD are usually required only with special or "problem" data. Do not assume that you must use any of these options.

The syntax of LRECL, RECFM, and PAD is presented below:

```
LRECL=length
RECFM=F|D
[NO]PAD
```

LRECL sets the column limit that can be accessed in the INPUT statement: if it is longer than the actual record length, SAS fills the remaining positions with blanks or special characters.

The *length* is a positive integer constant between 1 and 32,767. It indicates how many bytes, or columns, will be available to the INPUT statement when SAS actually reads the data.

In RECFM, *F* indicates that data from the raw dataset will be read in fixed "chunks" whose size is indicated by the LRECL option. *D*, the default value, indicates that SAS is left to figure out where one record ends and another begins.

PAD ensures that the padding character will be a blank. If it is less than the actual record length, the remaining, rightmost columns will not be copied into memory or be available to the INPUT statement.

Data Presentation Options.   Options to control how the data are presented to the INPUT statement include LINESIZE, LENGTH, and TABS. LINESIZE controls how many columns are actually made available to INPUT. LENGTH stores the length of the line just read. TABS determines whether to expand tab stops in the data. Their syntax is presented below:

```
LINESIZE=n_cols
LENGTH=line_len
TABS
```

In LINESIZE, *n_cols* is a positive integer constant specifying the largest column number accessible by the INPUT statement. This option is useful if the editor used for entering the data inserts sequence numbers to the rightmost columns. Rather than reading sequence numbers in columns 73 through 80 as "real" data, for example, the LINESIZE option could specify an *n_cols* value of 72.

In LENGTH, *line_len* is a numeric variable whose value contains the length of the current record read from the raw dataset. This option can be used for identifying "short" records. It is often used to read records from datasets with several types of records, each with a different length (see Exhibit 19.13).

TABS instructs SAS to expand any tab-stop characters in the raw data to the computer system's default settings. These default settings vary from system to system. Most microcomputers, for example, are set to eight-column intervals beginning in column 9. The default, NOTABS, tells SAS to ignore the tab settings and simply treat them as nonblank data.

The END Option.   An INFILE option not related to data movement or presentation is END. Its function is exactly the same as the END option of the SET, MERGE, and UPDATE statements described in Chapter 13: it resets a specified variable from 0 to 1 when the end of the raw data is reached. Keep in mind that if the OBS option is used to limit the number of data lines read, the END variable may not be set to 1 since the physical end of the raw data may not be reached. Its syntax is presented below:

```
END=end_var
```

The *end_var* is a numeric variable whose value is 1 when the last data line has been read, and 0 otherwise.

Examples.   Exhibit 19.12 demonstrates the impact of the LRECL and PAD options. The INPUT statement uses fixed length ($CHAR10.) fields to read four variables. When all four variables are not in the line, the line is "short" and SAS begins reading the next line to satisfy the current INPUT statement. (Notice the NOTE in the "Without LRECL" listing and that the dataset has only two observations.) Adding PAD and a large LRECL ensures that SAS will have enough space to read all four variables from a single data line.

EXHIBIT 19.12   LRECL and PAD options used in INFILE

```
Raw data, read with
INPUT NAME $CHAR10. (TRAIT1-TRAIT3)($CHAR10.);
-------------------------------------------------
Arnold     short     flabby
Mary       energetic lean       type a
Lou        feisty    balding
Michael    athletic  golfer


-------------------------------------------------------------------------

Without LRECL and PAD options in INFILE

NOTE: 4 records were read from the infile C:\CHAP18\PROGS\18P150.DAT.
      The minimum record length was 26.
      The maximum record length was 36.
NOTE: SAS went to a new line when INPUT statement reached past the end
      of a line.
NOTE: The data set WORK.TRAITS has 2 observations and 4 variables.

NAME       TRAIT1    TRAIT2    TRAIT3
Arnold     short     flabby    Mary
Lou        feisty    balding   Michael


-------------------------------------------------------------------------

Using LRECL=60 and PAD options in INFILE

NOTE: 4 records were read from the infile C:\CHAP18\PROGS\18P150.DAT.
      The minimum record length was 60.
      The maximum record length was 60.
NOTE: The data set WORK.TRAITS has 4 observations and 4 variables.
NOTE: The DATA statement used 9.00 seconds.

NAME       TRAIT1    TRAIT2    TRAIT3
Arnold     short     flabby
Mary       energetic lean      type a
Lou        feisty    balding
Michael    athletic  golfer
```

Exhibit 19.13 demonstrates the effective use of the LENGTH option. The dataset contains two types of data, one for a household and one for people within the household. Because the variables are different for each record type, different INPUT statements should be used. Because there is no record-type indicator in the data itself, the only way to distinguish one type of record from another is via the line length (100 for households, 25 for people). This is not the most elegant way of handling this type of file, but it is often the only way to read many public use and commercial datasets.

Exhibit 19.14 uses INFILE options END and LINESIZE. The input dataset BATCH1.DAT was entered with a program that marked the last eight positions of the fixed-length 80-column records with a line number. LINESIZE=72 is used to avoid treating these columns as data. To track the number of observations with excessive numbers of missing values, the NMISS function is used, a value

EXHIBIT 19.13   LENGTH option used in INFILE to determine how to read a line of data

```
filename survey '\survdata\mar91';

data people hholds;
infile survey length=linelen;
retain hh_id state county n_in_hh hh_inc;
if linelen = 100 then do;
    input hh_id $7. state $2. county $3. @25 n_in_hh 2. @43 hh_inc 7. ;
    output hholds;
    end;
    else if linelen = 25 then do;
        input hh_id $7. personid $2. (age income marstat)(3. 6. $1.);
        output people;
        end;
    else delete;
run;
```

EXHIBIT 19.14   END and LINESIZE options used in INFILE

```
libname dataout '\survey\round1\';
filename datain '\survey\round1\batch1.dat';

data dataout.surv(drop=_misshi)
    manymiss(keep=_misshi) ;
infile datain linesize=72 end=eod;
input id q1-q20;
if nmiss(of q1-q20) > 10 then _misshi + 1;
output dataout.surv;
if eod = 1 then output manymiss;
run;
```

RETAINed, and an observation output with this value only when at the end of the raw data, when EOD equals 1.

INPUT Extensions

This section describes two additional features of the INPUT statement. The colon (:) format modifier provides some flexibility by permitting data to be read with a combination of list and formatted input. An asterisk (*) subscript allows specifying array elements in the INPUT statement.

The : Format Modifier.   The colon (:) format modifier combines several SAS input styles. It is placed before a format (e.g., : 3., : $char20., : date7.) and tells SAS to read the variable beginning in the next nonblank column and ending when either a blank column is read, the length of the variable (if character) is reached, or the end of the data line is reached. This feature frees the user from lining up data values in specific columns yet keeps the advantages of using special formats such as COMMA and DATE.

Exhibit 19.15 illustrates some raw data entered with commas and date values.

EXHIBIT 19.15   Using the colon (:) format modifier

```
Raw data
-------------------------
J Andrews  15,000 04feb76
W Scott   25,380 05feb87
C Thorpe  43000 03dec84

DATA step
----------------------------------------------------------
data employee;
length name $15;
input name & $ salary : comma7. hiredate : date7. ;

PRINT output
----------------------------------------
NAME        SALARY   HIREDATE
J Andrews    15000      5878
W Scott      25380      9897
C Thorpe     43000      9103
```

INPUTting Array Elements.   SAS allows all or some of an array's elements in the INPUT statement to be read. To read all elements, specify the array name and use an asterisk (*) subscript. If a specific element will be read, enter the index value directly in the subscript. The subscript may be a numeric constant, variable, or expression.

In Exhibit 19.16 the reading of all elements of an array is streamlined. Notice that not too much effort is saved, since entering "score1-score5" in the INPUT

EXHIBIT 19.16   All array elements with asterisk (*) notation read

```
Raw data
-----------------------------------
cd9 235.1 221.5 225
nb0 280 290.1 200.2 276.4 188.0
nb1 301.1 301.4 300.3 310. 299.8

DATA step
-----------------------------
data rates;
infile cards missover;
array scores{*} score1-score5;
input id $3. scores{*};
cards;

PRINT output
-----------------------------------------------------
```

SCORE1	SCORE2	SCORE3	SCORE4	SCORE5	ID
235.1	221.5	225.0	.	.	cd9
280.0	290.1	200.2	276.4	188.0	nb0
301.1	301.4	300.3	310.0	299.8	nb1

statement is not a great hardship. Consider the savings, however, if there were 50 uniquely named elements in the array.

Exhibit 19.17 complicates things a bit by reading the eight-element array one element at a time. We assume that there are no gaps in the array, that if element n is missing, elements $n + 1$ through 8 are also missing. Missing elements are set to zero. This piece-by-piece reading can also be used for other activities such as comparing successive elements or stopping reading when an element or the entire array exceeds a critical value.

EXHIBIT 19.17   Individual array elements read

```
------------------- Raw data ----------------------

pgbl 109.1 109.3 110.1 110.1 112.9 113.0 113.0 116.8
anva 105.0       .      .      .      .      .      .
rrpd 123.4 127.2 118.0       .      .      .      .
vbch 110.2 106.8 107.8 109.0 100.7       .      .      .


------------------- Program listing --------------------

data closings;
infile cards;
array close{8} close1-close8;
input co_id $4. close{1} 6.1 @;
list;
do i = 2 to 8;
   input close{i} 6.1 @ ;
   if close{i} = . then do;
      do j = i to 8;
         close{j} = 0 ;
      end;
      return;
      end;
end;
cards;


------------------- Output listing --------------------
```

CO_ID	CLOSE1	CLOSE2	CLOSE3	CLOSE4	CLOSE5	CLOSE6	CLOSE7	CLOSE8
pgbl	109.1	109.3	110.1	110.1	112.9	113	113	116.8
anva	105.0	0.0	0.0	0.0	0.0	0	0	0.0
rrpd	123.4	127.2	118.0	0.0	0.0	0	0	0.0
vbch	110.2	106.8	107.8	109.0	100.7	0	0	0.0

Exercises

19.17. A raw dataset contains two types of records. The only way to distinguish between the records is by their line length. Records 40 bytes long have the variable DIRNAME in columns 1 through 40. Records 27 bytes long have the following format:

```
file name: columns 1-8, character
extension: columns 10-12, character
size: columns 14-21, numeric
```

Write the INFILE and INPUT statements required to read this data. Use a "ref" of RAWDATIN.

19.18. A SAS program containing in-stream data was written with a mainframe text editor that inserts line numbers in columns 73 through 80. Write the INFILE statement that ensures that SAS will not treat the line numbers as data.

19.19. Refer to Exhibit 19.13. Assume that there is, in fact, a record-type indicator located in column 80, coded as either a 'P' or an 'H'. Change the DATA step to take advantage of this new variable.

19.20. You want to read the following data (the column ruler is not part of the data):

```
----+----1----+----2----+----3----+----4
Erebus    b-2   15,123 16,150   05sep72
```

The INPUT statement that follows is incomplete, since it needs colon (:) format modifiers. Insert the colons where necessary.

```
input name $10. grade $ & alt comma6.
      mean_alt comma5. disc date7.;
```

19.7 Familiar Statements in Unfamiliar Places

At this point in the book we have covered quite a lot of ground with PROCs and, in particular, DATA steps. If you have actually been programming the examples or using the features as you have been reading, you have probably found yourself developing some coding habits. One of the easiest and most beneficial habits to acquire is to consistently order the statements in the DATA step. For example, you may always put SET and MERGE statements after the DATA statement, and place LABEL, DROP, and KEEP statements at the bottom of the DATA step.

These habits permit faster coding of programs, easier debugging, and over time give your programs a characteristic look and feel. However, these habits also belie the DATA step's flexibility: *any* executable (or action) statement can be placed *anywhere* in the DATA step. Executable statements used so far in this book include those that control execution flow (IF, ELSE, DO, STOP, RETURN, DELETE) and move data (INPUT, SET, MERGE, UPDATE, OUTPUT).

POINTERS AND PITFALLS

The DATA step is much more flexible than both SAS Institute manuals and this book have led you to believe. If a task becomes too awkward or unwieldy using traditional statement ordering, experiment with different arrangements, multiple SET, MERGE, and INFILE statements, and so on. As your SAS skills increase, so will your experience base and intuition about what is possible, legal, and implausible.

If placing statements consistently is desirable, why put them in different places and potentially confuse yourself? Such "clever coding" often makes potentially cumbersome DATA steps more compact and easier to read. They can also significantly reduce computer resources required to perform a task. There is also the more ephemeral issue of program elegance: many of the features employed later in this section are neat, succinct means of solving unwieldy problems.

The purpose of this section is not to present the definitive grab bag of legitimate-yet-clever DATA step tricks. It simply illustrates how you can increase the power of the DATA step without having to learn any new statements.

Two caveats must be pointed out. First, unusual placement, especially for data movement statements, can have unanticipated effects. *Run test programs on small datasets and check the results carefully.* Second, not all techniques work exactly the same in all programming environments. If a program will be developed on a PC and run on a mainframe, make sure early on in the process that it behaves identically in both locations.

Examples

The three examples in this section focus on data movement applications. In the first example population figures are computed with MEANS, then compared to individuals in a DATA step. The step reads (via SET) the only observation in SAS dataset NATIONAL, then loops while merging the three regional datasets and computing a few variables. Once the END statement is reached, execution jumps back to the DO and performs the merge, calculation, and output. The process is repeated until the end of the last merged dataset.

The DATA statement is executed only once. This means that the step's non-RETAINed variables will not be set to missing each time an observation is read. In effect, all variables are RETAINed. The result can be mixed. Computer resource use often drops drastically once DATA statement activity is bypassed. The downside is manifest in the IF statement. Because values are not reset, once an observation's value of DEVIATE is set to 'Y,' all remaining observations will keep that value. The situation could be avoided if an unconditional ELSE statement assigned an 'N' to DEVIATE, thus ensuring all situations are covered and nothing will be unintentionally held over to the next observation.

```
proc means data=in.all;
var rate;
output out=national(keep=natl_avg natl_sd)
       mean=natl_avg std=natl_sd;
run;

data deviate;
set national;
enddata = 0;
do while (enddata=0);
   merge reg1(rename=(rate=rate1))
         reg2(rename=(rate=rate2))
         reg3(rename=(rate=rate3))
         end=enddata;
   by regionid;
   diff = rate - natl_avg;
   if abs(diff) > 1.5 * natl_sd then deviate = 'Y';
   output;
end;
run;
```

The second example takes every third observation from a SAS dataset, aborting if POINT accidentally specifies an invalid observation number (automatic variable _ERROR_ equals 1). Just as the first example highlighted the DATA statement's variable maintenance, this one shows how _N_'s behavior is different when reading within a loop. Variable COUNT will always be 1: automatic variable _N_ is "stuck" at 1 since it counts the number of times the DATA statement is executed. In this program the DATA statement is, in fact, executed only once.

The traditional approach is to read every observation, compute MOD(N,3), and output when the remainder is zero. The execution speed improves dramatically when we point directly to the required observations. Notice that a STOP statement is required. Without it we would be caught in an "infinite loop," continually POINTing but with no indication of when to STOP, since the physical end of the dataset may never be reached.

Notice that variable COUNT will always be 1. Since _N_ is the count of the number of times the DATA statement is executed, it will never have a chance to be reset: the DATA statement is executed once, at the beginning of the DATA step, the loop continues until all observations are read, and then the STOP statement is executed. _N_ is still 1! COUNT would be more meaningful if it were assigned as COUNT = COUNT + 1.

```
data sample;
do i = 1 to totnobs by 3;
    set indata.univ1 indata.univ2 point=i nobs=totnobs;
    if _error_ = 1 then abort abend;
    count = _n_;
    output;
end;
stop;
run;
```

The third example demonstrates different sampling rates within BY-groups. Dataset WHERE has one observation per BY-group of dataset MASTER. Variables BEGIN and END contain the observation numbers of the start and end of the group. The second DATA step reads an observation from WHERE, then determines how often to sample from the group. A DO-loop points from the beginning to the end of the group and selects observations using a random number generated each time a record is read from MASTER. A STOP statement is not required since the DATA step will stop "naturally" at the end of WHERE.

```
data where(keep=begin end);
set   master;
by    class;
retain begin end;
if first.class then begin = _n_;
if last.class then do;
    end = _n_;
    output;
    end;
run;

data sample(drop=begin end rate);
set where;
if end - begin < 100 then rate = .50 ;
    else                   rate = .25 ;
do go = begin to end;
    set master point=go;
    if ranuni(54321) <= rate then output;
end;
run;
```

Exercises

19.21. Refer to the first example above and the accompanying discussion about implicitly RETAINed variables. Insert the statement that will avoid the potential problem with variable DEVIATE.

19.22. The following program uses multiple INFILE statements to read two different raw datasets. You want to read 50 observations from each dataset (dataset ONE, then, should contain 100 observations). The ellipses (...) indicate parts of the program that are not germane to the exercise.

```
data one;
infile in1 ... ;
do i = 1 to 50;
   input ... ;
end;
infile in2 ... ;
do i = 1 to 50;
   input ... ;
end;
run;
```

a. Will the program work correctly? Justify your answer.

b. Rewrite the program using only one DO-loop to read the two files.

19.23. Describe the contents of OUT.POINTER. What does each observation represent? What do variables ST and END contain?

```
data out.pointer;
set in.accts;
by branch;
retain st end;
if first.branch then st = _n_;
if last.branch then do;
   end = _n_;
   output;
   end;
keep branch st end;
```

19.24. What does the following program accomplish? What is contained in datasets SELECT and SUBSET?

```
data select;
input branch @@;
cards;
100 120 122 180 200 230 240
run;

data subset;
merge out.pointer(in=ptr) select=(in=sel);
by branch;
if ptr = 1 & sel = 1 then do;
   do i = st to end;
      set in.accts point=i;
      output;
   end;
   end;
```

Summary

The SAS data handling language has many powerful features that simplify the coding of even the most demanding tasks. Chapter 19 continues the discussion of the DATA step's syntax and logic. The RETAIN statement may be used to override the DATA step's usual action of setting all variables to MISSING at the beginning of each execution of the step. RETAIN is sometimes used with the FIRST and LAST prefixes to BY-group variables. FIRST and LAST allow the programmer to take actions when the DATA step is processing the beginning and end of a unique level of the BY-group variables.

The MISSING statement is used to allow fine-tuning of missing values. Up to 28 separate missing values may be specified for a dataset's numeric values. This flexibility, for example, allows the analyst to distinguish "don't know" values from "unavailable."

Automatic variables, those maintained by SAS but accessible to the programmer, are discussed. The END, NOBS, and POINT options in the SET, MERGE, and UPDATE statements allow greater control over identification of the end of the input datasets, supplying the number of observations in the datasets, and accessing a particular observation in a dataset. Also discussed are some extensions to the reading of raw data. Options in the INFILE and INPUT statements extend the DATA step's ability to read "problem" data.

Following standard coding practices makes programming more reliable but may also develop habits that stifle creative programming. The last section in the chapter presents a few examples of using familiar statements in unfamiliar locations in the DATA step. Programming efficiency and effectiveness may be improved with such unorthodox solutions but may bear the cost of being difficult to understand and modify.

20 Custom Reports

The SAS PROCs discussed so far display results in a reasonably appealing format. Statistics and graph axes can be clearly labelled, and columns of numbers can be printed with headings. There are situations, however, when the user needs more control over how the data are presented. For example, page headings may need variable values to make them more descriptive, and reports may require very specific layouts. And in some circumstances a raw dataset must be written. For example, some microcomputer-based programs cannot read SAS datasets. You must "export" the data from SAS into a format readable by the program. This is done with custom report features described in this chapter. This chapter presents the basics of "custom" report writing. It discusses the pros and cons of such reports, a general strategy for report design, and DATA step extensions, and provides complete examples of simple to complex reports. It draws heavily on the tools presented in Chapter 19.

20.1 Custom Report Pros and Cons

Before jumping into the thick of custom report writing, it is a good idea to determine whether a custom report is actually needed. Ask yourself the following questions:

- Is the desired output a raw dataset rather than a printed table?
- Is the printed report arranged so that the PROCs cannot display the numbers satisfactorily (even with the help of system options to control centering, page numbering, and so on)?
- Are messages needed within the DATA step to highlight data out of range, intermediate calculations, and the like?

- Have other procedures to achieve the desired results been tried? (See Chapter 25 for examples of procedures not discussed in detail elsewhere in the book and Appendix C for information on locating new procedures.)

If the answer to any of these questions is "yes," then a custom report is warranted.

There are some advantages to writing custom reports rather than accepting PROC output. First, they give the user complete control over what goes where. This enables the creation of reports that would be impossible with the PROCs described earlier in the book. Second, this flexibility permits creative display of the data. For example, subtotals can be identified more clearly, values of interest can be underlined, and some graphic display capability is possible.

There are also several drawbacks to creating custom reports. They may require a significant amount of time to write. Poorly written report-writing programs can be difficult to modify: it is much easier to add a variable to the PRINT procedure than it is to add a column to a custom report. Finally, custom reports cannot take advantage of the automatic calculation of interval midpoints, statistics, and page control options that PROCs perform. Nearly all of this dirty work must be handled by the programmer.

POINTERS AND PITFALLS

Custom reports can be attractive and persuasive but often at the cost of considerable programming effort. Carefully consider the costs and benefits before attempting complex reports.

As the examples throughout this chapter will demonstrate, report writing is an extremely powerful and often demanding process. It probably exercises more SAS statements and options than any other activity undertaken by the novice SAS programmer. Because of the number and variety of SAS features used, writng custom reports is a good way to learn the SAS DATA step language and produce useful, interesting, and, occasionally, unintentionally humorous output.

20.2 Guidelines for Designing, Programming, and Testing Custom Reports

Designing, programming, and testing custom reports is considerably easier if some general guidelines are kept in mind:

- *Combine forms of output.* Any combination of printed reports, raw datasets, and SAS datasets may be created by a single DATA step. There is no need to use one step to create the SAS dataset, another to print a report, another to log data errors, and so on. This support of different types of output makes SAS a very efficient and powerful tool.

- *Sketch, then program.* Rather than jumping in and improvising the program as you go, first make a rough sketch of how the report will appear. Do as much redesigning and tinkering as possible on paper rather than in front of the computer. Try to address the basic elements of the report in different, physically separate sections of the DATA step. These elements usually include page and column headings, format of detail lines, subtotals, and summary information at the end of the report. Not all reports, of course, require all elements.

- *Order statements correctly.* Statement order is critical. The executable, or action, statements of the DATA step are sensitive to statement order. Report lines are written in the order they are found. This obvious point becomes important when conditionally printing at the beginning and end of BY-groups and at the end of the dataset. For a dataset processed with BY-variables GROUP and TEST, for example, lines written at FIRST.GROUP should usually precede those written at FIRST.TEST. Likewise, LAST.TEST activity should be carried out prior to that of LAST.GROUP. Any end-of-file activity should take place after *any* FIRST. and LAST. activity.

- *Get it right, then pretty.* Once a preliminary report is produced, suppress the urge to dress it up until the substance of the report, the text and numbers, is accurate. A visually appealing report with faulty information is worthless.

- *Minimize arithmetic.* Try to minimize the amount of arithmetic operations in the report. There is no reason to worry about calculating summary statistics, regression coefficients, standard scores, and the like when SAS procedures can save this information in a dataset. See if you can capture the required data in a procedure's output dataset. Work with this dataset, possibly merging it with the original data. Focus on the arrangement and presentation of the data, and leave as many calculations as possible to SAS.

20.3   Report-Writing Statements

This section describes statements and options commonly used to write custom reports. Some of them should be familiar. Others, although new, simply reverse familiar processes and write, rather than read, data.

NULL Datasets

The special SAS dataset name _NULL_ is used when the DATA step's sole purpose is report writing. SAS assumes that all you want to do in the DATA step is write a report and that you do not want to save anything in a SAS dataset for later use. Using _NULL_ usually results in much faster program execution since SAS does not have to perform much of the housekeeping normally associated with the DATA statement. _NULL_'s syntax is illustrated below.

```
DATA _NULL_;
```

Keep in mind the following points about null datasets:

- When _NULL_ is used, no other dataset names may appear in the DATA statement.

- Dataset options (KEEP, DROP, RENAME, and so on) have no meaning with null datasets. They are ignored if specified.

The FILE Statement

Just as INFILE identifies where raw data are *read*, the FILE statement identifies where raw data are *written*. FILE has options that help control the format and presentation of the report. More than one FILE statement may appear in the DATA step. This flexibility enables multiple reports to be written simultaneously.

For example, you may specify one destination for a formal report containing verified data and another for data that failed one or more validity checks. This multiple file capability is the key to writing programs of virtually limitless power and, sometimes, equally limitless confusion.

The INFILE-FILE parallels can also cause semantic confusion, possibly with dire results. If you accidentally specify FILE when you mean *IN*FILE, you may inadvertently overwrite your data. It is worth the effort to learn how to protect INFILE datasets: most operating systems allow you to designate files "read only."

The syntax of the FILE statement is presented below. Integrated examples of FILE's use are found in Section 20.4.

```
FILE dest [HEADER=label] [NOTITLES] [N=lines]
    [COL=colnum] [LINE=linenum] [LINESLEFT=remlines]
    [PAGESIZE=plines]
    [LINESIZE=length] [RECFM=format]
    [LRECL=ncols] [PAD|NOPAD] [MOD|OLD]
    [PRINT|NOPRINT];
```

Some common applications of the FILE statement follow:

```
file '\export.raw' notitles;
file p1 linesleft=ll pagesize=60 n=ps;
file rawdata recfm=80 mod;
```

The FILE statement has 12 components:

1. The *dest* identifies the location of a report file. It can take one of four forms: The first is the file reference of the file containing the report. This should have been identified earlier via a FILENAME statement. The second form is 'filename,' which directly identifies the file name. The third form is LOG, which writes the report to the SAS Log file. Output will appear after the DATA step and before any NOTEs about computer resources used by the step. LOG is the default destination. The fourth form, PRINT, routes the report to the OUTPUT file, the location of output for most of the PROCs described in previous chapters.

2. *HEADER=label* identifies where SAS should look for page header statements. The *label* is the name of a statement *label* identifying the beginning of the top-of-page statements. The syntax of *label* is discussed on page 401. A RETURN statement must precede the labelled statement, and another RETURN statement must be the last statement in the page heading group. Exhibit 20.1 presents a generalized form of HEADER use. The first RETURN in the exhibit tells SAS to stop the execution of the dataset for the current observation and return to the DATA statement, ready to read the next observation. The second RETURN instructs SAS to go back to wherever it was executing before the page heading was written.

───────

EXHIBIT 20.1   Custom report DATA step schematic

───────────────────────────────

```
data _null_;
...
file print header=pagetop;
...
return;
pagetop: [page heading statements] ;
        return;
```

───────────────────────────────

3. *NOTITLES*, also written as *NOTITLE*, suppresses the printing of TITLE statements in the report. This option is useful when writing raw datasets. FOOTNOTES are always ignored when writing custom reports.

4. *N=lines* tells SAS to use *lines* lines at a time when writing the report file. A special value of *lines*, *PS*, indicates that the entire page can be written to. The default for *lines* is the maximum value of the # line pointer found in any statement.

5. *COL=colnum* and *LINE=linenum* create numeric variables named *colnum* and *linenum*, which contain the current location of the column and line pointers. These variables are altered automatically by SAS and are not kept in the SAS dataset(s) created in the DATA step.

6. *LINESLEFT=remlines* identifies a numeric variable *remlines*, which contains the number of lines available for printing on the current page. This option is useful when writing subheadings: if you do not want to begin a new category near the bottom of a page, use *remlines* in an IF statement to tell SAS to begin a new page, *then* begin the subhead. *LINESLEFT* may be abbreviated *LL*.

7. *PAGESIZE=plines* and *LINESIZE=length* establish the maximum number of lines and columns, respectively, that can be written in the report. If you attempt to write beyond *plines* or *length*, SAS goes to a new page or new line, respectively. *PAGESIZE*, abbreviated as *PS*, ranges from 20 to 500 lines. *LINESIZE*, abbreviated as *LS*, ranges from 64 to 256 columns. Defaults vary from system to system but are usually 66 and 132 columns for batch processing and 25 and 80 columns for interactive.

8. *RECFM=format* controls the format of the output file's records. Its more common values are *F* (fixed length, which requires the *LRECL* option be entered as well) or *D* (data sensitive, where lines are written as long as necessary to accommodate the data).

9. *LRECL=ncols* controls the record length of the output file. If not specified, SAS uses the *LINESIZE* option's value. Use this option if *RECFM=F* is specified.

10. *PAD* and *NOPAD* control whether SAS pads with blanks if a record is shorter than the *ncols* value of *LRECL*. The default is *NOPAD* (avoid padding, keep record length to a minimum).

11. *MOD* and *OLD* control whether SAS begins writing the report at the end of an already existing file (*MOD*) or starts writing at the beginning of the file (*OLD*, the default).

12. *PRINT* and *NOPRINT* control whether page feed and other special formatting characters should be added to the report file. This is usually needed only if the report's *dest* is not *LOG* or *PRINT*.

The PUT Statement

Just as INFILE and FILE are complementary, so are INPUT and PUT. INPUT tells SAS how to *read* data, and PUT tells how to *write* it. PUT has additional features that let you annotate output, overprint parts of a line, and quickly display variables. Multiple PUT statements may appear in the DATA step. This flexibility enables writing reports to multiple destinations and/or multiple types of lines to the same report.

The syntax of the PUT statement is virtually the same as INPUT. The +, @ (@column and trailing @), /, and # pointers may be used [specifying +(−n) allows *left* movement *n* positions]. List, column, and formatted input styles are available in PUT and may be used in the same statement. User-written formats may be used as well as SAS System formats. Refer to Exhibit 9.3 for a list of valid SAS System formats that may be used in the PUT statement.

PUT also has capabilities beyond those paralleling the INPUT statement. These features permit quick display of variables, control page feeds and over-printing, and allow annotation of output by using character constants.

Variable Display. The DATA step stores the raw data line read by the most recent INPUT statement in the special variable name _INFILE_. To display all variables in the DATA step and their current values using their default formats, use the special variable name _ALL_. SAS writes the variables in the order in which they were defined in the DATA step: it prints their name, an equals sign (=), and the value. To display a single variable, specify *varname=*. An entire array may be printed by specifying *arrayname{*}=*, while an array element may be specified *arrayname{n}=*. The _ALL_ and *varname=* forms are commonly used when trying to diagnose problems in a DATA step, and less often for more formal reports or datasets. The different forms of these variable display options are compared in Exhibit 20.2.

EXHIBIT 20.2   Annotated display of variables and data with PUT statements

```
Raw Data           put _infile_;          put char= var1= var2=;
-----------        ----------------       --------------------
   1 1 . 8 9          1 1 . 8 9           CHAR=ab VAR1=1 VAR2=2
cd 1 . . . .        cd 1 . . . .          CHAR=cd VAR1=1 VAR2=.
ab 1 2 3 4 5        ab 1 2 3 4 5          CHAR=  VAR1=1 VAR2=1
-- 9 8 . . .        -- 9 8 . . .          CHAR=-- VAR1=9 VAR2=8
!! 8 9 7 6 2        !! 8 9 7 6 2          CHAR=!! VAR1=8 VAR2=9
 put _all_;
                   -----------------------------------------------------------
VAR3=3 VAR4=4 VAR5=5 CHAR=ab VAR1=1 VAR2=2 _ERROR_=0 _N_=1
VAR3=. VAR4=. VAR5=. CHAR=cd VAR1=1 VAR2=. _ERROR_=0 _N_=2
VAR3=. VAR4=8 VAR5=9 CHAR=  VAR1=1 VAR2=1 _ERROR_=0 _N_=3
VAR3=. VAR4=. VAR5=. CHAR=-- VAR1=9 VAR2=8 _ERROR_=0 _N_=4
VAR3=7 VAR4=6 VAR5=2 CHAR=!! VAR1=8 VAR2=9 _ERROR_=0 _N_=5
 array v{3} v3-v5;
 put v{*}=;                          put v{3}=;
-----------------------              -------------
VAR3=3 VAR4=4 VAR5=5                 VAR5=5
VAR3=. VAR4=. VAR5=.                 VAR5=.
VAR3=. VAR4=8 VAR5=9                 VAR5=9
VAR3=. VAR4=. VAR5=.                 VAR5=.
VAR3=7 VAR4=6 VAR5=2                 VAR5=2
```

Page Feeds and Overprinting. Normally, SAS writes lines until it reaches the limit imposed by the PAGESIZE value. Some conditions, however, require a new page prior to the physical end of the current page. For example, if a report on sales force performance will be distributed to regional managers, then a page feed must occur when SAS encounters the beginning of a new region (FIRST.REGION = 1). This prevents overlap of different regions on the same page.

To force SAS to begin a new page, use the _PAGE_ keyword. Note that if a HEADER= option was specified in the FILE statement, the top-of-page statements will be executed.

Overprinting is often helpful to emphasize fields or to create unusual printer graphics. The special keyword OVERPRINT instructs SAS to display the rest of the PUT statement's output on top of what was already written. This allows you to overprint underscores (_), enabling underlining of titles, column headings, important data, and the like. If OVERPRINT begins a PUT statement, it applies to the previous PUT statement for the same file. OVERPRINT may be used only with PRINT and LOG files, and only when the N= option in FILE has a value of 1.

POINTERS AND PITFALLS

Not all printers allow overprinting. Test your printer with a short program to determine if you will have overprint capability. If not, compromise is necessary. Rather than underlining, for example, you could skip to the next line and write dashes or equal signs.

The program and output in Exhibit 20.3 (see p. 402) give a feel for the effect of the impact of _PAGE_ and OVERPRINT.

Annotating the Output. A report filled with variable values but having no description of their meaning is of little use. Character constants may be used in the PUT statement to enhance the report. Their syntax is exactly the same as in assignment statements. Section 20.4 contains many examples that use character constants.

Statement Labels

The HEADER= option of the file statement identifies a new feature of SAS syntax, the statement label. This is simply a tag, or identifier, at the beginning of a statement that indicates where the page header processing should begin. The label follows the same rules of length and characters as SAS variables, and it must be followed by a colon (:). As with variables, the name should communicate the statement's purpose: NEWPAGE and PAGETOP convey meaning; GOHERE and START do not.

Exercises

20.1. Write FILE statements to handle the following situations:
   a. Output is directed to file name RPT. It is a maximum of 130 columns wide and does not have any default TITLEs or FOOTNOTEs included in it. The page header lines are identified by statement label PAGETOP.
   b. Output is sent to the SAS Log. Variables are set up to identify the line and column numbers for PUT statement variables and constants. The maximum width of a page is 78 columns.
   c. A print file has to be sensitive to the possibility of multiple lines for a single observation being spread across two or more pages. Set the page size to 55 lines and set up a variable that contains the number of remaining lines on the current page. The print file identified by file reference RPTOUT1 should contain page feed characters.

20.2. Write PUT statements to meet the following requirements:
   a. Print NAME in columns 1 through 20 and SCORE1 through SCORE5 with format COMMA10. starting in column 23.
   b. Quickly list all the variables and values in the DATA step.
   c. Echo the input line just read. Each line should begin with the character string "Line just read ==>."

EXHIBIT 20.3   PUT statement's OVERPRINT option used

```
------------------- Program listing -------------------

data _null_;
set  areas;
by region;
file print notitles;
if first.region then do;
   put _page_;
   put 'Report for Region: ' region $1. // ;
   end;
put rep @15 sales comma8.2;
if sales < 0 then put overprint @15 '_____';
run;


------------------- Input dataset AREAS ---------------

REGION     REP           SALES

   e       Adams         45.2000
   e       Smith         88.3000
   e       Welch         -7.2000
   e       Williams      78.0000
   s       Andrews       -1.2000
   s       Marshall      33.4000


------------------- Output listing --------------------

Report for Region: e
Adams           45.20
Smith           88.30
Welch           -7.20
                -----
Williams        78.00


- - - - - - - - - - New page of output - - - - - - - - -

Report for Region: s
Andrews         -1.20
                -----
Marshall        33.40
```

Exercises
(continued)

d. Display NAME ($20.) on the first line, then the constant "Scores" starting at column 25 of the next line, followed by variables TEST1 through TEST5, each using a format of 5.2.

e. Variable INDENT is numeric. It contains the number of columns used to indent the current line. Write variables NAME and DEPT, each with format $20., starting at column number INDENT. Skip two columns between the variables.

20.3. What, if anything, is wrong with the following statements?

```
a. put 'Dept' @10 'Level' @1 '____' @10 '_____' ;
b. put _page // "Radon Survey - Summary Results' ;
c. put _infile_ / repeat('*', 80) ;
d.     file print header=top;
   put '| ' state $15. ' | ' fips 2. ' |';
        top: put 'State FIPS Code Listing' // ;
           return;
```

20.4   Examples of Custom Reports

Most of the examples in this section use data on national parks in the United States. Observations are parks, variables are park name, state postal code (if the park extends across states, only one state is used), coast ("W" or "E," relative to the Mississippi River), year established, and size in acres.

Key features and options used in this section are summarized in the following list:

Exhibit	Features/Options
20.4	First five lines dumped: problems in the data identified
20.5	PUT with column pointers; end-of-file processing
20.6	Multiple raw data files written in a single DATA step
20.7	Rough listing of park data
20.8	HEADER routine used to insert page title and column headings
20.9	Subtotals calculated and printed; date and page number inserted in title line
20.10	Appearance enhanced with LINESLEFT, LINE, PS, and N options
20.11	Comprehensive report-writing example
20.12	PUT for debugging: setup
20.13	PUT for debugging: PUT statements inserted at critical points in the DATA step
20.14	Final, debugged program and output
20.15	Report using generated data; PUT statement sensitivity to location

Exhibit 20.4 (see p. 404) reads the park raw data. Observations with missing acreage or short/missing state abbreviations are identified. To get a feel for what the raw data looks like, the first five lines are listed.

Exhibit 20.5 (see p. 405) is similar to Exhibit 20.4, but is a bit easier to read because the messages and data values are aligned with column pointers. When the last observation is processed (ENDDATA = 1), the number of records read, the number of records with bad data, and the distribution of bad data types are printed.

Exhibit 20.6 (see p. 407) demonstrates PUT's flexibility for writing raw data files. Three files (identified by different FILE statements) are written in a single DATA step: formatted (consistent location of data items across all records), listed (separate items by one or more spaces), and comma-delimited (suitable for processing by most microcomputer software). Notice the separation of park name and state abbreviation in the list example by two spaces (+2 column pointer) instead of one. This allows another SAS program to safely read the embedded blank in the park name field by using the & format modifier (discussed in Chapter 18).

Exhibit 20.7 (see p. 408) is a "bare bones" listing of the data. Variables are labelled on each line with character constants. Although the program is quickly written, the overall effect is cluttered.

Exhibit 20.8 (see p. 409) communicates the same information as Exhibit 20.7 but presents it more effectively by using a page header (PAGEHEAD statement). The routine prints a report title and supplies column headings for each variable. Notice the specificity required when locating the headings in PAGEHEAD's PUT statement: an adjustment in each park's PUT locations would require a similar modification of the column header locations in the PAGEHEAD routine's PUT statement.

Exhibit 20.9 (see p. 410) is a report with subtotals. The data are sorted by coast prior to the DATA step. At the beginning of a new coast (FIRST.COAST),

EXHIBIT 20.4   Lines dumped; problems in the data identified

```
-------------------- SAS Log ---------------------------

   5     filename datain 'c:\data\raw\nprkdirt.v2';
   6
   7     data _null_;
   8     infile datain missover;
   9     input park $20. @22 st $2. @25 coast $1. @27 yrestab 4.
  10          @32 acres 8.;
  11     *
  12     | List the first five records just to get a feel for what the input
  13     | data looks like. ;
  14     *;
  15     if _n_ <= 5 then put 'Input record' _n_ 3. +2 _infile_;
  16     *
  17     | Check for two forms of invalid values: short state abbreviation,
  18     | missing acreage.
  19     *;
  20     if length(st) < 2 then put 'Short state abbreviation '
  21         _n_ = st= ;
  22     if acres = . then put 'Missing acreage '
  23         _n_ = park= ;
  24     run;
```

```
Input record  1  Acadia             M  E 1919
Short state abbreviation _N_=1 ST=M
Missing acreage _N_=1 PARK=Acadia
Input record  2  Arches             UT W 1971    73378
Input record  3  Badlands           SD W 1978   243302
Input record  4  Big Bend           TX W 1935   735416
Input record  5  Biscayne           FL E 1980   173039
Missing acreage _N_=6 PARK=Canyonlands
Missing acreage _N_=8 PARK=Carlsbad Caverns
Short state abbreviation _N_=13 ST=
Short state abbreviation _N_=14 ST=M
Missing acreage _N_=15 PARK=Glacier Bay
Short state abbreviation _N_=19 ST=N
Missing acreage _N_=39 PARK=Sequoia
Missing acreage _N_=40 PARK=Shenandoah
```

> PUT statement output is embedded in SAS Log.

```
NOTE: The infile DATAIN is file C:\DATA\RAW\NPRKDIRT.V2.
NOTE: 49 records were read from the infile C:\DATA\RAW\NPRKDIRT.V2.
      The minimum record length was 30.
      The maximum record length was 39.
NOTE: The DATA statement used 12.00 seconds.
```

its name is printed and the coastal acreage total is reset to zero. At the end of the coast (LAST.COAST), the acreage total is printed. This report also demonstrates techniques for inserting dates and page numbers (variables DATE and PG) in the report header. To avoid unnecessary calculations, today's date is computed at the beginning of the DATA step (when _N_ = 1), and this value (statements 14 and 15) is retained. Notice the statement order: FIRST. processing, each park's PUT, LAST. processing, then end-of-file processing (when EOF is true, or 1).

In Exhibit 20.10 (see p. 412) the dataset is sorted by state name and the number of parks in each state is reported. Totals are printed only if the state has more than one park. This exhibit illustrates some of the FILE statement options that enhance report appearance: the LINESLEFT variable prevents a state header line or trailer being "orphaned." An orphan occurs when a header or trailer is written at the bottom of one page but space constraints force the actual data

---

EXHIBIT 20.5   Columns aligned with column pointer; end-of-file processing

---

```
------------------- SAS Log ---------------------------

 5     filename datain 'c:\data\raw\nprkdirt.v2';
 6
 7     data _null_;
 8     infile datain end=enddata missover;
 9     input park $20. @22 st $2. @25 coast $1. @27 yrestab 4.
10           @32 acres 8.;
11     *
12     | Check for two forms of invalid values: short state abbreviation,
13     | missing acreage.
14     *;
15     if length(st) < 2 then do;
16        short_st + 1;  * Increment bad abbreviation counter ;
17        badobs = 'y';  * Mark an observation as having invalid data ;
18        put 'Short state abbreviation '
19            @30 _n_ = st= ;
20        end;
21     if acres = . then do;
22        missacre + 1;  * Increment missing acreage counter ;
23        badobs = 'y';  * Mark an observation as having invalid data ;
24        put 'Missing acreage '
25            @30 _n_ = park= ;
26        end;
27     *
28     | Avoid double-counting observations: if either or both of short
29     | state and missing acres is missing, increment # bad obs. counter.
30     *;
31     if badobs = 'y' then nbadobs + 1;
32     if enddata = 1 then
33        put / '------------------------------------'
34            / _n_      3. ' records were read'
35            / nbadobs  3. ' obs. with bad data'
36            / short_st 3. ' short state abbreviations'
37            / missacre 3. ' missing acreages'
38            ;
39     run;
```

```
Short state abbreviation      _N_=1 ST=M
Missing acreage               _N_=1 PARK=Acadia
Missing acreage               _N_=6 PARK=Canyonlands
Missing acreage               _N_=8 PARK=Carlsbad Caverns
Short state abbreviation      _N_=13 ST=
Short state abbreviation      _N_=14 ST=M
Missing acreage               _N_=15 PARK=Glacier Bay
Short state abbreviation      _N_=19 ST=N
Missing acreage               _N_=39 PARK=Sequoia
Missing acreage               _N_=40 PARK=Shenandoah
```

Observation-level PUT statement output

EXHIBIT 20.5 *(continued)*

```
------------------------------------------
   49 records were read
    9 obs. with bad data
    4 short state abbreviations
    6 missing acreages
```
◄─────────────────────── End-of-file DO loop output

```
NOTE: The infile DATAIN is file C:\DATA\RAW\NPRKDIRT.V2.
NOTE: 49 records were read from the infile C:\DATA\RAW\NPRKDIRT.V2.
      The minimum record length was 30.
      The maximum record length was 39.
NOTE: The DATA statement used 17.00 seconds.
```

to be written at the beginning of the next page. The LINE indicator avoids overwriting the footnote written by the page header routine. Notice that the page header writes to line 60 (#60), then jumps to the top of the page (#1) and writes the header. This is perfectly legal, provided the pagesize system option (PS, line 4) is at least 60 and the N=PS INFILE option is specified.

Exhibit 20.11 (see p. 415) tracks the political nature of park additions. Two formats, PARTY and PRES, are created that identify the president's name and party during the years in which the national parks were established. The report prints Democrats on the left side and Republicans on the right. Park listings are interrupted when a new president is associated with park creation. Each president's name is printed at the start of his group. A park count and acreage total are printed (line 60) at the end of his group only if he added more than one park. At the end of the dataset a total for each party of number of parks and acres added is printed (lines 73–83).

This example uses the @variable form of pointer control to shift between columns. It also demonstrates the creation of a SAS dataset (OUT.PARKPRES) at the same time reporting takes place. The output is attractive and effective but carries the cost of having to write a fairly lengthy program. Notice how many features of the SAS System are brought together in this program: user-written formats, functions, retained variables, dataset options, and formatted output are all needed for this relatively elaborate output.

PUT statements are often helpful when debugging a program. The next three exhibits illustrate this use. The input data is a series of test scores sorted by an id number and test name (geology, English, and so on). The output dataset contains one observation per id-test name combination and holds up to four scores. If *any* test was taken prior to 1982, the entire set of scores for that person's test is discarded. Exhibit 20.12 (see p. 419) presents the input data and the initial program.

Although syntactically correct, the program's results are suspect since SAS dataset TWO has no observations. Exhibit 20.13 (see p. 420) shows the program with some debugging PUT statements. Any time something important happens in the program (beginning or end of a BY-group, comparisons, change of subscripts, and so on), the values of the appropriate variables are written and line feeds, character constants, and pointer control (@col for indentation) are used to present the information effectively.

Notice that GRP and YEAR are not entered as expected. GRP values begin with capital letters and so do not compare correctly to "eng." YEAR is entered with four digits instead of two, thus setting all values of BAD_GRP to missing. Since the array subscript value goes to its maximum (4), it may be prudent to

EXHIBIT 20.6   Multiple raw data files written in a single DATA step

```
-------------------- SAS Log ----------------------------

    4      filename datain 'c:\data\raw\nparks';
    5
    6      data _null_;
    7      infile datain missover;
    8      input park $20. @22 st $2. @25 coast $1. @27 yrestab 4.
    9           @32 acres 8.;
   10      file '\data\raw\19fmt' notitles;
   11      put park $20. st $2. coast $1. yrestab 4. acres 8.;
   12      file '\data\raw\19lst' notitles;
   13      put park +2 st coast yrestab acres;
   14      file '\data\raw\19com' notitles;
   15      put '"' st +(-1) '","' park +(-1) '","' coast +(-1)
   16           '",' yrestab +(-1) ',' acres ;
   17      run;
NOTE: The infile DATAIN is file C:\DATA\RAW\NPARKS.
NOTE: 49 records were read from the infile C:\DATA\RAW\NPARKS.
      The minimum record length was 39.
      The maximum record length was 39.
NOTE: The file '\data\raw\19fmt' is file C:\DATA\RAW\19FMT.
NOTE: 49 records were written to the file C:\DATA\RAW\19FMT.
      The minimum record length was 35.
      The maximum record length was 35.
NOTE: The file '\data\raw\19lst' is file C:\DATA\RAW\19LST.
NOTE: 49 records were written to the file C:\DATA\RAW\19LST.
      The minimum record length was 23.
      The maximum record length was 39.
NOTE: The file '\data\raw\19com' is file C:\DATA\RAW\19COM.
NOTE: 49 records were written to the file C:\DATA\RAW\19COM.
      The minimum record length was 27.
      The maximum record length was 43.
NOTE: The DATA statement used 17.00 seconds.

-------------------- Dataset listings --------------------
 C:\DATA\RAW\19FMT.
Acadia              MEE1919   41365
Arches              UTW1971   73378
Badlands            SDW1978  243302
Big Bend            TXW1935  735416
Biscayne            FLE1980  173039
 C:\DATA\RAW\19LST.
Acadia    ME E 1919 41365
Arches    UT W 1971 73378
Badlands  SD W 1978 243302
Big Bend  TX W 1935 735416
Biscayne  FL E 1980 173039
 C:\DATA\RAW\19COM.
"Acadia","ME","E",1919,41365
"Arches","UT","W",1971,73378
"Badlands","SD","W",1978,243302
"Big Bend","TX","W",1935,735416
"Biscayne","FL","E",1980,173039
```

EXHIBIT 20.7   Rough listing of park data (no column headings)

```
-------------------- SAS Log ----------------------------

    3    libname  out '\data\sas\';
    4    options ps=60 ls=120;
    5
    6    data _null_;
    7    set  out.natlpark;
    8    file print notitles;
    9    put 'Park: ' park $20. ' state: ' st $2.
   10        ' coast: ' coast $1. ' year estab.: ' yrestab 4.
   11        ' acres: ' acres comma9.;
   12    run;
NOTE: 49 lines were written to file PRINT.
NOTE: The DATA statement used 13.00 seconds.
Park: Acadia               state: ME coast: E year estab.: 1919 acres:    41,365
Park: Arches               state: UT coast: W year estab.: 1971 acres:    73,378
Park: Badlands             state: SD coast: W year estab.: 1978 acres:   243,302
Park: Big Bend             state: TX coast: W year estab.: 1935 acres:   735,416
Park: Biscayne             state: FL coast: E year estab.: 1980 acres:   173,039
.

.

.
Park: Wind Cave            state: SD coast: W year estab.: 1903 acres:    28,292
Park: Wrangell-St. Elias   state: AK coast: W year estab.: 1980 acres: 8,331,604
Park: Yellowstone          state: WY coast: W year estab.: 1872 acres: 2,219,784
Park: Yosemite             state: CA coast: W year estab.: 1980 acres:   761,170
Park: Zion                 state: UT coast: W year estab.: 1919 acres:   146,597
```

add an IF statement checking its value prior to assignment in the S array. The new statement assigns SCORE to an array element only if the input data has four or fewer tests for a subject-group. The modified program and the "correct" output data are shown in Exhibit 20.14 (see p. 421).

Once the program functions correctly, the PUT statements can either be removed or commented out. If you anticipate modifying the program later, it would be wise to leave the PUT statements in the DATA step as comments. This allows you to reuse the statements without much effort should the need arise. Notice how the indentation and labelling of the output assists in resolving the problems. Also notice that this debugging output can become voluminous!

Exhibit 20.15 (see p. 423) demonstrates a report using data values generated entirely within the DATA step. The MORT function is used to calculate monthly payments on a $50,000 loan at a variety of terms and interest rates. Notice the effect of the PUT statements' placement: the first PUT occurs prior to the DO-loops generating the data and is executed only once. The second is located between the END statements of the nested DO's and is executed once for each value of TERM, the loop index variable. The inner loop fills the PAY array, and the numbers are displayed *after* the loop is complete and *before* the END closing off the outer loop.

Exercises

20.4. Refer to Appendix E's description of the NATLPARK dataset. Write DATA steps to meet the following requirements:

EXHIBIT 20.8 HEADER routine used to insert page title, column headings

```
------------------- SAS Log --------------------------

  3     libname  out '\data\sas\';
  4     options ps=60 ls=120;
  5
  6     data _null_;
  7     set  out.natlpark;
  8     file print notitles header=pagehead;
  9     put @1 park $20.  @25 st $2.  @32 coast $1.  @38 yrestab 4.
 10        @45 acres comma9. ;
 11     return;
 12     pagehead: put 'National Park Data, Sorted by Park Name' //
 13                 @38 'Year' /
 14                 @1 '|------ Park ------|'
 15                 @23 'State'   @30 'Coast'   @37 'Estab.'
 16                 @45 '| Acres |' ;
 17     run;
NOTE: 52 lines were written to file PRINT.
NOTE: The DATA statement used 13.00 seconds.

------------------- Output listing --------------------

National Park Data, Sorted by Park Name
                                 Year
|------ Park ------| State Coast Estab.  | Acres |
Acadia                 ME    E    1919      41,365
Arches                 UT    W    1971      73,378
Badlands               SD    W    1978     243,302
Big Bend               TX    W    1935     735,416
Biscayne               FL    E    1980     173,039
 .
 .
 .
Wind Cave              SD    W    1903      28,292
Wrangell-St. Elias     AK    W    1980   8,331,604
Yellowstone            WY    W    1872   2,219,784
Yosemite               CA    W    1980     761,170
Zion                   UT    W    1919     146,597
```

a. List, for the first 20 observations, the name and value of each variable in the dataset. Pay no attention to how the display is formatted; just display the information with a minimum of effort.

b. Write the dataset to a "flat" file (i.e., a dataset you can read with your text editor). Do not separate variables with blank columns.

c. Sort the dataset by variable ST. Then list the data using a layout similar to this one:

```
State name
          park name    # of acres
```

Column headings are not necessary; just print the data itself.

d. Expand the report created in part c. Add a title to the report (use the HEADER option of the FILE statement, not a TITLE statement). If a state has more than one park, add the acreage for all its parks (variable TOTACRES) and print subtotals. Clearly label the subtotal lines.

(*continued on p. 414*)

EXHIBIT 20.9   Subtotals printed; date and page number inserted in title line

```
------------------- SAS Log ----------------------------

3     libname  out '\data\sas\';
4     options ps=60 ls=120;
5
6     proc sort data=out.natlpark out=sorted;
7     by coast;
8     run;
NOTE: The data set WORK.SORTED has 49 observations and 5 variables.
NOTE: The PROCEDURE SORT used 6.00 seconds.
9
10    data _null_;
11    set  sorted end=eof;
12    by coast;
13    file print notitles header=pagehead;
14    retain date;
15    if _n_ = 1 then date = today();
16    if first.coast then do;
17       coastacr = 0;
18       put / 'Coast: ' coast / ;
19       end;
20    coastacr + acres;
21    totacres + acres;
22    put @1 park $20.  @25 st $2.  @31 yrestab 4.
23        @38 acres comma10. ;
24    if last.coast then put @38 '==========' /
25                          @1 'Total acres, coast ' coast
26                          @38 coastacr comma10. ;
27    if eof then put // @38 '==========' /
28                          @1 'Grand total'
29                          @38 totacres comma10. ;
30    return;
31    pagehead: pg + 1;
32           put 'National Park Data, Sorted by Park Name'
33               @66 date date7. @74 'Page ' pg //
34               @31 'Year' /
35               @1 '|------ Park ------|'
36               @23 'State'   @30 'Estab.'
37               @38 '|- Acres |' ;
38           return;
39    run;
NOTE: 63 lines were written to file PRINT.
NOTE: The DATA statement used 19.00 seconds.


------------------- Output listing --------------------
National Park Data, Sorted by Park Name          03MAY89 Page 1


                                  Year
|------ Park ------| State  Estab.  |- Acres |
Coast: E
Acadia                ME     1919       41,365
Biscayne              FL     1980      173,039
Everglades            FL     1934    1,398,938
Great Smoky Mts.      NC     1926      520,269
Isle Royale           MI     1931      571,790
```

EXHIBIT 20.9 *(continued)*

```
Mammoth Cave              KY    1926        52,420
Shenandoah                VA    1926       195,346
Virgin Islands            VI    1956        14,695
                                          ==========
Total acres, coast E                     2,967,862
Coast: W
Arches                    UT    1971        73,378
Badlands                  SD    1978       243,302
Big Bend                  TX    1935       735,416
Bryce Canyon              UT    1924        35,835
Canyonlands               UT    1964       337,570
Capitol Reef              UT    1971       241,904
Carlsbad Caverns          NM    1930        46,755
Channel Islands           CA    1980       249,353
Crater Lake               OR    1902       183,224
Denali                    AK    1917     4,716,726
Gates of the Arctic       AK    1980     7,523,888
Glacier                   MT    1910     1,013,572
Glacier Bay               AK    1980     3,225,284
Grand Canyon              AR    1919     1,218,375
Grand Teton               WY    1929       310,521
Great Basin               NV    1986        77,109
Guadalupe Mts.            TX    1966        76,293
Hakeakala                 HI    1960        28,655
Hawaii Volcanoes          HI    1916       229,177
Hot Springs               AK    1921         5,839
Katmai                    AK    1980     3,716,000
Kenai Fjords              AK    1980       670,000
Kings Canyon              CA    1940       461,901
Kobuk Valley              AK    1980     1,750,421
Lake Clark                AK    1980     2,636,839
Lassen Volcanic           CA    1916       106,372
Mesa Verde                CO    1906        52,085
Mount Rainier             WA    1899       235,404
North Cascades            WA    1968       504,780
Olympic                   WA    1938       914,818
Petrified Forest          AR    1962        93,532
Redwood                   CA    1968       110,178
Rocky Mountain            CO    1915       265,200
Sequoia                   CA    1890       402,482
Theodore Roosevelt        ND    1978        70,416
Voyageurs                 MN    1971       218,059
Wind Cave                 SD    1903        28,292
Wrangell-St. Elias        AK    1980     8,331,604
Yellowstone               WY    1872     2,219,784
Yosemite                  CA    1980       761,170
National Park Data, Sorted by Park Name          03MAY89 Page 2
                                 Year
|------ Park ------|  State  Estab.  |- Acres |
Zion                      UT    1919       146,597
                                          ==========
Total acres, coast W                    44,268,110
                                          ==========
Grand total                             47,235,972
```

EXHIBIT 20.10   Appearance enhanced with LINESLEFT, LINE, PS, and N options

```
------------------- SAS Log --------------------------

    3     libname  out '\data\sas\';
    4     options ps=60 ls=120;
    5
    6     proc sort data=out.natlpark out=sorted;
    7     by st park;
    8     run;
NOTE: The data set WORK.SORTED has 49 observations and 5 variables.
NOTE: The PROCEDURE SORT used 6.00 seconds.
    9
   10     data _null_;
   11     set  sorted end=eof;
   12     by st park;
   13     file print notitles header=pagehead linesleft=lines2go
   14          line=currline n=ps;
   15     retain date;
   16     if _n_ = 1 then date = today();
   17     if first.st then do;
   18       _nstates + 1;
   19       _ninst = 0;
   20       if lines2go < 4 then put _page_;
   21       put / st @ ;
   22       end;
   23     _ninst + 1;
   24     _npark + 1;
   25     if currline >= 58 then put _page_;
   26     put @7 park $20.  @31 coast $1.  @37 yrestab 4.
   27        @44 acres comma10. ;
   28     if last.st & _ninst > 1 then do;
   29        if lines2go < 3 then put _page_;
   30        put @7 '--------------------' /
   31           @7 '# parks in state = ' _ninst ;
   32        end;
   33     if eof then put // _nstates 2. ' states with parks'
   34                      / _npark  2. ' parks' ;
   35     return;
   36     pagehead: pg + 1;
   37             put #60 'Source: Dept. of the Interior'
   38                 #1  'National Park Data, Sorted by Park Name'
   39                 @66 date date7. @74 'Page ' pg //
   40                 @37 'Year' /
   41                 @1 'State'
   42                 @7 '|------ Park ------|'
   43                 @29 'Coast'    @36 'Estab.'
   44                 @44 '|- Acres |' ;
   45          return;
   46     run;
NOTE: 81 lines were written to file PRINT.
NOTE: The DATA statement used 21.00 seconds.
```

EXHIBIT 20.11   Comprehensive report-writing example

```
-------------------- SAS Log ----------------------------

  3     libname  out '\data\sas\';
  4     options ps=60 linesize=80;
  5
  6     proc format;
  7     value party  1869-1885, 1889-1893, 1897-1913,
  8                   1921-1933, 1953-1961, 1969-1977,
  9                   1981-1989 = 'Rep'
 10                   other     = 'Dem';
NOTE: Format PARTY has been output.
 11     value pres   1869-<1877 = 'Grant'       1877-<1881 = 'Hayes'
 12                   1881-<1885 = 'Arthur'
 13                   1885-<1889, 1893-<1897 = 'Cleveland'
 14                   1889-<1893 = 'B. Harrison'   1897-<1901 = 'McKinley'
 15                   1901-<1909 = 'T. Roosevelt'  1909-<1913 = 'Taft'
 16                   1913-<1921 = 'Wilson'        1921-<1923 = 'Harding'
 17                   1923-<1929 = 'Coolidge'      1929-<1933 = 'Hoover'
 18                   1933-<1945 = 'F. Roosevelt'  1945-<1953 = 'Truman'
 19                   1953-<1961 = 'Eisenhower'    1961-<1963 = 'Kennedy'
 20                   1963-<1969 = 'L. Johnson'    1969-<1974 = 'Nixon'
 21                   1974-<1977 = 'Ford'          1977-<1981 = 'Carter'
 22                   1981-<1989 = 'Reagan'  ;
NOTE: Format PRES has been output.
 23     run;
NOTE: The PROCEDURE FORMAT used 10.00 seconds.
 24
 25     proc sort data=out.natlpark out=sorted;
 26     by yrestab park;
 27     run;
NOTE: The data set WORK.SORTED has 49 observations and 5 variables.
NOTE: The PROCEDURE SORT used 5.00 seconds.
 28
 29     data complete;
 30     set  sorted;
 31     length presname $15 party $3;
 32     presname = put(yrestab,pres.);
 33     party = put(yrestab,party.);
 34     run;
NOTE: The data set WORK.COMPLETE has 49 observations and 8 variables.
NOTE: The DATA statement used 13.00 seconds.
 35
 36     data out.parkpres(keep=presname party _presprk _presacr);
 37     set  complete end=eof;
 38     by presname notsorted;
 39     file print notitles header=pagehead linesleft=left;
 40     length dashes $79;
 41     retain dashes;
 42     if _n_ = 1 then dashes = repeat('-',79);
 43     _npark + 1;
 44     if party = 'Dem' then do;
 45        _ndem + 1;
 46        _demacre + acres;
 47        col = 3;
```

EXHIBIT 20.11 *(continued)*

```
48          end;
49          else do;
50              _nrep + 1;
51              _repacre + acres;
52              col = 42;
53              end;
54      if left < 4 then
55          put @1 dashes @1 '+' @40 '+' @79 '+' _page_ ;
56      if first.presname then do;
57          put   @1 '|' @40 '|' @79 '|'
58              / @1 '|' @40 '|' @79 '|' @col '**** ' presname ;
59          _presprk = 0;
60          _presacr = 0;
61          end;
62      _presprk + 1;
63      _presacr + acres;
64      put @col yrestab 4. +1 park $20. +1 acres comma10.
65          @1 '|' @40 '|' @79 '|'   ;
66      if last.presname then do;
67          output;
68          if _presprk > 1 then
69            put @col +5 '**** Parks=' _presprk 'Acres= '
70                @col +26 _presacr comma10.
71                @1 '|' @40 '|' @79 '|'   ;
72          end;
73      if eof then do;
74          if left < 6 then put _page_ ;
75          put    @1 dashes @1 '+' @40 '+' @79 '+'
76              // @3  'Democrat totals:'
77                 @42 'Republican totals:'
78               / @3  '... parks added: ' _ndem
79                 @42 '... parks added: ' _nrep
80               / @3  '... acres added: ' _demacre comma10.
81                 @42 '... acres added: ' _repacre comma10.
82             ;
83          end;
84      return;
85      pagehead: put   @1 dashes @1 '+' @40 '+' @79 '+'
86                    / @1 '|' @40 '|' @79 '|'
87                      @3 'D e m o c r a t s'
88                      @42 'R e p u b l i c a n s'
89                    / @1 '|' @40 '|' @79 '|'
90                    / @1 '|' @40 '|' @79 '|'
91                      @3 'Year ' @42 'Year'
92                    / @1 '|' @40 '|' @79 '|'
93                      @3 'Est. Park Name' @29 'Acres'
94                      @42 'Est. Park Name' @68 'Acres'
95                    / @1 dashes @1 '+' @40 '+' @79 '+'
96                    ;
97              if first.presname = 0 then
98                  put   @1 '|' @40 '|' @79 '|'
99                      / @1 '|' @40 '|' @79 '|'
100                       @col '**** ' presname '(cont.)' ;
101             return;
102     run;
```

EXHIBIT 20.11 *(continued)*

```
NOTE: 110 lines were written to file PRINT.
NOTE: The data set OUT.PARKPRES has 16 observations and 4 variables.
NOTE: The DATA statement used 29.00 seconds.

-------------------- Output listing --------------------

+----------------------------------+--------------------------------------+
| D e m o c r a t s                | R e p u b l i c a n s                |
|                                  |                                      |
| Year                             | Year                                 |
| Est. Park Name         Acres     | Est. Park Name         Acres         |
+----------------------------------+--------------------------------------+
|                                  |                                      |
|                                  | **** Grant                           |
|                                  | 1872 Yellowstone      2,219,784      |
|                                  |                                      |
|                                  | **** B. Harrison                     |
|                                  | 1890 Sequoia            402,482      |
|                                  | 1890 Yosemite           761,170      |
|                                  |    **** Parks=2 Acres= 1,163,652     |
|                                  |                                      |
|                                  | **** McKinley                        |
|                                  | 1899 Mount Rainier      235,404      |
|                                  |                                      |
|                                  | **** T. Roosevelt                    |
|                                  | 1902 Crater Lake        183,224      |
|                                  | 1903 Wind Cave           28,292      |
|                                  | 1906 Mesa Verde          52,085      |
|                                  |    **** Parks=3 Acres=  263,601      |
|                                  |                                      |
|                                  | **** Taft                            |
|                                  | 1910 Glacier          1,013,572      |
|                                  |                                      |
| **** Wilson                      |                                      |
| 1915 Rocky Mountain    265,200   |                                      |
| 1916 Hawaii Volcanoes  229,177   |                                      |
| 1916 Lassen Volcanic   106,372   |                                      |
| 1917 Denali          4,716,726   |                                      |
| 1919 Acadia             41,365   |                                      |
| 1919 Grand Canyon    1,218,375   |                                      |
| 1919 Zion              146,597   |                                      |
|    **** Parks=7 Acres= 6,723,812 |                                      |
|                                  |                                      |
|                                  | **** Harding                         |
|                                  | 1921 Hot Springs          5,839      |
|                                  |                                      |
|                                  | **** Coolidge                        |
|                                  | 1924 Bryce Canyon        35,835      |
|                                  | 1926 Great Smoky Mts.   520,269      |
|                                  | 1926 Mammoth Cave        52,420      |
|                                  | 1926 Shenandoah         195,346      |
|                                  |    **** Parks=4 Acres=  803,870      |
|                                  |                                      |
|                                  | **** Hoover                          |
|                                  | 1929 Grand Teton        310,521      |
|                                  | 1930 Carlsbad Caverns    46,755      |
```

EXHIBIT 20.11   *(continued)*

```
|                                    | 1931 Isle Royale               571,790 |
|                                    |      **** Parks=3 Acres=       929,066 |
|                                    |                                        |
| **** F. Roosevelt                  |                                        |
| 1934 Everglades         1,398,938  |                                        |
| 1935 Big Bend             735,416  |                                        |
| 1938 Olympic              914,818  |                                        |
+-----------------------------------+----------------------------------------+

     - - - - - - - - - - - New page of output - - - - - - - - -


+-----------------------------------+----------------------------------------+
| D e m o c r a t s                  | R e p u b l i c a n s                  |
|                                    |                                        |
| Year                               | Year                                   |
| Est. Park Name          Acres      | Est. Park Name          Acres          |
+-----------------------------------+----------------------------------------+
|                                    |                                        |
| **** F. Roosevelt (cont.)          |                                        |
| 1940 Kings Canyon         461,901  |                                        |
|      **** Parks=4 Acres= 3,511,073 |                                        |
|                                    |                                        |
|                                    | **** Eisenhower                        |
|                                    | 1956 Virgin Islands         14,695     |
|                                    | 1960 Haleakala              28,655     |
|                                    |      **** Parks=2 Acres=    43,350     |
|                                    |                                        |
| **** Kennedy                       |                                        |
| 1962 Petrified Forest      93,532  |                                        |
|                                    |                                        |
| **** L. Johnson                    |                                        |
| 1964 Canyonlands          337,570  |                                        |
| 1966 Guadalupe Mts.        76,293  |                                        |
| 1968 North Cascades       504,780  |                                        |
| 1968 Redwood              110,178  |                                        |
|      **** Parks=4 Acres= 1,028,821 |                                        |
|                                    |                                        |
|                                    | **** Nixon                             |
|                                    | 1971 Arches                 73,378     |
|                                    | 1971 Capitol Reef          241,904     |
|                                    | 1971 Voyageurs             218,059     |
|                                    |      **** Parks=3 Acres=   533,341     |
|                                    |                                        |
| **** Carter                        |                                        |
| 1978 Badlands             243,302  |                                        |
| 1978 Theodore Roosevelt    70,416  |                                        |
| 1980 Biscayne             173,039  |                                        |
| 1980 Channel Islands      249,353  |                                        |
| 1980 Gates of the Arctic 7,523,888 |                                        |
| 1980 Glacier Bay        3,225,284  |                                        |
| 1980 Katmai             3,716,000  |                                        |
| 1980 Kenai Fjords         670,000  |                                        |
| 1980 Kobuk Valley       1,750,421  |                                        |
| 1980 Lake Clark         2,636,839  |                                        |
| 1980 Wrangell-St. Elias 8,331,604  |                                        |
|      **** Parks=11 Acres= 28,590,146 |                                      |
```

EXHIBIT 20.11 *(continued)*

```
|                                    |                                            |
|                                    | **** Reagan                                |
|                                    | 1986 Great Basin              77,109 |
+------------------------------------+--------------------------------------------+
  Democrat totals:                      Republican totals:
  ... parks added: 27                   ... parks added: 22
  ... acres added: 39,947,384           ... acres added:  7,288,588
```

EXHIBIT 20.12   PUT used in debugging: setup

```
------------------- SAS Log ----------------------------

    3     options ps=60 linesize=80;
    4
    5     data one ; input id grp $ score year ; list ; cards ;
RULE:----+----1----+----2----+----3----+----4----+----5----+----6----+----7
   6 110 Geog 82 1981
   7 110 Geol 90 1982
   8 110 Geol 92 1982
   9 110 Geol 91 1983
  10 110 Geol 90 1980
  11 119 Eng  90 1981
  12 119 Engl 95 1982
  13    run;
NOTE: The data set WORK.ONE has 7 observations and 4 variables.
NOTE: The DATA statement used 8.00 seconds.
  14
  15    * Select only those groups of tests that were ALL taken in 1982 or
  16      later. If ANY test in the group is pre-1982, do not output. ;
  17    data two;
  18    set  one;
  19    by id grp;
  20    array s{3} s1-s3;
  21    if first.grp then do;
  22       do i = 1 to 3; s{i} = .; end;
  23       end;
  24    sub = 0;
  25    if year < 82 then bad_grp = 'y';
  26    sub + 1;
  27    s{sub} = score;
  28    if last.grp then do;
  29      if bad_grp = 'n' then do;
  30        if grp = 'eng' then quant = 'y' ;
  31          else            quant = 'n' ;
  32        output;
  33        end;
  34      end;
  35    run;
NOTE: The data set WORK.TWO has 0 observations and 11 variables.
NOTE: The DATA statement used 11.00 seconds.
```

EXHIBIT 20.13   PUT statements inserted at critical points in the DATA step

---

```
-------------------- SAS Log ----------------------------

15    * Select only those groups of tests that were ALL taken in 1982 or
16       later. If ANY test in the group is pre-1982, do not output. ;
17    data two;
18    set  one;
19    by id grp;
20    array s{3} s1-s3;
21    if first.grp then do;
22       put // 'Begin group ' grp 'at obs ' _n_ ;
23       do i = 1 to 3; s{i} = .; end;
24       end;
25    sub = 0;
26    if year < 82 then bad_grp = 'y';
27    put @3 'Year is ' year 'bad_grp flag set to ' bad_grp ;
28    sub + 1;
29    put @5 'Score array subscript is ' sub ;
30    s{sub} = score;
31    if last.grp then do;
32       put @7 'End of group';
33       if bad_grp = 'n' then do;
34          put @10 '*** output ***' / _all_;
35          if grp = 'eng' then quant = 'y' ;
36             else          quant = 'n' ;
37          output;
38          end;
39       end;
40    run;
Begin group Geog at obs 1
Year is 1981 bad_grp flag set to
  Score array subscript is 1
    End of group
Begin group Geol at obs 2
Year is 1982 bad_grp flag set to
  Score array subscript is 1
Year is 1982 bad_grp flag set to
  Score array subscript is 1
Year is 1983 bad_grp flag set to
  Score array subscript is 1
Year is 1980 bad_grp flag set to
  Score array subscript is 1
    End of group
Begin group Eng at obs 6
Year is 1981 bad_grp flag set to
  Score array subscript is 1
    End of group
Begin group Engl at obs 7
Year is 1982 bad_grp flag set to
  Score array subscript is 1
    End of group
NOTE: The data set WORK.TWO has 0 observations and 11 variables.
NOTE: The DATA statement used 14.00 seconds.
```

EXHIBIT 20.14   Final, debugged program and output

```
------------------- SAS Log --------------------------

17    * Select only those groups of tests that were ALL taken in 1982 or
18       later. If ANY test in the group is pre-1982, do not output. ;
19    data two;
20    set  one;
21    by id grp;
22    retain bad_grp ' ';  *--> Keep BAD_GRP from obs to obs ;
23    retain s1-s4;  *--> Scores must be held across observations so we
24                        can output at LAST.GRP ;
25    array s{4} s1-s4;  *--> Previous size was 3 ;
26    if first.grp then do;
27       put / 'Begin group ' grp 'at obs ' _n_ ;
28       do i = 1 to 4; s{i} = .; end;
29       sub = 0;  *--> Previously was executed in EVERY obs ;
30       bad_grp = 'n';  *--> Reset at start of each GRP level ;
31       end;
32    if year < 1982 then bad_grp = 'y';  *--> Add 19 prefix to year # ;
33    put @3 'Year is ' year 'bad_grp flag set to ' bad_grp ;
34    sub + 1;
35    put @5 'Score array subscript is ' sub ;
36    if sub <= 4 then s{sub} = score;  *--> Add IF-THEN to ensure that
37                                           we will have a valid
                                               subscript. ;
38    if last.grp then do;
39       put @7 'End of group';
40       if bad_grp = 'n' then do;
41          put @10 '*** output ***' / _all_;
42          if grp = 'Eng ' then quant = 'y' ; /* EXACT comparison of */
43             else            quant = 'n' ; /*   test GRP name      */
44          output;
45          end;
46       end;
47    keep id quant grp s1-s4;
48    run;
Begin group Geog at obs 1
 Year is 1981 bad_grp flag set to y
   Score array subscript is 1
     End of group
Begin group Geol at obs 2
 Year is 1982 bad_grp flag set to n
   Score array subscript is 1
 Year is 1982 bad_grp flag set to n
   Score array subscript is 2
 Year is 1983 bad_grp flag set to n
   Score array subscript is 3
 Year is 1983 bad_grp flag set to n
   Score array subscript is 4
     End of group
       *** output ***
```

EXHIBIT 20.14   *(continued)*

```
ID=110 GRP=Geol SCORE=90 YEAR=1983 FIRST.ID=0 LAST.ID=1 FIRST.GRP=0
LAST.GRP=1
BAD_GRP=n S1=90 S2=92 S3=91 S4=90 I=. SUB=4 QUANT=  _ERROR_=0 _N_=5
 Begin group Hist at obs 6
  Year is 1983 bad_grp flag set to n
    Score array subscript is 1
  Year is 1984 bad_grp flag set to n
    Score array subscript is 2
      End of group
       *** output ***
ID=115 GRP=Hist SCORE=70 YEAR=1984 FIRST.ID=0 LAST.ID=1 FIRST.GRP=0
LAST.GRP=1
BAD_GRP=n S1=74 S2=70 S3=. S4=. I=. SUB=2 QUANT=  _ERROR_=0 _N_=7
 Begin group Eng at obs 8
  Year is 1981 bad_grp flag set to y
    Score array subscript is 1
      End of group
 Begin group Engl at obs 9
  Year is 1982 bad_grp flag set to n
    Score array subscript is 1
      End of group
       *** output ***
ID=119 GRP=Engl SCORE=95 YEAR=1982 FIRST.ID=0 LAST.ID=1 FIRST.GRP=1
LAST.GRP=1
BAD_GRP=n S1=95 S2=. S3=. S4=. I=5 SUB=1 QUANT=  _ERROR_=0 _N_=9
NOTE: The data set WORK.TWO has 3 observations and 7 variables.
NOTE: The DATA statement used 17.00 seconds.
   49
   50    proc print;
   51    run;
NOTE: The PROCEDURE PRINT used 8.00 seconds.

------------------- Output listing --------------------

OBS    ID    GRP    S1    S2    S3    S4    QUANT

 1    110    Geol    90    92    91    90    n
 2    115    Hist    74    70    .     .     n
 3    119    Engl    95    .     .     .     n
```

EXHIBIT 20.15   Report with generated data; PUT statement's sensitivity to location

---

```
-------------------- Program listing --------------------

data _null_;
file print notitles;
array pay{6};
dashes = repeat('=======  ',6);
put 'Repayment Scenarios for a Loan of $50,000' //
    ' Term    Monthly payments at different, fixed interest rates' /
    '(years)      7%       8%       9%      10%      11%      12%' /
    dashes;
do term = 15 to 30 by 5;
   element = 0;
   do rate = .07 to .12 by .01;
      element = element + 1;
      pay{element} = mort(50000, ., rate/12, term*12);
   end;
   put term 7. +2 (pay{*}) (comma7. +2);
end;
run;

-------------------- Output listing --------------------

Repayment Scenarios for a Loan of $50,000

Term    Monthly payments at different, fixed interest rates
(years)    7%       8%       9%      10%      11%      12%
=======  =======  =======  =======  =======  =======  =======
   15     449      478      507      537      568      600
   20     388      418      450      483      516      551
   25     353      386      420      454      490      527
   30     333      367      402      439      476      514
```

Summary

Usually the default or option-altered output of statistical and reporting procedures is adequate for analysis and presentation requirements. Occasionally, however, the required degree of control over placement of text and data calls for custom report writing. Chapter 20 discusses the rationale and mechanics of report writing. It briefly outlines some of the criteria for determining whether a custom report is actually needed, it then steps through a series of points to consider when designing and writing report programs.

Several DATA step extensions are presented. The special dataset name _NULL_ improves processing efficiency during report writing. FILE identifies the location of the report and sets options that control the report header, the length of the output line, and other aesthetic considerations. The PUT statement actually writes the report line. It allows you to mix variables and constants and offers complete control of their placement on the page or screen. These statements are powerful tools that are sometimes difficult to put together successfully. As the example programs demonstrate, however, the impact and visual appeal of the final product can make the effort worthwhile.

PART FIVE

Statistics Revisited:
Procedures for Bivariate
and Multivariate Analysis

21 Two-Variable Graphics

This chapter shifts the focus away from SAS data management features and moves it to data analysis. Although statistics have been used throughout the book, they have been simple univariate measures. Single-variable characteristics such as category frequency, central tendency, dispersion, rank, and standardized values were covered by the FREQ, MEANS, UNIVARIATE, RANK, and STANDARD procedures. This chapter begins the discussion of calculating and displaying bivariate and multivariate statistics. This family of statistics is concerned with measuring the nature and strength of the relationship between two or more variables. The techniques range from simple bivariate measures such as cross tabulations, correlations, and nonparametric measures of association to predictive, multivariable techniques such as linear regression. More complex, multivariate techniques such as MANOVA are beyond the scope of this book and are only mentioned briefly.

One of the simplest means of assessing the relationship between two variables is graphical representation. This chapter presents the PLOT procedure, which generates scatterplots. Scatterplots may be used to supplement the more powerful and sophisticated analytical procedures presented in later chapters.

21.1 Scatterplots: The PLOT Procedure

Scatterplots are a visually appealing way to examine the relationship between two variables. They help you "eyeball" your data and answer questions such as these:

- Does there appear to be a straight line, tightly clustered association between the two variables?

- Does the display suggest that a higher-order term (quadratic) or a transformation (logarithm) may give a better fit?
- Is there any trend to the display of predicted values from a regression equation when plotted against actual values?
- Are there outlier values for either variable?

These sorts of questions are given at least a preliminary answer by PLOT. The relationship between two variables of any data type (character or numeric) and scale (nominal, ordinal, interval, or ratio) may be represented. Typically, both variables are numerics and are scaled interval or ratio. Nominal and ordinal data are usually more effectively presented with the CHART procedure (Chapter 6).

How PLOT Works

The mechanics of the PLOT procedure are straightforward. SAS reads an observation's values for a pair of analysis variables. It determines the values' intersection along the plot's x and y axis and indicates this point with a nonblank character. By default, the intersection is marked by an A. If two observations have identical values or if there is not enough room on the display to distinguish values, a B is printed. The letters C, D, and so on through Z are used to indicate the piling up of value pairs. If a value falls outside a user-specified range for an axis, it is not printed and a message is printed on the plot. The area available for the actual plot is the space left over once TITLEs, FOOTNOTEs, LEGENDs, and messages are displayed.

PLOT Features and Flexibility

Exhibit 21.1 presents a simple plot of two variables from the STATES dataset. No PLOT options are specified; all default values are used. The y-axis value is each state's percentage of college-educated residents. The x axis measures the states' per capita expenditure on teacher salaries. There is a moderately strong association: as the proportion of college-educated people rises, so does the expenditure for education. Notice the two outliers: their identity could be revealed by using the UNIVARIATE procedure with an ID statement.

Exhibit 21.1 illustrates several of PLOT's features, which are flexible enough to allow a variety of presentation styles. These options, which are described as they apply to Exhibit 21.1, cover such aspects of plots as content displayed, number of plots per page, the plot's general appearance, and axis scaling. These options are covered in detail in the following Syntax section.

POINTERS AND PITFALLS

Although the user has control over many of PLOT's features, usually only a few are needed to accurately portray the variables' distribution. Many of the options are concerned with aesthetic, rather than statistical or computational, aspects of the plot.

Content.   All nonmissing pairs of variables in the entire dataset are displayed. This option is overridden by using BY-group processing (BY statement). Axis specifications may be used to "window" a portion of the plot. These windows restrict the display to a particular range of values along the x and y axes, thus allowing a closeup look at the data points. By enlarging the physical space allowed for the plot, windows often eliminate the piling up of observations at an intersection on the display.

EXHIBIT 21.1   Simple plot: All default settings used

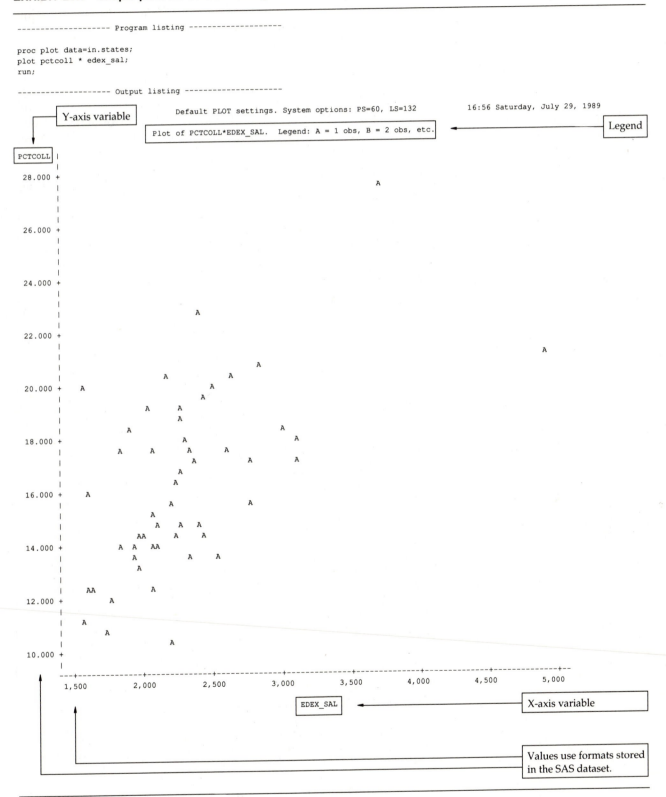

Plots per Page. Each plot request, that is, each pair of variables, possibly for each BY-group, is displayed on one page of the output file. This feature is overridden by using the HPCT and VPCT options (PROC statement) or the OVERLAY option (PLOT statement). These options allow multiple plots per page or the extension of a single plot over several pages of output.

Appearance. A legend is printed at the top of the plot, and the x and y axes are drawn. The plot is confined to the default system page length and line width settings. These defaults are overridden by using the NOLEGEND option (PROC statement), the BOX, HREF, VREF, HREFCHAR, and VREFCHAR options (PLOT statement), and the PS and LS system options (OPTIONS statement).

Axis Scaling. Axis value ranges are determined by the minimum and maximum for each variable in the dataset or BY-group. Tick marks are evenly spaced between these extremes. The default scaling is overridden by using the UNIFORM, NOMISS, and VTOH options (PROC statement), and the HAXIS, VAXIS, HZERO, VZERO, HREVERSE, VREVERSE, HSPACE, and VSPACE options (PLOT statement).

Syntax

The PLOT procedure's syntax is described below. In addition to the procedure-specific statements described, the BY, LABEL, ATTRIB, FORMAT, TITLE, and FOOTNOTE statements may also be used. These statements were discussed in Chapter 5.

```
PROC PLOT [DATA=dset] [UNIFORM] [NOLEGEND]
          [NOMISS] [VTOH=ratio]
          [VPCT=pct_list] [HPCT=pct_list] ;
PLOT request [/ [OVERLAY] [BOX]
              [HREF=axis_spec] [HREFCHAR='char']
              [VREF=axis_spec] [VREFCHAR='char']
              [HAXIS=axis_spec] [VAXIS=axis_spec]
              [HZERO] [VZERO] [HREVERSE] [VREVERSE]
              [HSPACE=gap_len] [VSPACE=gap_len]
              [HPOS=positions] [VPOS=positions] ] ;
[RUN;]
[QUIT;]
```

POINTERS AND PITFALLS

Start using PLOT with as many defaults as possible. It is usually best to get the axis scale and orientation correct, then work on spacing, reference lines, and other, more cosmetic features.

Some common applications of the PLOT procedure follow:

```
proc plot data=m1;
plot supp1*debt1 / box hzero yzero;

proc plot data=bball nolegend;
options ps=24 linesize=78;
plot winpct*(reb pts assist) / overlay;
run;
options ps=60 linesize=120;
plot winpct*reb ;
quit;
```

The PLOT procedure has 18 components:

1. The *dset* is the name of the dataset to be analyzed. If the *DATA* option is not entered, PLOT uses the most recently created dataset.

2. *UNIFORM* forces axes to be scaled identically across all levels of BY-groups. This uniform scaling permits direct comparison of plots for different BY-group levels. By default, axis ranges are determined by each BY-group's minimum and maximum values. *UNIFORM* has no effect if specified without a BY statement.

3. *NOLEGEND* suppresses the printing of the legend at the top of each plot.

4. *NOMISS* requests that observations with missing values not be used when determining axis ranges. If a plot of *A* against *B* was requested and an observation's value of *A* was not specified, the plot's *B* value might still replace the current *B* axis minimum or maximum values. *NOMISS* prevents this replacement. Axis-scaling options in the PLOT statement override *NOMISS*. This option is available only in version 6 of the SAS System.

5. *VTOH* specifies the *ratio* of vertical axis length to horizontal. The number of spaces between tick marks is adjusted to match or approximate *ratio*. This option is available only in version 6 of the SAS System.

6. *VPCT* and *HPCT* request overriding of the usual one page per plot. The values in *pct_list*, integers separated by one or more blanks, indicate the percentage of a page used by the vertical (*VPCT*) and horizontal (*HPCT*) axes. Values greater than 100 force the plot to extend over more than one page; values under 100 enable multiple plots per page; and the special value 0 forces PLOT to a new page even if space is available on the current page. Either or both *VPCT* and *HPCT* may be specified. Exhibit 21.2 (see p. 432) presents some values for these options and describes the layout of the resulting plots. These options are available only in version 6 of the SAS System.

7. The *request* specifies which variables will be plotted. Its use is required and takes three general forms. The first specifies the vertical (*y*-axis) variables and the horizontal (*x*-axis) variables. The *y*-axis variables are followed by an asterisk (*). Valid plot requests and the resulting output are presented in Exhibit 21.3. The second form of *request* specifies how the scatterplot point should be displayed. It takes either of two forms. The first form is

```
y * x = 'char'
```

The *char* is the character PLOT will use to identify points. When this option is selected, PLOT cannot identify locations where observations pile up: the *A*, *B*, *C*, . . ., *Z* notation is suppressed. The second form is

```
y * x = var
```

In this form *var* is a numeric or character variable. The first (leftmost) position of the formatted value of the variable is used to locate the observation's *x-y* pair on the plot. If the plot request in Exhibit 21.1 were reworded to PCTCOLL * EDEX_SAL = REGION, the points displayed would be *N*, *M*, *S*, *W*, and *P* rather than the default *A*, *B*, and so on.

8. The / signals the end of the plot request and the beginning of a list of options.

EXHIBIT 21.2   Displaying partial or multiple plots per page

Option	Plot
vpct=50	Up to two plots on a page, each half a page long and a full page wide.

```
+-----------------------,--+
|          Plot  1         |
+--------------------------+
|          Plot  2         |
+--------------------------+
```

Option	Plot
vpct=50 hpct=50	Up to four plots per page, each taking a quarter of the page.

```
+------------+------------+
|  Plot  1   |  Plot  2   |
+------------+------------+
|  Plot  3   |  Plot  4   |
+------------+------------+
```

Option	Plot
vpct=50 25 25 hpct=50	A maximum of six plots on a page. The first two each take a quarter of the page, the remaining four each use an eighth (25% of the vertical by 50% of the horizontal).

```
+------------+------------+
|  Plot  1   |  Plot2     |
|            |            |
+------------+------------+
|  Plot  3   |  Plot  4   |
+------------+------------+
|  Plot  5   |  Plot  6   |
+------------+------------+
```

Option	Plot
hpct=200	The plot is displayed in sections over two pages.

```
+----------------+   +----------------+
| Left side      |   | Right side     |
|                |   |                |
| (page 1)       |   | (page 2)       |
+----------------+   +----------------+
```

9. *OVERLAY* displays all requests in the PLOT statement in a single plot. The axes are scaled to accommodate the widest x and y variable range in *any* pair of plot variables. Only the first form of plot specifications ($y * x = $ 'char') may be used with the *OVERLAY* option.

10. *BOX* draws a border around the entire plot rather than just on the bottom and left side. Tick marks, but not values at those marks, are also drawn. This gives plots a more "finished" effect. The use of *BOX* is purely cosmetic. BOX is available only in version 6 of the SAS System.

EXHIBIT 21.3 Plots produced from different plot requests

Plot Request	Variable Pairs Plotted
y * x	y-x
(y1 y2) * x	y1-x y2-x
y * (x1 x2)	y-x1 y-x2
y * x1 y * x2	y-x1 y-x2
(y1 y2) * (x1 x2)	y1-x1 y1-x2 y2-x1 y2-x2
y * (x1-x3)	y-x1 y-x2 y-x3
(x y z)	x-y x-z y-z

11. *HREF* and *VREF* identify points along the *x* (horizontal) and *y* (vertical) axes where PLOT will draw reference lines. Reference lines help emphasize data trends, answering questions such as, Do values for family income tend to fall below the poverty level? Do closing prices for a stock fall between the most recent year's high and low? The syntax for the *axis_refs* is discussed in "Specifying Axis Reference Points" later in the chapter.

12. *HREFCHAR* and *VREFCHAR* specify the characters used to display the horizontal and vertical reference lines. The *char* is a single character. The default *HREFCHAR* value is "¦." The default *VREFCHAR* value is "-."

13. *HAXIS* and *VAXIS* describe the location of labelled tick marks along the horizontal and vertical axes, respectively. By default, PLOT chooses values in equal increments along each axis. The syntax for *axis_refs* is discussed in the "Specifying Axis Reference Points" section.

14. *HZERO* and *VZERO* instruct PLOT to begin the horizontal and vertical axes, respectively, with a value of 0. *HZERO* is ignored if *HAXIS* specifies a non-0 starting value, and *VZERO* is similarly overridden by *VAXIS*.

15. *HREVERSE* and *VREVERSE* reverse the order of the horizontal and vertical axes. Rather than displaying *y*-axis values from 20 at the bottom to 100 at the top, *VREVERSE* prints 20 at the top, ending with 100 at the bottom.

16. *HSPACE* and *VSPACE* specify the spacing between tick marks on the axis. When used with *HSPACE*, *gap_len* indicates print positions. When used with *VSPACE*, it indicates print lines.

17. *HPOS* and *VPOS* request that SAS use *positions* print positions when drawing the horizontal and vertical axes, respectively. The *positions* is ignored if it conflicts with other page-sizing specifications such as the LINESIZE and PAGESIZE system options or the HPCT and VPCT options in the PROC statement.

18. *RUN* and *QUIT* control the timing of plots in both batch and interactive environments. *RUN* instructs PLOT to execute the PLOT statements entered since the previous *RUN* statement. *QUIT* executes the most recently specified plots (like *RUN*) and then ends the procedure. If you attempt to end an interactive SAS session while in PLOT and you have not *QUIT*, SAS will ask you to verify the end of the session.

Specifying Axis Reference Points

PLOT's default axis reference points are adequate to encompass all values of the observations being plotted but often lack a logical or aesthetic sense. Fortunately, there are several simple, powerful features in the VAXIS, HAXIS, VREF, and

HREF statements that allow quick coding of user-friendly axis labels. Axis tick mark values and reference points may be specified directly or indirectly.

Direct Specification.   Direct specification simply involves listing values. Numeric variable values must be in a constant ascending or descending order; character values may be in any order. The tick marks will be evenly spaced along the axis. The following specifications are valid:

```
haxis = 0 5 10 15 20 25 30
haxis = 0 5 10
vaxis = -5 0 5
```

Indirect Specification.   Indirect specification may be used with numeric variables. It takes the same form as the iterative DO-loops discussed in Chapter 11. If date values are plotted, they may be entered as date constants. BY values for date specifications may take the same form as the increments used in the INTCK function: DAY, WEEK, MONTH, QTR, and YEAR. Formats should be used with date values to make the axis labelling presentable. Examples of valid specifications follow:

```
haxis = '01feb89'd to '31dec91'd by qtr
vaxis = 0 to 30 by 5
href = 0, 4 to 8, 12
vref = 0, 20 to 50 by 10
```

Usage Notes

Keep in mind the following points about PLOT's features:

- Axis-scaling options (HAXIS and VAXIS) may unintentionally exclude observations from the display by specifying a too-narrow range of values for an axis. Likewise, the options may err in the opposite direction, specifying a range that is inappropriately wide for the data values. This will cluster the observations into a corner of the plot. Examples of both errors are presented later in the chapter (see Exhibits 21.8 and 21.13).

- If axis scaling excludes observations from a plot, the message "[n] obs were out of range" is printed in a note below the graph.

- If the $y * x = $ 'char' or $y * x = var$ plot request is used or the OVERLAY option is selected, the message "[n] obs hidden" is printed below the graph if two points occupy the same plot location. In this case, SAS has no way to distinguish the number of occurrences of a pair of values from the pair of variables being plotted (see Exhibit 21.4).

- Overlaying PLOT with wide-ranging variables (e.g., state population) and comparatively narrow ranges (percent unemployed) often renders the display virtually useless. A more effective strategy is to present the plots side by side using the HPCT and VPCT options of the PROC statement.

- Any number of plot requests may appear in a single execution of PLOT. Remember that options specified in the PROC statement (UNIFORM, NOLEGEND, NOMISS, and the scaling features) are in effect for *all* PLOT statements.

- The RUN statement gives you the flexibility to change titles, formats, labels, and system options during the procedure. Exhibit 21.4 illustrates this flexibility. The title "Sales by Date" and the DATE format are used in the first plot, and the title "Sales by Quarter" and the YYQ format are used in the second plot. The OPTIONS statements specify a line width of 78 columns for the first plot and 132 columns for the second.

EXHIBIT 21.4 Using RUN to separate plots with different OPTIONs, FORMATs, and TITLEs

```
proc plot data=salesumm;
plot sale_vol * date / haxis = '01jan90'd to '31dec91'd by month;
options ls=78;
format date date7.;
title "Sales by Date";
run;
```
Run plots with current options, formats, titles. Remain in PLOT.

```
plot sale_vol * qtr / haxis = '01jan90'd to '31dec91'd by qtr;
options ls=132;
format qtr yyq6.;
title "Sales by Quarter";
quit;
```
Run plots with altered display parameters. Then exit PLOT.

Examples

Key features and options used in this section are summarized in the following table:

Exhibit	Features/Options
21.5	Plot style $x*y = z$; hidden observation message
21.6	OVERLAY plots; distinguished with different symbols
21.7	Axis annotation options (VAXIS, HAXIS, VREFCHAR, and HREFCHAR) used
21.8	Reduced display area size increases number of hidden observations
21.9	Plot appearance enhanced with BOX and NOLEGEND options and ATTRIB statement
21.10	Multiple plots per page
21.11	Invalid HAXIS specification: no values in range
21.12	A group unintentionally excluded from a plot
21.13	Outliers distorting axis scale
21.14	Axis specifications used to display only a portion of the display space

This section illustrates appropriate, inappropriate but syntactically correct, and erroneous uses of PROC PLOT. Unless otherwise specified, the system options used to specify page dimensions were LS = 132 and PS = 60.

Exhibit 21.5 (see p. 436) plots two variables, marking their location on the page by the value of a third variable. All defaults are taken. A note is printed at the bottom of the page indicating that an observation was hidden: two observations had the same pair of values but only one value could be printed at that location.

Exhibit 21.6 (see p. 437) overlays two plots (OVERLAY option), using different symbols ($ and c) to distinguish them. The meaning of the two plotting symbols is explained in the legend at the top of the page.

Exhibit 21.7 (see p. 438) is the same as Exhibit 21.6 but with axis annotation (VAXIS and HAXIS options) controlled. Both the vertical and horizontal axes use reference lines (VREF and HREF options) and specific reference line characters (VREFCHAR and HREFCHAR options). Notice that there are five hidden observations in this plot even though the plot in Exhibit 21.6 had only two. This seeming discrepancy was caused by the TITLE statements: Exhibit 21.7

EXHIBIT 21.5   Two variables plotted by a third; hidden observation message

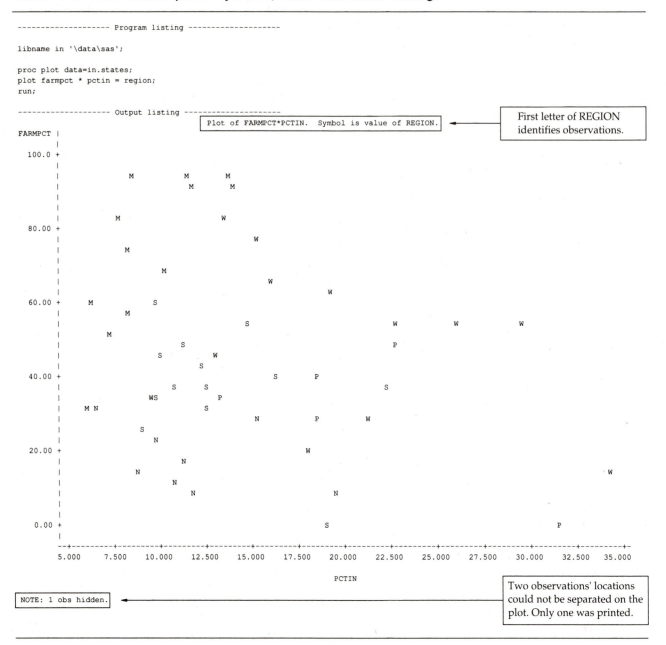

```
------------------- Program listing -------------------

libname in '\data\sas';

proc plot data=in.states;
plot farmpct * pctin = region;
run;

------------------- Output listing -------------------
```

First letter of REGION identifies observations.

NOTE: 1 obs hidden.

Two observations' locations could not be separated on the plot. Only one was printed.

has four TITLE lines, while Exhibit 21.6 had only two. Thus Exhibit 21.6 had more space available for the display of the plot and did not have to hide as many observations.

Exhibit 21.8 (see p. 439) is the same as Exhibit 21.7 except the dimensions of the display space are changed. The system options PS = 21 and LS = 78 are used by most interactive, full-screen sessions. The constrained display space forces PLOT to "hide" 30 observations: since the symbols '$' and 'c' cannot suggest the piling up of value pairs, PLOT simply notes that there were 30 instances of duplicate locations. The NOTE about hidden observations is moved below the legend.

EXHIBIT 21.6   OVERLAY plots; distinguished with different symbols

```
------------------- Program listing -------------------

libname in '\data\sas';

proc plot data=in.states;
plot pcturb*pcapinc='$' pcturb*crimert='c'/ overlay ;
run;

------------------- Output listing -------------------
```

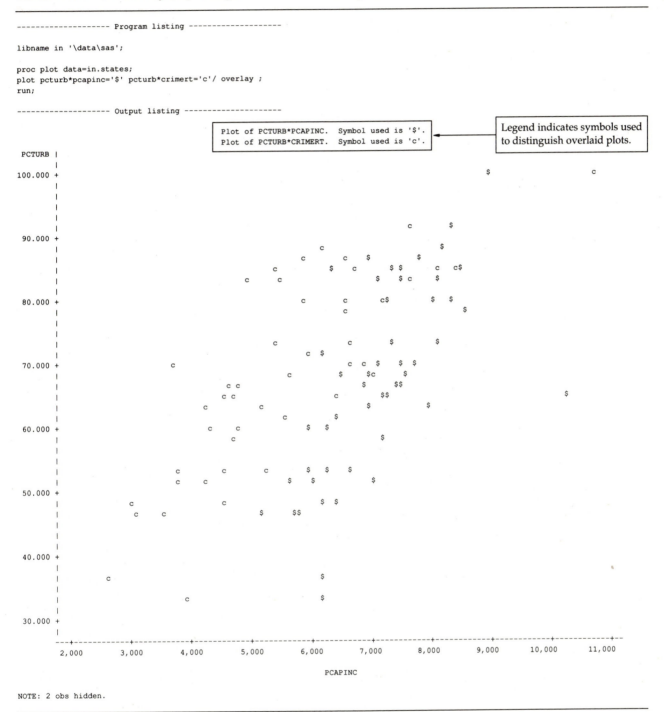

Plot of PCTURB*PCAPINC.   Symbol used is '$'.
Plot of PCTURB*CRIMERT.   Symbol used is 'c'.

Legend indicates symbols used to distinguish overlaid plots.

NOTE: 2 obs hidden.

In Exhibit 21.9 (see p. 440) a simple plot is dressed up. The printing of the legend (NOLEGEND option) is suppressed, parallel *x* and *y* axes (BOX option) are drawn, and the ATTRIB statement is used to assign temporary LABELs and FORMATs to the variable.

EXHIBIT 21.7   Axis intervals controlled; reference lines specified

```
------------------- Program listing --------------------

libname in '\data\sas';

proc plot data=in.states;
plot pcturb*pcapinc='$' pcturb*crimrt='c'
     / overlay vaxis=0 to 100 by 25 vref=25 75 vrefchar='-'
       haxis=2000 to 12000 by 2000   href=6000 hrefchar='.';
format pcturb 4.;
run;

------------------- Output listing --------------------
```

Line drawn with VREF option

```
                                  Plot of PCTURB*PCAPINC.  Symbol used is '$'.
                                  Plot of PCTURB*CRIMERT.  Symbol used is 'c'.

    PCTURB |                              .
           |                              .
      100 +                               .                        $           c
           |                              .
           |                              .
           |                              .              c      $
           |                            . c           $
           |                           c.    c    $      $     c$
           |               c    cc    .    $   c  $  $ $c    c$
           |                         c.    c    $      $ $
      75.0 +----------------------------------------------------------------------------
           |                           c   c$    c       $
           |                  c            c   .   $ c c$ $  $ $
           |                     c c       .          $c   $ $$
           |                  c c c   c        .   c    $ $$      $              $
           |                  c   c      c    $. $  $
           |                     c            .        $
           |
           |           c    c   c      c  $ $. $  $  $
      50.0 +        c                         $   $
           |           c   c        c     $   $$ .$
           |                              .
           |                              .
           |              c               .$
           |                              .
           |        c                     . $
           |           c                  .
           |                              .
      25.0 +--------------------------------------------------------------------------
           |                              .
           |                              .
           |                              .
           |                              .
           |                              .
           |                              .
      .000 +                              .
           |                              .
           ---+--------------+--------------+--------------+--------------+--------------+--
           2,000          4,000          6,000          8,000         10,000         12,000

                                              PCAPINC
```

Line character specified by VREFCHAR option

Line character specified by HREFCHAR option

Line drawn with VREF option

Line drawn with HREF option

```
NOTE: 5 obs hidden.
```

Reference lines cause three additional observations to be hidden (compare to Exhibit 21.6).

EXHIBIT 21.8 Hidden observations increase as display space decreases

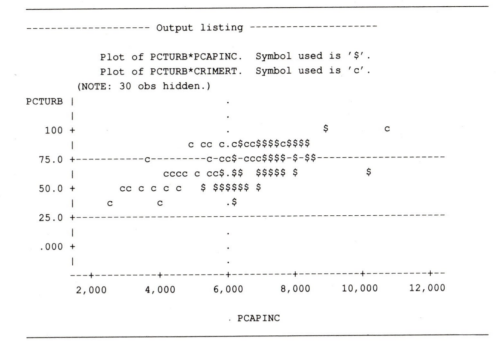

```
------------------- Output listing --------------------

            Plot of PCTURB*PCAPINC.  Symbol used is '$'.
            Plot of PCTURB*CRIMERT.  Symbol used is 'c'.
        (NOTE: 30 obs hidden.)
PCTURB |                          .
       |                          .
  100  +                     .            $          c
       |                 c cc c.c$cc$$$$c$$$$
 75.0  +-----------c---------c-cc$-ccc$$$$-$-$$--------------------
       |              cccc c cc$.$$  $$$$$ $               $
 50.0  +        cc c c c    $ $$$$$$ $
       |      c        c         .$
 25.0  +--------------------------------------------------------------
       |                          .
 .000  +                          .
       |                          .
       ---+----------+----------+----------+----------+----------+--
        2,000      4,000      6,000      8,000     10,000     12,000

                            . PCAPINC
```

Exhibit 21.10 (see p. 441) displays two plots on the same page. The VPCT = 50 option splits the vertical dimension of the page equally and leaves the horizontal dimension unchanged. The HPOS option in each PLOT statement ensures that the same number of print positions (or columns) is used in each plot. Formats are used to enhance the presentation of the tick marks. Notice that four observations had missing values and thus were excluded from the plot.

Exhibit 21.11 (see p. 442) displays nominal, rather than interval, scale data along the *x* axis. REGION has values of "Northeast", "South", "Midwest", and so on. An attempt was made to restrict the display to regions east of the Mississippi River by using the HAXIS option. Since no observations have values of "N", "S", or "M", none are in range and the PLOT request causes an error.

Exhibit 21.12 (see p. 443) changes the HAXIS specifications used in Exhibit 21.11, but mistakenly specifies "North" rather than "Northeast". A tick mark is displayed at "North" with no observations plotted above it. Character axis specifications require exact matches!

Exhibit 21.13 (see p. 445) shows how two outliers can distort an entire plot. Since two observations have very high POPSQMI values, the vertical axis is "stretched" out to include them. This makes distinguishing among the other observations difficult.

Exhibit 21.14 (see p. 446) uses the VAXIS option to "window" the bottom portion of the plot in Exhibit 21.12. Notice the "2 obs were out of range" message: these are the observations that were deliberately excluded by choosing a narrow VAXIS value range. This is a good example of the benefits of running PLOT interactively. You could have run the statements in Exhibit 21.13, seen the distortion, then rerun the PLOT with the new VAXIS specification. The same problem-solving process is, of course, possible in batch mode but would take more elapsed time.

EXHIBIT 21.9   Plot appearance enhanced with BOX and NOLEGEND options and ATTRIB statement

```
------------------- Output listing --------------------

libname in '\data\sas';

proc plot data=in.states nolegend;
attrib farmpct format=3.0 label='% land in farms'
       pctin   format=3.0 label='% in migration' ;
plot farmpct*pctin / box;
run;

------------------- Output listing --------------------
```

Border produced by BOX option in PLOT statement

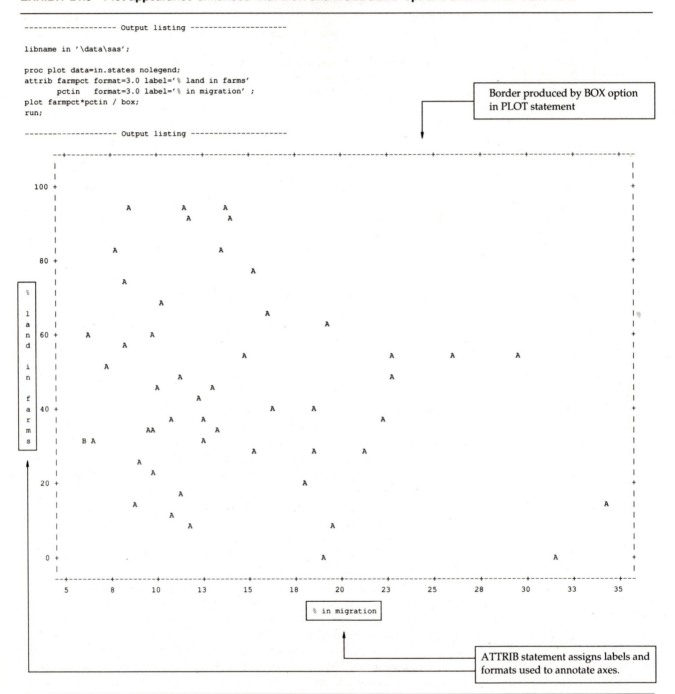

% in migration

ATTRIB statement assigns labels and formats used to annotate axes.

EXHIBIT 21.10   Multiple plots per page

```
------------------- Program listing -------------------
libname in '\data\sas';

proc plot data=in.states vpct=50;
plot pcturb*pcapinc / haxis=2000 to 12000 by 2000 box hpos=110;
plot pcturb*crimert / box hpos=110;
run;

------------------- Output listing -------------------
```

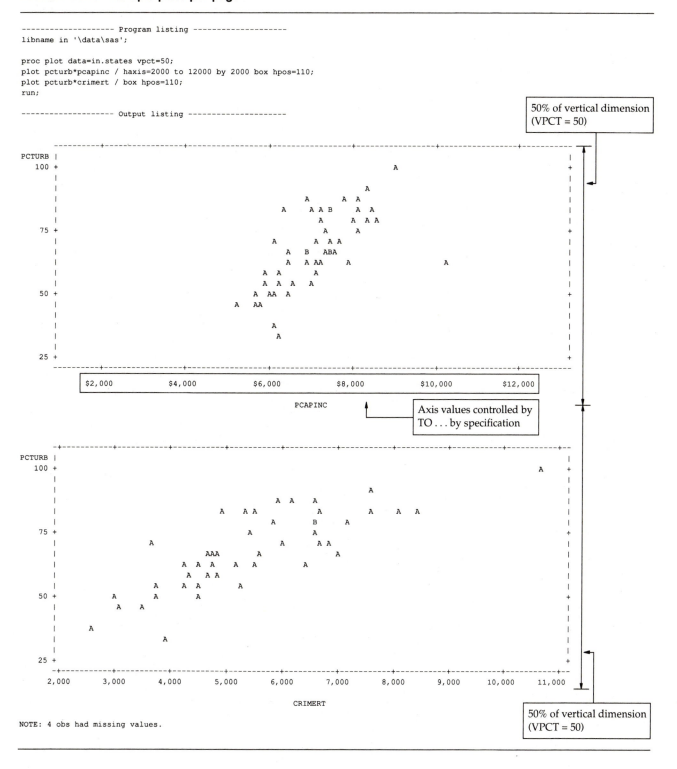

50% of vertical dimension (VPCT = 50)

Axis values controlled by TO . . . by specification

50% of vertical dimension (VPCT = 50)

```
NOTE: 4 obs had missing values.
```

EXHIBIT 21.11   Invalid HAXIS specification: no values in range

```
------------------- SAS Log --------------------------
  1    libname in '\data\sas';
  2
  3    proc plot data=in.states;
  4    plot pctcrowd*region / haxis = 'N' 'S' 'M';
  5    title1 "pctcrowd*region / haxis = 'N' 'S' 'M';" ;
  6    run;
ERROR: All values missing or out of range for PCTCROWD*REGION. Check VAXIS= and HAXIS= specifications.
```

Exercises

21.1. Specify VAXIS values to meet the following requirements:
   a. Tick marks at 1, 3, 5, and so on, up to 21.
   b. Tick marks from "01mar91" through "01apr92" in increments of one month.
   c. Tick marks starting at 0, ending at 50, in increments of 5.
   d. Tick marks of odd values in range 1 through 11, even values in range 12 through 20.

21.2. Write PLOT statements to meet the following requirements. All variables are interval scale unless otherwise specified. All answers should use a single PLOT statement. Vertical axis variables are specified first.
   a. Plot NREPS by SALES. Draw a box around the plot and suppress the legend.
   b. Produce three plots: NREPS by SALES, PROFITS, and NETCHG. All plots should be displayed on the same page, each taking a third of the page height. Begin each x-axis tick mark at 0 even if the minimum x-axis value is greater than 0. Format the x-axis variables with DOLLAR10.
   c. Plot NREPS by SALES for each level of REGION. Keep axis specifications identical across all levels of REGION.
   d. Produce two plots: NREPS by SALES and PROFITS. Print the plots on top of each other. Indicate the SALES-NREPS points by the character "S." Indicate the PROFITS-NREPS points by the character "P."
   e. Follow the instructions in part d, but draw a box around the plot and add a vertical reference line at x-axis values 100,000 and 500,000 (use the character "¦" to draw the line). A reference line at the y-axis should be drawn with character "-" at value 250.

21.3. What plots are produced by each of the following specifications (indicate y-axis variable first)?
   a. (y1 - y3) * x
   b. y1 * (x1 - x3)
   c. (y1 - y3)
   d. y1 * x y2 * x
   e. (y1 y2) * x
   f. (y1 y2) * (x1 - x3)
   g. y * x = '1'

21.4. Variable INCZSCOR ranges from –3 to 3. Variable INPTILE ranges from 0 to 100. What, if anything, is wrong with the following statements?
   a. `plot yrs_exp * (inczscor inptile) / overlay ;`
   b. `plot yrs_exp * inczscor / haxis = 0 to 100`
             `by 10;`

21.5. The y-axis variable Y varies from –5 to 5, x-axis variable X from 0 to 100. Write PLOT statements to display only a portion of the full plot (the one where the y axis varies from –5 to 5 and the x axis from 0 to 100). The statements must meet the following requirements. Choose appropriate tick mark increments.
   a. Y values: any
       X values: 0 through 50

EXHIBIT 21.12   Group unintentionally excluded from a plot

```
------------------- Program listing -------------------

libname in '\data\sas';

proc plot data=in.states;
plot pctcrowd*region / haxis = 'North' 'South' 'Midwest';
run;

------------------- Output listing -------------------

                    Plot of PCTCROWD*REGION.   Legend: A = 1 obs, B = 2 obs, etc.

          16.000 +
                 |
                 |
                 |
                 |
                 |
                 |
          14.000 +
                 |
                 |
                 |
                 |
                 |
                 |
          12.000 +
                 |
                 |
        PCTCROWD |
                 |
                 |
                 |
          10.000 +
                 |
                 |
                 |
                 |
                 |
                 |
           8.000 +                                B
                 |
                 |
                 |
                 |
                 |
                 |
           6.000 +                                A
                 |                                C
                 |
                 |
                 |                                B
                 |                                A                     A
           4.000 +                                A
                 |                                                      A
                 |                                A                     A
                 |                                A                     B
```

EXHIBIT 21.12   *(continued)*

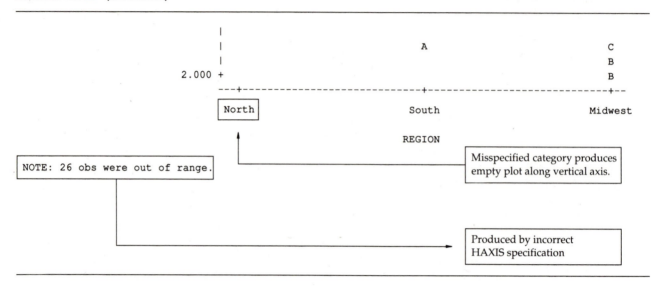

Exercises

(continued)

  b. *Y* values: –5 through 0
       X values: 0 through 50
  c. *Y* values: –10 through 10
       X values: 50 through 100

21.6. Write PLOT procedure statements to plot *y*1 * *x* with the title "Measure 1 by Income," then plot *y*2 by *x* with the title "Measure 2 by Income." Use only one PROC PLOT statement.

21.7. Refer to Exhibit 21.5. How would you describe the regional patterns revealed by the plot? How would you clarify the individual REGIONs' patterns using PLOT?

21.8. Refer to Exhibit 21.8. What does the note "30 obs hidden" mean? Is this an error? How might you change the plot statement or SAS System options to eliminate this message? Under what circumstances may a plot this size be acceptable?

21.9. Refer to Exhibit 21.10. How would you change the PROC PLOT statement to allow three plots on the page? What statement would you add to plot PCTURB * PCAPINC overlaid on PCTURB * CRIMERT? What is the advantage of displaying all three plots on a single page, since information is lost (there are "hidden" observations)?

21.10. Refer to Exhibit 21.13. Use the UNIVARIATE procedure to identify the value of the variable POPSQMI outliers as well as their associated NAME value.

EXHIBIT 21.13   Outliers distorting plot scaling

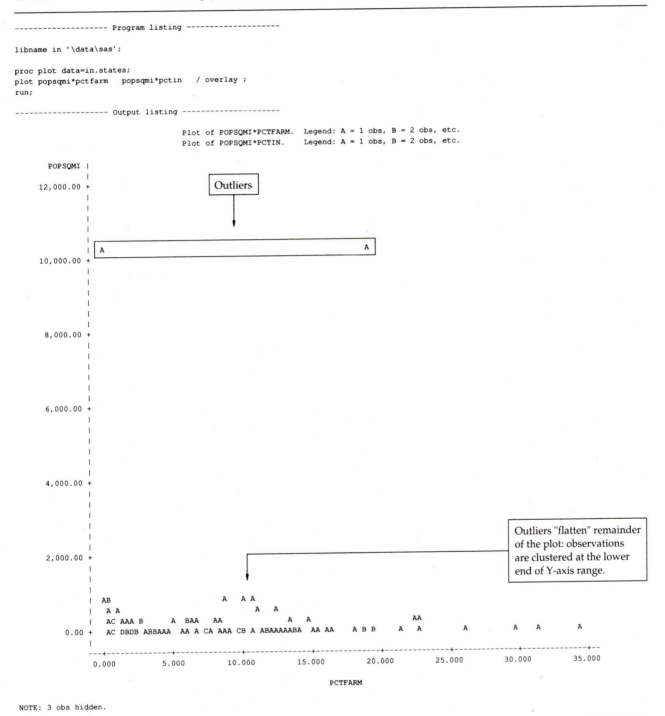

```
-------------------- Program listing --------------------

libname in '\data\sas';

proc plot data=in.states;
plot popsqmi*pctfarm   popsqmi*pctin   / overlay ;
run;

-------------------- Output listing --------------------

                         Plot of POPSQMI*PCTFARM.   Legend: A = 1 obs, B = 2 obs, etc.
                         Plot of POPSQMI*PCTIN.     Legend: A = 1 obs, B = 2 obs, etc.
```

NOTE: 3 obs hidden.

EXHIBIT 21.14   Axis specifications used to display only a portion of the display space

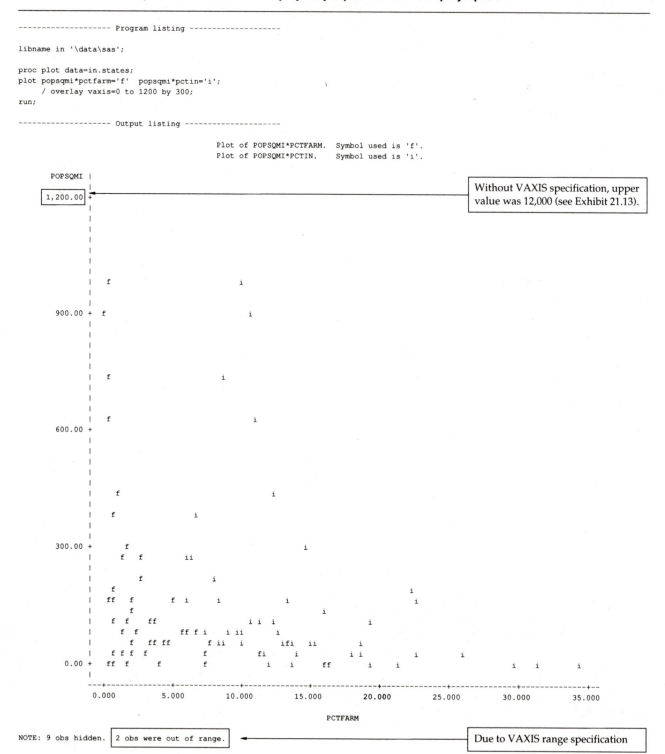

```
------------------ Program listing --------------------

libname in '\data\sas';

proc plot data=in.states;
plot popsqmi*pctfarm='f'  popsqmi*pctin='i';
    / overlay vaxis=0 to 1200 by 300;
run;

------------------ Output listing --------------------
```

```
                              Plot of POPSQMI*PCTFARM.   Symbol used is 'f'.
                              Plot of POPSQMI*PCTIN.     Symbol used is 'i'.


   POPSQMI |
           |
  1,200.00 +◄────────────────────────────────────────────
           |
           |
           |
           |
           |
           |
           |
           |
           |   f                          i
           |
           |
   900.00 +   f                          i
           |
           |
           |
           |
           |   f                      i
           |
           |
           |
           |
           |   f                           i
   600.00 +
           |
           |
           |
           |
           |
           |     f                         i
           |
           |     f                    i
           |
           |
   300.00 +       f                                i
           |       f   f        ii
           |
           |         f                i
           |   f
           |  ff   f        f  i    i            i               i                   i
           |       f                           i
           |   f f    ff           i i i             i
           |    f  f         ff f i  i ii       i
           |      f   ff ff      f ii  i     ifi ii       i
           |   f f f  f          f      fi    i      i i       i        i
     0.00 +  ff  f     f         f         i   i    ff     i    i            i    i       i
           |
           ---+--------------+--------------+--------------+--------------+--------------+--------------+--------------+--
              0.000          5.000         10.000         15.000         20.000         25.000         30.000         35.000

                                                      PCTFARM
```

```
NOTE: 9 obs hidden.   2 obs were out of range. ◄───────────────────────
```

Without VAXIS specification, upper value was 12,000 (see Exhibit 21.13).

Due to VAXIS range specification

Summary

The PLOT procedure is a straightforward way to display the relationship between pairs of variables. In its simplest form the intersection of a pair of values is plotted with the letter A. Two pairs sharing the same plot location are plotted with a B, and so on. Many options are available for improving the informative and visual content of the output. Axis scaling and origin may be controlled, reference lines may be drawn along the axes, plots can be overlaid, and multiple plots may be displayed on a single page. The examples in the chapter highlight these features and also point out how plots can be rendered uninformative by incorrect specification of axis values or the presence of outliers along one or both axes.

22 Bivariate Statistics

The plotting techniques discussed in Chapter 21 were fine for quick visual displays of the relationship between variables. Usually this graphic presentation is only the beginning of the data analysis process. This chapter presents some simple empirical measures of the strength and nature of the relationship between two variables: the FREQ, TTEST, and CORR procedures.

22.1 Contingency Tables: PROC FREQ

Chapter 6 discussed the single-variable use of the FREQ procedure. With just a few simple additions to the procedure's statements, two- or higher dimensional contingency tables can be produced. These tables and their associated statistics show the strength and pattern of relationships between two or more variables. They answer questions such as, Is there any relationship between race and level of education? and, Does the association between income group and education level vary by region of the country?

Notice the scale of the variables in these questions. Race and region use *nominal* scales: there is no implied magnitude between the codes. You cannot say that the region value of "North" is intrinsically larger or smaller than that of region "West". The education-level variable may use an *ordinal* scale: a coding scheme of "1" for only grade school completed, "2" for some high school, "3" for completed high school, and so on does imply that the larger the value of the variable, the greater the underlying concept being measured (i.e., education level). Finally, income group is measured on an *interval* scale. It may have any number of negative and positive values measured by a common standard (i.e., the local currency). The standard enables comparisons that are not possible in other scales. You can say that an income of $20,000 is twice as large as that of

$10,000, but you cannot say that an education level of "2" is twice that of "1." Notice that the sample research question posed above uses income *group:* interval scales may be reduced to ordinal scales via recoding (discussed in Chapter 15).

The FREQ procedure will accept any form of character or numeric values for its tables. It will produce more meaningful output when it uses nominal- and ordinal-scaled data.

In Exhibit 22.1 FREQ is used to produce cross-tabulations. The analysis dataset's unit of observation is state and state-equivalents in the United States. The percent of adults 25 years and older graduating from college (variable PCT-COLL) is cross-tabulated with per capita educational expenditures for teacher salaries (EDEX_SAL) in each state. Both variables are interval-level data. They are transformed into ordinal data by two user-written formats defined prior to FREQ (see Chapter 15 for a discussion of user-written formats). This grouping reduces what could be a 51 row by 51 column table of individual values into a 2 by 2 table of grouped data. Each observation's (i.e., state's) value of these variables is likely to be unique, so without the formats FREQ would assign each value to a row or column in the table. The output would be difficult to interpret and not credible from a statistical viewpoint.

In Exhibit 22.1 college percent (PCTCOLL) is the row variable and per capita expenditure (EDEX_SAL) the column variable. The formatted values of each table dimension are displayed in ascending order (top to bottom in the rows and left to right in the columns). Each cell contains four rows of information. The meaning of each row is identified in the cell at the top left corner of the table:

1. *Frequency.* The frequency is how many times that particular row-column combination occurred in the dataset.

2. *Percent.* The percent is what percentage of the total number of observations used in the table is contained in the cell.

3. *Row Pct.* The row percent is the percent of the row total found in the cell going across all *columns* of the table.

4. *Col pct.* The column percent is the percent of the column total found in the cell going down all *rows* of the table.

The "Total" column and row headings are the marginal totals and their percent of the dimension total. For example, the first row's "Total" value of 26 represents the sum of the frequencies of that row across all columns (20 plus 6 equals 26), and the 50.98 indicates that the 26 represents 50.98% of all observations used in the table. The next row's total of 25 and its 49.02 percentage bring the grand total to 51 observations and, of course, 100%.

FREQ Options

Exhibit 22.1 presented default output. FREQ has many options that control the content and presentation of the tables. This section gives a brief overview of options that may be used to customize output. The options are described in more detail in the "Syntax" section below.

Altering Table Contents. Cell contents can be altered. The cell's contribution to the table's chi-square value, expected value, and deviation from expected and cumulative percentage across rows may be displayed using the CELLCHISQ, EXPECTED, DEVIATION, and CUMCOL options. Missing values and nonoccurring row-column combinations may be displayed with the MISSPRINT and SPARSE options.

All or part of the default cell contents may be suppressed. The NOROW, NOCOL, NOPERCENT, and NOFREQ options control the four default items. NOPRINT completely avoids printing the table.

EXHIBIT 22.1   Default output for cross-tabulation

```
-------------------- Program listing --------------------

libname x 'c:\data\sas\';
options ps=60;

proc format;
value coll(max=8) 0-16= '<= 16%' 16<-high= '> 16%' ;
value edsal(max=8) 0-2250= '$2.25k-' 2250<-high='>$2.25k';
run;

proc freq data=x.states;
tables pctcoll * edex_sal ;
format pctcoll coll8. edex_sal edsal8.;
run;

-------------------- Output listing --------------------
```

TABLE OF PCTCOLL BY EDEX_SAL — Row variable

```
PCTCOLL      EDEX_SAL                       ← Column variable

Frequency
Percent
Row Pct
Col Pct      $2.25k- |>$2.25k |   Total
          ---------+--------+--------+
<= 16%          20 |      6 |      26    ← Row "marginal" count
          +  39.22 |  11.76 |   50.98    ← Percent of table count
          +  76.92 |  23.08 |              accounted for by this row
          +  71.43 |  26.09 |
          ---------+--------+--------+
> 16%        |      8 |     17 |      25
             |  15.69 |  33.33 |   49.02
             |  32.00 |  68.00 |
             |  28.57 |  73.91 |
          ---------+--------+--------+
Total           28       23       51    ← Table count

             54.90    45.10   100.00
```

Row variable

Column variable

Row "marginal" count

Percent of table count accounted for by this row

Table count

Column "marginal" count

Percent of table count accounted for by this column

Displaying Statistics.   By default, FREQ displays only raw counts and percentages. A variety of correlation coefficients and measures of association are available using the CHISQ, CMH, MEASURES, and ALL options.

Using Other Options.   The user can also control other FREQ features: a list format of the table variables may be selected instead of the default row and column table (LIST); missing values may be treated as valid and nonmissing (MISSING); and a single table's values may be saved in a SAS dataset (OUT).

Syntax

FREQ's syntax is described below. In addition to these procedure-specific statements, BY, LABEL, FORMAT, ATTRIB, TITLE, FOOTNOTE, and WEIGHT may be used. These statements were presented in Chapter 5. Options not in boldface were explained in the discussion of FREQ in Chapter 6.

```
PROC FREQ [DATA=dataset] [ORDER=FREQ|DATA|INTERNAL]
        [PAGE] ;
TABLES requests [ / [CELLCHISQ] [CUMCOL] [DEVIATION]
      [EXPECTED] [MISSPRINT] [SPARSE]
      [NOCOL] [NOROW] [NOCUM] [NOFREQ] [NOPERCENT]
      [NOPRINT]
      [CHISQ] [CMH] [CMH1] [CMH2] [EXACT] [MEASURES]
      [ALL] [ALPHA=prob]
      [LIST] [MISSING] [OUT=dataset] ;
```

Some common applications of the FREQ procedure follow:

```
proc freq data=in.harris;
tables region*q1 race*sex / chisq;
tables region*(q1-q5) / all;
tables region*race*(q23a q23b) / norow nocol nopercent ;
```

General Options.   FREQ has three general options:

1. *PAGE* allows only one table to be printed per page of output. The default action is to display as many tables as possible on a single page. PAGE is available only in version 6 of the SAS System.

2. The *requests* specifies which variables will be cross-tabulated. Its use is required for producing cross-tabulation: if it is omitted, FREQ produces single-variable frequency distributions of all variables in the dataset. The general form of *requests* is the specification of the table row variable followed by an asterisk (*) and the column variable. Valid table requests and the resulting output are shown in the following list:

Table Request	Table(s) Produced
$y * x$	$y–x$
$z * y * x$	$y–x$ for each value of z
$y * (x1\ x2)$	$y–x1\ y–x2$
$y * x1\ y * x2$	$y–x1\ y–x2$
$(y1\ y2) * x$	$y1–x\ y2–x$
$(y1\ y2) * (x1\ x2)$	$y1–x1\ y1–x2\ y2–x1\ y2–x2$
$y * (x1–x3)$	$y–x1\ y–x2\ y–x3$
$z * y * (x1\ x2)$	$y–x1$ and $y–x2$ for each value of z

3. The / signals the end of the plot request and the beginning of a list of options.

Options Increasing Table Detail. The following six options increase the amount of detail in a table:

1. *CELLCHISQ* requests printing of each cell's contribution to the table's chi-square statistic. The value is (frequency − expected)^2 / expected.

2. *CUMCOL* requests printing of cumulative column percentages in each cell. As you read down the rows of the table, the percent of each column's frequency accounted for up to and including that row is displayed.

3. *DEVIATION* requests printing of the difference between each cell's expected and actual frequencies.

4. *EXPECTED* requests printing of each cell's expected frequency under the hypothesis of row-column variable independence. The *LIST* option (see the "Other Options" section below) suppresses *EXPECTED*'s output.

5. *MISSPRINT* requests printing of missing values in the tables. To include missing values in statistic calculations, use the *MISSING* option (see the "Other Options" section).

6. *SPARSE*, when used with the *LIST* option, enumerates all possible combinations of the table variables even if some combinations do not actually occur in the dataset. *SPARSE* cannot be used if any statistics are requested.

Options Decreasing Table Detail. The following five options decrease the amount of detail in a table:

1. *NOCOL* and *NOROW* suppress the printing of each cell's contribution to its column and row frequency, respectively.

2. *NOCUM*, when used with the *LIST* option, suppresses the printing of cumulative frequencies and percentages.

3. *NOFREQ* suppresses the printing of each cell's frequency.

4. *NOPERCENT* suppresses the printing of each cell's share of the number of observations used in the table.

5. *NOPRINT* suppresses the printing of the table but does permit display of requested statistics (see the following "Statistics Options" section for details).

Statistics Options. The following seven options request statistics for the table. These statistics are computed for each "layer" of the table: for three- (or higher) dimensional tables, for example, $z * y * x$, the statistic is computed for $y * x$ at each level of z. Likewise, a BY statement computes statistics for each BY-group.

1. *CHISQ* requests a chi-square test of homogeneity. Other chi-square measures of association are also produced: Pearson, likelihood ratio, and Mantel–Haenszel. Phi and contingency coefficients and Cramer's V are also calculated.

2. *CMH* requests Cochran–Mantel–Haenszel (CMH) statistics, testing for association between row and column variables after adjusting for other variables in the *TABLES* statement. For 2×2 tables, FREQ also produces relative-risk estimates and tests for homogeneity of odds ratios.

3. *CMH1* and *CMH2* resemble the *CMH* option but restrict the amount of statistics generated, possibly saving a great deal of computer time. *CMH1* produces a correlation statistic with one degree of freedom, and *CMH2* produces this correlation and mean score (ANOVA) statistics. These options are available only in version 6 of the SAS System.

4. *EXACT* requests calculation of Fisher's exact test for tables larger than 2×2. When the number of observations in the dataset divided by the degrees

of freedom in the table exceeds five, the calculation becomes extremely computer-intensive and may fail. EXACT is available only in version 6 of the SAS System.

5. *MEASURES* requests the calculation of a set of measures of association not covered in the preceding options. Included are correlation coefficients (Pearson and Spearman), gamma, Kendall's tau-*b*, Stuart's tau-*c*, Somer's *D*, lambda, and uncertainty coefficients. For 2 × 2 tables, FREQ also produces odds and risk ratios.

6. *ALL* requests calculation of all statistics described in the *CHISQ, CMH,* and *MEASURES* options. Specifying *ALL* and either *CMH1* or *CMH2* will restrict the number of CMH statistics produced, possibly saving great amounts of computer resources.

7. The *prob* specifies a width for confidence intervals. It is a decimal value representing 100*(1 − *prob*), thus .05 requests a 95% confidence interval, and .01 a 99% confidence interval. The *prob* must be between .0001 and .9999. The default is .05.

Other Options.   Several other options to control the presentation, content, and retention of tables are available:

1. *LIST* displays tables as a list of paired values rather than a grid. One line of output is used for each combination of *x-y* variable levels in the table. The output line contains the values of a combination of table variables and their frequency, percent, cumulative frequency, and cumulative percent. Selecting the *CELLCHISQ, EXPECTED,* and *DEVIATION* options has no effect on the presentation of *LIST* output.

2. *MISSING* requests that missing values be treated as valid, nonmissing categories when calculating statistics. Use the *MISSPRINT* option to include missing values in the printed table but not in the calculation of statistics.

3. *OUT=dataset* creates a SAS dataset containing counts and percentages from the most recent (i.e., rightmost) request in the *TABLES* statement. The *dataset* can specify a temporary or permanent SAS dataset name. Observations in *dataset* are unique combinations of the table's variables. The *dataset*'s variables include the table variable name(s) and the BY variable(s), if appropriate. The variables FREQ and PERCENT represent the number of occurrences of the observation's table variable combination (i.e., cell frequency) and the percent of total observations represented by this combination. (See Exhibit 22.8 for an example of its use.)

Usage Notes

Keep the following points in mind when using PROC FREQ:

- More than one TABLES statement may be used with a single PROC statement. Any options specified in the PROC statement (ORDER, PAGE, or DATA) apply to *all* tables specified by *all* TABLES statements.

- Single-variable and multivariable tables may be combined in the same TABLES statement. Options unique to multivariable tables are simply ignored for single-variable tables: no errors, warnings, or notes are produced.

- Table categories are determined by the 16 leftmost formatted positions. Values longer than 16 are truncated and may be grouped together. FREQ prints a message if this grouping takes place.

- Since cross-tabulations are oriented toward nominal- and discrete-scale variables, care should be taken when using interval- or ratio-scale data. Usually such data should be grouped into a manageable number of categories. The regrouping can be performed with a series of IF-THEN statements in a DATA step or with user-written formats (see Chapter 15 for details). FREQ uses each formatted value of the TABLES variables for its rows and columns. Thus if an interval-level variable is ungrouped or only partially grouped by the format, the table may become quite large and, typically, useless from an analytical standpoint. (See Exhibit 22.10.)

- Interactive sessions typically use a page size of 21 lines (system options PS=21, TPS=21). Non-LIST tables in these environments that display many pieces of information in each cell are usually restricted to one row per "page," or screen, of output. To make the table more readable, the page size can be enlarged or the table information reduced. Multidimensional tables and those using ungrouped interval-scale data can produce large amounts of output. It may be prudent to test the table request with a subset of the data (possibly using the OBS dataset or system options described in Chapter 12).

Examples

This section shows examples of FREQ used correctly as well as inappropriately. Unless otherwise noted, all the examples using continuous-scale data were grouped into a smaller number of categories with user-written formats (discussed in Chapter 15).

Key features and options used in this section are summarized in the following list:

Exhibit Features/Options

22.2	All defaults used for a two-way table
22.3	NOROW, NOCOL, and NOPERCENT used to reduce cell information
22.4	DEVIATION and CELLCHISQ options; no overall chi-square statistic generated
22.5	ALL option selected for statistics
22.6	Table listing suppressed; chi-square and related statistics generated
22.7	LIST format used
22.8	Three-way table; output dataset generated
22.9	WEIGHT used to change table's unit of observation
22.10	Table with ungrouped interval-scale variable

Exhibit 22.2 is a simple cross-tabulation of the percent of adults with four years of college by per capita expenditure on education. No options are taken.

Suppose only the cell frequencies are of interest. Exhibit 22.3 (see p. 457) uses the NOROW, NOCOL, and NOPERCENT options to suppress extraneous cell information. The row and column marginal totals do not display percentages when these options are used.

Exhibit 22.4 (see p. 458) requests information about each cell's deviation from its expected value (DEVIATION) and its contribution to the chi-square statistic (CELLCHISQ). Notice that CELLCHISQ generates only the cells' share of the test statistic and does *not* generate a table-level chi-square test.

Exhibit 22.5 (see p. 459) takes default table presentation but specifies that all appropriate statistics be calculated (ALL).

Exhibit 22.6 (see p. 461) suppresses the display of the table (NOPRINT) and asks for chi-square and related measures of association (CHISQ).

Exhibit 22.7 (see p. 462) requests the same table as previous exhibits in this chapter but displays it in list format (LIST). The information in this listing differs from the default format: there are no "rows" and "columns" as there were in the table.

EXHIBIT 22.2　Simple cross-tabulation; all defaults taken

```
------------------- Program listing -------------------

proc freq data=college;
tables pctcoll * edex_sal;
format pctcoll coll8. edex_sal edsal8.;
run;

------------------- Output listing -------------------

TABLE OF PCTCOLL BY EDEX_SAL

PCTCOLL     EDEX_SAL

Frequency|
Percent  |
Row Pct  |
Col Pct  |$2.25k- |>$2.25k |  Total
---------+--------+--------+
<= 16%   |    20  |     6  |    26
         |  39.22 |  11.76 | 50.98
         |  76.92 |  23.08 |
         |  71.43 |  26.09 |
---------+--------+--------+
>  16%   |     8  |    17  |    25
         |  15.69 |  33.33 | 49.02
         |  32.00 |  68.00 |
         |  28.57 |  73.91 |
---------+--------+--------+
Total         28       23       51
            54.90    45.10   100.00
```

Exhibit 22.8 (see p. 463) requests a three-way table: the college-by-expenditure table used in Exhibit 22.7 is calculated for each level of EASTWEST, an east/west of the Mississippi River indicator. Chi-square statistics are generated for each level, or strata, of the table, that is, each value of EASTWEST. Notice that although the FREQOUT dataset is sorted by EASWEST, the input dataset is not: the SAS dataset does not have to be sorted. The first table, that for EASWEST level "e" (East Coast), is shown in the Exhibit.

The exhibit also demonstrates creation of an output dataset (FREQOUT). Exhibit 22.8 shows a listing (PROC PRINT) and the directory contents (PROC CONTENTS) of the dataset. Notice in the contents listing that FREQ creates variable labels for PERCENT and COUNT. The output dataset's directory also contains the names of the formats used to group the variables.

Exhibit 22.9 (see p. 465) is the default table used in Exhibit 22.1 but with a twist. Rather than have each observation contribute once to each cell (one observation, one state), each state's contribution to the table is weighted by its population (in millions). Thus we see that low college rate (<= 16%), low expenditure ($2.25k-) states account for 59.703 million people. The total frequency of table cells (the sum of the weights) is 226.546, the population, in millions, of the United States in 1980.

EXHIBIT 22.3   NOROW, NOCOL, and NOPERCENT used to reduce cell information

```
------------------- Program listing -------------------

proc freq data=college;
tables pctcoll * edex_sal / norow nocol nopercent;
format pctcoll coll8. edex_sal edsal8.;
run;

------------------- Output listing -------------------

TABLE OF PCTCOLL BY EDEX_SAL

PCTCOLL      EDEX_SAL

Frequency|$2.25k- |>$2.25k |   Total
---------+--------+--------+
<= 16%   |   20 |     6 |      26
---------+--------+--------+
> 16%    |    8 |    17 |      25
---------+--------+--------+
Total         28       23       51
```

Exhibit 22.10 (see p. 466) runs the table in Exhibit 22.1 but omits the format for interval-scale variable PCTCOLL. Since there is no format to do grouping, FREQ treats *each* value as a category. Thus 10.417 is distinct from 10.820 and 11.077. Since no states have duplicate PCTCOLL values, the table has 51 rows, one for each state and state-equivalent. Statistics generated from this table would probably be meaningless. Exhibit 22.10 presents the first few rows of this accidentally long table.

Exercises

22.1. Refer to Exhibit 22.1. In the questions that follow, "college graduates" refers to variable PCTCOLL (the percent of the population over 25 years of age with four or more years of college education). "Salaries" refers to variable EDEX_SAL (per capita education expenditures for teacher salaries). The unit of observation in the table is U.S. states.

    a. What percentage of states have lower rates of college graduates and lower salaries? Express this percent as a fraction: what numbers were used for the numerator and denominator?

    b. How many states have salaries less than $2,500 per capita?

    c. Which is the smallest cell in the table? What percent of the total number of tabled values is this?

    d. What does the "51" at the bottom right of the table represent? The "100.00"?

    e. EDEX_SAL and PCTCOLL are interval-scale variables grouped into two levels with user-written formats. Do you need to perform such grouping with ordinal- or nominal-scale data? Justify your answer. What would happen to the table's statistical reliability and size if the user-written formats were not used to group the variables?

EXHIBIT 22.4   DEVIATION and CELLCHISQ increase cell information

```
------------------- Program listing -------------------

proc freq data=college;
tables pctcoll * edex_sal / deviation cellchisq;
format pctcoll coll8. edex_sal edsal8.;
run;

------------------- Output listing -------------------

TABLE OF PCTCOLL BY EDEX_SAL
```

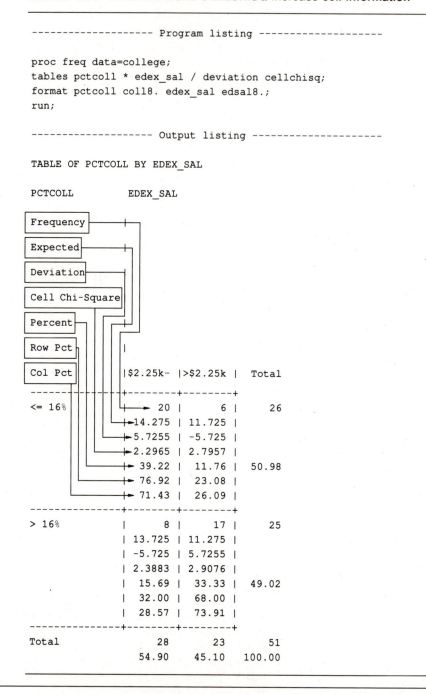

PCTCOLL	EDEX_SAL		
Frequency Expected Deviation Cell Chi-Square Percent Row Pct Col Pct	|$2.25k-	|>$2.25k |	Total
<= 16%	20	6	26
	14.275	11.725	
	5.7255	-5.725	
	2.2965	2.7957	
	39.22	11.76	50.98
	76.92	23.08	
	71.43	26.09	
> 16%	8	17	25
	13.725	11.275	
	-5.725	5.7255	
	2.3883	2.9076	
	15.69	33.33	49.02
	32.00	68.00	
	28.57	73.91	
Total	28	23	51
	54.90	45.10	100.00

Exercises
(*continued*)

22.2. Write TABLES statements to meet the following requirements. If there is more than one way to specify the table request, choose the most succinct.
   a. A by B
   b. A by B for each level of C
   c. A by B, A by C, both tables by each level of D
   d. A by B, A by C, A by D
   e. B, C, and D by A
   f. A by B; B by C; D by E by each level of F
   g. A by B, A by C, D by B, D by C

Exercises
(continued)

22.3. Which options in the TABLES statement would you use to accomplish the following?
   a. Display only cell counts in the table (no percentages).
   b. Display only the actual and expected cell counts, along with their difference.
   c. Display all possible combinations of cell values in a list rather than in tabular format.
   d. Treat missing values as a valid, nonmissing level of the analysis variable(s).

(continued on p. 462)

EXHIBIT 22.5   ALL option selected for statistics

```
------------------- Program listing -------------------

proc freq data=college;
tables pctcoll * edex_sal / all;
format pctcoll coll8. edex_sal edsal8.;
run;


------------------- Output listing -------------------

TABLE OF PCTCOLL BY EDEX_SAL

PCTCOLL      EDEX_SAL

Frequency|
Percent  |
Row Pct  |
Col Pct  |$2.25k- |>$2.25k |   Total
---------+--------+--------+
<= 16%   |    20 |      6 |     26
         | 39.22 |  11.76 |  50.98
         | 76.92 |  23.08 |
         | 71.43 |  26.09 |
---------+--------+--------+
> 16%    |     8 |     17 |     25
         | 15.69 |  33.33 |  49.02
         | 32.00 |  68.00 |
         | 28.57 |  73.91 |
---------+--------+--------+
Total         28       23       51
            54.90    45.10   100.00
```

STATISTICS FOR TABLE OF PCTCOLL BY EDEX_SAL

Statistic	DF	Value	Prob
Chi-Square	1	10.388	0.001
Likelihood Ratio Chi-Square	1	10.776	0.001
Continuity Adj. Chi-Square	1	8.653	0.003
Mantel-Haenszel Chi-Square	1	10.184	0.001
Fisher's Exact Test (Left)			1.000
(Right)			1.44E-03
(2-Tail)			1.91E-03
Phi Coefficient		0.451	
Contingency Coefficient		0.411	
Cramer's V		0.451	

CHISQ option output (also produced by ALL option)

EXACT option output (also produced by ALL option)

EXHIBIT 22.5   *(continued)*

```
Statistic                              Value       ASE
-------------------------------------------------------
Gamma                                  0.753       0.137
Kendall's Tau-b                        0.451       0.125
Stuart's Tau-c                         0.449       0.125

Somers' D C|R                          0.449       0.125
Somers' D R|C                          0.453       0.125

Pearson Correlation                    0.451       0.125
Spearman Correlation                   0.451       0.125

Lambda Asymmetric C|R                  0.391       0.170
Lambda Asymmetric R|C                  0.440       0.144
Lambda Symmetric                       0.417       0.149

Uncertainty Coefficient C|R            0.153       0.088
Uncertainty Coefficient R|C            0.152       0.088
Uncertainty Coefficient Symmetric      0.153       0.088


Estimates of the Relative Risk (Row1/Row2)


                                    95%
Type of Study        Value    Confidence Bounds
-------------------------------------------------------

Case-Control          7.083    2.049      24.486
Cohort (Col1 Risk)    2.404    1.307       4.420
Cohort (Col2 Risk)    0.339    0.160       0.720


Sample Size = 51
```

◄─── MEASURES option output (also produced by ALL option)

```
SUMMARY STATISTICS FOR PCTCOLL BY EDEX_SAL


Cochran-Mantel-Haenszel Statistics (Based on Table Scores)

Statistic   Alternative Hypothesis    DF     Value     Prob
-------------------------------------------------------------
    1       Nonzero Correlation        1     10.184    0.001
    2       Row Mean Scores Differ     1     10.184    0.001
    3       General Association        1     10.184    0.001
```

◄─── CMH option output (also produced by ALL option)

EXHIBIT 22.5 *(continued)*

```
Estimates of the Common Relative Risk (Row1/Row2)
                                                     95%
Type of Study   Method             Value    Confidence Bounds
----------------------------------------------------------------
Case-Control    Mantel-Haenszel    7.083     2.128    23.573
  (Odds Ratio)  Logit              7.083     2.049    24.486


Cohort          Mantel-Haenszel    2.404     1.403     4.119
  (Col1 Risk)   Logit              2.404     1.307     4.420


Cohort          Mantel-Haenszel    0.339     0.175     0.659
  (Col2 Risk)   Logit              0.339     0.160     0.720


The confidence bounds for the M-H estimates are test-based.

Total Sample Size = 51
```

EXHIBIT 22.6 **Table listing suppressed; chi-square and related statistics generated**

```
-------------------- Program listing --------------------

proc freq data=college;
tables pctcoll * edex_sal / noprint chisq ;
format pctcoll col18. edex_sal edsal8.;
run;

-------------------- Output listing --------------------

STATISTICS FOR TABLE OF PCTCOLL BY EDEX_SAL

Statistic                      DF    Value     Prob
----------------------------------------------------------
Chi-Square                      1    10.388    0.001
Likelihood Ratio Chi-Square     1    10.776    0.001
Continuity Adj. Chi-Square      1     8.653    0.003
Mantel-Haenszel Chi-Square      1    10.184    0.001
Fisher's Exact Test (Left)                     1.000
                    (Right)                    1.44E-03
                    (2-Tail)                   1.91E-03
Phi Coefficient                       0.451
Contingency Coefficient               0.411
Cramer's V                            0.451

Sample Size = 51
```

EXHIBIT 22.7   LIST format used

```
------------------ Program listing ------------------

proc freq data=college;
tables pctcoll * edex_sal / list;
format pctcoll coll8. edex_sal edsal8.;
run;


------------------ Output listing -------------------
```

| | | | | Cumulative | Cumulative |
PCTCOLL	EDEX_SAL	Frequency	Percent	Frequency	Percent
<= 16%	$2.25k-	20	39.2	20	39.2
<= 16%	>$2.25k	6	11.8	26	51.0
> 16%	$2.25k-	8	15.7	34	66.7
> 16%	>$2.25k	17	33.3	51	100.0

Exercises
(continued)

    e. Just display missing values. Do not include them in the calculation of statistics, counts, and percentages.

    f. Compute all statistics available in FREQ for two-way tables.

    g. Compute only a chi-square statistic.

22.4. Compare the default FREQ table output of Exhibit 22.2 to the list-style output of Exhibit 22.7.

    a. What information is lost by going to list style? What information is gained?

    b. Suppose Exhibit 22.2 had zero-count, or empty, cells. How would you ensure that this entry would be present in the list-style output?

22.2   Comparing Distributions: PROC TTEST

The TTEST procedure computes (not surprisingly) *t*-tests. These are statistics that calculate the probability that the means of two groups of observations are equal assuming the groups have equal variances. It also provides an option to compare means using the assumption of unequal variances.

Syntax

The syntax of PROC TTEST is described below. In addition to these procedure-specific statements, BY, LABEL, FORMAT, ATTRIB, TITLE, and FOOTNOTE may be used. These statements were presented in Chapter 5.

```
PROC TTEST [DATA=dataset] [COCHRAN];
CLASS stratifier;
[VAR analysis_vars;]
```

Some common applications of the TTEST procedure follow:

```
proc ttest data=x.survey;
class sex;
var p1_q1--p2_q3;
```

The TTEST procedure has four components:

    1. The *dataset* is the name of the dataset to be analyzed. If the *DATA* option is not entered, the most recently created dataset is used.

2. *COCHRAN* requests the Cochran and Cox *t*-test when variances are *unequal*.

3. *CLASS* specifies a single *stratifier* variable. This variable must have exactly two nonmissing levels. These levels form the two groups whose means and variances are compared by TTEST.

EXHIBIT 22.8   Three-way table; output dataset generated

```
------------------- Program listing -------------------

proc freq data=college;
tables eastwest * pctcoll * edex_sal / out=freqout chisq;
format pctcoll coll8. edex_sal edsal8.;
run;

------------------- FREQ output --------------------
```

```
TABLE 1 OF PCTCOLL BY EDEX_SAL
CONTROLLING FOR EASTWEST=e
```
◄─────────────────────   Message indicating level of stratifier variable EASTWEST

```
PCTCOLL      EDEX_SAL

Frequency|
Percent  |
Row Pct  |
Col Pct  |$2.25k- |>$2.25k |  Total
---------+--------+--------+
<= 16%   |    10  |     6  |    16
         | 37.04  | 22.22  | 59.26
         | 62.50  | 37.50  |
         | 71.43  | 46.15  |
---------+--------+--------+
> 16%    |     4  |     7  |    11
         | 14.81  | 25.93  | 40.74
         | 36.36  | 63.64  |
         | 28.57  | 53.85  |
---------+--------+--------+
Total         14       13       27
            51.85    48.15   100.00
```

```
STATISTICS FOR TABLE 1 OF PCTCOLL BY EDEX_SAL
CONTROLLING FOR EASTWEST=e
```
◄─────────────────────

```
Statistic                       DF    Value      Prob
-------------------------------------------------------
Chi-Square                       1    1.784      0.182
Likelihood Ratio Chi-Square      1    1.802      0.179
Continuity Adj. Chi-Square       1    0.890      0.345
Mantel-Haenszel Chi-Square       1    1.718      0.190
Fisher's Exact Test (Left)                       0.959
                    (Right)                      0.173
                    (2-Tail)                     0.252
Phi Coefficient                       0.257
Contingency Coefficient               0.249
Cramer's V                            0.257

Sample Size = 27
```

EXHIBIT 22.8 *(continued)*

```
------------------- FREQOUT listing --------------------

OBS    EASTWEST    PCTCOLL    EDEX_SAL    COUNT    PERCENT

 1        e        <= 16%     $2.25k-      10      19.6078
 2        e        <= 16%     >$2.25k       6      11.7647
 3        e         > 16%     $2.25k-       4       7.8431
 4        e         > 16%     >$2.25k       7      13.7255
 5        w        <= 16%     $2.25k-      10      19.6078
 6        w         > 16%     $2.25k-       4       7.8431
 7        w         > 16%     >$2.25k      10      19.6078

------------------- FREQOUT contents --------------------

CONTENTS PROCEDURE

   Data Set Name:   WORK.FREQOUT          Type:
   Observations:    7                     Record Len: 37
   Variables:       5
   Label:

-----Alphabetic List of Variables and Attributes-----

   #   Variable   Type   Len   Pos   Format     Label
   4   COUNT      Num     8    21              FREQUENCY COUNT
   1   EASTWEST   Char    1     4
   3   EDEX_SAL   Num     8    13    EDSAL8.
   2   PCTCOLL    Num     8     5    COLL8.3
   5   PERCENT    Num     8    29              PERCENT OF TOTAL FREQUENCY
```

LABELs generated by FREQ

4. If the *VAR* statement is not present, TTEST compares the means of all numeric variables not used elsewhere in the procedure. Control the variables analyzed by specifying *analysis_vars*.

Usage Notes

Keep in mind the following points when using TTEST:

- TTEST does not know about the scale of an analysis variable, but *you* do. Results on *analysis_vars* other than interval- or ratio-scale data are suspect.

- If BY-group processing is used, the analyses are performed independently for each distinct, formatted level of the BY variables. This may create situations where the number of observations and/or variances in the BY-groups yields unreliable results.

- Because the procedure compares *observations*, not *variables*, it is inappropriate for paired comparisons. Handling paired comparisons was discussed in the description of PROC MEANS in Chapter 6.

- CLASS variables may be either numeric or character. Only the first (leftmost) 16 positions of character variables are used to assign groups. The two-group restriction applies to the *formatted* values of *stratifier*: nondichotomous numeric variables may be grouped using user-written formats. See Chapters 14 and 15 for a description of user-written formats.

EXHIBIT 22.9   WEIGHT used to change unit of observation (from states to people)

```
------------------- Program listing -------------------

proc freq data=x.states;
tables  pctcoll * edex_sal ;
weight pop;
format pctcoll coll8. edex_sal edsal8.;
run;

------------------- Output listing -------------------

TABLE OF PCTCOLL BY EDEX_SAL

PCTCOLL    EDEX_SAL

Frequency|
Percent  |
Row Pct  |
Col Pct  |$2.25k- |>$2.25k |  Total
---------+--------+--------+
<= 16%   | 59.703 | 47.323 | 107.03
         |  26.35 |  20.89 |  47.24
         |  55.78 |  44.22 |
         |  70.78 |  33.28 |
---------+--------+--------+
> 16%    | 24.652 | 94.867 | 119.52
         |  10.88 |  41.88 |  52.76
         |  20.63 |  79.37 |
         |  29.22 |  66.72 |
---------+--------+--------+
Total      84.3557  142.19  226.546
           37.24    62.76   100.00
```

Examples

All examples in this section use the STATES dataset in Appendix E. In Exhibit 22.11 (see p. 467) the means and variance of record high temperatures (HIGHTEMP) and average unemployment benefits (UNEMPBEN) are compared. The comparison groups are formed by each state's level of EASTWEST, a character variable indicating the state's location east or west of the Mississippi River.

Exhibit 22.12 (see p. 468) compares crime rates (variable CRIMERT) by level of population density (CLASS variable POPSQMI). Because the density variable is continuous, it cannot be used as a CLASS variable until it is dichotomized. PROC FORMAT is used to create a format (DENSITY) that groups states at or below 100 into one category (<= 100/sq mi) and other, more densely populated states into a second category (> 100/sq mi). The FORMAT statement in the PROC tells TTEST to group POPSQMI using the format DENSITY. Labels are used to annotate the output. The COCHRAN option generates additional hypothesis-testing output.

Exhibit 22.13 (see p. 469) uses the same TTEST program statements as Exhibits 22.11 and 22.12 but with different formats. The DENSITY format was defined incorrectly: rather than beginning the second group at 100, an extra 0 was inadvertently added that caused the group to begin at 1,000. This groups only states with 100 or fewer and greater than 1,000 per square mile. Levels of

---

EXHIBIT 22.10   Table with ungrouped interval-scale variable

```
------------------- Program listing --------------------

proc freq data=college;
tables pctcoll * edex_sal;
format edex_sal edsal8.;
run;

------------------- Output listing --------------------

TABLE OF PCTCOLL BY EDEX_SAL

PCTCOLL      EDEX_SAL

Frequency|
Percent  |
Row Pct  |
Col Pct  |$2.25k- |>$2.25k |   Total
---------+--------+--------+
  10.417 |    1 |      0 |     1
         |  1.96 |  0.00 |  1.96
         | 100.00 |  0.00 |
         |  3.57 |  0.00 |
---------+--------+--------+
  10.820 |    1 |      0 |     1
         |  1.96 |  0.00 |  1.96
         | 100.00 |  0.00 |
         |  3.57 |  0.00 |
---------+--------+--------+
  11.077 |    1 |      0 |     1
         |  1.96 |  0.00 |  1.96
         | 100.00 |  0.00 |
         |  3.57 |  0.00 |
---------+--------+--------+
  12.180 |    1 |      0 |     1
         |  1.96 |  0.00 |  1.96
         | 100.00 |  0.00 |
         |  3.57 |  0.00 |
(Continued)
```

POPSQMI for other states are left as is, thus creating more than two categories. Exhibit 22.13 shows the errant FORMAT and how TTEST communicates the error.

Exercises

22.5. Write TTEST statements to meet the following requirements. The dataset name is HUMANRES.

   a. Compare the means of variables SALARY, SOCSEC, and FICA for groups classified by variable REGION.

   b. For each level of variable YEAR, compare means for groups classified by variable EARN_GRP. Analyze all numeric variables not used elsewhere by the procedure.

22.6. Variable DECADE represents the decade of first exposure to a toxic chemical: 50 = 1950s, 60 = 1960s, and so on. What is wrong with the following TTEST procedures?

  a. proc ttest data=expose;
     class decade;
     var icd174 icd209;

  b. proc ttest data=expose;
     class type;
     var icd174 icd209;
     by type;

22.7. How can you supplement TTEST output with the PLOT or CHART procedures? Use the analyses in Exhibit 22.12 as a starting point. You might want to chart the differences in the two groups' size (N's), plot the variables to highlight clusters of observations, and so on.

EXHIBIT 22.11   Character CLASS variable

```
------------------- Program listing -------------------

proc ttest data=book.states;
var hightemp unempben;
class eastwest;
run;

------------------- Output listing --------------------

TTEST PROCEDURE

Variable: HIGHTEMP
```

EASTWEST	N	Mean	Std Dev	Std Error
e	26	110.38461538	3.58972894	0.70400377
w	24	117.62500000	6.88294995	1.40497628

Variances	T	DF	Prob>|T|
Unequal	-4.6073	34.0	0.0001
Equal	-4.7164	48.0	0.0000

```
For H0: Variances are equal, F' = 3.68    DF = (23,25)    Prob>F' = 0.0020


Variable: UNEMPBEN
```

EASTWEST	N	Mean	Std Dev	Std Error
e	27	203.18518519	54.93985315	10.57317967
w	24	190.91666667	33.46499361	6.83101322

Variances	T	DF	Prob>|T|
Unequal	0.9746	43.6	0.3351
Equal	0.9482	49.0	0.3477

```
For H0: Variances are equal, F' = 2.70    DF = (26,23)    Prob>F' = 0.0189
```

EXHIBIT 22.12   Grouped interval-scale variable in CLASS

```
------------------- Program listing -------------------

proc format;
value density low  -100  = '<= 100/sq mi'
              100 <-high = '> 100/sq mi'  ;
run;

proc ttest data=book.states cochran;
var crimert;
class popsqmi;
label popsqmi = 'Population density'
      crimert = 'Crimes per 100,000 pop'
      ;
format popsqmi density.;
run;

------------------- Output listing -------------------

TTEST PROCEDURE

Variable: CRIMERT      Crimes per 100,000 pop

  POPSQMI      N              Mean           Std Dev          Std Error
--------------------------------------------------------------------------
<= 100/sq     30       5018.30000000     1428.60561750      260.82650749
 > 100/sq     21       5967.85714286     1593.70013760      347.77388169


Variances      T    Method            DF    Prob>|T|
--------------------------------------------------------------
Unequal    -2.1843  Satterthwaite    40.1    0.0348
                    Cochran           .      0.0396
Equal      -2.2276                   49.0    0.0305

For H0: Variances are equal, F' = 1.24   DF = (20,29)   Prob>F' = 0.5789
```

22.3   Correlations: PROC CORR

Another group of common two-variable statistics are correlation coefficients. These measures indicate the direction and strength of association between a pair of numeric variables, optionally controlling for the effects of a third variable. Correlations may be computed for both continuous, interval-scale data and ordinal, ranked data.

The CORR procedure calculates both parametric and nonparametric correlations. Printed output includes descriptive statistics of the analysis variables and the correlation matrix. Each cell in the matrix contains the correlation coefficient, the number of observations used in its calculation, and the probability that the association is zero. One or more output datasets may be created and used later as input to regression procedures (these are discussed in Chapter 23).

EXHIBIT 22.13   Attempt to use a CLASS variable that is not dichotomous

```
    1    proc format;
    2    value density low  -100  = '<= 100/sq mi'
    3                       1000 <-high = '> 100/sq mi'  ;
NOTE: Format DENSITY has been output.
    4    run;
NOTE: The PROCEDURE FORMAT used 3.00 seconds.
    5
    6    proc ttest data=book.states cochran;
    7    var crimert;
    8    class popsqmi;
    9    label popsqmi = 'Population density'
   10          crimert = 'Crimes per 100,000 pop'
   11          ;
   12    format popsqmi density.;
   13    run;
ERROR: The CLASS variable has more than two levels.
NOTE: The PROCEDURE TTEST used 6.00 seconds.
```

POINTERS AND PITFALLS

A particularly useful application of CORR is its ability to create an output dataset that may be used as input to regression and analysis of variance procedures. Analyzing the matrix rather than the original data often saves large amounts of computer resources and time.

Exhibit 22.14 (see p. 470) shows the output produced by a simple use of PROC CORR. Pearson product-moment correlations are calculated.

The analysis first displays the names of the variables. The output contains a square, symmetric matrix of coefficients: row 1 and column 1 of the matrix contain the same variable. This setup continues through row n and column n. Some "simple statistics" are displayed for each variable: the count of nonmissing values, the mean, standard deviation, sum, and minimum and maximum values. Some of these values are used in computing the correlation, while others are displayed simply to assist other parts of the analysis.

Printed below the descriptive statistics are the correlations themselves. Each cell of the correlation matrix has three lines:

1. The correlation value
2. The probability of accepting the null hypothesis of no relationship
3. The number of observations with nonmissing values for the pair of variables.

CORR Options

Informative as it was, Exhibit 22.14 was only the default display. CORR has options that may be used to customize output. Much more information can be printed, all printing can be suppressed, and new datasets can be created.

Adding to the Display.   Normally, matrices that are by-products of the analysis are not printed. The covariance matrix and the corrected and uncorrected sums

EXHIBIT 22.14   Pearson correlation matrix; default output

```
------------------- SAS Log --------------------------

  1    options ps=60 ls=132;
  2
  3    proc corr data=book.states ;
  4    var pctin pctmfg  pctcoll pctfarm pcturb btus pcapinc;
  5    label pctin = '% in-migration'
  6          pctmfg = '% labor force in manufacturing'
  7          pctcoll = '% over 25 w. college ed'
  8          pcapinc = 'per capita income'
  9          pctfarm = '% land in farms'
 10          pcturb = '% pop in urban areas'
 11          btus  = 'BTUs used per capita'
 12          ;
 13    run;
NOTE: The PROCEDURE CORR used 14.00 seconds.

------------------- Output listing ---------------------

CORRELATION ANALYSIS

    7 'VAR' Variables:  PCTIN    PCTMFG   PCTCOLL  PCTFARM  PCTURB   BTUS     PCAPINC
```

Simple Statistics

Variable	N	Mean	Std Dev	Sum	Minimum	Maximum	Label
PCTIN	51	14.229673	6.578015	725.713331	5.936137	34.252274	% in-migration
PCTMFG	51	19.944283	8.307113	1017.158417	4.519854	32.767392	% labor force in manufacturing
PCTCOLL	51	16.312528	3.334439	831.938933	10.417055	27.461978	% over 25 w. college ed
PCTFARM	51	3.493370	3.868253	178.161856	0	16.337468	% land in farms
PCTURB	51	67.597185	14.996714	3447.456432	33.773189	100.000000	% pop in urban areas
BTUS	50	331.820000	128.819394	16591	190.000000	858.000000	BTUs used per capita
PCAPINC	51	7092.803922	973.557908	361733	5183.000000	10193	per capita income

Pearson Correlation Coefficients / Prob > |R| under Ho: Rho=0 / Number of Observations						
	PCTIN	PCTMFG	PCTCOLL	PCTFARM	PCTURB	BTUS	PCAPINC

	PCTIN	PCTMFG	PCTCOLL	PCTFARM	PCTURB	BTUS	PCAPINC
PCTIN	1.00000	-0.66318	0.41549	-0.19791	0.18470	0.32635	0.37423
% in-migration	0.0	0.0001	0.0024	0.1639	0.1945	0.0207	0.0068
	51	51	51	51	51	50	51
PCTMFG	-0.66318	1.00000	-0.38332	-0.22057	-0.16553	-0.43722	-0.26405
% labor force in manufacturing	0.0001	0.0	0.0055	0.1199	0.2457	0.0015	0.0612
	51	51	51	51	51	50	51
PCTCOLL	0.41549	-0.38332	1.00000	-0.32882	0.57895	-0.03918	0.67624
% over 25 w. college ed	0.0024	0.0055	0.0	0.0185	0.0001	0.7871	0.0001
	51	51	51	51	51	50	51
PCTFARM	-0.19791	-0.22057	-0.32882	1.00000	-0.49659	0.08301	-0.34348
% land in farms	0.1639	0.1199	0.0185	0.0	0.0002	0.5666	0.0136
	51	51	51	51	51	50	51
PCTURB	0.18470	-0.16553	0.57895	-0.49659	1.00000	-0.14449	0.65746
% pop in urban areas	0.1945	0.2457	0.0001	0.0002	0.0	0.3168	0.0001
	51	51	51	51	51	50	51
BTUS	0.32635	-0.43722	-0.03918	0.08301	-0.14449	1.00000	0.22323
BTUs used per capita	0.0207	0.0015	0.7871	0.5666	0.3168	0.0	0.1192
	50	50	50	50	50	50	50
PCAPINC	0.37423	-0.26405	0.67624	-0.34348	0.65746	0.22323	1.00000
per capita income	0.0068	0.0612	0.0001	0.0136	0.0001	0.1192	0.0
	51	51	51	51	51	50	51

of squares and cross-product matrices are controlled by the COV, CSSCP, and SSCP options in the PROC statement. The RANK option adds interpretive power to the display by arranging the rows of the matrix in order of descending absolute value of the coefficients.

Reducing the Display.   All or part of the listing may be suppressed by the NOCORR, NOSIMPLE, NOPROB, NOPRINT, and BEST options in the PROC statement. These are particularly useful when CORR is being used only to generate output datasets containing the correlation matrices.

Selecting the Type of Correlation.   By default, CORR computes Pearson product-moment correlations. Hoeffding's D statistic, Kendall's tau-b, and Spearman's rank-order correlation may also be requested via the HOEFFDING, KENDALL, and SPEARMAN options in the PROC statement. Conbach's alpha (ALPHA option in the PROC) and partial correlations (PARTIAL statement) are also available.

Creating Output Datasets.   Output datasets containing the Hoeffding, Spearman, Kendall, and Pearson correlation matrices are created by specifying the OUTH, OUTS, OUTK, and OUTP options in the PROC statement.

Missing Values.   Normally each cell of the correlation matrix uses all available nonmissing pairs of variables. The NOMISS option of the PROC statement eliminates an observation from the analysis if *any* of the analysis variables are missing from an observation. This ensures that the number of observations used for each pair of variables is equal. This is particularly useful if the correlation matrix will be used as input to a linear regression procedure, since equal n's are preferred by many statisticians.

Syntax

The syntax for the CORR procedure is described below. In addition to procedure-specific statements BY, FREQ, ATTRIB, LABEL, FORMAT, TITLE, and FOOT-NOTE may be used. These statements were presented in Chapter 5.

```
PROC CORR [DATA=dataset] [SSCP] [COV] [CSSCP]
          [NOSIMPLE] [NOPROB] [NOCORR] [NOPRINT]
          [RANK] [BEST=bestn]
          [PEARSON] [KENDALL] [HOEFFDING] [SPEARMAN]
          [NOMISS]
          [OUTH=outdset_h] [OUTK=outdset_k]
          [OUTP=outdset_p] [OUTS=outdset_s]
          ;
[VAR var_list;]
[WITH with_list;]
[PARTIAL partial_list;]
```

Some common applications of the CORR procedure follow:

```
proc corr data=in.measures pearson kendall;
var m1-m15;

proc corr data=in.measures;
var m1-m5;
with t1-t5;

proc corr data=summ;
var wgt1--pulse;
partial age;
```

The CORR procedure has the following elements:

1. The *dataset* is the name of the analysis dataset. If the *DATA* option is not entered, the most recently created dataset is used.

2. *SSCP* requests printing of the sums of squares and cross-products. If the *PARTIAL* statement (see #16 below) is also entered, the partialled SSCP matrix is printed. If the *OUTP* option (see #13) in the PROC statement is also entered, the SSCP matrix is included in the output dataset. Using *SSCP* implies use of the *PEARSON* option (the default correlation method, see #11).

3. *COV* requests printing of the covariance matrix. If the *PARTIAL* statement is also entered, the partial covariance matrix is printed. If the *OUTP* option in the PROC statement is also entered, the covariance matrix is included in the output dataset. If both *PARTIAL* and *OUTP* are entered, the output dataset contains the partialled covariances. Using *COV* implies use of the *PEARSON* option.

4. *CSSCP* requests printing of the corrected sums of squares and cross-products. If the *PARTIAL* statement is also entered, the partialled CSSCP matrix is printed. If the *OUTP* option in the PROC statement is also entered, the CSSCP matrix is included in the output dataset. Using *CSSCP* implies use of the *PEARSON* option.

5. *NOSIMPLE* suppresses display of descriptive statistics for the analysis variables.

6. *NOPROB* suppresses display of the correlations' significance probabilities.

7. *NOCORR* suppresses calculation and display of Pearson correlations. Other correlations and matrices will be printed, however.

8. *NOPRINT* suppresses all printed output from CORR. Use this option only when you want CORR to create one or more output datasets.

9. *RANK* prints correlations in order of descending absolute value. This usually means that the contents of a given column in the table vary from row to row. CORR prints the name of each column variable in each cell of the printed matrix.

10. *BEST* prints only a specific number (*bestn*) of the largest absolute value correlations for each variable. The default is display of all correlations regardless of magnitude and significance.

11. *PEARSON, KENDALL, HOEFFDING,* and *SPEARMAN* request the types of correlations to display: Pearson's product-moment correlation, Kendall's tau-*b*, Hoeffding's *D*, and Spearman's rank-order coefficients, respectively. If none are specified, the default is *PEARSON. PEARSON* must be specified if any of the other three options is requested.

12. *NOMISS* allows CORR to exclude an observation from the analysis if *any* of its analysis variables contain missing values. By default, CORR uses any observation with pairs of nonmissing values. This "pairwise" calculation often creates matrices with unequal numbers of observations per cell and may use significantly more computer resources than if *NOMISS* were used.

13. *OUTP, OUTK, OUTH,* and *OUTS* request generation of one or more SAS datasets containing Pearson, Kendall, Hoeffding, and Spearman correlations, respectively. The *outset_p, outset_k, outset_h,* and *outset_s may* be either temporary or permanent SAS datasets. Their typical use is as input to SAS regression procedures.

14. *VAR* specifies the variables for the analysis. If omitted, CORR uses all numeric variables not used elsewhere in the PROC. If the *VAR* statement is specified, CORR displays a square matrix using *var_list* for its rows and columns.

15. *WITH* restricts the contents of the correlation matrices. By default, CORR prints a square matrix using *var_list* or all available numeric variables in the dataset. *WITH* restricts output by requesting correlations between pairs of variables. The *with_list* variables define the rows of the output matrix and *var_list* specifies the columns. A *VAR* statement must be entered if *WITH* is specified. Exhibit 22.15 illustrates the appearance of the matrices produced by different *VAR* and *WITH* statements.

EXHIBIT 22.15   Form of correlation matrix using different VAR-WITH combinations

---

```
var a b c;                      var d e;
with d e;                       with a b c;


   | a | b | c |                    | d | e |
--+---+---+---+                 --+---+---+
d |   |   |   |                 a |   |   |
--+---+---+---+                 --+---+---+
e |   |   |   |                 b |   |   |
--+---+---+---+                 --+---+---+
                                c |   |   |
                                --+---+---+
```

---

16. *PARTIAL* requests partial correlations, controlling for the variables specified in *partial_list*. All types of correlations except *HOEFFDING* may produce partial correlations. The *NOMISS* option is automatically used throughout the analysis if *PARTIAL* is specified. This option is available only in version 6 of the SAS System.

Usage Notes

Keep in mind the following points when using PROC CORR:

- WEIGHT statements may be used only when computing Pearson correlations. FREQ statements may be used with any correlation method.

- If a BY statement is used, the statistics are calculated independently for each level of the BY variable(s). BY variables are included in CORR's output datasets (created by the OUTP, OUTK, OUTH, and OUTS options).

- More than one output dataset may be created in a single run of PROC CORR. Options such as MISSING, SSCP, and COV and the WITH, VAR, and PARTIAL statements will affect the contents of each dataset. You cannot, for example, have a covariance in the OUTP dataset and not the OUTK dataset created in the same run of CORR.

- If an option is invalid or inappropriate, CORR usually prints a message on the Log but not on the output itself. Exhibit 22.18 demonstrates a request for a partial tau-*b*. A NOTE is printed in the Log indicating that probability values cannot be calculated for these values.

- A correlation matrix dataset used as input to regression procedures should be square and symmetric; that is, the CORR statements creating them should not include the WITH statement. If the correlation dataset is not square, most regression procedures will note that the "Correlation matrix is incomplete" and terminate.

- Using the RANK and BEST options does not affect the contents of datasets created by CORR. If 10 variables were specified in the VAR statement and BEST=5 was specified in the PROC statement, the output dataset would contain the full 10 × 10 matrix.

Examples

All examples below use the system options PS=60 and LS=132. They also use LABEL statements. Recall that this variable annotation is automatically used if LABELs are stored in the dataset.

Key features and options used in this section are summarized in the following list:

Exhibit	Features/Options
22.16	SSCP, COV matrices; PEARSON, KENDALL statistics printed; WITH statement
22.17	Univariate statistics suppressed; format of printed correlations (BEST, RANK options) modified
22.18	Partial correlations (PARTIAL statement); descriptive statistics suppressed (NOSIMPLE)
22.19	Printed output suppressed; Pearson output dataset created
22.20	Covariances (COV option) added to Pearson output dataset

Exhibit 22.16 requests printing of additional matrices (SSCP and COV options in the PROC statement) and statistics (PEARSON and KENDALL). Notice that since a nondefault statistic (KENDALL) is being printed, *all* statistics must be specified (PEARSON, the default, is explicitly noted). Only certain pairs of correlations (WITH statement) rather than the default square matrix are requested.

In Exhibit 22.17 (see p. 478) the display of univariate statistics (NOSIMPLE option in the PROC statement) is suppressed. Output is restricted to the four most highly correlated variables with each analysis variable (BEST option). These values are arranged in descending order (RANK option). Notice how the appearance of the matrix is changed by the ranking and selection options: each cell of the table must be individually labelled since there is a variable-to-variable difference in strength of association.

Partial correlations (PARTIAL statement) are computed in Exhibit 22.18 (see p. 480). The descriptive statistics are suppressed (NOSIMPLE option) and two types of correlations are requested (PEARSON and KENDALL options). Notice that probability values for the Kendall correlations are not in the output. This is not an error or omission, simply a calculation that SAS cannot carry out.

The omission of the probabilities on the output file emphasizes the importance of always looking at the Log, even when the output "looks right." The Log often notes computational and other problems that may affect your interpretation of the output. (In this case, you would not be able to determine p values for Kendall's partial tau-*b*.)

In Exhibit 22.19 (see p. 481) an output dataset of Pearson correlations is created (OUTP option), possibly for use in a regression procedure later in the SAS session or job. Since all we want from CORR is the output matrix, the listing of the descriptive statistics and the correlation matrix is not printed (NOPRINT option). The exhibit's output shows a PRINT and CONTENTS of the output dataset PEARSAVE.

Exhibit 22.20 (see p. 482) is nearly the same as Exhibit 22.19. This time, however, covariances are included (COV option) in the output dataset. Aside from a different number of observations, the CONTENTS listing (not shown) is identical to that in Exhibit 22.19. Notice how the structure of the dataset changes: covariances are printed prior to the standard OUTP options. Be aware of the difference if you are actually manipulating these output datasets and not simply passing them to regression procedures.

EXHIBIT 22.16   SSCP, COV matrices; Pearson and Kendall statistics printed; WITH statement

```
------------------- SAS Log ---------------------------

    4     proc corr data=states nosimple sscp cov
    5           pearson kendall ;
    6     var pctin pctmfg pctcoll;
    7     with pctfarm pcturb btus pcapinc;
    8     label pctin = '% in-migration'
    9           pctmfg = '% labor force in manufacturing'
   10           pctcoll = '% over 25 w. college ed'
   11           pcapinc = 'per capita income'
   12           pctfarm = '% land in farms'
   13           pcturb = '% pop in urban areas'
   14           btus  = 'BTUs used per capita'
   15           ;
   16     run;


------------------- Output listing --------------------
```

CORRELATION ANALYSIS

```
    4 'WITH' Variables:   PCTFARM   PCTURB    BTUS      PCAPINC
    3 'VAR'  Variables:   PCTIN     PCTMFG    PCTCOLL
```

Sum-of-Squares and Crossproducts

'W'\'V'	PCTIN	PCTMFG	PCTCOLL
PCTFARM	2283	3199	2694
% land in farms	1371	1371	1371
	12490	23737	14127
PCTURB	49967	67726	57684
% pop in urban areas	244283	244283	244283
	12490	23737	14127
BTUS	248092	313684	266210
BTUs used per capita	6318353	6318353	6318353
	12127	23716	13373
PCAPINC	5267173	7107731	6010542
per capita income	2613091991	2613091991	2613091991
	12490	23737	14127

EXHIBIT 22.16   *(continued)*

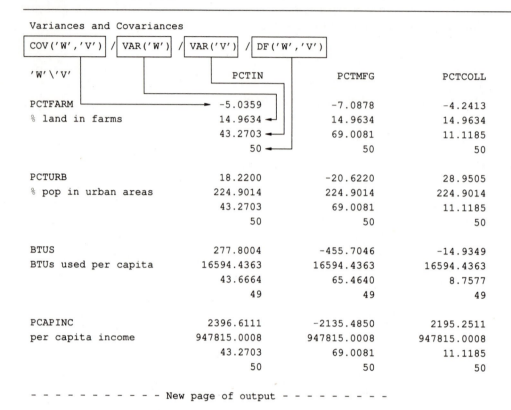

```
Variances and Covariances

┌──────────────┐ ┌───────────┐ ┌───────────┐ ┌────────────────┐
│ COV('W','V') │/│ VAR('W')  │/│ VAR('V')  │/│ DF('W','V')    │
└──────────────┘ └───────────┘ └───────────┘ └────────────────┘

 'W'\'V'                    PCTIN            PCTMFG           PCTCOLL

PCTFARM                  -5.0359          -7.0878           -4.2413
% land in farms          14.9634          14.9634           14.9634
                         43.2703          69.0081           11.1185
                              50               50                50


PCTURB                   18.2200         -20.6220           28.9505
% pop in urban areas    224.9014         224.9014          224.9014
                         43.2703          69.0081           11.1185
                              50               50                50


BTUS                    277.8004        -455.7046          -14.9349
BTUs used per capita  16594.4363       16594.4363        16594.4363
                         43.6664          65.4640            8.7577
                              49               49                49


PCAPINC                2396.6111       -2135.4850         2195.2511
per capita income    947815.0008      947815.0008       947815.0008
                         43.2703          69.0081           11.1185
                              50               50                50
```

- - - - - - - - - - - - - New page of output - - - - - - - - -

CORRELATION ANALYSIS

Pearson Correlation Coefficients / Prob > |R| under Ho: Rho=0 / Number of Observations

```
                         PCTIN            PCTMFG           PCTCOLL

PCTFARM                -0.19791         -0.22057          -0.32882
% land in farms         0.1639           0.1199            0.0185
                            51               51                51


PCTURB                  0.18470         -0.16553           0.57895
% pop in urban areas    0.1945           0.2457            0.0001
                            51               51                51


BTUS                    0.32635         -0.43722          -0.03918
BTUs used per capita    0.0207           0.0015            0.7871
                            50               50                50


PCAPINC                 0.37423         -0.26405           0.67624
per capita income       0.0068           0.0612            0.0001
                            51               51                51
```

Exercises

22.8. Refer to Exhibit 22.14.

   a. Ignoring the main diagonal (whose correlations are always 1.0), identify the
   three highest correlations. Then identify the three lowest. Remember to look

EXHIBIT 22.16 *(continued)*

```
Kendall Tau-b Correlation Coefficients / Prob > |R| under Ho:
Rho=0 / Number of Observations
```

	PCTIN	PCTMFG	PCTCOLL
PCTFARM	-0.13098	-0.03686	-0.32235
% land in farms	0.1750	0.7027	0.0008
	51	51	51
PCTURB	0.07922	-0.09961	0.38667
% pop in urban areas	0.4120	0.3023	0.0001
	51	51	51
BTUS	0.07029	-0.24683	-0.22395
BTUs used per capita	0.4719	0.0115	0.0219
	50	50	50
PCAPINC	0.11216	-0.14510	0.49804
per capita income	0.2454	0.1329	0.0001
	51	51	51

for the high and low values ignoring the sign of the correlation (i.e., look for absolute values).

b. Choose one of the high correlations from part a. In one or two sentences, describe what the correlation represents. Refer to the description of the STATES dataset in Appendix E for variable labels.

22.9. Write CORR procedure statements to meet the following requirements. All exercises use dataset IN.PERFORM.

a. Print PEARSON correlations for all numeric variables in the dataset. Suppress the printing of the descriptive statistics.

b. Print a correlation matrix of the largest five correlations for each variable. Display them in descending order of absolute value.

c. Calculate Kendall and Spearman correlations for variables RATE80Q1--RATE85Q4. Ensure that the statistics are based on observations with complete data for every analysis variable. Print only the Kendall and Spearman correlations.

d. Calculate Pearson correlations and store the results in SAS dataset PCORR for use later in a regression procedure. Use variables ASSESS01 through ASSESS25 and OVERALL. Do not print any printed output.

e. Print Pearson correlations creating a matrix of the form ASSESS01--ASSESS25 as column variables and RATE80Q1--RATE85Q4 as row variables. Also print the sums of squares and covariance matrices.

f. Print the Spearman rank-order coefficients for DEPRNK1–DEPRNK5 with PERANK1–PERANK10. Repeat this for each level of variable DISTRICT (the dataset is already sorted by DISTRICT).

22.10. What, if anything, is wrong with the following uses of PROC CORR?

a.
```
proc corr outp=out.lab1pear data=in.lab1
        noprint nosimple nomissing;
var bmass1-bmass10;
with o2ing1-o2ing10;
```

LAB1PEAR is later used as input to a regression procedure.

(continued on p. 479)

EXHIBIT 22.17 Suppress univariate statistics; modify format of printed correlations (BEST, RANK options)

```
20      proc corr data=states nosimple rank best=4 ;
21      var pctin pctmfg pctcoll pctfarm pcturb btus pcapinc;
22      label pctin = '% in-migration'
23            pctmfg = '% labor force in manufacturing'
24            pctcoll = '% over 25 w. college ed'
25            pcapinc = 'per capita income'
26            pctfarm = '% land in farms'
27            pcturb = '% pop in urban areas'
28            btus  = 'BTUs used per capita'
29            ;
31      run;
```

> Variables with the strongest association with PCTIN (number of variables set by BEST option, their arrangement in descending order set by RANK option)

CORRELATION ANALYSIS

   7 'VAR' Variables: PCTIN   PCTMFG   PCTCOLL   PCTFARM   PCTURB   BTUS   PCAPINC

Pearson Correlation Coefficients / Prob > |R| under Ho: Rho=0 / Number of Observations

PCTIN
% in-migration

PCTIN	PCTMFG	PCTCOLL	PCAPINC
1.00000	-0.66318	0.41549	0.37423
0.0	0.0001	0.0024	0.0068
51	51	51	51

PCTMFG
% labor force in manufacturing

PCTMFG	PCTIN	BTUS	PCTCOLL
1.00000	-0.66318	-0.43722	-0.38332
0.0	0.0001	0.0015	0.0055
51	51	50	51

PCTCOLL
% over 25 w. college ed

PCTCOLL	PCAPINC	PCTURB	PCTIN
1.00000	0.67624	0.57895	0.41549
0.0	0.0001	0.0001	0.0024
51	51	51	51

PCTFARM
% land in farms

PCTFARM	PCTURB	PCAPINC	PCTCOLL
1.00000	-0.49659	-0.34348	-0.32882
0.0	0.0002	0.0136	0.0185
51	51	51	51

PCTURB
% pop in urban areas

PCTURB	PCAPINC	PCTCOLL	PCTFARM
1.00000	0.65746	0.57895	-0.49659
0.0	0.0001	0.0001	0.0002
51	51	51	51

EXHIBIT 22.17 *(continued)*

BTUS
BTUs used per capita

	BTUS	PCTMFG	PCTIN	PCAPINC
	1.00000	-0.43722	0.32635	0.22323
	0.0	0.0015	0.0207	0.1192
	50	50	50	50

PCAPINC
per capita income

	PCAPINC	PCTCOLL	PCTURB	PCTIN
	1.00000	0.67624	0.65746	0.37423
	0.0	0.0001	0.0001	0.0068
	51	51	51	51

Exercises

(continued)

b. `proc corr data=in.lab1;`
   `var bmass1-bmass10;`
   `with bmass1-bmass10;`

c. `proc corr data=in.lab1 out=out.lab1p;`
   `var bmass1-bmass10 o2ing1-o2ing10;`

   LAB1P will be used as input to a regression procedure. Among the linear models to be tested is one that proposes that variable DECAY is predicted by variables BMASS1 and O2ING1.

22.11. Refer to the output reproduced in Exhibit 22.17.

  a. Which variable appears most frequently in the table (do not count correlations of variables with themselves)?

  b. What option could have been used to ensure that all correlations are based on the same number of observations (i.e., those with complete data for all analysis variables)?

22.12. Refer to Exhibit 22.16. The "W"\"V" labels in the output refer to the "with" and "variables" lists, respectively. For the variable pair PCTURB-PCTCOLL, identify the following items on the output:

  a. Sum of squared cross-products

  b. Variance of each variable

  c. Number of nonmissing pairs of observations

EXHIBIT 22.18   Partial correlations (PARTIAL statement); descriptive statistics suppressed (NOSIMPLE)

```
------------------- SAS Log ---------------------------

  34     proc corr data=states nosimple pearson kendall ;
  35     var pctin pctmfg pctcoll pcapinc;
  36     partial pcturb;
  37     label pctin = '% in-migration'
  38           pctmfg = '% labor force in manufacturing'
  39           pctcoll = '% over 25 w. college ed'
  40           pcapinc = 'per capita income'
  41           pcturb = '% pop in urban areas'
  42           ;
  43     run;
NOTE: The probability values for Kendall's partial tau-b are not available.
NOTE: The PROCEDURE CORR used 16.00 seconds.

------------------- Output listing --------------------
```

CORRELATION ANALYSIS

```
  1 'PARTIAL' Variables:  PCTURB
  4 'VAR'     Variables:  PCTIN    PCTMFG    PCTCOLL   PCAPINC
```

Pearson Partial Correlation Coefficients / Prob > |R| under Ho: Partial Rho=0 / N = 51

	PCTIN	PCTMFG	PCTCOLL	PCAPINC
PCTIN	1.00000	-0.65268	0.38506	0.34138
% in-migration	0.0	0.0001	0.0058	0.0153
PCTMFG	-0.65268	1.00000	-0.35752	-0.20888
% labor force in manufacturing	0.0001	0.0	0.0108	0.1455
PCTCOLL	0.38506	-0.35752	1.00000	0.48115
% over 25 w. college ed	0.0058	0.0108	0.0	0.0004
PCAPINC	0.34138	-0.20888	0.48115	1.00000
per capita income	0.0153	0.1455	0.0004	0.0

Kendall Partial Tau-b Correlation Coefficients / N = 51

	PCTIN	PCTMFG	PCTCOLL	PCAPINC
PCTIN	1.00000	-0.45145	0.34121	0.08329
% in-migration				
PCTMFG	-0.45145	1.00000	-0.21872	-0.10972
% labor force in manufacturing				
PCTCOLL	0.34121	-0.21872	1.00000	0.37665
% over 25 w. college ed				
PCAPINC	0.08329	-0.10972	0.37665	1.00000
per capita income				

EXHIBIT 22.19   Printed output suppressed; Pearson output dataset created

---

```
------------------- SAS Log --------------------------

47    proc corr data=states noprint outp=pearsave ;
48    var pctin pctmfg pctcoll pcapinc;
49    label pctin = '% in-migration'
50          pctmfg = '% labor force in manufacturing'
51          pctcoll = '% over 25 w. college ed'
52          pcapinc = 'per capita income'
53          pcturb = '% pop in urban areas'
54          ;
55    run;
NOTE: The data set WORK.PEARSAVE has 7 observations and 6 variables.
NOTE: The PROCEDURE CORR used 7.00 seconds.
```

```
------------------- PEARSAVE listing --------------------
```

OBS	_TYPE_	_NAME_	PCTIN	PCTMFG	PCTCOLL	PCAPINC
1	MEAN		14.230	19.944	16.313	7,093
2	STD		6.578	8.307	3.334	974
3	N		51.000	51.000	51.000	51
4	CORR	PCTIN	1.000	-0.663	0.415	0
5	CORR	PCTMFG	-0.663	1.000	-0.383	-0
6	CORR	PCTCOLL	0.415	-0.383	1.000	1
7	CORR	PCAPINC	0.374	-0.264	0.676	1

Variables printed using formats stored in SAS dataset STATES

Descriptive statistics

Correlation matrix

```
------------------- PEARSAVE contents --------------------
```

CONTENTS PROCEDURE

Data Set Name:	WORK.PEARSAVE	Type:	CORR
Observations:	7	Record Len:	52
Variables:	6		
Label:	Pearson Correlation Matrix		

```
-----Alphabetic List of Variables and Attributes-----
```

#	Variable	Type	Len	Pos	Format	Label
6	PCAPINC	Num	8	44	COMMA7.	per capita income
5	PCTCOLL	Num	8	36	7.3	% over 25 w. college ed
3	PCTIN	Num	8	20	7.3	% in-migration
4	PCTMFG	Num	8	28	7.3	% labor force in manufacturing
2	_NAME_	Char	8	12		
1	_TYPE_	Char	8	4		

EXHIBIT 22.20   Covariances (COV option) added to Pearson output dataset

```
-------------------- SAS Log --------------------------

65     proc corr data=states noprint outp=pearsave cov;
66     var pctin pctmfg pctcoll pcapinc;
67     label pctin = '% in-migration'
68           pctmfg = '% labor force in manufacturing'
69           pctcoll = '% over 25 w. college ed'
70           pcapinc = 'per capita income'
71           pcturb = '% pop in urban areas'
72           ;
73     run;
NOTE: The data set WORK.PEARSAVE has 11 observations and 6 variables.
NOTE: The PROCEDURE CORR used 7.00 seconds.

-------------------- PEARSAVE listing ----------------------
```

OBS	_TYPE_	_NAME_	PCTIN	PCTMFG	PCTCOLL	PCAPINC
1	COV	PCTIN	43.270	-36.239	9.113	2,397
2	COV	PCTMFG	-36.239	69.008	-10.618	-2,135
3	COV	PCTCOLL	9.113	-10.618	11.118	2,195
4	COV	PCAPINC	2396.61	-2135.5	2195.25	947,815
5	MEAN		14.230	19.944	16.313	7,093
6	STD		6.578	8.307	3.334	974
7	N		51.000	51.000	51.000	51
8	CORR	PCTIN	1.000	-0.663	0.415	0
9	CORR	PCTMFG	-0.663	1.000	-0.383	-0
10	CORR	PCTCOLL	0.415	-0.383	1.000	1
11	CORR	PCAPINC	0.374	-0.264	0.676	1

Summary

Chapter 22 presents some of the SAS System's procedures for analyses comparing two numeric variables, possibly controlling for one or more other numeric or character variables. The FREQ procedure produces a wide variety of measures of association. Its tabular output detail may be altered as needed. FREQ may produce an output dataset containing basic items such as cell counts and percentages. Examples of most common options, illustrations of the format of output when controlling for a third variable, and lists of the contents and directory of an output dataset are given.

The TTEST and CORR procedures are also described. CORR's ability to produce an output dataset containing descriptive statistics and correlations is particularly useful to analysts who will be using the more sophisticated multivariable methods described later in the book.

23 Linear Regression

Linear regression is one of the most popular analytical tools in the social sciences. This technique allows prediction of a variable's value based on the values of one or more other variables, tests for the statistical significance of the contribution one or more variables makes to the prediction, and has many other useful properties.

This chapter reviews some of the statistical theory underlying linear regression, describes how to identify and react to the most common complications in regression analysis, and implements these discussions via use of the REG procedure. Many people spend a good part of their professional lives exploring the linear models upon which REG is based. This chapter should in no way be regarded as anything but a cursory introduction to the topic.

23.1 Regression Basics

This section briefly outlines the basic issues addressed by linear regression. It discusses the concept of least squares estimation of parameters and other basic regression terminology.

Questions and Terminology

Linear regression answers a deceptively simple research question: How well do the values of one or more continuous variables predict the most likely value of another continuous variable? As Section 23.2 demonstrates, regression output answers the question with both a linear equation showing the best estimate of the relationship plus various diagnostics showing how well the equation fits the observed data.

The predicted variable is commonly referred to as the **dependent variable**. Its values are thought to depend on the values of other variables, the so-called regressors or **independent variables**. Both types of variables are usually continuous: items such as height, weight, salary, percentages, and income are valid, while region numbers and other nominal-scale data are not. These restrictions may be relaxed only with cautions and justifications beyond the scope of this overview.

Statement of the Problem

The Bivariate Case.   The simplest form of regression is the bivariate case, where a single independent variable is used to predict the value of the dependent variable. Exhibit 23.1 illustrates a hypothetical x-y plot for the data points of two variables. It uses a traditional axis arrangement: the y axis is used for the dependent variable, the x axis for the independent.

The scatterplot in Exhibit 23.1 illustrates several fundamental points. First, the variables are positively related; that is, as one variable's value increases, so does the other's. Second, the form of the relationship may be expressed algebraically in the form of a straight line equation such as

$$y = a + bx$$

where a is the value of y at which the line crosses the y axis. This is also called the **intercept**. The b is the slope of the line, the amount of change in y (CRIMERT) for a unit change in x (PCTURB).

This simple notation is complicated by the "real world": it would only be a valid expression of the relationship if all the data points fell exactly on the line defined by the equation. It is obvious from the scatterplot that although the points are clustered around a hypothetical line, they do not all lie on it. Once we recognize that there will usually be some prediction error at any given point, we arrive at the basic "least squares" regression equation for the bivariate case:

$$y = a + bx + e$$

where a and b have the same meaning as before and e is the error term, a correction for the fact that the *predicted* value of y for a particular observation will not always match its *actual* value. Thus e_i for a given observation is calculated as the predicted value (sometimes called "$\hat{y}$" or "y hat") minus the actual value. The e_i's are also referred to as the **residuals**.

The least squares regression estimate for a and b is the line that minimizes the sum of e_i^2 for all cases. An infinite number of lines could be drawn estimating the relationship, but only one will minimize the sum of e_i^2. In geometric terms, the least squares regression estimate is also the line that minimizes the sum of the vertical distances drawn from the points to the regression line.

The regression line is overlaid on the scatterplot in Exhibit 23.2 (see p. 486).

Extension to Several Independent Variables.   The simple bivariate case may be extended to include more than one independent variable:

$$y_i = a + b_1 x_{i1} + b_2 x_{i2} + \cdots + b_j x_{ij} + e_i$$

The predicted value of a given y (y_i) is determined by the intercept (a) plus the b_1 slope times the x_1 value of the observation, plus the b_2 slope times the x_2 value of the observation, and so on through the jth b times the jth x.

While the bivariate case estimates a least squares line, two independent variables fit a least squares plane. Multiple independent variables in general fit a least squares surface, minimizing the squared error terms in an n-dimensional space.

EXHIBIT 23.1   Scatterplot of interval-scale variables

```
------------------- Program listing -------------------

libname x '\data\sas';

proc plot data=x.states;
plot crimert*pcturb;
run;

------------------- Output listing -------------------

          Plot of CRIMERT*PCTURB.  Legend: A = 1 obs, B = 2 obs, etc.


     CRIMERT |
             |
    15,000 +
             |
             |
             |
             |
    12,000 +
             |
             |                                                              A
             |
             |
     9,000 +
             |                                                A
             |                                              B     A
             |                                 A          A
             |                        A   AAA   AA  A A
     6,000 +                                     A    A    AA
             |                  A    B  AA  A         B
             |            A    A   AA  BABA         A
             |      A     A    A    A A
             |            A   AA            A
     3,000 +            AA  A
             |      A
             |
             |
             |
         0 +
             |
             ---+-------------+-------------+-------------+-------------+--
          20.000       40.000       60.000       80.000      100.000

                                    PCTURB
```

23.2   Simple Diagnostics

Sometimes the least squares regression parameters (the intercept and b's) are all that are required for an analysis. Typically, however, empirical measures of the goodness of fit of the regression equation are required. This section discusses

EXHIBIT 23.2 Scatterplot with least squares regression line

```
------------------- Program listing -------------------

libname x '\data\sas';

proc plot data=x.states;
plot crimert*pcturb;
run;

------------------- Output listing -------------------

        Plot of CRIMERT*PCTURB.   Legend: A = 1 obs, B = 2 obs, etc.
```

Plot of CRIMERT*PCTURB with regression line $y = -291 + 84x$

these and other simple diagnostic tests that can be easily carried out using the regression procedure described later in Section 23.4.

Pearson Correlation

In the bivariate case (one dependent, one independent variable), the Pearson r (discussed in Chapter 22) measures the goodness of fit of the regression line to the data points. The sign of the r corresponds to the slope of the regression line: if the line's slope (the b coefficient) is positive, r is positive; if the slope is negative, indicating an inverse relationship, r is negative as well.

The r may range from -1 (indicating a perfect negative relationship) to 0 (no relationship) to $+1$ (perfect positive relationship). The closer its value to the extreme (-1 and $+1$) values, the better the correspondence between the observed and predicted y's. Another measure, r^2, is a more commonly used indicator of strength of relationship. It is the proportion of the variance in y accounted for or "explained" by the regression equation. The error or "unexplained" variance in y is $1 - r^2$.

R (Coefficient of Determination)

Regressions with two or more independent variables use R rather than r as a measure of goodness of fit. R, also known as the **coefficient of determination**, has no sign, since some slopes may be positive and others negative. Its value ranges from 0 to 1. A value of 0 indicates no relationship between any of the independent variables and the dependent variable. An R of 1, on the other hand, suggests a perfect prediction of the dependent variable by the independent variables. A related measure, R^2, is the more commonly used measure of goodness of fit and is interpreted in a manner similar to r^2.

Significance Tests

Usually the first hypothesis examined in a regression equation is whether the multiple R is significantly different from zero. This may also be stated as the null hypothesis:

$$H_0 : b_1 = b_2 = \cdots b_j = 0$$

The **null hypothesis** states that all slopes are zero and have no predictive value. If the significance levels are such that the null hypothesis cannot be rejected, the lack of relationship is accepted.

Assuming the null hypothesis is rejected, the next step in the research process often involves testing each of the b's difference from zero. Rejecting the null hypothesis that a particular b is different from zero means that its corresponding independent variable contributes to the prediction of the dependent variable after taking into account the contribution of the other independent variables. This is sometimes referred to as *controlling for* the effects of the other variables in the equation.

The impact of a set of variables on a model may be assessed using a test of the differences between two R^2's, one for the "full" or unrestricted equation, containing all independent variables, and a second for the "restricted" model's variables, containing a subset of the full model's independent variables. If the proportion of explained variance (R^2) for the full model is not significantly different from that of the restricted model, then the additional variables make no real difference in predicting the values of the dependent variable. The F statistic for the comparison is computed as follows:

$$F_{\text{diff}, N - Kf - 1} = [(R^2_{\text{Full}} - R^2_R)/(K_{\text{Full}} - K_R)]/[(1 - R^2)_{\text{Full}}/N - K_{\text{Full}} - 1]$$

where Full represents the full model, R the reduced model, N the number of observations used in the model, and K the number of regressors in the model.

23.3 Complications

The least squares solution for regression estimates is based on several assumptions about the data and errors (e_i's). Violation of these assumptions can lead to misleading results that are often not detectable from the basic SAS regression output. This section discusses several common problem areas in regression analysis: outliers, specification errors, heteroscedasticity, autocorrelation, and multicollinearity. All these problems can be identified using the PLOT procedure and the options for the regression procedure described in Section 23.4.

Outliers

An **outlier** is an observation in the dataset that is characterized by an unusually large difference between predicted and actual values. This discrepancy may be due to a data input error or the inclusion of an observation from a portion of the population not suitable for the model. It may also simply be a legitimate anomalous value for which there is no satisfactory explanation. Exhibit 23.3 illustrates a hypothetical dataset. It has several obvious outliers as well as several points in the gray area between outlier and clustered. The temptation to discard these cases is strong, sometimes warranted, and ultimately a substantive, not a statistical, decision.

EXHIBIT 23.3 Distribution with outliers

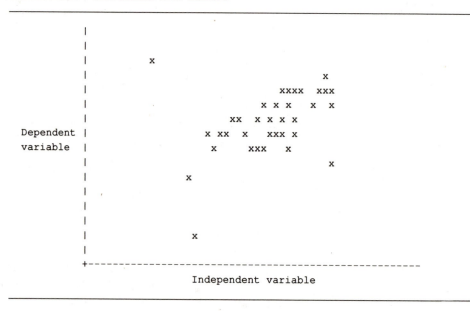

What to Do About Outliers. If the number of observations in the dataset is relatively small, use the R option in PROC REG's MODEL statement to produce a graphic display of the residuals. REG's ID statement assists in the identification of individual observations. Studentized residuals greater than an absolute value of 2.5 and those with high values of Cook's D warrant closer investigation.

Analyses with a large number of observations require bulky listings of individual residuals. In these cases the more appropriate strategy is to use the PLOT procedure to display a scatterplot of residual (y axis) and predicted (x axis) values. The dataset containing these values is specified in REG's OUTPUT statement. Since the mean of the residuals is 0, a reference line can be drawn parallel to the x axis intersecting the y axis at 0 using the VREF=0 option in the PLOT statement. In the ideal case, the residuals should form a "cloud" clustered more or less evenly around the reference line.

Specification Errors

The general model for linear regression assumes that the independent variables are linearly related to the dependent variable and that all relevant independent variables are included in the model. A **specification error** is made when these conditions are not met.

To detect this type of error, plot the residuals against the predicted value of the dependent variable. If the points form a diagonal (rather than cloudlike, random) band, a predictive variable may have been omitted from the model. A hypothetical example is presented in Exhibit 23.4. If the points display a curvilinear pattern, there is a possible nonlinear relationship of one or more of the independent variables to the dependent. A hypothetical example is presented in Exhibit 23.5.

EXHIBIT 23.4   Possible specification error: variable missing from the model

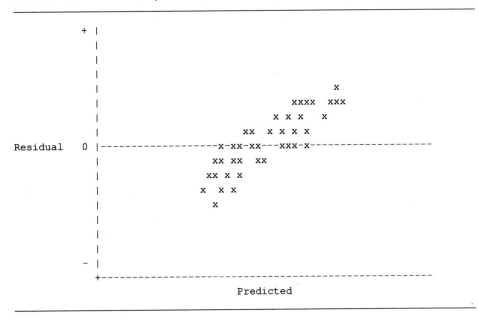

What to Do About Specification Errors.   If you suspect the specification error is one of omission, consider which variables should be added to the model. This process should include variables in the dataset not used as well as items that may need to be added to the data.

Errors of nonlinearity are more problematic, since they require the identification of the curvilinear independent variable. Use REG's OUTPUT statement to create a dataset containing residuals and actual values, then PLOT each independent variable against the residuals, looking for curvilinear patterns. Based on the shape of the relationship, additional independent variables may need to be created in a DATA step, then added to the model. Common independent variable transformations that correct for nonlinearity include squaring, cubing, and taking logarithms.

Heteroscedasticity

Another assumption of the general model for linear regression assumes that the variances of the dependent variable around each point on the regression line or surface are equal. A condition of **heteroscedasticity**, or nonconstant variance,

EXHIBIT 23.5   Specification error: possible nonlinear relationship between *x* and *y*

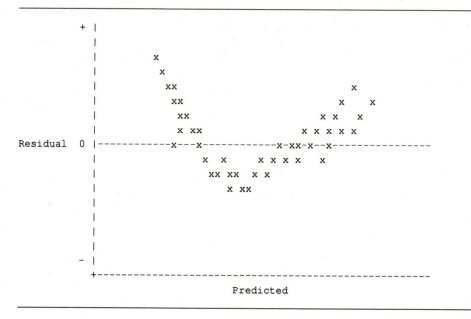

exists when there is an appreciable trend in the plot of residuals versus predicted values. This can mean that the standard errors of the *b*'s and hence their tests of significance will be incorrect. "Appreciable" is usually safely judged by examining a residual-predicted scatterplot, one like that shown in Exhibit 23.6. The exhibit reveals a pronounced funneling of values from right to left. This pattern, its reverse, or a bulge somewhere along the *x* axis suggests the presence of heteroscedasticity.

EXHIBIT 23.6   Scatterplot revealing heteroscedasticity

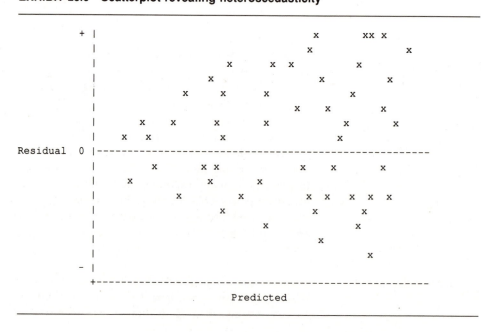

What to Do About Heteroscedasticity. Two solutions are typically employed. The first is to transform the dependent variable. This activity, frequently a logarithmic transformation, can be easily performed in a DATA step prior to the regression procedure. A variation of this technique is to redefine the dependent variable. For example, rather than use a highly variable measure such as sales volume by region, a more appropriate field may be per capita sales.

The second technique, weighted least squares, provides a more accurate estimate of the regression coefficients (the b's) but entails more programming effort. It requires creating a weight for each observation. Details about the circumstances under which this is possible are found in many econometrics texts. The weight is the reciprocal of the expected variance of the residual for each observation. The SAS regression procedures allow a WEIGHT statement to be used to calculate weights.

Autocorrelation

The general model for linear regression also assumes that the residuals (the e_i's) are not correlated. **Autocorrelation** occurs when the observations for a unit of analysis are serially or temporally ordered, and an observation's value of a variable at time t influences the value at time $t+1$. Monthly economic indicators, for example, often reveal cyclical patterns.

Autocorrelation may be detected either with statistical or visual tests. If it is reasonable to assume that time t values are only influenced by the previous period ($t-1$) and no further back, the Durbin–Watson statistic may be requested in the definition of the regression model. Longer cycles (i.e., two or more periods influencing values at time t) require more sophisticated methods beyond the scope of this book.

A visual test for the presence of autocorrelation is an ordered listing of residuals. Observations are listed in temporal order along the y axis, and residuals are plotted along the x axis. Exhibit 23.7 illustrates a clear case of autocorrelation.

What to Do About Autocorrelation. Part of the advice given to remedy specification errors applies here as well: the autocorrelation of residuals may be reduced by introducing a new, theoretically justifiable, independent variable. Other techniques such as differences and the Cochran–Orcutt method are beyond the scope of this book. SAS's AUTOREG procedure, not described in this book, may be used to estimate data that are influenced by two or more time periods.

Multicollinearity

Regression may be used to predict values of a dependent variable. Another area of inquiry focuses on the slopes of the independent variable(s) as structural parameters of a causal process. This is often the case in sociology and economics, and is known as the problem of **multicollinearity**, sometimes simply referred to as **collinearity**.

The problem is caused by highly correlated independent variables. Ideally, none of the independent variables will be correlated. In practice this is seldom achieved, however, and the regression model and its tests are sufficiently robust to accommodate small correlations. Once the independent variables associate too strongly, the standard errors of their b coefficients become very large, producing inaccurate estimates of the coefficients. One manifestation of the problem occurs when a significant R value is accompanied by b's that are not significant and have large standard errors.

What to Do About Collinearity. Identification of the correlated variables is often problematic. Highly related pairs of variables are easily spotted in a standard

EXHIBIT 23.7   Residuals revealing autocorrelation

```
Obser-   t     |                    x    |
vation   t+1   |                     x   |
time     t+2   |                     x   |
         t+3   |                        |x
         t+4   |                        | x
         t+5   |                        |     x
         t+6   |                        |     x
         t+7   |                        |   x
         t+8   |                        |     x
         t+9   |                        |        x
         t+10  |                        |     x
         t+11  |                        | x
         t+12  |                 x      |
         t+13  |            x           |
         t+14  |         x              |
         t+15  |             x          |
         t+16  |             x          |
         t+17  |                  x     |
         t+18  |                  x     |
         t+19  |                        |x
         t+20  |                        |x
               +--+----+----+----+----+----+----+----+----+
                 -4   -3   -2   -1    0    1    2    3    4

                              Residual
```

Pearson correlation matrix. If the problem lies with more than two intercorrelated variables, detection becomes difficult.

SAS provides two options (VIF and COLLINOINT) that help target the source of the collinearity. These options produce a variance inflation factor and condition number for each independent variable. The inflation factor is a function of each independent variable regressed on all others. A relatively large value for a particular independent variable suggests that the associated variables are intercorrelated.

The condition number produced by the COLLINOINT option also aids identification of the collinearity sources. A large value may have several high variance proportions in two or more variables. These variables are the source of the collinearity.

The collinearity may be reduced by dropping one or more of the variables identified as correlated. There should be a rationale for dropping them, however, other than improving the regression diagnostics. If the variables cannot be dropped, more sophisticated techniques that are beyond the scope of this book must be employed.

23.4   The REG Procedure

SAS offers many procedures that perform linear regression. This section describes REG, a procedure that is powerful, interactive, and easy to use. Many users will find its defaults adequate for their analyses. Others will take advantage of

REG's ability to quickly modify models, plot residuals, create output datasets, and perform tests to attack the complications noted in Section 23.3. Regardless of the level of complexity, be aware that REG, like any SAS procedure, will analyze any type of data. It is up to the analyst to ensure that the correct type of data is being used.

Syntax

The syntax for the REG procedure is outlined below. The number and variety of features should not be intimidating: many analyses use only the PROC and MODEL statements with no options selected.

```
PROC REG [DATA=dset] [OUTEST=param_est]
         [OUTSSCP=sscp] [SIMPLE] [CORR] [NOPRINT] ;
[MODEL dep_var = indep_var [...]
       [ / [NOPRINT] [STB] [VIF] [COLLIN] [COLLINOINT]
           [TOL] [P] [R] [CLM] [CLI] [DW] [PARTIAL] ];]
[TEST restriction [,...] ;]
[VAR varlist;]
[ID idvar;]
[OUTPUT [OUT=out_dset] [P=pred] [R=resid]
        [L95M=low_mean] [U95M=high_mean]
        [L95=low_pred] [U95=high_pred]
        [STDP=std_pred] [STDR=std_resid]
        [STDI=std_indiv]
        [COOKD=influ]
        [STUDENT=stu_resid] [RSTUDENT=stu_resid_del]
        ;]
[FREQ freqvar;]
[WEIGHT wgtvar;]
[REWEIGHT wgtvar;]
[RUN;]
[QUIT;]
```

POINTERS AND PITFALLS

Begin your analysis with a simple model. Add terms to the model, options, and other tests gradually so you can easily identify the source of incremental changes in significance levels, standard errors, and other statistics.

Some common applications of the REG procedure follow:

```
proc reg data=datain.trade;
model cpi = t_vol / p r ;

proc reg data=datain.trade corr;
model gnp = cpi_tr t_vol ;
test t_vol;

proc reg data=datain.trade simple;
model gnp = tr1 tr2 tr3 tr4 / dw;
id cntry_id;
```

REG's *PROC* statement has six features:

1. The *dset* indicates the name of the SAS dataset to be analyzed. If the *DATA* option is not specified, the most recently created SAS dataset is used. For additional information, see the "Efficiency Tips" section later in the chapter.

2. The *param_est* indicates the name of a SAS dataset containing parameter estimates and other statistics for each model run during the procedure. For additional information, see the "Types of Output Datasets" section later in the chapter.

3. The *sscp* indicates the name of a SAS dataset containing sums of squares and cross-products. This dataset is particularly useful when the input (DATA=) dataset has many observations and/or many variables. The SSCP dataset may be used as input to subsequent REGs, thus saving potentially large amounts of computer resources. For additional information, see the "Efficiency Tips" section.

4. *SIMPLE* prints descriptive statistics for each variable used in REG.

5. *CORR* requests printing of the Pearson correlation matrix for each variable used in REG. CORR is available only in version 6 of the SAS System.

6. *NOPRINT* suppresses all printed output. This option is usually taken when REG is being run to create *OUTEST* and/or *OUTSSCP* datasets. The *NOPRINT* option in the *MODEL* statement suppresses printed output on a model-by-model basis.

The REG *MODEL* statement has nine features:

1. The *dep_var* specifies the predicted, or dependent, variable; the *indep_var* specifies one or more predictor, or independent, variables.

2. *NOPRINT* suppresses printed output for the model. This option is useful when REG is run to produce only output datasets.

3. *STB* requests printing of standardized regression coefficients.

4. *TOL* and *VIF* request tolerance values and variance inflation factors for the estimates, respectively. *TOL* is defined as $1 - R^2$, where R^2 is determined by regressing a given independent variable on all other independent variables. VIF is the reciprocal of tolerance.

5. *COLLIN* and *COLLINOINT* request analyses related to multicollinearity among the independent variables. *COLLIN* includes the intercept variable in the analysis; *COLLINOINT* does not.

6. *CLI* and *CLM* print 95% upper and lower confidence limits for individual predicted values and expected values of the dependent variable, respectively. *CLI* requests a prediction interval; *CLM* does not because it does not account for variation in the error term.

7. *P* and *R* request analyses of residuals. For each observation used in the analysis, *P* displays the observation number in the dataset, the value of the *ID* variable (if specified), actual and predicted values, and the residual. *R* displays everything that *P* does plus the standard errors of the predicted and actual values, the Studentized residual, and Cook's D statistic.

8. *DW* is appropriate only when analyzing time-series data. It computes a Durbin–Watson statistic testing for first-order autocorrelation.

9. *PARTIAL* produces partial regression leverage plots for each independent variable in the full model. It plots residuals for the equation of the dependent variable with the independent variable omitted on the *y* axis and residuals for the independent variable regressed against all other independent variables on the *x* axis.

In the REG *TEST* statement, *restriction* specifies a constraint on the model specified in the preceding *MODEL* statement. This is a useful means of testing the influence of adding additional variables, determining whether certain parameters equal 0, and so on. The *restriction* takes the form of an equation. If no equals sign appears, the equation is set to 0 (i.e., a test for the variable's significance in the equation). The special value *INTERCEPT* may be used to indicate the model's intercept. Multiple restrictions are separated with a comma. Valid *restrictions* are shown in Exhibit 23.8

EXHIBIT 23.8   TEST statements and their interpretation

```
b for HEIGHT = 0
test height;

test height=0;

b's for HEIGHT and WEIGHT = 0
test height, weight;

b's for HEIGHT and WEIGHT and the intercept = 0
test height=weight=intercept;

b for HEIGHT is twice that of WEIGHT
test 2*height=weight;
```

In the REG VAR statement, *varlist* identifies numeric variables in the dataset that may be used by *any MODEL* statement during execution of REG. It must appear before the first *RUN* statement. A *VAR* statement without a *MODEL* statement implies that REG is being run to create an SSCP dataset for use by subsequent SAS sessions and jobs. See Exhibit 23.8 for an illustration of this use of *VAR*.

In the REG *ID* statement, *idvar* indicates a single character or numeric variable used to identify an observation. Character *idvars* are limited to the first (leftmost) eight characters. If any of the *P*, *R*, *CLI*, or *CLM* options are taken, statistics for individual observations are printed. The *idvar* simply aids identification of these observations. An *idvar* might be the name of a survey respondent, a company number, date value, or an industrial or occupational code.

The REG *OUTPUT* statement creates an output SAS dataset containing statistics for each observation used in the most recent *MODEL* statement. More than one *OUTPUT* statement may be used in a single execution of REG. *OUTPUT* is useful when the PLOT procedure will be used later to display predicted versus actual values. The *OUTPUT* statement has six features:

1. The *out_dset* indicates the name of the SAS dataset to be created. It may be either a one-level (temporary) or two-level (permanent) dataset. If *OUT* is not specified, SAS names the dataset DATAn, where *n* is the next available integer (e.g., if DATA3 had already been created, SAS will select DATA4).

2. The *P* and *R* options identify variables containing the predicted and residual values, respectively. *STDP, STDR,* and *STDI* name variables for the standard errors of the mean predicted value, the residual, and the individual predicted values, respectively.

3. *L95M* and *U95M* represent the lower and upper bounds of the 95% confidence interval for the expected value of the dependent variable. These values correspond to those printed by the *CLM* option.

4. *L95* and *U95* represent the lower and upper bounds of the 95% confidence interval for an individual observation's prediction. These values correspond to those printed by the CLI option.

5. *COOKD* names a variable representing the observation's value for Cook's D influence statistic.

6. STUDENT = *stu_resid* and RSTUDENT = *stu_resid_del* name variables representing Studentized residuals (residuals divided by their standard errors). The *stu_resid* represents the residual for all observations in the dataset, while *stu_resid_del* is the residual for all but the current observation.

In the REG FREQ statement, *freqvar* indicates a single, positive numeric variable whose integer value represents the observation's actual occurrence in the population from which the sample was drawn. The *freqvar*'s value influences all calculations used in the analysis: degrees of freedom, means, variances, and so on. If *freqvar* is missing or less than 1, the observation is not used in the analysis.

In the REG *WEIGHT* and *REWEIGHT* statements, *wgtvar* indicates a single, positive numeric variable whose value represents a relative weight for a weighted least squares model. Use the *REWEIGHT* statement during an interactive analysis to change *wgtvar*.

Using REG Interactively

Like DATASETS and PLOT, discussed in Chapters 10 and 21, respectively, REG is an interactive procedure. This section discusses some features to keep in mind when using REG interactively.

Starting and Stopping REG.   Entering the RUN statement begins execution of a group of REG statements: it does not terminate REG. Once the statements are executed, you can either enter another model specification or QUIT to leave the procedure.

Multiple Titles and Footnotes.   A useful feature of REG is its ability to accept more than one set of TITLEs and FOOTNOTEs in a single execution of the procedure. This is illustrated in the following example:

```
proc reg data=test1;
model income = ed_level age work_yrs;
title "Full Model -> education plus experience";
title2 "Harris dataset";
footnote "Homework for Sept 24th";
run;
test work_yrs=0;
title "Test for work experience effect";
run;
```

The first analysis will have two title lines and a footnote. The second has one title (TITLE2 is overwritten) and uses the same FOOTNOTE.

Statement Order.   Interactive use of REG requires that some care be exercised over statement order. Exhibit 23.9 shows the order in which statements may be entered. Note that using BY-group processing (the BY statement) disables some interactive features such as TEST and REWEIGHT.

Types of Output Datasets

REG can produce three types of output datasets. One, described in the OUTPUT statement above, contains one observation for each observation used in the anal-

---

EXHIBIT 23.9 REG statement order

```
MODEL

Before first RUN statement:
FREQ
WEIGHT
ID
VAR
BY

Anywhere between MODEL and RUN statements:
TITLE
FOOTNOTE
OUTPUT
REWEIGHT
TEST
```

---

ysis. The variables contain such information as actual, predicted, and residual values, confidence limits, and standard errors.

The other output datasets—OUTEST and OUTSSCP—are specified in the PROC statement and do not have a one-to-one correspondence with observations in the input dataset.

OUTEST Datasets. Datasets created by the OUTEST option in the PROC statement contain parameter estimates and other statistics for each model in the PROC. The dataset holds these items for each MODEL run for each BY-group level, if specified. OUTEST datasets are useful when you want to display the estimates with procedures such as PLOT or CHART.

Exhibit 23.18 (in Section 23.5) illustrates the creation and contents of OUTEST datasets. Among the variables in the dataset are the following:

- Values of the BY variables, if specified
- MODEL: a character variable of the form MODELn, where *n* indicates the model number (1, 2, and so on) of the PROC
- _DEPVAR_: a character variable indicating the name of the dependent variable
- _RMSE_: the root mean squared error (the estimate of the standard deviation of the error term)
- INTERCEP: the estimated intercept. This is included only if NOINT is not specified
- All variables used in all MODEL statements. If a variable was not used in a particular model, it is assigned a missing value. If the variable is the dependent variable for a model, it is assigned a value of -1. The values for the rest of the variables are the regression coefficients for the model.

OUTSSCP Datasets. Datasets created by the OUTSSCP option contain sums of squares (*SS*) and cross-products (*CP*) used in calculating parameter estimates and other regression statistics. OUTSSCP datasets can be used as input to subsequent uses of REG. This often saves large amounts of computer resources since the individual observations do not have to be read and their statistics recalculated. The "Efficiency Tips" section discusses this point in greater detail.

Exhibit 23.17 (in Section 23.5) illustrates the creation and contents of OUT-SSCP datasets. Among the variables in the dataset are the following:

- _NAME_: the name of the variable in the MODEL or VAR statements, or INTERCEP, which contains the sum of the analysis variables
- All variables used in all MODEL statements plus any additional variables specified by the VAR statement
- INTERCEP: a variable containing sums of the analysis variables and the count of nonmissing observations

Special Note: The VAR Statement.   A VAR statement may be used to supplement or replace the variables specified in the MODEL statements. In either case, only numeric variables may be specified in the VAR statement. This feature is useful if all you want to do in the REG procedure is create an SSCP dataset for variables that will be used in subsequent analyses.

Missing Data

An important consideration of REG's use is its insistence on having "complete" data; that is, only observations with nonmissing values for *all* values of *all* variables used in *every* MODEL are included in the analysis. This "listwise" deletion gives good estimates for the observations used for the analysis but is subject to two drawbacks. First, the cases used may be representative of the population of all cases, including those not used. Second, as the number of variables in the MODEL statements increases, so usually does the proportion of cases deleted from the analysis. What seemed to be a large dataset can shrink rapidly if the missing values are spread evenly throughout the dataset.

A viable, if not satisfactory, solution to part of this problem is to run separate REGs, one for each model. This avoids one model's *n* being penalized for the lack of complete data for another that uses a variable with a high proportion of missing values.

Efficiency Tips

The power of the REG procedure is not without a price. Calculating the required sums of squares, correlation, and other matrices may require significant (and costly) computer resources. Even if you are running SAS on a microcomputer and resources are "free," the time spent waiting for the calculations may become considerable and tiresome.

A simple solution to this problem is to use correlation and SSCP datasets as input to REG. This bypasses the reading of each observation from the input dataset: REG immediately goes to work with the required matrices, often saving vast amounts of time and money. Correlation datasets were discussed in Chapter 22; SSCP datasets in Section 23.3. No changes are required when using either one with REG: an indicator in the dataset directory tells REG that it is reading a CORR or SSCP rather than a "regular" dataset.

Two cautions when using these datasets. First, you should take care to include all required variables in the dataset. It is usually prudent to include too many variables in the dataset rather than too few—otherwise the efficiency gain will be wiped out when you have to create a new, larger CORR or SSCP dataset. Second, use of these summary-level datasets as input means individual observations are, in effect, lost. Creation of confidence limits, prediction intervals, residuals, and the like is impossible. If these observation-level analyses are necessary, you must work with the original dataset.

> **POINTERS AND PITFALLS**
>
> Use a correlation matrix as input to REG when you do not require observation-level detail from the procedure (no residuals, confidence intervals, and so on).

Exercises

23.1. Refer to the description of the STATES dataset in Appendix E. Propose a model whose dependent variable is PCTIN, the proportion of the population five years or older that lived in a different state in 1975. PCTIN is a rough indicator of in-migration and, implicitly, an area's attractiveness as a place to live. Possible independent variables include, but are not limited to the following:

PCTURB	The proportion of the population living in urban areas (Urban areas have more employment opportunities and thus would be positively related to PCTIN.)
PCTFARM	The proportion of the population living on farms (This follows the same logic as PCTURB but would suggest a negative relation to PCTIN.)
PCTMFG	Proportion of the labor force in manufacturing jobs (Given the shift to service-sector jobs, you would expect a negative relationship with PCTIN—"undesirable" jobs are not a source of population growth.)
PCAPINC	Per capita income (Areas with the highest incomes are often magnets for growth. This suggests a positive relation with PCTIN.)

   a. Write the REG procedure statements to specify a model with one independent variable.

   b. Add two independent variables to the model.

   c. Test the contribution to the model made by the variables added in part b.

   d. Can you do parts a, b, and c in a single execution of REG? If not, why not? If you can, write the necessary statements.

   e. Assume that region of the country (variable REGION) has an effect on in-migration. Rather than use REGION directly as an independent variable, create a series of dummy variables. Write the necessary DATA step statements.

23.2. Describe how to use SAS to identify, and possibly correct, the following problem areas. Your answer need not be confined to the REG procedure.

   a. Outliers

   b. Specification errors

   c. Heteroscedasticity

   d. Autocorrelation

   e. Collinearity

23.3. Refer to the full model in Exercise 23.1b. Use the CORR procedure to create a correlation matrix of the variables used in the model. Rerun the model using the correlation matrix, rather than individual (state-level) observations, as input. Note the execution time (printed in the SAS Log) of this method. Is it less than Exercise 23.1b? Why (or why not)?

23.4. Rerun the model in Exercise 23.1b. Add the statements to save an output dataset containing the predicted and residual values. Also, suppress printed output from the procedure. Using the PLOT procedure, produce a scatterplot of residual versus predicted values. Draw a reference line at residual value 0.

23.5. What, if anything, is wrong with the following REG program?

```
proc reg data=test;
model y = x1 x2;
model y = x1;
run;
model y = x1 x2 x3;
model y = x1 x2 x1*x2;
quit;
```

23.5 Examples of the REG Procedure

The examples in this section use the STATES dataset in Appendix E. They highlight some of the more common uses of the REG procedure and are easily extended to more complicated models and a broader selection of options.

Key features and options used in this section are summarized in the following list:

Exhibit	Features/Options
23.10	Two independent variables; all defaults taken
23.11	Dummy variable added to model in Exhibit 23.10
23.12	Interaction term specified; multiple RUN statements; VAR statement used
23.13	Univariate statistics added (SIMPLE); MODEL used STB, P, R, CLI options; ID statement to aid observation identification
23.14	Collinearity options specified: VIF, TOL, COLLIN
23.15	Output dataset created; output suppressed (NOPRINT)
23.16	TEST used to assess variables' contribution to model
23.17	OUTSSCP dataset created for use in later PROCs
23.18	OUTEST dataset containing model information created
23.19	Input to REG is CORR output dataset

Exhibit 23.10 specifies a simple model. Two independent variables are used and all defaults in the PROC and MODEL statements are taken.

Exhibit 23.11 (see p. 502) tests the significance of geographic region in the model. A DATA step is used to create dummy variable DUMY_REG, whose value is 1 if the state is on the East coast, 0 otherwise. Since the EASTWEST variable used to create the dummy has two levels, only $2 - 1$, or 1, variable is needed to distinguish regions in the model.

Exhibit 23.12 (see p. 503) investigates the significance of the PCTURB–PCTCOLL interaction. A DATA step is used prior to REG to create URB_COLL. Notice that the interactions must be specified as a variable in the MODEL statement: entries of the form PCTURB*PCTCOLL are not allowed. The exhibit also demonstrates REG's ability to run several models. The VAR statement appears prior to the first RUN statement and tells REG which variables will be used in *all* models in the PROC. TITLEs will be changed with each model: a model is requested, a title specified, a RUN encountered, and SAS immediately displays the results with the current title.

Exhibit 23.13 (see p. 504) adds some options to the basic model of Exhibit 23.10. In the PROC statement, display of univariate statistics for variables in the model is requested (the SIMPLE option). In the MODEL statement, standardized coefficients (the STB option), predicted values (P), residuals (R), and 95% prediction intervals (CLI) are requested. The ID statement is also used to

EXHIBIT 23.10   Two independent variables; all defaults taken

---

```
------------------ SAS Log ----------------------

    1     libname x 'c:\book\data\sas';

    2

    3     options ls=120 ps=60 nocenter;

    4

    5     proc reg data=x.states;
    6     model crimert = pcturb pctcoll;
NOTE: 51 observations read.
      51 observations used in computations.
```

No observations excluded due to missing values

```
------------------ Output listing --------------------
```

Models are numbered automatically by REG.

```
Model: MODEL1
```

Dependent Variable: CRIMERT

Analysis of Variance

Source	DF	Sum of Squares	Mean Square	F Value	Prob>F
Model	2	85111147.389	42555573.694	56.723	0.0001
Error	48	36011099.199	750231.23332		
C Total	50	121122246.59			

Root MSE	866.15890	R-square	0.7027	
Dep Mean	5409.29412	Adj R-sq	0.6903	
C.V.	16.01242			

Parameter Estimates

Variable	DF	Parameter Estimate	Standard Error	T for H0: Parameter=0	Prob > |T|
INTERCEP	1	-1188.574891	661.17422659	-1.798	0.0785
PCTURB	1	69.158407	10.01760164	6.904	0.0001
PCTCOLL	1	117.882121	45.05439171	2.616	0.0118

help identify observations. Notice the Log message that the ID variable was truncated to eight characters.

Exhibit 23.14 (see p. 506) adds a variable (POPSQMI) that is somewhat correlated with PCTURB. This is done to demonstrate fruitful use of the collinearity options in the MODEL statement (VIF, TOL, COLLIN).

Exhibit 23.15 (see p. 508) demonstrates the use of the OUTPUT statement. A dataset DIAGS, containing various predicted and residual values from the model, is created. (If there were multiple MODEL statements, only the one immediately preceding the OUTPUT statement would be captured in the output dataset.) REG's usual output is suppressed by using the NOPRINT option in the PROC statement.

The exhibit also presents a CONTENTS procedure listing of the output dataset, followed by a PRINT of the first 10 observations. Finally, it displays a scatterplot of the residual and predicted values. A vertical axis reference line

EXHIBIT 23.11   Dummy variable added to the model in Exhibit 23.10

```
------------------- SAS Log --------------------------

    1      libname x 'c:\book\data\sas';
    2
    3      options ls=120 ps=60 nocenter;
    4
    5      data temp;
    6      set  x.states;
    7      if eastwest = 'e' then dumy_reg = 1;
    8        else            dumy_reg = 0;
    9      run;
NOTE: The data set WORK.TEMP has 51 observations and 65 variables.
NOTE: The DATA statement used 9.00 seconds.
   10
   11      proc reg data=temp;
   12      model crimert = pcturb pctcoll dumy_reg;
   13      run;
NOTE: 51 observations read.
      51 observations used in computations.

------------------- Output listing --------------------
```

Model: MODEL1
Dependent Variable: CRIMERT

Analysis of Variance

Source	DF	Sum of Squares	Mean Square	F Value	Prob>F
Model	3	85139805.260	28379935.087	37.070	0.0001
Error	47	35982441.328	765583.85805		
C Total	50	121122246.59			

Root MSE	874.97649	R-square	0.7029
Dep Mean	5409.29412	Adj R-sq	0.6840
C.V.	16.17543		

Parameter Estimates

Variable	DF	Parameter Estimate	Standard Error	T for H0: Parameter=0	Prob > |T|
INTERCEP	1	-1229.852205	701.15187861	-1.754	0.0859
PCTURB	1	69.098986	10.12424137	6.825	0.0001
PCTCOLL	1	119.099937	45.94624606	2.592	0.0127
DUMY_REG	1	48.031310	248.25548490	0.193	0.8474

EXHIBIT 23.12   Interaction term; VAR statement and multiple RUN, MODEL statements used

------------------- SAS Log ----------------------------

```
    1      libname x 'c:\book\data\sas';
    2
    3      options ls=120 ps=60 nocenter;
    4
    5      data temp;
    6      set  x.states;
    7      if eastwest = 'e' then dumy_reg = 1;
    8         else                dumy_reg = 0;
    9      urb_coll = pcturb * pctcoll ;          ◄——————  Create interaction term
   10      run;
NOTE: The data set WORK.TEMP has 51 observations and 65 variables.
NOTE: The DATA statement used 9.00 seconds.
   11                                         ┌─── Variables used in all models
   12      proc reg data=temp;
   13      var pcturb pctcoll dumy_reg urb_coll;  ◄──────
   14      model crimert = pcturb pctcoll dumy_reg;   ◄───  First MODEL specification
   15      title 'Dummy variable example';                  and title
   16      run;
NOTE: 51 observations read.
      51 observations used in computations.
   17      model crimert = pcturb pctcoll urb_coll;   ◄───  Second MODEL specification
   18      title 'Interaction example';                     and title
   19      run;
   20      quit;
NOTE: The PROCEDURE REG used 15.00 seconds.
```

------------------- Output: model with interaction -----------

< First MODEL output is not shown >

┌─────────────────────┐   TITLE used in second
│ Interaction example │ ◄──────────────────────────────   MODEL specification
└─────────────────────┘

Model: MODEL2
Dependent Variable: CRIMERT

Analysis of Variance

Source	DF	Sum of Squares	Mean Square	F Value	Prob>F
Model	3	85576180.825	28525393.608	37.717	0.0001
Error	47	35546065.763	756299.27156		
C Total	50	121122246.59			

Root MSE	869.65469	R-square	0.7065
Dep Mean	5409.29412	Adj R-sq	0.6878
C.V.	16.07705		

EXHIBIT 23.12   *(continued)*

Parameter Estimates

Variable	DF	Parameter Estimate	Standard Error	T for H0: Parameter=0	Prob > |T|
INTERCEP	1	618.998917	2398.8419732	0.258	0.7975
PCTURB	1	43.468176	34.27133291	1.268	0.2109
PCTCOLL	1	5.434313	150.36788568	0.036	0.9713
URB_COLL	1	1.558994	1.98814991	0.784	0.4369

EXHIBIT 23.13   SIMPLE, STB, P, R, CLI options; ID statement to aid observation identification

```
------------------ SAS Log ----------------------

   1    libname x 'c:\book\data\sas';
   2
   3    options ls=120 ps=60 nocenter;
   4
   5    proc reg data=x.states simple;
   6    model crimert = pcturb pctcoll / stb p r  cli;
   7    id name;
   8    run;
NOTE: 51 observations read.
      51 observations used in computations.
NOTE: Some ID variables have been truncated to 8 characters.

------------------ Output listing --------------------
```

Descriptive Statistics

Variables	Sum	Mean	Uncorrected SS	Variance	Std Deviation
INTERCEP	51	1	51	0	0
PCTURB	3447.4564322	67.597184946	244283.42103	224.90141975	14.996713632
PCTCOLL	831.938933	16.312528098	14126.951261	11.118480814	3.3344386055
CRIMERT	275874	5409.2941176	1613405852	2422444.9318	1556.4205511

Produced by SIMPLE option in PROC statement

```
- - - - - - - - - - New page of output - - - - - - - - -
```

Model: MODEL1
Dependent Variable: CRIMERT

Analysis of Variance

Source	DF	Sum of Squares	Mean Square	F Value	Prob>F
Model	2	85111147.389	42555573.694	56.723	0.0001
Error	48	36011099.199	750231.23332		
C Total	50	121122246.59			

Root MSE	866.15890	R-square	0.7027
Dep Mean	5409.29412	Adj R-sq	0.6903
C.V.	16.01242		

Parameter Estimates

Variable	DF	Parameter Estimate	Standard Error	T for H0: Parameter=0	Prob > |T|	Standardized Estimate
INTERCEP	1	-1188.574891	661.17422659	-1.798	0.0785	0.00000000
PCTURB	1	69.158407	10.01760164	6.904	0.0001	0.66636798
PCTCOLL	1	117.882121	45.05439171	2.616	0.0118	0.25254787

EXHIBIT 23.13 *(continued)*

```
- - - - - - - - - - - New page of output - - - - - - - - -
```

Obs	NAME	Dep Var CRIMERT	Predict Value	Std Err Predict	Lower95% Predict	Upper95% Predict	Residual	Std Err Residual	Student Residual	-2-1-0 1 2	Cook's D
1	Alabama	4801.0	4399.2	196.945	2613.2	6185.1	401.8	843.471	0.476	| | |	0.004
2	Alaska	6410.0	5745.8	264.443	3925.0	7566.7	664.2	824.804	0.805	| |* |	0.022
3	Arizona	7614.0	6661.9	185.145	4881.0	8442.7	952.1	846.140	1.125	| |** |	0.020
4	Arkansas	3743.0	3654.8	236.038	1849.7	5459.8	88.2487	833.377	0.106	| | |	0.000
5	Californ	7592.0	7435.9	228.677	5634.7	9237.1	156.1	835.427	0.187	| | |`	0.001
6	Colorado	7189.0	7093.4	275.986	5265.6	8921.2	95.6	821.013	0.116	| | |	0.001
7	Connecti	6552.0	6698.7	200.528	4911.1	8486.3	-146.7	842.627	-0.174	| | |	0.001
8	Delaware	6644.0	5755.7	128.505	3995.1	7516.3	888.3	856.573	1.037	| |** |	0.008
9	DC	10692.0	8964.5	428.498	7021.6	10907.5	1727.5	752.742	2.295	| |**** |	0.569
10	Florida	8048.0	6393.1	243.294	4584.2	8202.0	1654.9	831.288	1.991	| |*** |	0.113

```
... other observations are not listed ...
```

```
Sum of Residuals           -7.95808E-11
Sum of Squared Residuals    36011099.199
Predicted Resid SS (Press)  42458759.150
```

First eight characters of ID variable

Residual/Std Err Residual

at zero helps to visualize the balance in number and magnitude of positive and negative residuals.

Exhibit 23.16 (see p. 510) uses the TEST statement to assess the contribution of POPSQMI and PCTCOLL to the model. The full model is specified in the MODEL statement, and the null hypothesis of zero slope is posed in the TEST statement.

Exhibit 23.17 (see p. 511) creates an output dataset with the information needed for later use of REG. The OUTSSCP dataset CROSPROD may be a one- or two-level (temporary or permanent) SAS dataset name and may be used as input to later executions of REG. VAR, together with the omission of a MODEL statement, tells REG that the purpose of the procedure is dataset creation: CROSPROD will contain information required for using any variable in the VAR statement in a model. The exhibit also displays a CONTENTS listing, PRINTs the output dataset, and then uses the dataset as input to REG. Notice that the only indication that CROSPROD is not a "regular" SAS dataset is the "Type" indication in the CONTENTS listing: most SAS datasets discussed in this book are type "Data"; the OUTSSCP option produces a specially formatted dataset with a type of "SSCP". When REG reads the input dataset and detects an SSCP format, it goes through a special, faster-executing series of calculations to produce the regression output. The drawback to this efficiency is the loss of observation-level information necessary for display of residuals, predicted values, and confidence intervals. The P and R options requesting such detail were specified, so REG printed a NOTE that the options were ignored.

An OUTEST dataset is created in Exhibit 23.18 (see p. 513). Four models are run, and their parameter estimates and other identifying information stored in dataset ESTIMATE. As it did with the OUTSSCP dataset in Exhibit 23.17, SAS gave the dataset a special "type" (EST) and supplied labels for the variables it created (INTERCEP, _DEPVAR_, _TYPE_, and so on). The exhibit shows a CONTENTS and PRINT of the dataset.

The REG statements in Exhibit 23.19 (see p. 514) are nearly identical to those of Exhibit 23.13. Only the PROC statement differs: its input dataset is a correlation matrix created in PROC CORR. The execution speed gained by using the correlation matrix is somewhat offset by the loss of observation-level

EXHIBIT 23.14 Collinearity diagnostics specified: VIF, TOL, and COLLIN

```
------------------- SAS Log ----------------------------

   1     libname x 'c:\book\data\sas';
   2
   3     options ls=120 ps=60 nocenter;
   4
   5     proc reg data=x.states corr ;
   6     model crimert = pcturb pctcoll popsqmi / vif tol collin ;
   7     run;
NOTE: 51 observations read.
      51 observations used in computations.


------------------- Output listing --------------------
```

Correlation

CORR	PCTURB	PCTCOLL	POPSQMI	CRIMERT
PCTURB	1.0000	0.5789	0.3752	0.8126
PCTCOLL	0.5789	1.0000	0.5019	0.6383
POPSQMI	0.3752	0.5019	1.0000	0.5100
CRIMERT	0.8126	0.6383	0.5100	1.0000

```
- - - - - - - - - - New page of output - - - - - - - - - -
```

Model: MODEL1
Dependent Variable: CRIMERT

Analysis of Variance

Source	DF	Sum of Squares	Mean Square	F Value	Prob>F
Model	3	88026672.954	29342224.318	41.670	0.0001
Error	47	33095573.634	704161.14115		
C Total	50	121122246.59			

Root MSE	839.14310	R-square	0.7268
Dep Mean	5409.29412	Adj R-sq	0.7093
C.V.	15.51299		

Parameter Estimates

Variable	DF	Parameter Estimate	Standard Error	T for H0: Parameter=0	Prob > |T|	Tolerance	Variance Inflation
INTERCEP	1	-507.642988	722.69808981	-0.702	0.4859		0.00000000
PCTURB	1	66.771707	9.77577133	6.830	0.0001	0.65525114	1.52613241
PCTCOLL	1	81.771563	47.11882279	1.735	0.0892	0.57051588	1.75279959
POPSQMI	1	0.197901	0.09725817	2.035	0.0475	0.73736084	1.35618811

VIF option

TOL option

EXHIBIT 23.14 *(continued)*

```
CollinearityDiagnostics

                    Condition   Var Prop   Var Prop   Var Prop   Var Prop
Number   Eigenvalue    Number    INTERCEP    PCTURB     PCTCOLL    POPSQMI

   1      3.08598     1.00000    0.0026     0.0031     0.0023     0.0129
   2      0.87840     1.87435    0.0012     0.0005     0.0002     0.7404
   3      0.02078    12.18698    0.4093     0.9155     0.0497     0.0583
   4      0.01484    14.41907    0.5869     0.0810     0.9478     0.1884
```

← COLLIN option

detail. Notice that the P, R, and CLI options cannot be satisfied since there is no data at the observation level. The ID statement is also problematic since the variable NAME, while on the *original* dataset (STATES), is not on the correlation matrix dataset. If you can live with these restrictions, storing the matrix as a SAS dataset and using it with one or more PROC REGs can dramatically conserve computer resources. Notice that you can store the correlation matrix as a permanent SAS dataset and realize the resource savings over more than one SAS session.

Exercises

23.6. Refer to Exhibit 23.10. Identify or compute the following items:
   a. The number of observations in input dataset STATES with complete (nonmissing) data for all analysis variables.
   b. Value of R-square, accounting for the number of independent variables in the model.
   c. Mean of CRIMERT.
   d. The equation represented by the model. Your answer should be similar to:

   ```
   CRIMERT = intercept + b₁(PCTURB) + b₂(PCTCOLL)
   ```

   Supply values for "intercept," "b_1," and "b_2."

23.7. Refer to Exhibit 23.11.
   a. Review the coding scheme for DUMY_REG. Will missing values be assigned to the 0's group? If so, how would you rewrite the ELSE statement to exclude missing values?
   b. Suppose the basis for the dummy variable (EASTWEST) had three levels rather than two. Rewrite the IF statement to handle EAST-WEST values of 'E,' 'W,' and 'N.'
   c. Does DUMY_REG make a significant contribution to the model?

23.8. Refer to Exhibit 23.12.
   a. Why is the VAR statement needed?
   b. None of the parameters is significant at the .05 level, but the *overall* model is significant (p value of .0001). Explain.

23.9. Refer to Exhibit 23.13's display of observations and their residual and predicted values and associated statistics.
   a. Of the observations listed, which have the greatest difference between actual and predicted values? the smallest?
   b. Which columns can be used to compute "Residual"? "Student Residual"?

23.10. Refer to Exhibit 23.15.
   a. Why might NOPRINT have been used in the PROC statement?
   b. What problem in the data is suggested by the scatterplot?
   c. Why does output dataset DIAGS have formats of COMMA9. and 7.3 assigned to its variables?

EXHIBIT 23.15   Output dataset created (OUTPUT statement); printed output suppressed (NOPRINT)

```
------------------ SAS Log ----------------------

    1     libname x 'c:\book\data\sas';
    2
    3     options ls=120 ps=60 nocenter;
    4
    5     proc reg data=x.states(keep=crimert pcturb pctcoll popsqmi)
    6             noprint;
    7     model crimert = pcturb pctcoll popsqmi ;
    8     output out=diags p=pred r=resid student=st_res rstudent=r_st_res
    9                     stdr=std_res;
   10     quit;
NOTE: 51 observations read.
      51 observations used in computations.
NOTE: The data set WORK.DIAGS has 51 observations and 9 variables.
NOTE: The PROCEDURE REG used 13.00 seconds.
   11
   12     proc contents;
   13     run;
NOTE: The PROCEDURE CONTENTS used 4.00 seconds.
   14
   15     proc print data=diags(obs=10);
   16     run;
NOTE: The PROCEDURE PRINT used 5.00 seconds.
   17
   18     proc plot data=diags;
   19     plot resid*pred='x' / vref=0 ;
   20     run;
```

```
------------------ DIAGS Contents --------------------

CONTENTS PROCEDURE

    Data Set Name:  WORK.DIAGS          Type:
    Observations:   51                  Record Len: 76
    Variables:      9
    Label:

-----Alphabetic List of Variables and Attributes-----

    #  Variable  Type  Len  Pos  Format    Label

    2  CRIMERT   Num   8    12   COMMA9.
    4  PCTCOLL   Num   8    28   7.3
    3  PCTURB    Num   8    20   7.3
    1  POPSQMI   Num   8    4    COMMA9.2

    5  PRED      Num   8    36            Predicted Value of CRIMERT
    6  RESID     Num   8    44            Residual
    9  R_ST_RES  Num   8    68            Studentized Residual without Current Obs
    7  STD_RES   Num   8    52            Standard Error of Residual
    8  ST_RES    Num   8    60            Studentized Residual
```

Formats stored in permanent dataset STATES

LABELS added by REG

```
------------------ Listing of DIAGS ---------------------

OBS   POPSQMI     CRIMERT   PCTURB    PCTCOLL     PRED       RESID     STD_RES    ST_RES    R_ST_RES

  1     76.70       4,801    60.035    12.180    4512.16    288.84    815.274    0.35428    0.35096
  2      0.70       6,410    64.344    21.076    5512.27    897.73    790.790    1.13523    1.13881
  3     23.90       7,614    83.832    17.414    6518.62   1095.38    816.721    1.34119    1.35299
  4     43.90       3,743    51.589    10.820    3830.52    -87.52    802.749   -0.10903   -0.10788
  5    151.40       7,592    91.295    19.602    7221.11    370.89    802.454    0.46220    0.45830
  6     27.90       7,189    80.619    22.959    6758.36    430.64    778.178    0.55339    0.54927
  7    637.90       6,552    78.832    20.659    6571.72    -19.72    813.955   -0.02423   -0.02397
  8    307.60       6,644    70.636    17.468    5698.16    945.84    829.374    1.14042    1.14417
  9 10,180.70      10,692   100.000    27.462   10429.91    262.09    114.921    2.28062    2.39249
 10    180.00       8,048    84.261    14.882    6371.17   1676.83    805.287    2.08228    2.16215
```

MODEL variables

Variables specified in OUTPUT statement

EXHIBIT 23.15 *(continued)*

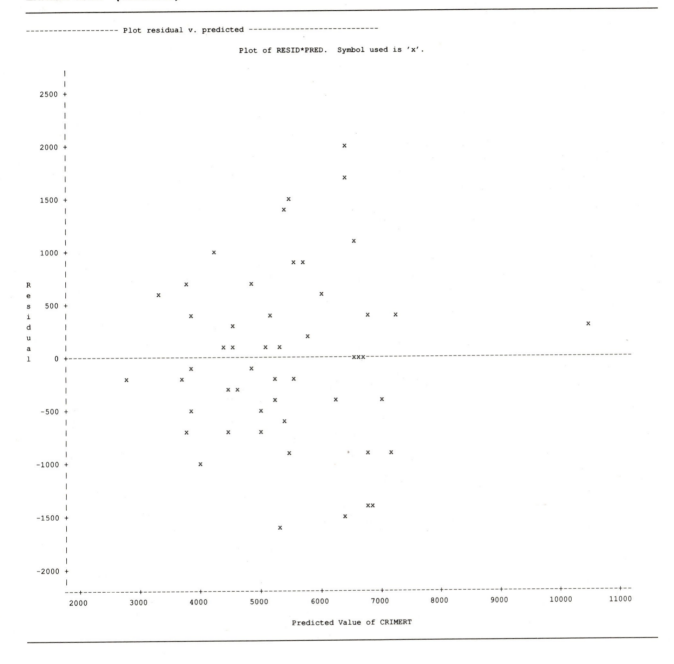

-------------------- Plot residual v. predicted ----------------------------

Plot of RESID*PRED. Symbol used is 'x'.

Predicted Value of CRIMERT

EXHIBIT 23.16   TEST statement used to assess variables' contribution to model

```
------------------- SAS Log -------------------------

    1     libname x 'c:\book\data\sas';
    2
    3     options ls=120 ps=60 nocenter;
    4
    5     proc reg data=x.states;
    6     model crimert = pcturb pctcoll popsqmi ;
    7     test pctcoll=popsqmi=0;
    8     run;
NOTE: 51 observations read.
      51 observations used in computations.


------------------- Output listing --------------------
```

Model: MODEL1
Dependent Variable: CRIMERT

Analysis of Variance

Source	DF	Sum of Squares	Mean Square	F Value	Prob>F
Model	3	88026672.954	29342224.318	41.670	0.0001
Error	47	33095573.634	704161.14115		
C Total	50	121122246.59			

Root MSE	839.14310	R-square	0.7268	
Dep Mean	5409.29412	Adj R-sq	0.7093	
C.V.	15.51299			

Parameter Estimates

Variable	DF	Parameter Estimate	Standard Error	T for H0: Parameter=0	Prob > |T|
INTERCEP	1	-507.642988	722.69808981	-0.702	0.4859
PCTURB	1	66.771707	9.77577133	6.830	0.0001
PCTCOLL	1	81.771563	47.11882279	1.735	0.0892
POPSQMI	1	0.197901	0.09725817	2.035	0.0475

```
- - - - - - - - - - New page of output - - - - - - - - -

SAS
```

```
┌─────────────────────────────────────────────────────────┐
│ Dependent Variable: CRIMERT                              │
│ Numerator:4025713.9849  DF:    2   F value:   5.7170     │ ◄──────  TEST statement output
│ Denominator:  704161.1  DF:   47   Prob>F:   0.0060      │
└─────────────────────────────────────────────────────────┘
```

EXHIBIT 23.17   Creation and use of OUTSSCP dataset

```
------------------- SAS Log ----------------------------

    1     libname x 'c:\book\data\sas';

    2

    3     options ls=120 ps=60 nocenter;

    4

    5     proc reg data=x.states outsscp=sscp noprint;
    6     var crimert pcturb pctcoll popsqmi;
    7     run;
NOTE: 51 observations read.
      51 observations used in computations.
NOTE: The data set WORK.CROSPROD has 6 observations and 7 variables.
NOTE: The PROCEDURE REG used 7.00 seconds.
    8
    9     proc contents;
   10     run;
NOTE: The PROCEDURE CONTENTS used 3.00 seconds.
   11
   12     proc print data=sscp;
   13     run;
NOTE: The PROCEDURE PRINT used 5.00 seconds.
   14
   15     proc reg data=sscp;
   16     model crimert = pcturb / p r;
   17     run;
```

```
NOTE: No raw data are available. Some options are ignored.
```
← SSCP dataset has only summary data. No observation-level ("raw") data are available.

```
NOTE: The PROCEDURE REG used 9.00 seconds.

------------------- SSCP Contents --------------------------

CONTENTS PROCEDURE

  Dataset Name:   WORK.CROSPROD        Type:        SSCP
  Observations:   6                    Record Len:  60
  Variables:      7
  Label:          Sum of Squares and Crossproducts

-----Alphabetic List of Variables and Attributes-----

  #   Variable   Type   Len   Pos   Format    Label
  4   CRIMERT    Num     8    28    COMMA9.
  3   INTERCEP   Num     8    20
  6   PCTCOLL    Num     8    44    7.3
  5   PCTURB     Num     8    36    7.3
  7   POPSQMI    Num     8    52    COMMA9.2
  2   _NAME_     Char    8    12
  1   _TYPE_     Char    8     4
```

EXHIBIT 23.17 *(continued)*

```
------------------- SSCP listing --------------------

OBS   _TYPE_    _NAME_    INTERCEP    CRIMERT    PCTURB    PCTCOLL    POPSQMI

 1    SSCP    INTERCEP      51.00    275,874    3447.46    831.939    17,899.30
 2    SSCP    CRIMERT    275874.00   1.61341E9    1.96E7    4665845    153216016
 3    SSCP    PCTURB       3447.46   19596634    244283     57684.3    1609685.1
 4    SSCP    PCTCOLL       831.94   4,665,845   57684.3    14127.0    410878.19
 5    SSCP    POPSQMI     17899.30   153216016   1609685    410878    107239839
 6    N                      51.00         51    51.000     51.000        51.00

------------------- SSCP used as input to REG -------
```

Model: MODEL1
Dependent Variable: CRIMERT

Analysis of Variance

Source	DF	Sum of Squares	Mean Square	F Value	Prob>F
Model	1	79975244.984	79975244.984	95.239	0.0001
Error	49	41147001.604	839734.72661		
C Total	50	121122246.59			

Root MSE	916.37041	R-square	0.6603
Dep Mean	5409.29412	Adj R-sq	0.6534
C.V.	16.94067		

Parameter Estimates

Variable	DF	Parameter Estimate	Standard Error	T for H0: Parameter=0	Prob > |T|
INTERCEP	1	-291.367979	598.06978866	-0.487	0.6283
PCTURB	1	84.332833	8.64151635	9.759	0.0001

EXHIBIT 23.18   OUTEST dataset containing model information created

```
------------------- SAS Log ----------------------------

    1      libname x 'c:\book\data\sas';
    2
    3      options ls=120 ps=60 nocenter;
    4
    5      proc reg data=x.states outest=estimate noprint;
    6      model crimert = pcturb pctcoll popsqmi;
    7      model crimert = pcturb;
    8      model crimert = pctcol;
    9      model crimert = popsqmi;
   10      quit;
NOTE: 51 observations read.
      51 observations used in computations.
NOTE: The data set WORK.X has 4 observations and 9 variables.
NOTE: The PROCEDURE REG used 11.00 seconds.
   11
   12      proc contents data=estimate;
   13      run;
NOTE: The PROCEDURE CONTENTS used 4.00 seconds.
   14
   15      proc print data=estimate;
   16      run;
NOTE: The PROCEDURE PRINT used 5.00 seconds.

------------------- ESTIMATE Contents ----------------------

CONTENTS PROCEDURE

  Data Set Name:  WORK.X              Type:        EST
  Observations:   4                   Record Len:  76
  Variables:      9
  Label:          Parameter Estimates and Statistics


-----Alphabetic List of Variables and Attributes-----

  #   Variable  Type  Len  Pos  Format    Label
  9   CRIMERT   Num    8   68   COMMA9.
  5   INTERCEP  Num    8   36             Intercept
  7   PCTCOLL   Num    8   52   7.3
  6   PCTURB    Num    8   44   7.3
  8   POPSQMI   Num    8   60   COMMA9.2
  3   _DEPVAR_  Char   8   20             Dependent variable
  1   _MODEL_   Char   8    4             Label of model
  4   _RMSE_    Num    8   28             Root mean squared error
  2   _TYPE_    Char   8   12             Type of statistics
```

```
------------------- ESTIMATE listing --------------------
```

OBS	_MODEL_	_TYPE_	_DEPVAR_	_RMSE_	INTERCEP	PCTURB	PCTCOLL	POPSQMI	CRIMERT
1	MODEL1	PARMS	CRIMERT	839.14	-507.64	66.772	81.772	0.20	-1
2	MODEL2	PARMS	CRIMERT	916.37	-291.37	84.333	.	.	-1
3	MODEL3	PARMS	CRIMERT	1210.23	548.84	.	297.958	.	-1
4	MODEL4	PARMS	CRIMERT	1352.41	5213.25	.	.	0.56	-1

EXHIBIT 23.19   Input to REG is CORR output dataset

```
------------------- SAS Log ---------------------------

    1     libname x 'c:\book\data\sas';
    2
    3     options ls=120 ps=60 nocenter;
    4
    5     proc corr data=x.states out=corrmtrx noprint pearson;
    6     var crimert pcturb pctcoll;
    7     run;
NOTE: The data set WORK.CORRMTRX has 6 observations and 5 variables.
NOTE: The PROCEDURE CORR used 7.00 seconds.
    8
    9     proc reg data=corrmtrx simple;
   10     model crimert = pcturb pctcoll / stb p r  cli;
   11     id name;
ERROR: Variable NAME not found.
NOTE: The previous statement has been deleted.
   12     run;
NOTE: No raw data are available. Some options are ignored.
NOTE: The PROCEDURE REG used 11.00 seconds.

------------------- Output listing ---------------------
```

Descriptive Statistics

Variables	Sum	Mean	Uncorrected SS	Variance	Std Deviation
INTERCEP	51	1	51	0	0
PCTURB	3447.4564322	67.597184946	244283.42103	224.90141975	14.996713632
PCTCOLL	831.938933	16.312528098	14126.951261	11.118480814	3.3344386055
CRIMERT	275874	5409.2941176	1613405852	2422444.9318	1556.4205511

- - - - - - - - - - - New page of output - - - - - - - - -

SAS

Model: MODEL1
Dependent Variable: CRIMERT

Analysis of Variance

Source	DF	Sum of Squares	Mean Square	F Value	Prob>F
Model	2	85111147.389	42555573.694	56.723	0.0001
Error	48	36011099.199	750231.23332		
C Total	50	121122246.59			

Root MSE	866.15890	R-square	0.7027
Dep Mean	5409.29412	Adj R-sq	0.6903
C.V.	16.01242		

Parameter Estimates

Variable	DF	Parameter Estimate	Standard Error	T for H0: Parameter=0	Prob > |T|	Standardized Estimate
INTERCEP	1	-1188.574891	661.17422659	-1.798	0.0785	0.00000000
PCTURB	1	69.158407	10.01760164	6.904	0.0001	0.66636798
PCTCOLL	1	117.882121	45.05439171	2.616	0.0118	0.25254787

Summary

Linear regression is one of the most popular forms of multivariable analysis. Chapter 23 presents an overview of the topic, describes some of the model's complications normally encountered in real-world applications, and discusses REG, the procedure of choice for all but the most complex models. Regression enables the researcher to predict a variable's value based on the values of one or more other variables. The strength of the relationship may be measured by the Pearson r and the coefficient of determination. Significance tests may indicate the departure zero relationship of the model as a whole as well as the individual predictor, or independent, variables.

The chapter describes some of the complications to the model typically encountered by the researcher. Outliers are observations in the dataset with an unusually large discrepancy between actual and predicted values. Specification errors are committed when not all relevant predictor variables are included in the model. Heteroscedasticity exists when the variances of the dependent variable around the line drawn by the regression equation are not equal. Autocorrelation is present in the model when observations are serially ordered, that is, when an observation at time t influences values of an observation at time $t + 1$. Finally, multicollinearity exists when independent variables in the model are themselves intercorrellated.

The PLOT procedure can be used to diagnose some of the model complications. This chapter describes the basic syntax of the REG procedure. REG produces a wide variety of diagnostic and significance testing measures, some of them with accompanying graphics. It may be used interactively, and produces output datasets useful when processing efficiency is a concern.

24 Analysis of Variance

Analysis of variance (ANOVA) techniques allow testing of hypotheses about the equality of means of a continuous dependent variable for different subpopulations of one or more independent, categorical variables. The topic's complexity is compounded by SAS's unique approach to calculating results. This chapter first presents some background ANOVA material, emphasizing how SAS's approach affects the presentation of procedure output. The rest of the chapter describes ANOVA and GLM, the two SAS procedures used for most analysis of variance tasks. Syntax descriptions and examples are presented for both PROCs. The analysis of variance is a very complex topic. The chapter covers only the basics of ANOVA and is not intended to be a substitute for more thorough and in-depth treatments of the topic found elsewhere.

24.1 One-Way ANOVA

Analysis of variance resembles linear regression in its attempt to predict the value of a continuous, dependent variable given the value of one or more categorical independent variables. One-way ANOVA techniques attempt to estimate the mean of the dependent variable for each level of the independent variable using a prediction equation that looks similar to regression models. Common ANOVA independent variables include region of the country, race, and treatment versus control group for experimental studies.

SAS's one-way ANOVA model scores the categorical (independent) variable by creating two or more indicator variables, each scored as a 0 or 1, for each level of the independent variable. These variables are used internally by the procedure and are not added to the SAS dataset. To illustrate this scoring process, consider the variable REGION, which has four levels. The ANOVA and GLM procedures create the indicator variables shown in Exhibit 24.1. Each observation in the analysis dataset has a score for all four indicator variables. The pattern of the 0's and 1's in these variables shows which category of REGION the observation belongs to: a 1 indicates that the observation is in a particular category, and a 0 indicates that it is not. A model for predicting per capita income (y) for region of the country would closely resemble the regression model discussed in Chapter 23:

$$y_i = a + b_1 x_{1i} + b_2 x_{2i} + b_3 x_{3i} + b_4 x_{4i} + e_i$$

where the x's are the indicator variables for each region and e_i is the error term.

EXHIBIT 24.1   Creation of indicator variables

Level of REGION	Region Indicator Variable			
	North	South	Midwest	West
North	1	0	0	0
South	0	1	0	0
Midwest	0	0	1	0
West	0	0	0	1

Although the one-way ANOVA model is similar to the regression model, it presents a unique computational problem. There are only four REGION categories but there are five parameters being estimated in the model: $a, b_1, b_2, b_3, b_4, b_5$. Thus the model is *overspecified*.

Most statistics texts on ANOVA solve the overspecification problem by placing a restriction on the b's. One of the possible restrictions is to set one of the b's, usually the last, to 0. In practice this is often done by not including one of the indicator variables in the regression equation to be estimated. When this is done and least squares estimates are obtained for the remaining parameters, the estimate of "a" will be the mean of the omitted category and the estimate of b will be the differences between the mean of the omitted category and the mean of the category represented by the indicator variable for that b.

SAS's ANOVA and GLM procedures produce b's that, for the most straightforward situations, are identical to those produced by omitting the indicator variable for the highest-valued category. However, for reasons beyond the scope of this chapter, SAS does this by solving for the least squares estimates of the b's using a computational method known as a **generalized inverse**. It is for this reason that, when asked to print the b's for an ANOVA model with the SOLUTION option on the MODEL statement, SAS always issues the message, "The matrix has been found to be singular." This outcome is almost always a result of how SAS solves the model, not a problem with the data.

POINTERS AND PITFALLS

SAS's ANOVA calculations and terminology sometimes diverge from those used by other statistical packages. The resulting numbers, however, are consistent with those produced by the other packages' methods.

There are two types of tests for differences between specific category means in ANOVA: planned and unplanned. These tests are also referred to as planned contrasts and post hoc comparisons. In **planned tests** the researcher states the hypotheses before running the analysis program about differences in group means. The hypotheses are expressed as *contrasts* among the b's for the indicator variable categories. The values of a hypothesis' contrasts sum to zero, suggesting that there were, in fact, no differences between the means they represent. Consider the planned hypotheses and their associated contrasts described in Exhibit 24.2.

EXHIBIT 24.2   Contrast examples

Hypothesis	Contrast Values			
	North	South	Midwest	West
North differs from South	1	-1	0	0
North and Midwest differ from South and West	1	-1	1	-1
North differs from the average of South, Midwest, and West	1	-.33	-.33	-.33

It is important to remember that although many planned contrasts may be specified to SAS, there are only $C - 1$ mathematically independent contrasts for a single variable with C categories. Planned contrasts can be performed only with the CONTRAST statement in the GLM procedure (described later in this chapter).

In **unplanned tests** the researcher does not have a set of planned comparisons of means but may still wish to know which categories differ from each other. Simply performing a *t*-test for each pair of means is *not* an appropriate strategy, since the probability of a Type I error increases dramatically as the number of comparisons tested increases—if you make enough comparisons you are almost certain to find significant differences even if the differences are due just to chance.

24.2   *n*-Way ANOVA

Analyses of variance become more difficult when there is more than one independent variable. Such analyses are referred to by the number of dependent variables used, hence the terms two-, three-, or *n*-way ANOVA. One complication of the additional predictor variables is assessing their impact on the prediction of the dependent variable after the effect of the other independent variables has been controlled (so-called **main effects**). In addition to having main effects, independent variables may also have a synergistic effect known as an interaction.

An **interaction** occurs when the relationship between one of the independent-dependent variable relationships is modified by another independent variable. For example, the relationship between level of education and income may be modified by a person's gender. The difference in average income between high school graduates and college graduates may vary by gender—the difference may be greater for males than for females. The explanation of interactions becomes more complicated as the number of independent variables involved increases. However, if significant interaction effects are found, the variables involved in those interactions are generally not considered to leave any interpretable main effects.

The tests for the significance of main effects and interactions in n-way ANOVAs are based on the partitioning of the total variation in the dependent variable. This variance is divided into main, interaction, and residual (or error) components. For the two-way ANOVA case with independent variables X_1 and X_2:

$$SS\,(\text{total}) = SS\,(\text{Model}) \qquad\qquad + SS\,(\text{residual})$$
$$SS\,(\text{total}) = SS\,(x_1) + SS\,(x_2) + SS\,(x_1 * x_2) + SS\,(\text{residual})$$

Notice that the explained, or model, sum of squared deviations (SS) is partitioned into three components: main effects x_1 and x_2 and the interaction effect ($x_1 * x_2$). Each of these effects has an F statistic that tests its statistical significance.

An important concept in n-way ANOVA is that of balanced designs. When all of the cell means for the cross-classification of the independent variables, or factors, are based on equal numbers of cases, the analysis is called a **balanced** ANOVA. The calculation of the SS for the model components (main effects and interaction) is straightforward. Balanced ANOVAs are usually performed on preplanned experiments in which equal numbers of units of analysis are randomly assigned to treatment groups. *SAS's PROC ANOVA calculates valid SS and tests only for balanced designs.*

For **unbalanced** designs, where the number of cases for each cell mean differs, the SS are not unique, several tests for an effect are possible. Two general types of tests are available within the GLM procedure. The first is a sequential test, in which any effect is tested for significance, controlling only for the effects *preceding* it in the model. GLM refers to these SS and associated tests as TYPE I sums of squares. The second test for unbalanced designs tests for the significance of an effect, controlling for all other effects in the model. GLM labels these as TYPE III sums of squares, also referred to as regression sums of squares. The choice of appropriate test and thus the type of sum of squares used depends on the substantive question the investigator has in mind.

When an independent variable has a significant effect, this implies that one or more *pairs of means* for that effect differ. Just as one-way ANOVA's overall test of significance for an effect does not indicate which means differ, n-way ANOVA must use either preplanned contrasts or post hoc tests to determine which means differ from one another. Post hoc comparisons may be obtained for any effect using the MEANS statement.

Specifying preplanned contrasts for main effects follows logic similar to that of one-way ANOVA: the scoring of any contrast simply compares the means for the subcategories of the variable of interest, disregarding the categories of the other variables in the model. Also in keeping with the one-way ANOVA logic, the contrasts must be specified according to how SAS sorts the levels of the categorical variable: the score for the lowest level of the variable must be specified first, followed by the next largest, and so on. The first page of GLM output usually supplies univariate, diagnostic statistics that make this sort order explicit, should there be any question. Specifying contrasts for interaction effects is more complicated and depends on how SAS sorts the levels of the variables in the interaction.

24.3 MANOVA

It is not unusual in experimental work or social science research to be concerned with several dependent, or outcome, variables in an investigation. ANOVA techniques may be applied to these situations. These techniques are commonly referred to as **multivariate analyses of variance**, or MANOVA. MANOVA tests for the effects of a particular independent variable by providing an overall test of significance for all of the dependent variables. This test accounts for the fact that two or more dependent variables are being examined simultaneously. An obvious alternative, simply performing separate ANOVAs for each dependent variable, does not consider the multivariate nature of the analysis and is not usually an appropriate strategy.

The MANOVA statement in the ANOVA and GLM procedures requests multivariate tests. The general null hypothesis for these tests states that none of the subcategory means differ for any of the dependent variables. If the multivariate test for an effect is significant, the ANOVAs for the individual variables may then be examined to determine which of the individual dependent variables the effect applies to. SAS prints these individual, univariate ANOVAs *prior to* the multivariate tests: so remember to check the MANOVA test results before examining the ANOVAs.

Both the ANOVA and GLM procedures produce several varieties of multivariate tests for a particular effect. In most situations these tests have similar significance levels. However, occasionally the tests may disagree. The reasons for this disagreement are beyond the scope of this brief overview. The best solution is to seek the advice of a statistician skilled in multivariate analysis!

24.4 Repeated Measures

A special case of MANOVA occurs when the multiple dependent variables are the result of using the same measure taken at multiple time periods on the same cases. Such *within group* analyses occur in both designed experiments and observational studies. When dependent variables are measures across time periods, they are analyzed using a *repeated measures* ANOVA that can test for the difference in mean level of the dependent variable across time.

Both the ANOVA and GLM procedures use the REPEATED statement for such analyses. REPEATED prints multivariate tests for mean differences over time as well as whether the time effect differs for any groups of cases used, and differences among categories of any independent variables as well as whether time differences vary by categories of the independent variables. The latter tests are labeled in the output as TIME by Independent Variable interactions.

It is possible to have very complicated repeated measures designs. One of the more common complications arises when the dependent variables are not measured only across time but also under different experimental conditions at each time period.

24.5 Cautions and Caveats

It would be wise to reiterate a point made at the beginning of this chapter: the material covered here only sketches the most significant aspects of ANOVA and is by no means a substitute for a thorough understanding of the statistics involved. It is also important to identify some of the assumptions on which the ANOVA model depends.

The basic assumptions of the general linear model for ANOVA are similar to those for linear regression. The assumptions of **homoscedasticity** and **normality of errors** are particularly important (refer to Chapter 23 for a discussion of these concepts). ANOVA techniques are relatively robust when the distribution of errors is not perfectly normal. Homoscedasticity implies that the population variances for each of the subcategories are equal. Lack of homoscedasticity can have a deleterious impact on inferences about differences between category means, especially in unbalanced designs.

In some situations if the data seriously violate either or both of these assumptions, PROC RANK can be used to convert the dependent variable to ranks and then the NPAR1WAY or ANOVA procedures can be used. These PROCs are described in SAS Institute publications (the "SAS/Stat" guides).

Another assumption critical to the foregoing discussion is that the categorical variables have a *fixed number of categories* and that occurrences of each category are found in the data. These are referred to as **fixed effect** variables. It is also possible to have **random effect** variables, in which the number and type of categories found in the data are a random sample of those occurring in the population of interest. Partitioning sums of squares and significance testing for random effect models differs from partitioning sums of squares automatically provided by the GLM and ANOVA procedures. Significance testing is also beyond the scope of this chapter.

24.6   SAS Procedures for ANOVA

SAS offers a near-bewildering variety of PROCs and options to perform analysis of variance. This section describes part of the ANOVA and GLM procedures. Both may be used interactively (see the "Using the PROCs Interactively" section later in the chapter for details). They also have many statements in common, so you can easily switch between them.

How do you choose which of the two procedures to use? ANOVA executes faster and requires less memory than GLM, in part because it does not have as many options. It produces valid results only for balanced designs (equal cell sizes). For analyses requiring preplanned contrasts and output of model results, or having unbalanced designs, GLM is the only practical choice.

POINTERS AND PITFALLS

The ANOVA procedure runs faster than GLM but has fewer options and may reliably analyze only balanced designs. Both procedures may use correlation datasets as input (see Chapter 22 for CORR details).

ANOVA

If the number and complexity of statements and options described here seem daunting, remember that most analyses need only the PROC, MODEL, and CLASS statements. Refer to the "Statement Order" section later in this chapter for guidelines on grouping statements.

```
PROC ANOVA [DATA=dataset] [MANOVA] [OUTSTAT=outdset];
MODEL dep ... [= indep ...] [/ [INT] [NOUNI]] ;
CLASS classvar ... ;
[MANOVA [M=xform_matrix] [E=err_effect]
       [/ [PRINTE] [SUMMARY]]; ]
[MEANS var ... [/ [BON] [DUNCAN] [SCHEFFE] [T]
       [TUKEY] [WALLER] [CLM] [ALPHA=prob] [CLDIFF]
       [E=effect] [NOSORT]]; ]
[REPEATED factor nlevels (values)
       [CONTRAST[(reflevel)]|PROFILE] ...
       [/ [NOM NOU SUMMARY]; ]
[RUN;]
[QUIT;]
```

Some common applications of PROC ANOVA follow:

```
proc anova data=measures;
class grp race;
model rating = grp race;

class rating = grp race grp*race;
means grp race / bon duncan;
```

The ANOVA PROC statement has three features:

1. The *dataset* is the name of the dataset to be analyzed. If the *DATA* option is not entered, the most recently created dataset is used.

2. *MANOVA* requests that ANOVA eliminate observations if *any* of the dependent variables in a multivariate analysis have missing values. This option is available only in version 6 of the SAS System.

3. *OUTSTAT = outdset* identifies an output dataset containing statistics for each effect in the model. Among these are sums of squares, *F* statistics, and probability levels. See Exhibit 24.8 for an example of its use. This option is available only in version 6 of the SAS System.

A single *MODEL* statement is required. *MODEL* has three elements:

1. The *dep* specifies one or more dependent variables; *indep* specifies the independent variables or effects. If *indep* is not entered, ANOVA tests the hypothesis that the mean of *dep* is zero (i.e., it only fits the intercept). Some valid model specifications and their interpretation are listed in Exhibit 24.3.

2. *INT* forces the intercept to be treated as an effect in the model. By default, ANOVA includes the intercept in the model but does not use it in hypothesis testing.

3. *NOUNI* suppresses the univariate statistics normally produced by ANOVA.

The *CLASS* statement is required. In *CLASS*, *classvar* specifies one or more variables to be used as classification variables. These variables identify the subgroups for the analysis. They can be either numeric or character variables, and may be grouped by using user-written formats (see Chapter 15 for recoding with user-written formats). Character variables may use only the leftmost 16 positions to determine groups.

The *MANOVA* statement has four components:

1. The *xform_matrix* indicates a transformation matrix for the *dep* variables listed in the *MODEL* statement. Using the M option requests an analysis that may differ from that specified in the *MODEL* statement. The matrix may be specified as several equations or as a list of numbers. If numbers are

EXHIBIT 24.3 **Interpretation of MODEL statements**

```
class x y;
model w = x y;
```

Two-factor main effects model. Factors are x and y, and the dependent variable is w class x y;
```
model w = x y x*y;
```

Add a crossed factor, or interaction effect, to the model specified above.

```
class x y z;
model w = x y x*y x*z z*y x*y*z;
```

"Full factorial" model, with all possible interactions.

```
class x y z;
model w = x|y|z;
```

"Full factorial" model, using "bar notation". Bars signify all possible interactions.

```
class x y z ;
model w = x (y) z;
```

```
class x y z;
model w = x y z(x y);
```

"Nested" model, where the levels of z were not the same for variables x and y. The z is said to be nested within x and y. The nested effect z(x y) is not treated as a main effect.

```
class x y x;
model w = x y(x) y*x;
```

Nesting and crossed factors may be combined in a single model.

---

specified, there must be as many columns as there are dependent variables. Refer to Exhibit 24.4 for examples and their interpretation.

2. The *err_effect* specifies an error effect term. By default, the sum of squared cross-products (SSCP) matrix is used.

3. *PRINTE* requests printing of the error matrix. If the E option was not specified, partial correlations of the dependent variables given the independent variables are also printed.

4. *SUMMARY* prints analysis of variance tables for each dependent variable. The content of the table is affected by the specification, if any, of the M option. If specified, the table reflects the transformation of the dependent variable. If M was not specified, the original dependent variables are analyzed.

The *MEANS* statement computes means and comparison tests for one or more effects specified in the most recent *MODEL* statement. The *MEANS* statement has 12 elements:

EXHIBIT 24.4   MANOVA transformation matrices

```
class x y;
model s1-s4 = x y;
manova m = s1-s2, s2-s3, s3-s4;

/* M matrix transforms the dependent variables into pairs of differences */

manova m = (1 -1  0  0,
            0  1 -1  0,
            0  0  1 -1);

/* The matrix can be specified directly. Note how spacing the matrix across
three lines contributes to the statement's purpose and readability. */

manova m = (2 -1 -1 0,
            0  2 -2 0,
            0  0 -1 1)

/* The sum of the coefficients in each row (each group of
     numbers delimited by a comma) must equal zero. */
```

1. The *var* specifies one or more effects in the previous *MODEL* statement's model.

2. *BON* requests Bonferroni *t*-tests. These test the differences in means for all main effect means included in *var*. Also see the *CLM* and *CLDIFF* options, discussed in #8 and #9.

3. *DUNCAN* requests Duncan's multiple-range test for all main effect means included in *var*.

4. *SCHEFFE* requests Scheffe's multiple-comparison test on all main effect means included in *var*. Also see the *CLM* and *CLDIFF* options.

5. *T* requests pairwise *t*-tests for all main effect means included in *var*. This test is identical to Fisher's least-significant difference test for the case of equal cell sizes. Also see the *CLM* and *CLDIFF* options.

6. *TUKEY* requests Tukey's Studentized range test on all main effect means included in *var*. Also see the *CLM* and *CLDIFF* options.

7. *WALLER* requests Waller–Duncan *k*-ratio *t*-tests for all main effect means included in *var*.

8. *CLM* presents output from the *BON, SCHEFFE,* and *T* options as confidence intervals for the mean of each level of *var*. Also see the *NOSORT* option in #12.

9. The *prob* specifies a significance level for the comparisons. The default value of *ALPHA* is .05. *ALPHA* may be overridden with any value between .0001 and .9999. When the *DUNCAN* option is specified, *prob* may only take values of .01, .05, or .10.

10. *CLDIFF* presents output from the *BON, SCHEFFE, T,* and *TUKEY* options as confidence intervals for pairwise differences between the means of *var*. Also see the *NOSORT* option.

11. The *effect* identifies the error mean square to use in the tests. The residual mean square is used by default. The *effect* must be a term in the most recent *MODEL* statement.

12. *NOSORT* avoids the default sorting when *CLM* or *CLDIFF* are specified. If it is not specified, the largest means are displayed first.

The *REPEATED* statement enables testing of hypotheses in which the dependent variables represent repeated measurements on the same experimental unit. The statement allows testing of both within-subject (dependent variables) and between-subject (dependent-independent variable) factors. The *REPEATED* statement has six features:

1. The *factor* names a factor associated with the dependent variable. The *factor* should not already be in the dataset being analyzed.

2. The *nlevels* specifies the number of levels associated with *factor*. For a single within-subject factor, *nlevels* equals the number of dependent variables. In all cases, single and multiple within-subject factors, the product of all values of *nlevels* should equal the number of dependent variables. The order in which the *factors* are specified is important! See the last examples in Exhibit 24.5 for details.

3. The *values* assigns values corresponding to levels of a *factor*. The number of *values* specified should equal *nlevels* for that factor.

4. *CONTRAST* generates contrasts between levels of the factor. Optionally, an ordinal value, *reflevel*, may be specified as a reference level (thus a value of 2 would use the second level of the factor rather than the number 2). If *reflevel* is omitted, the last, or highest, value of *factor* is used as a reference level.

5. *PROFILE* generates contrasts between successive, adjacent levels of *factor*.

6. *NOM* requests printing of only univariate analyses; *NOU* prints only multivariate analyses; and *SUMMARY* prints analysis of variance tables for each within-subject factor. *SUMMARY* also produces a test of the contrast's mean being equal to zero.

Exhibit 24.5 contains some valid REPEATED statements and their interpretation.

Statements common to all PROCs that can be used with ANOVA include BY, FREQ, TITLE, FOOTNOTE, FORMAT, LABEL, and ATTRIB. See Chapter 5 for a complete description of these features.

GLM

The preliminary note in the ANOVA description is also appropriate here: do not be put off by the complexity and number of statements and options. Many analyses require simple uses of only the PROC, MODEL, and CLASS statements. Refer to the "Statement Order" section later in this chapter for guidelines on grouping statements.

```
PROC GLM [DATA=dset] [OUTSTAT=outdset]
      [ORDER=FREQ|DATA|FORMATTED|INTERNAL]
      [MANOVA] [NOPRINT];
CLASS classvar ... ;
MODEL dep ... = indep ... [/ [[NO]INT] [NOUNI]] ];
[CONTRAST 'contrast_label' effect values [, ...]
        [/ [E=contr_err]]; ]
[LSMEANS effect ... [/ [COV] [NOPRINT]
      [OUT=lsdset]; ]
[MANOVA [M=xform_matrix] [E=err_effect]
        [/ [PRINTE] [SUMMARY]]; ]
```

EXHIBIT 24.5   REPEATED statements and their interpretation

Four measurements, equally spaced (equal spacing is implied by 0, 1, 2, and 3 being equally separated). By default, the last (fourth) level of the factor is used as a reference level.

```
class x y;
model t0-t3 = x y;
repeated times;
```

The default values are explicitly entered.

```
class x y;
model t0-t3 = x y;
repeated times 4 (0 1 2 3) contrast(4);
```

Unevenly spaced levels. Specify them in parentheses. Notice that *nlevels* does not need to be specified since there is only one factor: the procedure assumes there are as many levels as there are dependent variables.

```
class x y;
model t0 t2 t5 t10 = x y;
repeated times (0 2 5 10);
```

Two factors, whose 10 (5 TIMES * 2 TREAT) levels equals the number of dependent variables. Notice the comma separating the information for the individual factors. Also note that *nlevels* (5 and 2) *must* be specified. The order and number of levels of the factors implies the following relationship between factor levels and the dependent variables:

Dep. Var.	TIMES	TREAT
m1	1	1
m2	1	2
m3	2	1
m4	2	2
m5	3	1
m6	3	2
m7	4	1
m8	4	2
m9	5	1
m10	5	2

```
class x y;
model m1-m10 = x y;
repeated times 5 (0 2 5 10),
         treat 2 ;
```

```
[MEANS var ... [/ [BON] [DUNCAN] [SCHEFFE] [T]
        [TUKEY] [WALLER] [CLM] [ALPHA=prob] [CLDIFF]
        [E=effect] [NOSORT]]; ]
[REPEATED factor nlevels (values)
        [CONTRAST[(reflevel)]|[PROFILE] ...
        [/ [NOM NOU SUMMARY]; ]
[RUN;] [QUIT;]
```

The GLM PROC statement has five elements:

1. The *dset* is the name of the dataset to be analyzed. If the *DATA* option is not entered, the most recently created dataset is used.

2. OUTSTAT= *outdset* identifies an output dataset containing statistics for each effect in the model and, if used, each user-specified contrast. Among the items included are sums of squares, *F* statistics, and probability levels. See Exhibit 24.16 for an example of its use. This option is available only in version 6 of the SAS System.

3. *ORDER* specifies the sort order of the classification variable(s). The option indicates how the levels of the model's parameters will be assigned. Thus the only time *ORDER* usually needs to be stated explicitly is when user-supplied contrasts (*CONTRAST* statement) are also entered. *FREQ* sorts in descending category frequency, *DATA* in the order in which levels were found in the dataset, *FORMATTED* in the external, formatted value, and *INTERNAL* in the internal, unformatted value. *FORMATTED* is the default.

4. *MANOVA* requests that GLM eliminate observations if *any* of the dependent variables in a multivariate analysis have missing values. This option is available only in version 6 of the SAS System.

5. *NOPRINT* suppresses the creation of the usual output, or listing, files. This option is usually taken when GLM is being run solely to create output datasets with the PROC's *OUTSTAT* option or the *OUT* statement.

The *CLASS* statement is required. It identifies variables to be used as classification variables in the analysis. It is identical in form and function to the ANOVA procedure's CLASS statement.

The *MODEL* statement identifies the dependent variables, main effects, and, possibly, interactions. There may be only one *MODEL* statement. It is identical in form and function to the ANOVA procedure's *MODEL* statement.

The *CONTRAST* statement enables you to construct hypotheses (planned contrasts) about the relationships between effect levels. The default GLM output tests the model's overall predictive power; *CONTRAST* lets you ask specific questions about differences of means. The regional differences discussed in Section 24.1 are typical of the kinds of questions addressed by contrasts. Multiple *CONTRAST* statements may be entered. They must appear *after* the *MODEL* statement. The *CONTRAST* statement has four components:

1. The *contrast_label* is a label used to identify contrast-related statistics on the printout. It cannot exceed 20 characters and must be enclosed in single quotes (').

2. The *effect* is an effect specified in the *MODEL* statement. The special effect *INTERCEPT* may be used if the intercept was fitted as part of the model (INT option in the *MODEL* statement).

3. The *values* are coefficients associated with each level of *effect*. If there are fewer *values* than there are levels of *effect*, the remaining coefficients are set to zero. If there are more *values* than there are *effects*, the extra entries are ignored and GLM prints a warning message. The *values* are applied against *effect* levels in the level order specified in the *ORDER* option in the PROC statement. Exhibit 24.6 illustrates some *CONTRAST* statements and their interpretation.

EXHIBIT 24.6   CONTRAST statements and their interpretation

Create contrasts necessary to test the planned hypotheses in Section 24.1.
Effect REGION has four levels. Notice that the first contrast could have
been written simply as 1 -1, since GLM fills short lists with 0's.

```
model pc_inc = region ;
contrast 'Regional differences'
        region 1 -1 0 0,
        region 1 -1 1 -1,
        region 1 -.33 -.33 -.33  ;
```

REGION is mistakenly assumed to have had five levels. The third contrast
(level 1 differs from the average of the other levels) is respecified. GLM
prints a warning that the number of levels of the effect REGION (4) is less
than the number of coefficients (5) specified in the CONTRAST matrix.

```
model pc_inc = region ;
contrast 'Regional differences'
        region 1 -1 0 0,
        region 1 -1 1 -1,
        region 1 -.25 -.25 -.25 -.25 ;
```

Test the linearity of LAB, which has five levels.

```
model rate = lab;
contrast 'LAB linear?' lab -2 -1 0 1 2 ;
```

Test hypotheses in a two-way ANOVA. LAB has five levels, METHOD two.

```
model rate = lab method;
contrast 'LAB-METHOD tests'
        lab -2 -1 0 1 2 method -1 1 ;
```

Include intercept as a model effect in the preceding two-way ANOVA.

```
model rate = lab method / int ;
contrast 'LAB-METH. w. intercept'
        intercept 1 lab -2 -1 0 1 2  method -1 1 ;
```

---

4. E=*contr_err* identifies an effect in the model that will be used as an error
   term. The default is the mean square (*MS*) error.

   The *LSMEANS* statement computes least squares estimates of class and sub-
class marginal means in unbalanced designs. These are the means of an effect in
the model that would be expected if the design were, in fact, balanced. *LSMEANS*
has three elements:

1. The *effect* identifies an effect (main or interaction) specified in the *MODEL*
   statement.

2. The *lsdset* identifies an output dataset containing the least square means and
   related statistics. Covariances are included if the *COV* option is specified.

3. *NOPRINT* suppresses creation of the output listing. It is usually used only
   when *OUT* is used to create an output dataset.

The *MANOVA* statement requests multivariate analyses when *MODEL* statements have more than one dependent variable. It is identical in form and function to the ANOVA procedure's MANOVA statement.

The *MEANS* statement requests calculation of the observed means and standard deviations for any continuous variables in the model. These statistics may be printed for each level or combination of levels of *var*, an effect in the model that does not use continuous variables. It is identical in form and function to the ANOVA procedure's MEANS statement. Statements common to all PROCs that can be used with GLM include BY, REQ, TITLE, FOOTNOTE, FORMAT, LABEL, and ATTRIB. See Chapter 5 for a complete description of these features.

The *REPEATED* statement allows specification of a repeated measures design and provides a variety of multivariate and univariate tests. It is identical in form and function to the ANOVA procedure's REPEATED statement.

Statements common to all PROCs that can be used with GLM include BY, FREQ, TITLE, FOOTNOTE, FORMAT, LABEL, and ATTRIB. See Chapter 5 for a complete description of these features.

Statement Order

Both ANOVA and GLM can handle multiple occurrences of some statements but are sensitive to their order and may have only one MODEL statement. The following chart illustrates the order in which each procedure expects to find its statements.

First statement after PROC
    CLASS
Next statement
    MODEL
Before first RUN statement
    BY, FREQ
Anywhere after MODEL
    TITLE, FOOTNOTE, MANOVA, MEANS, REPEATED
    (GLM only: CONTRAST, LSMEANS)

Using the PROCs Interactively

Like the DATASETS, PLOT, and REG procedures discussed in earlier chapters, ANOVA and GLM are interactive. A RUN statement begins execution of a group of procedure statements. Once the statements execute, the procedure attempts to process the next group of statements. This statement grouping enables you to change titles and footnotes between RUN groups. The QUIT statement functions identically to RUN except that once the statements complete execution, the procedure terminates.

Exercises

24.1. **a.** Variable TESTGRP has three levels: "RDU", "QDR", and "XYC". Describe how their values indicator variables will be set. Use Exhibit 24.1 as a model.

b. INCOME is the dependent variable in a model using TESTGRP. Rather than write the model as

$$\text{INCOME}_i = a + b\text{TESTGRP}_i + e_i$$

substitute the indicator variables for TESTGRP. (This is similar to the equation in the discusssion accompanying Exhibit 24.1.)

24.2. Refer to Exhibit 24.2.
   a. Expand on the comments accompanying the contrasts. Describe each hypothesis in one or two sentences.
   b. Write a new contrast and explain the hypothesis.

24.3. The preceding sections have described several variations of ANOVA techniques. Among these were one-way ANOVA, *n*-way ANOVA, MANOVA, and repeated measures designs. In the following procedures, identify the technique(s) being used.

   a.
```
proc anova data=marketing;
class testgrp;
model income = testgrp;
```
   b.
```
proc anova data=marketing;
class testgrp t_slice;
model income = testgrp t_slice;
```
   c.
```
proc anova data=marketing;
class testgrp;
model income1-income4 = testgrp;
repeated inc;
```
   d.
```
proc anova data=marketing;
class testgrp;
model con1-con10 = testgrp;
repeated inc 2, occ 5;
```
   e.
```
proc anova data=marketing;
class testgrp;
model income = testgrp;
means testgrp / bon tukey;
```
   f.
```
proc anova data=marketing;
class testgrp;
model income1-income4 = testgrp;
manova m=(-3 -1 -1 -1,
          1 -1 0 0);
```

24.4. MASS0 through MASS3 are the dependent variables in a model. Write MODEL statements to meet the following requirements:
   a. MASS0 is predicted by variable LAB.
   b. The independent variable STYLE is added to the model in part a.
   c. The interaction of STYLE and LAB is considered. Specify the independent variables two ways.
   d. MASS0 through MASS3 are predicted by STYLE and LAB.

24.5. The following statement is used in an analysis of variance procedure:

```
model t1-t5 = grp;
```

Which variable is used as a reference level? Write the statement required to specify the reference level explicitly.

24.6. The following statement is used in an analysis of variance procedure:

```
model l1-l10 = contract;
```

   a. The following repeated measures design is specified:

```
repeated lab 2, rep 5;
```

How do values of LAB and REP correspond to L1 to L10? Arrange your answer like the display in Exhibit 24.5.
   b. Repeat part a using a different design:

```
repeated rep 5, lab 2;
```

24.7. Write a CONTRAST statement for the new contrast specified in Exercise 24.4. Refer to the first example in Exhibit 24.6 for other REGION contrasts.

24.8. What, if anything, is wrong with the following statements? An ellipsis (...) indicates portions of the statements that are not germane to the exercise.

    a. `proc ... ;`
       `class ... ;`
       `run;`
       `model ... ;`
       `quit;`

    b. `proc ... ;`
       `class ... ;`
       `model ... ;`
       `by ... ;`
       `run;`

    c. `proc ... ;`
       `class ... ;`
       `model ... ;`
       `run;`

    d. `proc ... ;`
       `class ... ;`
       `model ... ;`
       `quit;`

    e. `proc ... ;`
       `class ... ;`
       `model ... ;`
       `contrast ... ;`
       `run;`

24.9. Dataset FIN has variables DIV (sales division: 3 levels), STYLE (type of material fabricated: 4 levels), and MARGIN (profit per full-time employee).

    a. Assume that either GLM or ANOVA will be an appropriate procedure for your needs. How would you determine whether the one- and two-factor models will be balanced?

    b. Write the statements needed for two-way analysis of variance.

    c. Add the term that assesses the interaction of the two factors to the model in part b.

    d. Add the following items to part c: request pairwise t-tests, Duncan's multiple-range test, and Bonferroni's t-tests for independent variable STYLE.

24.7   Examples of ANOVA and GLM

The examples in this section use the CMI dataset described in Appendix E. ANOVA examples are presented first, then those for GLM. Key features and options used in this section are summarized in the following list:

Exhibit	Features/Options
24.7	All defaults for statistics and presentation of output
24.8	Two-way model; OUTSTAT option used to create dataset of model information
24.9	Interaction term added to a two-way model
24.10	MEANS statement; computational problem reported in output listing but not in Log
24.11	ANOVA used with an unbalanced design
24.12	GLM output for one-way ANOVA; all defaults for statistics and presentation of output

ANOVA

Exhibit 24.7 is a simple one-way analysis of a balanced design (each category of CLASS variable AREA occurs equally, 25 observations each, in the dataset). All defaults for statistics and presentation of output are taken. Notice the "Class Level Information" heading at the beginning of the output: it indicates how many levels were used for effect variable AREA and lists their values (1, 2, 3, and 4).

Exhibit 24.8 (see p. 535) is a two-way model. Both independent variables (RACIAL and HOUSING) are dichotomous. During the analysis their levels are arranged in alphabetical order ("Class Level Information"). An output dataset containing model information is also created (OUTSTAT option). CONTENTS procedure output and a listing of the OUTSTAT dataset follow the analysis.

Exhibit 24.9 (see p. 536) is similar to Exhibit 24.8. An interaction term is added (HOUSING*RACIAL). Using the "bar notation" described in Exhibit 24.3, the MODEL statement's independent variables could have been written as HOUSING|RACIAL. The procedure would have created the appropriate main effects and all possible interactions.

Exhibit 24.10 (see p. 537) adds a MEANS statement to a model resembling that used in Exhibit 24.7. The procedure appeared to complete normally: the Log did not contain an ERROR or other warning message. The output file, however, reveals a problem computing the Waller–Duncan k-ratio test. This exhibit, like others earlier in the book, emphasizes the importance of reading both the Log and output files to get a true understanding of how the procedure ran.

The description of the ANOVA procedure emphasized its use for balanced designs, those with equal numbers of observations in each category of the independent variable. It is also possible to use the procedure with unbalanced designs, as Exhibit 24.11 (see p. 540) shows. The results, especially with small and/or very different cell sizes, may be suspect.

GLM

Exhibit 24.12 (see p. 541) uses GLM for a balanced, one-factor analysis. This exhibit uses the same model and dataset as Exhibit 24.7. The format of the output and the results of the analysis are virtually identical to that of the ANOVA procedure.

In Exhibit 24.13 (see p. 542) the simple model from Exhibit 24.12 is rerun, this time using a dataset yielding an unbalanced design. The dataset size is 82 observations; only 69 observations have nonmissing levels of independent variable AREA. Notice that other than the Note about the number of usable observations, there is no explicit mention of the design being unbalanced.

To add more descriptive impact to AREA, a user-written format (discussed in Chapters 14 and 15) is added in Exhibit 24.14 (see p. 544). Value '1' becomes 'Cherryview,' '2' becomes 'Morningdale,' and so on (a FORMAT statement is added). A CONTRAST statement is used to compare the levels of levels '1' and '4'. No ORDER option is specified in the PROC statement, so the default ordering of AREA's levels is used. The formatted (alphabetical) values of AREA determine the arrangement of the levels, as shown in the Class Level Information

EXHIBIT 24.7   All defaults for statistics and presentation of output

```
------------------- Program listing --------------------

proc anova data=cmidata.cmi ;
class area ;
model cmi = area ;
title '1-way, balanced' ;
quit;

------------------- Output listing --------------------

1-way, balanced

Analysis of Variance Procedure
```

Class Level Information

Class	Levels	Values
AREA	4	1 2 3 4

$\longleftarrow$ Values of CLASS variable AREA

```
Number of observations in data set = 100

- - - - - - - - - - - - New page of output - - - - - - - - -

1-way, balanced

Analysis of Variance Procedure

Dependent Variable: CMI   CMI (well-being) score
```

Source	DF	Sum of Squares	Mean Square	F Value	Pr > F
Model	3	894.51000000	298.17000000	1.18	0.3223
Error	96	24301.60000000	253.14166667		
Corrected Total	99	25196.11000000			

R-Square	C.V.	Root MSE	CMI Mean
0.035502	72.88331	15.91042635	21.83000000

Source	DF	Anova SS	Mean Square	F Value	Pr > F
AREA	3	894.51000000	298.17000000	1.18	0.3223

portion of the listing. The CONTRAST, then, compares 'Cherryview' (level 1) with 'Easton' (level 4).

Exhibit 24.15 (see p. 545) uses the same user-written format as Exhibit 24.14 but overrides the level arrangement (ORDER option). By specifying ORDER=INTERNAL, the original classification of the data is preserved but the descriptive neighborhood names added by format $AREA are included. The CONTRAST statement to compare 'Cherryview' and 'Easton' is rewritten to reflect the internal, original coding of AREA rather than its alphabetical, formatted order. The analysis results are identical. It is wise to check the Class Level Information to make sure that your assumptions about level ordering agree with those actually used by GLM!

EXHIBIT 24.8 Two-way model; OUTSTAT used to create model-information dataset

---

```
------------------- Program listing -------------------

proc anova data=cmidata.cmi outstat=varout ;
class racial housing ;
model cmi = housing racial ;
title '2-way, main effects' ;
run ;

------------------- ANOVA Output -------------------

2-way, main effects

Analysis of Variance Procedure
Class Level Information

Class    Levels   Values

RACIAL      2     H L

HOUSING     2     H L


Number of observations in data set = 100

- - - - - - - - - - - New page of output - - - - - - - - -

2-way, main effects

Analysis of Variance Procedure

Dependent Variable: CMI    CMI (well-being) score
```

Source	DF	Sum of Squares	Mean Square	F Value	Pr > F
Model	2	490.50000000	245.25000000	0.96	0.3854
Error	97	24705.61000000	254.69701031		
Corrected Total	99	25196.11000000			

R-Square	C.V.	Root MSE	CMI Mean
0.019467	73.10687	15.95922963	21.83000000

Source	DF	Anova SS	Mean Square	F Value	Pr > F
HOUSING	1	110.25000000	110.25000000	0.43	0.5121
RACIAL	1	380.25000000	380.25000000	1.49	0.2247

EXHIBIT 24.8   *(continued)*

```
------------------- VAROUT Contents ----------------------------

Contents of OUTSTAT dataset

CONTENTS PROCEDURE

   Data Set Name:  WORK.VAROUT          Type:
   Observations:   3                    Record Len: 72
   Variables:      7
   Label:

-----Alphabetic List of Variables and Attributes-----

   #  Variable  Type   Len  Pos  Label
   4  DF        Num     8    40
   6  F         Num     8    56
   7  PROB      Num     8    64
   5  SS        Num     8    48
   1  _NAME_    Char    8     4
   2  _SOURCE_  Char   20    12
   3  _TYPE_    Char    8    32


------------------- VAROUT Listing ----------------------------

OBS   _NAME_    _SOURCE_    _TYPE_    DF      SS         F        PROB


1     CMI       ERROR       ERROR     97   24705.61      .          .
2     CMI       HOUSING     ANOVA     1      110.25    0.43287   0.51214
3     CMI       RACIAL      ANOVA     1      380.25    1.49295   0.22472
```

Exhibit 24.16 (see p. 546) analyzes a two-factor unbalanced design and includes a term for the interaction of the factors in the model. An LSMEANS statement computes least squares estimates for class means of both main effects. These estimates and other information are saved in an output dataset (OUT option). CONTENTS procedure output and a listing of the dataset are also included.

EXHIBIT 24.9   **Interaction term added to a two-way model**

```
------------------- Program listing -------------------

proc anova data=cmidata.cmi ;
class racial housing ;
model cmi = housing racial housing*racial ;
title1 '2-way, with an interaction term' ;
quit ;
```

EXHIBIT 24.9 *(continued)*

```
-------------------- Output listing --------------------
```

2-way, with an interaction term

Analysis of Variance Procedure

Dependent Variable: CMI   CMI (well-being) score

Source	DF	Sum of Squares	Mean Square	Value	Pr > F
Model	3	894.51000000	98.17000000	1.18	0.3223
Error	96	24301.60000000	53.14166667		
Corrected Total	99	25196.11000000			

R-Square	C.V.	Root MSE	CMI Mean
0.035502	72.88331	15.91042635	1.83000000

Source	DF	Anova SS	Mean Square	F Value	Pr > F
HOUSING	1	110.25000000	110.25000000	0.44	0.5109
RACIAL	1	380.25000000	380.25000000	1.50	0.2233
RACIAL*HOUSING	1	404.01000000	404.01000000	1.60	0.2095

EXHIBIT 24.10   MEANS statement; computational problem reported in output but not in Log

```
-------------------- Program listing --------------------
```

```
proc anova data=cmidata.cmi ;
class area ;
model cmi = area / int nouni ;
means area / bon duncan scheffe waller tukey t ;
title 'Intercept in model, MEANS statement added';
quit ;
```

```
-------------------- Output listing --------------------
```

Intercept in model, MEANS statement added

Analysis of Variance Procedure
Class Level Information

Class	Levels	Values
AREA	4	1 2 3 4

Number of observations in data set = 100

Intercept in model, MEANS statement added

Analysis of Variance Procedure

Waller-Duncan K-ratio T test for variable: CMI

NOTE: This test minimizes the Bayes risk under additive loss and certain other assumptions.

ERROR: Failure in Bayes T computation. F value too small.

EXHIBIT 24.10 *(continued)*

```
- - - - - - - - - - New page of output - - - - - - - - -

Intercept in model, MEANS statement added

Analysis of Variance Procedure

T tests (LSD) for variable: CMI

NOTE: This test controls the type I comparisonwise error rate not the experimentwise error rate.

Alpha= 0.05  df= 96  MSE= 253.1417
Critical Value of T= 1.98
Least Significant Difference= 8.9327

Means with the same letter are not significantly different.

T Grouping          Mean      N  AREA

            A       26.840    25  4
            A
            A       20.840    25  3
            A
            A       20.720    25  2
            A
            A       18.920    25  1

- - - - - - - - - - New page of output - - - - - - - - -

Intercept in model, MEANS statement added

Analysis of Variance Procedure

Duncan's Multiple Range Test for variable: CMI

NOTE: This test controls the type I comparisonwise error rate, not the experimentwise error rate.

Alpha= 0.05  df= 96  MSE= 253.1417

Number of Means     2     3     4
Critical Range  8.944 9.405 9.702

Means with the same letter are not significantly different.

Duncan Grouping       Mean      N  AREA

            A         26.840    25  4
            A
            A         20.840    25  3
            A
            A         20.720    25  2
            A
            A         18.920    25  1
```

EXHIBIT 24.10   *(continued)*

- - - - - - - - - - New page of output - - - - - - - - -

Intercept in model, MEANS statement added

Analysis of Variance Procedure

Tukey's Studentized Range (HSD) Test for variable: CMI

NOTE: This test controls the type I experimentwise error rate, but generally has a higher type II error rate than REGWQ.

Alpha= 0.05 df= 96 MSE= 253.1417
Critical Value of Studentized Range= 3.698
Minimum Significant Difference= 11.766

Means with the same letter are not significantly different.

Tukey Grouping		Mean	N	AREA
	A	26.840	25	4
	A			
	A	20.840	25	3
	A			
	A	20.720	25	2
	A			
	A	18.920	25	1

- - - - - - - - - - New page of output - - - - - - - - -

Intercept in model, MEANS statement added

Analysis of Variance Procedure

Bonferroni (Dunn) T tests for variable: CMI

NOTE: This test controls the type I experimentwise error rate, but generally has a higher type II error rate than REGWQ.

Alpha= 0.05 df= 96 MSE= 253.1417
Critical Value of T= 2.69
Minimum Significant Difference= 12.124

Means with the same letter are not significantly different.

Bon Grouping		Mean	N	AREA
	A	26.840	25	4
	A			
	A	20.840	25	3
	A			
	A	20.720	25	2
	A			
	A	18.920	25	1

EXHIBIT 24.10   *(continued)*

```
- - - - - - - - - - - New page of output - - - - - - - - -

Intercept in model, MEANS statement added

Analysis of Variance Procedure

Scheffe's test for variable: CMI

NOTE: This test controls the type I experimentwise error rate but generally has a higher type II error
rate than REGWF for all pairwise comparisons.

Alpha= 0.05  df= 96  MSE= 253.1417
Critical Value of F= 2.69939
Minimum Significant Difference= 12.806

Means with the same letter are not significantly different.

Scheffe Grouping          Mean       N  AREA

                 A       26.840     25  4
                 A
                 A       20.840     25  3
                 A
                 A       20.720     25  2
                 A
                 A       18.920     25  1
```

EXHIBIT 24.11   ANOVA used with an unbalanced design

```
------------------- SAS Log ----------------------------

   1     data unbal;
   2     set cmidata.cmi;
   3     if ranuni(7654321) <= .20 then delete;
   4     run;
NOTE: The data set WORK.UNBAL has 74 observations and 6 variables.
NOTE: The DATA statement used 8.00 seconds.
   5
   6     proc anova data=unbal ;
   7     class area ;
   8     model cmi = area ;
   9     title 'Unbalanced design';
  10     quit;
NOTE: The PROCEDURE ANOVA used 8.00 seconds.
```

EXHIBIT 24.11   (continued)

```
------------------- Output listing --------------------
```

Unbalanced design

Analysis of Variance Procedure
Class Level Information

Class   Levels   Values

AREA   4   1 2 3 4

Number of observations in data set = 74

```
- - - - - - - - - - - New page of output - - - - - - - - -
```

Unbalanced design

Analysis of Variance Procedure

Dependent Variable: CMI   CMI (well-being) score

Source	DF	Sum of Squares	Mean Square	F Value	Pr > F
Model	3	848.97009415	282.99003138	1.01	0.3944
Error	70	19647.08395990	280.67262800		
Corrected Total	73	20496.05405405			

R-Square	C.V.	Root MSE	CMI Mean
0.041421	73.35759	16.75328708	22.83783784

Source	DF	Anova SS	Mean Square	F Value	Pr > F
AREA	3	848.97009415	282.99003138	1.01	0.3944

EXHIBIT 24.12   GLM output for one-way ANOVA; all defaults for statistics and presentation of output

```
------------------- Program listing --------------------
```

```
proc glm data=cmidata.cmi ;
class area ;
model cmi = area ;
title '1-way, balanced' ;
```

```
------------------- Output listing --------------------
```

1-way, balanced

General Linear Models Procedure
Class Level Information

Class   Levels   Values

AREA   4   1 2 3 4

Number of observations in data set = 100

EXHIBIT 24.12   *(continued)*

- - - - - - - - - - - New page of output - - - - - - - - -

1-way, balanced

General Linear Models Procedure

Dependent Variable: CMI   CMI (well-being) score

Source	DF	Sum of Squares	Mean Square	F Value	Pr > F
Model	3	894.51000000	298.17000000	1.18	0.3223
Error	96	24301.60000000	253.14166667		
Corrected Total	99	25196.11000000			

R-Square	C.V.	Root MSE	CMI Mean
0.035502	72.88331	15.91042635	21.83000000

Source	DF	Type I SS	Mean Square	F Value	Pr > F
AREA	3	894.51000000	298.17000000	1.18	0.3223

Source	DF	Type III SS	Mean Square	F Value	Pr > F
AREA	3	894.51000000	298.17000000	1.18	0.3223

EXHIBIT 24.13   **One-factor, unbalanced design**

------------------- Program listing -------------------

```
proc glm data=unbal ;
class area ;
model cmi = area / e;
title '1-way, unbalanced' ;
quit;
```

------------------- Output listing --------------------

1-way, unbalanced

General Linear Models Procedure
Class Level Information

Class	Levels	Values
AREA	4	1 2 3 4

Number of observations in data set = 82

NOTE: Due to missing values, only 69 observations can be used in this analysis.

EXHIBIT 24.13 *(continued)*

```
- - - - - - - - - - New page of output - - - - - - - - -

1-way, unbalanced

General Linear Models Procedure
General Form of Estimable Functions

Effect          Coefficients

INTERCEPT       L1

AREA      1     L2
          2     L3
          3     L4
          4     L1-L2-L3-L4

- - - - - - - - - - New page of output - - - - - - - - -

1-way, unbalanced

General Linear Models Procedure

Dependent Variable: CMI    CMI (well-being) score
```

Source	DF	Sum of Squares	Mean Square	F Value	Pr > F
Model	3	119.36487414	39.78829138	0.17	0.9148
Error	65	15011.27280702	230.94265857		
Corrected Total	68	15130.63768116			

	R-Square	C.V.	Root MSE	CMI Mean
	0.007889	72.61628	15.19679764	20.92753623

Source	DF	Type I SS	Mean Square	F Value	Pr > F
AREA	3	119.36487414	39.78829138	0.17	0.9148

Source	DF	Type III SS	Mean Square	F Value	Pr > F
AREA	3	119.36487414	39.78829138	0.17	0.9148

EXHIBIT 24.14   CLASS-variable levels sorted by a user-written format

```
------------------- Program listing -------------------

proc format;
value $area '1' = 'Cherryview'   '2' = 'Morningdale'
            '3' = 'Northhills'   '4' = 'Easton'  ;
run;

proc glm data=unbal ;
class area ;
model cmi = area ;
contrast 'Compare (C and E)' area 1 -1 0 0 ;
format area $area.;
title 'Contrasts: 1 and 2, default ORDER (FORMATTED)';
quit;

------------------- Output listing -------------------

Contrasts: 1 and 2, default ORDER (FORMATTED)

General Linear Models Procedure
Class Level Information

Class     Levels    Values

AREA        4     │ Cherryview Easton Morningdale Northhills │  ◄──── Formatted values of AREA

Number of observations in data set = 82

NOTE: Due to missing values, only 69 observations can be used in this analysis.

- - - - - - - - - - New page of output - - - - - - - - -

Contrasts: 1 and 2, default ORDER (FORMATTED)

General Linear Models Procedure

Dependent Variable: CMI   CMI (well-being) score
```

Source	DF	Sum of Squares	Mean Square	F Value	Pr > F
Model	3	119.36487414	39.78829138	0.17	0.9148
Error	65	15011.27280702	230.94265857		
Corrected Total	68	15130.63768116			

	R-Square	C.V.	Root MSE	CMI Mean
	0.007889	72.61628	15.19679764	20.92753623

Source	DF	Type I SS	Mean Square	F Value	Pr > F
AREA	3	119.36487414	39.78829138	0.17	0.9148

Source	DF	Type III SS	Mean Square	F Value	Pr > F
AREA	3	119.36487414	39.78829138	0.17	0.9148

Contrast	DF	Contrast SS	Mean Square	F Value	Pr > F
Compare (C and E)	1	2.74189886	2.74189886	0.01	0.9136

EXHIBIT 24.15   ORDER option-user fomat interaction

```
------------------- Program listing -------------------

proc format;
value $area '1' = 'Cherryview'   '2' = 'Morningdale'
            '3' = 'Northhills'   '4' = 'Easton'  ;
run;

proc glm data=unbal order=internal;
class area ;
model cmi = area ;
contrast 'Compare (C and E)' area 1 0 0 -1 ;
format area $area.;
title 'Contrasts: 1 and 4, ORDER=INTERNAL';
quit;

------------------- Output listing --------------------

Contrasts: 1 and 4, ORDER=INTERNAL

General Linear Models Procedure
Class Level Information

Class     Levels    Values

AREA         4      Cherryview Morningdale Northhills Easton
```

> Formatted values are arranged by PROC option ORDER=INTERNAL. Compare to values in Exhibit 24.13.

```
Number of observations in data set = 82

NOTE: Due to missing values, only 69 observations can be used in this analysis.

- - - - - - - - - - New page of output - - - - - - - - -

Contrasts: 1 and 4, ORDER=INTERNAL

General Linear Models Procedure

Dependent Variable: CMI    CMI (well-being) score
```

Source	DF	Sum of Squares	Mean Square	F Value	Pr > F
Model	3	119.36487414	39.78829138	0.17	0.9148
Error	65	15011.27280702	230.94265857		
Corrected Total	68	15130.63768116			

	R-Square	C.V.	Root MSE	CMI Mean
	0.007889	72.61628	15.19679764	20.92753623

Source	DF	Type I SS	Mean Square	F Value	Pr > F
AREA	3	119.36487414	39.78829138	0.17	0.9148

Source	DF	Type III SS	Mean Square	F Value	Pr > F
AREA	3	119.36487414	39.78829138	0.17	0.9148

Contrast	DF	Contrast SS	Mean Square	F Value	Pr > F
Compare (C and E)	1	99.87150538	99.87150538	0.43	0.5131

EXHIBIT 24.16 **Two-factor, unbalanced model with interaction term; LSMEANS statement, OUT option**

```
------------------- Program listing --------------------

proc glm data=unbal ;
class racial housing;
model cmi = racial housing housing*racial;
lsmeans racial housing / out=temp;
title '2-factor, unbalanced';
quit;

------------------- Output listing --------------------

2-factor, unbalanced

General Linear Models Procedure
Class Level Information

Class     Levels    Values

RACIAL        2      H L
HOUSING       2      H L

Number of observations in data set = 82

NOTE: Due to missing values, only 69 observations can be used in this analysis.

- - - - - - - - - - New page of output - - - - - - - -

2-factor, unbalanced

General Linear Models Procedure
```

Dependent Variable: CMI   CMI (well-being) score

Source	DF	Sum of Squares	Mean Square	F Value	Pr > F
Model	3	119.36487414	39.78829138	0.17	0.9148
Error	65	15011.27280702	230.94265857		
Corrected Total	68	15130.63768116			

R-Square	C.V.	Root MSE	CMI Mean
0.007889	72.61628	15.19679764	20.92753623

Source	DF	Type I SS	Mean Square	F Value	Pr > F
RACIAL	1	77.40154671	77.40154671	0.34	0.5646
HOUSING	1	10.93728162	10.93728162	0.05	0.8284
RACIAL*HOUSING	1	31.02604582	31.02604582	0.13	0.7152

Source	DF	Type III SS	Mean Square	F Value	Pr > F
RACIAL	1	85.85859961	85.85859961	0.37	0.5442
HOUSING	1	10.28890831	10.28890831	0.04	0.8335
RACIAL*HOUSING	1	31.02604582	31.02604582	0.13	0.7152

EXHIBIT 24.16   *(continued)*

- - - - - - - - - - - New page of output - - - - - - - - -

2-factor, unbalanced

General Linear Models Procedure

Least Squares Means

RACIAL   CMI
                LSMEAN

H   22.0625000
L   19.8192982   Produced by LSMEANS statement

HOUSING   CMI
                LSMEAN

H   20.5526316
L   21.3291667

------------------- Output dataset contents --------------------

2-factor, unbalanced

CONTENTS PROCEDURE

   Data Set Name: WORK.TEMP   Type:
   Observations: 4   Record Len: 30
   Variables: 5
   Label:

-----Alphabetic List of Variables and Attributes-----

   # Variable Type Len Pos Label
   3 HOUSING   Char   1   13   > 100 h'holds in area = H, others = L
   4 LSMEAN   Num   8   14
   2 RACIAL   Char   1   12   > 50% black = H, others = L
   5 STDERR   Num   8   22
   1 _NAME_   Char   8   4

------------------- Output dataset listing --------------------

2-factor, unbalanced

OBS	_NAME_	RACIAL	HOUSING	LSMEAN	STDERR
1	CMI	H		22.0625	2.57822
2	CMI	L		19.8193	2.62445
3	CMI		H	20.5526	2.46525
4	CMI		L	21.3292	2.73085

Standard error was not in GLM output listing.

Summary

Analysis of variance (ANOVA) poses the question of equality of means of a dependent variable for different subpopulations of one or more categorical independent variables. The overall significance of the independent variables as well as differences between category means may be measured. Another feature of the model that is usually of interest is the prediction of the dependent variable based on the independent variables, the "main effects," versus prediction based on "interactions," or synergistic effects of the independent variables. Extensions to the basic ANOVA model include multiple dependent variables (MANOVA) and observations taken on the same subjects at different times (repeated measures designs). A basic assumption of ANOVA for all designs is an equal number of observations per cell, a "balanced" design.

SAS offers two procedures, both interactive, for ANOVA. The ANOVA procedure uses computer resources efficiently, may be used only for balanced designs, and offers a reasonable range of statistical tests. For more complex, unbalanced designs and more complete diagnostics, the GLM procedure should be used.

25 The Other 80%

The scope and content of this book has been confined to the needs of beginning to intermediate users. Given this restriction, only about 20% of the capability of the SAS System has been discussed. This chapter gives a quick tour of the highlights of the remaining 80% of the language. It covers features available to all SAS users in the SAS "Base product" as well as some of the specialized products sold separately.

The intent of this chapter is not to teach new features but to whet your appetite and curiosity to learn more about the SAS System. Thus it emphasizes capability rather than syntax. The purpose and more common uses of each feature are described, along with examples of its use.

25.1 Base Product Features

Every SAS installation has data management routines and PROCs that are the core of its SAS product base. All features discussed so far in this book have been in this Base product. Like many aspects of SAS Institute products, the content of the Base product is changing. In version 6 of the SAS System, statistical procedures such as GLM, REG, ANOVA, and TTEST will be separated from the Base product and licensed separately as "SAS/STAT." The distinction between Base and other products is important: some books and software products written for SAS require specialized extensions to SAS. Nothing is more frustrating than getting excited about a product and rushing to the terminal to try it only to discover that it is not installed at your site.

This section discusses DATA step options, PROCs, and user environments common to all SAS installations. These features dress up the appearance of interactive SAS, make routine tasks easier by using a powerful high-level language, and facilitate display of data.

Windows: Customized Data Entry Screens

A common programming task is to enable the computer to carry on a dialogue with the user. The user may need help filling out a form, enter bad data values that need to be corrected, want to browse through the contents of one or more datasets, want to see the results of calculations, and so on. An implicit requirement for these sorts of activities is that they take place using attractive, well-designed, user-friendly interactive environments.

Sometimes these requirements can be met outside SAS using the operating system's prompt and read facilities. Within SAS, however, there are several means of displaying messages and reading the user's response. This section describes one approach: SAS's windowing capability in the DATA step. Another approach to data entry, the SAS/FSP add-on product, is discussed in Section 25.2.

Two statements, WINDOW and DISPLAY, define screen layouts and activate different screens. WINDOW defines the screen. It specifies the size of the window, the location of text strings and/or variables to be displayed, and other visual and audio features. Multiple windows may be defined in a single DATA step. This enables display of, say, a data entry screen and several help and error message screens.

The DISPLAY statement initiates the presentation of the window. It may be part of an IF statement or executed unconditionally. DISPLAY has options that control how the window is overlaid on existing windows, whether the computer should beep when it displays the window, and so on.

Once a window is displayed on the screen, the user may have the option of entering data. The ENTER and TAB keys perform movement from field to field on the screen. Once all fields have been filled in, a special command or another ENTER transmits the contents of the variables just entered in the window back to the DATA step. The DATA step program can examine the variable contents and display another window or simply go on to the next observation in the dataset. All actions are determined by the programmer and depend on the requirements of the application.

The window facility is not only a very powerful tool but also one that can require significant amounts of programming effort to implement. The example in the "Using Windows" section demonstrates both sides of this coin.

For More Information.   The WINDOW and DISPLAY statements are discussed in some detail in the *SAS Language Guide for Personal Computers, Release 6.03 Edition*, published by SAS Institute. Version 6 updates to the *SAS User's Guide: Basics* also contains syntax descriptions and usage examples.

Using Windows.   The example developed in this section computes loan payment parameters. The initial window prompts the user for three of four possible items: loan amount, annual interest rate, number of years to repay, and monthly payment (the responses will be passed to the MORTGAGE function, discussed in Chapter 7). Entering INFO on the command line displays a short help file. If the user does not enter three items, an error message is displayed.

Once three of the four items are entered, the DATA step calculates the missing item and displays it on the right half of the screen. At this point the user has the option of saving the variables in an observation of a SAS dataset. Once that decision is made, a new observation is started. The process continues until STOP is entered on the COMMAND line at the top of the screen. As execution terminates, a message about number of models (observations) processed and number saved is displayed. At this point, pressing ENTER ends the DATA step and returns the user to the SAS session. The program is presented in Exhibit 25.1. The initial screen is shown in Exhibit 25.2 (see p. 553). Entering INFO on the COM-

MAND line and pressing ENTER displays the help file, as shown in Exhibit 25.3 (see p. 554).

What if two rather than three of the four parameters are accidentally entered? A small window at the bottom of the screen displays an error message (Exhibit 25.4, see p. 555). The window is written over the main screen. When the user presses ENTER to terminate the help screen, the main screen's original layout will be automatically restored.

The next step is to enter the correct number of parameters so that the remaining one may be calculated. When ENTER is pressed, the right side of the

---

EXHIBIT 25.1   DATA step using windows

```
data mortinfo(keep=amount rate term payment);
*
| Define all windows
*;
window readvals irow=1 icolumn=1
      #2   @2 'Enter values in three fields:'
      #5   @2 'Loan amount              ' amount
      #7   @2 'Annual interest rate     ' rate
      #9   @2 'Years to repay           ' term
      #11  @2 'Monthly payment          ' payment
      #16  @2 '*** Enter STOP on the COMMAND'
      #17  @2 '    line to end ***'
      #19  @2 '*** Enter INFO on the COMMAND'
      #20  @2 '    line for help ***'
      ;
window help irow=1 icolumn=40
      #2   @2 'LOAN AMOUNT should be entered'
      /    @2 '        without commas and dollar signs'
      /    @2 '        (e.g., 20000 NOT $20,000)'
      //   @2 'INTEREST RATE should be entered'
      /    @2 '        as an integer (e.g., 18 for 18%)'
      //   @2 'YEARS TO REPAY should be an integer'
      /    @2 '        (e.g., 20)'
      //   @2 'MONTHLY PAYMENT should be entered'
      /    @2 '        without commas and dollar signs'
      /    @2 '        (e.g., 1400 NOT $1,400)'
      ;
window badmiss irow=22 icolumn=1
      #1   @2 'Error: one field must be missing'
      ;
window goodvals irow=1 icolumn=40
      #2   @2 'Computed values:'
      #5   @2 'Loan amount              ' amount  dollar12.2  protect=yes
      #7   @2 'Annual interest rate     ' rate    12.2        protect=yes
      #9   @2 'Years to repay           ' term    12.2        protect=yes
      #11  @2 'Monthly payment          ' payment dollar12.2  protect=yes
      #15  @2 'Save in output dataset? Y or N ' commit $1.
      ;
window endstep irow=8 icolumn=1 rows=8
      #1   @2 'End of processing.'
      /    @5 'Obs. processed = ' nproc     protect=yes
      /    @5 'Records saved  = ' out_obs   protect=yes
      //   @2 '*** Press ENTER to continue ***'
```

EXHIBIT 25.1   *(continued)*

```
*
| First, display initial window.
*;
display readvals blank;
*
| What to do? STOP, INFO, or go ahead and process?
*;
if upcase(_cmd_) = 'STOP' then do;
   nproc = _n_ - 1;
   display endstep;
   stop;
   end;
   else if upcase(_cmd_) = 'INFO' then do;
       display help;
       display readvals;
       end;
*
| Display screen until correct number of missing values (1) shows.
*;
nmissing = nmiss(amount,rate,term,payment) ;
do while (nmissing ^= 1);
   display badmiss;
   display readvals;
   nmissing = nmiss(amount,rate,term,payment) ;
end;
*
| Now make adjustments to the input values.
*;
* Annualize value input on screen ;
if payment ^= . then payment = payment * 12 ;
* Adjust interest rate to be consistent with payment interval ;
if rate     ^= . then rate = rate / 100 ;
*
| Calculate the missing parameter.
*;
if amount  = . then amount  = mort(.,payment,rate,term);
if payment = . then payment = mort(amount,.,rate,term);
if rate    = . then rate    = mort(amount,payment,.,term);
if term    = . then term    = mort(amount,payment,rate,.);
*
| Convert units used in calculation back to their original units, then
| display them.
*;
payment = payment / 12 ;
rate    = rate * 100 ;
display goodvals ;
*
| Save this obs?
*;
if upcase(commit) = 'Y' then do;
   out_obs + 1 ;
   output;
   end;
run;
```

EXHIBIT 25.2   Initial screen seen by the user

```
┌ READVALS────────────────────────────────────────────────────┐
│ Command ===>                                                 │
│                                                              │
│                                                              │
│   Enter values in three fields:                              │
│                                                              │
│                                                              │
│   Loan amount              .                                 │
│                                                              │
│   Annual interest rate   .                                   │
│                                                              │
│   Years to repay           .                                 │
│                                                              │
│   Monthly payment          .                                 │
│                                                              │
│                                                              │
│                                                              │
│   *** Enter STOP on the COMMAND                              │
│        line to end ***                                       │
│                                                              │
│   *** Enter INFO on the COMMAND                              │
│        line for help ***                                     │
└──────────────────────────────────────────────────────────────┘
```

screen displays the three original values plus the fourth, computed one. The model does not have to be saved in the output dataset (a response of "n" to the prompt at the bottom of the right-hand window). The screen is reproduced in Exhibit 25.5 (see p. 556).

If you decide to exit after processing only one model, enter STOP on the COMMAND line, then press ENTER. The summary statistic window is displayed in the middle of the screen. Pressing ENTER again ends the DATA step and returns to interactive SAS. The exit screen is shown in Exhibit 25.6 (see p. 556).

FORMS: Quick, Compact Displays

Although the PRINT procedure is the simplest means to display data, it has several drawbacks. One is its inability to locate information at exact locations on the page. Another shortcoming is its inability to print several observations per line of output or print a single observation across several lines. A _NULL_ DATA step with PUT statements satisfies these requirements but can be time-consuming to write and debug.

The FORMS procedure addresses the multiple observation, multiple line, and exact placement issues and does so with a fairly straightforward syntax. Although it is usually thought of for use with mailing lists, anything that needs to be displayed compactly is suitable. For example, suppose you wanted to list several thousand student names and identification numbers. Using PRINT you would use one line per student, probably producing a wastefully thick print-out. FORMS, however, easily prints multiple groups of names and identification numbers across the page and can greatly reduce paper requirements.

EXHIBIT 25.3   Display help screen by entering INFO on command line

```
┌ READVALS ─────────────────────    ┌ HELP ──────────────────────────
Command ===>                        Command ===>


  Enter values in three fields:     LOAN AMOUNT should be entered
                                         without commas and dollar signs
                                         (e.g., 20000 NOT $20,000)
Loan amount          .
                                    INTEREST RATE should be entered
Annual interest rate .                   as an integer (e.g., 18 for 18%)

Years to repay       .              YEARS TO REPAY should be an integer
                                         (e.g., 20)
Monthly payment      .
                                    MONTHLY PAYMENT should be entered
                                         without commas and dollar signs
                                         (e.g., 1400 NOT $1,400)


*** Enter STOP on the COMMAND
    line to end ***

*** Enter INFO on the COMMAND
    line for help ***
```

FORMS' syntax is straightforward. Specify which variables will be placed on each line of the "form unit" (i.e., label), give the dimensions of the label, and tell how many labels go across and down each page. If more than one form unit will be printed on each line, FORMS prints them left to right as it processes the input SAS dataset. Titles and footnotes may be specified, and BY-group processing and FREQ statements may also be used.

The disadvantage of using this PROC purely for listing purposes is that you lose PRINT's and custom report's ability to add numeric variables. FORMS also does not label the variables on the form the way PRINT annotates its columns. If you can live with this restriction, consider using FORMS not only for obvious applications such as mailing lists but also for more subtle ones.

For More Information.   The FORMS procedure is described in the *SAS User's Guide: Basics* and *SAS Procedures Guide, Release 6.03 Edition*, both published by SAS Institute.

Using FORMS.   A simple, traditional application of FORMS is presented in this section. The input SAS dataset contains addresses and telephone numbers of organizations for a Christmas card mailing list.

Exhibit 25.7 (see p. 557) presents the program that reads the raw data, sorts it by ZIP code, and uses FORMS to print four labels across the page. Consider the level of difficulty involved if the listing had to be produced by a _NULL_ DATA step. Also think of the effort required if the specifications had to change (fewer spaces between labels, only three form units across, and so on). FORMS

EXHIBIT 25.4   Two rather than three values entered; error message displayed

```
┌ READVALS─────────────────────────────────────────────────────────────────┐
│ Command ===>                                                              │
│                                                                           │
│                                                                           │
│  Enter values in three fields:                                            │
│                                                                           │
│                                                                           │
│  Loan amount            20000                                             │
│                                                                           │
│  Annual interest rate   10                                                │
│                                                                           │
│  Years to repay          .                                                │
│                                                                           │
│  Monthly payment         .                                                │
│                                                                           │
│                                                                           │
│                                                                           │
│  *** Enter STOP on the COMMAND                                            │
│      line to end ***                                                       │
│ BADMISS ─────────────────────────────────────────────────────────────────│
│ Command ===>                                                              │
│                                                                           │
│  Error: one field must be missing                                         │
│                                                                           │
└───────────────────────────────────────────────────────────────────────────┘
```

permits quick initial specification of the page and lets you make modifications with equal ease.

The LINE statement has several options that improve display of each line: PACK removes unnecessary blanks between variables in the same line. REMOVE does not display a line if all its variables are missing (notice the uneven numbers of lines printed in the first row of forms).

TIMEPLOT: Displaying Temporal Data

A slowly evolving but important trend in SAS procedure capability is the integration of graphics into statistical reports. This capability was seen in the horizontal bar charts discussed in Chapter 6: bars and their corresponding frequency distributions were printed side by side. The TIMEPLOT procedure effectively blends hard numbers and a more visual, graphic approach to data presentation. It incorporates many features of both the PRINT and PLOT procedures: data points are positioned in a two-dimensional space, as in PLOT. The exact values of the data points are also displayed, along with optional identifying variables, as in PRINT.

TIMEPLOT listings may continue across more than one page if necessary. Observations are printed in the order they are read from the input SAS dataset. Thus if the data are sorted by a date variable, the listing gives a visual idea of trends of one or more variables over time. Another advantage of TIMEPLOT is its display of one observation per line. This type of display avoids "hidden" observation problems seen in the PLOT procedure (Chapter 21).

EXHIBIT 25.5   Calculated and original variables displayed in GOODVALS window

```
┌READVALS ──────────────────────┐ ┌GOODVALS──────────────────────┐
│Command ===>                   │ │Command ===>                  │
│                               │ │                              │
│                               │ │                              │
│ Enter values in three fields: │ │  Computed values:            │
│                               │ │                              │
│                               │ │                              │
│ Loan amount          20000    │ │  Loan amount       $20,000.00│
│                               │ │                              │
│ Annual interest rate 10       │ │  Annual interest rate   10.00│
│                               │ │                              │
│ Years to repay       25       │ │  Years to repay            25│
│                               │ │                              │
│ Monthly payment               │ │  Monthly payment      $183.61│
│                               │ │                              │
│                               │ │                              │
│                               │ │                              │
│                               │ │  Save in output dataset? Y or N n│
│ *** Enter STOP on the COMMAND │ │                              │
│     line to end ***           │ │                              │
│                               │ │                              │
│ *** Enter INFO on the COMMAND │ │                              │
│     line for help ***         │ │                              │
└───────────────────────────────┘ └──────────────────────────────┘
```

EXHIBIT 25.6   Termination window displays processing statistics

```
┌ READVALS──────────────────────────────────────────────────────┐
│Command ===>                                                    │
│                                                                │
│                                                                │
│ Enter values in three fields:                                  │
│                                                                │
│ENDSTEP─────────────────────────────────────────────────────── │
│Command ===>                                                    │
│                                                                │
│ End of processing.                                             │
│    Obs. processed = 1                                          │
│    Records saved  = 0                                          │
│                                                                │
│ *** Press ENTER to continue ***                                │
│                                                                │
│                                                                │
├────────────────────────────────────────────────────────────── ┤
│                                                                │
│ *** Enter STOP on the COMMAND                                  │
│     line to end ***                                            │
│                                                                │
│ *** Enter INFO on the COMMAND                                  │
│     line for help ***                                          │
└────────────────────────────────────────────────────────────────┘
```

TIMEPLOT's syntax closely resembles that of PLOT. A PLOT statement specifies the variables. Overlays of two or more plots are permitted. The variables are the *x*-axis values. The *y*-axis print positions represent individual observations. If plots are overlaid, the points along the *x* axis may be joined. This feature is useful, for example, when displaying stock market or other open/close, start/end data. Reference lines may be drawn along the *x* axis, and the default size of the plotting area may be overridden.

EXHIBIT 25.7   FORMS mailing list program and output

```
options ps=66 ls=132;

data assns;
infile '\xmas' missover;
length name add1 add2 $30 city_st $20 zip $5 phone $14;
input @1 name $char30. / @1 add1 $ & / @1 add2 $ &
      / @1 city_st $ &  zip $ / @1 phone $14.;
run;

proc sort;
by zip;
run;

proc forms data=assns
           width=30 lines=5 down=0 skip=2
           nacross=4 between=2 ndown=8;
line 1 name ;
line 2 add1 / remove ;
line 3 add2 / remove ;
line 4 city_st zip / pack ;
line 5 phone ;
title 'Christmas Card List';
run;

-------------------- Output listing ---------------------

Christmas Card List

American Society of Dowsers     National Assn. of Accountants   American Kennel Club            War Resisters League
Danville, VT 05828              10 Paragon Drive                51 Madison Ave.                339 Lafayette St.
(802) 684-3417                  PO Box 433                      New York, NY 10010             New York, NY 10012
                                Montvale, NJ 07645              (212) 696-8234                 (212) 228-0450
                                (201) 573-9000


CARE Inc.                       National Audubon Society        Amer. Civil Liberties Union    Alcoholics Anonymous
660 First Ave.                  950 Third Ave.                  132 W. 43rd St.                PO Box 459
New York, NY 10016              New York, NY 10022              New York, NY 10036             Grand Central Station
(212) 686-3110                  (212) 832-3200                  (212) 944-9800                 New York, NY 10163
                                                                                               (212) 686-1100


American Mensa, Ltd.            SCRABBLE Players                RID-USA                        American Philatelic Society
2626 E. 14th St.                PO Box 700                      Box 520                        PO Box 8000
Brooklyn, NY 11235              Front Street Garden             Schenectady, NY 12301          State College, PA 16803
(718) 934-3700                  Greenport, NY 11944             (518) 372-0034                 (814) 237-3803
                                (516) 477-0033


TOUGHLOVE                       American Friends Service Comm.  Aerospace Medical Assn.        Order of Sons of Italy
PO Box 1069                     1501 Cherry St.                 Washington National Airport    219 E. St., NE
Doylestown, PA 18901            Philadelphia, PA 19102          Washington, DC 20001           Washington, DC 20002
(215) 348-7090                  (215) 241-7060                  (703) 892-2240                 (202) 547-2900


National Space Society          Marine Technology Society       Zero Population Growth         National Geographic Society
922 Pennsylvania Ave., SE       1825 K St., NW                  1400 Sixteenth St., NW         17th and M Streets, NW
Washington, DC 20003            Suite 203                       Third Floor                    Washington, DC 20036
(202) 543-1900                  Washington, DC 20006            Washington, DC 20013           (202) 857-7000
                                (202) 775-5966                  (202) 332-2200
```

EXHIBIT 25.7 *(continued)*

American Veterans Comm. 1717 Massachusetts Ave., NW Suite 203 Washington, DC 20036 (202) 667-0090	World Wildlife Fund 1250 24th St., NW Washington, DC 20037 (202) 468-1800	4-H Program Room 3860-S US Dept. of Agriculture Washington, DC 20250 (202) 447-5853	World Future Society 4916 St. Elmo Ave. Bethesda, MD 20814 (301) 656-8274
Izaak Walton League of America 1701 N. Ft. Myer Drive Suite 1100 Arlington, VA 22209 (703) 528-1818	Muzzle Loading Rifle Assn. PO Box 67 Friendship, IN 47201 (812) 667-5131	Ducks Unlimited One Waterfowl Way Long Grove, IL 60047 (312) 438-4300	Modern Woodmen of America Mississippi River at 17th St. Rock Island, IL 61201 (309) 786-6481
Overeaters Anonymous PO Box 92870 Los Angeles, CA 90009 (213) 542-8363	Puppeteers of America 5 Cricklewood Path Pasadena, CA 91107 (818) 797-5748	Sierra Club 730 Polk St. San Francisco, CA 94109 (415) 776-2211	

For More Information.   The TIMEPLOT procedure is described in the *SAS User's Guide: Basics* and the *SAS Procedures Guide, Release 6.03 Edition*, both published by SAS Institute.

Using TIMEPLOT.   Exhibit 25.8 shows a simple use of TIMEPLOT. A sample of baseball players in the Hall of Fame is displayed, ordered by the first 16 characters of the player's last name. The beginning and ending dates of each player's career are plotted: beginning date is indicated in the plot area by a *b*, ending date by an *f*. The dates are joined by a dashed line.

The display gives a quick visual idea of when the player played, how long his career lasted, and (by comparing the overlap of the dashed lines) who his contemporaries were. More importantly, visual impressions about overlap and career duration are supported by the actual numbers printed on each line of the output.

TABULATE: Prettified, Complex Reports

Some circumstances require very "dressed up" presentation of data or moderately complex calculations. The TABULATE procedure addresses the appearance and complexity issues simultaneously. Output is automatically surrounded by boxes, and univariate statistics and complex row and column percentage definitions may be requested.

TABULATE does nothing that a PROC MEANS output dataset and a subsequent DATA step cannot do. The comments made during the FORMS discussion about ease of layout definition and modification also hold true for TABULATE. The choice to use TABULATE is simply a matter of convenience.

This convenience and simplicity are not immediately apparent, however. TABULATE has the distinction of being one of only two Base SAS procedures requiring a separate manual (see "For More Information" below). Some of the TABULATE syntax will be familiar to people who have mastered MEANS and other Base product PROCs, while some statements have rather obtuse syntax rules.

TABULATE produces tables with up to three dimensions (row, column, page), accumulating statistics and percentages by one or more stratifying (CLASS) variables. Output is neatly arranged in boxes with liberal amounts of default annotation. Options are available for substituting text for missing values, labelling statistical keywords and variable names, and overriding the default characters used for drawing the boxes. The complexity and confusion usually arise when trying to define subtotals and to do complex nesting of classification variables. The time spent learning TABULATE's syntax is worth the effort: the appearance of the output is impressive, and the alternative MEANS DATA step sequence is often not appealing.

EXHIBIT 25.8 TIMEPLOT program and output

```
------------------ Program listing ------------------

proc timeplot data=out.hallfame;
plot start='b' end='f'
     / overlay joinref axis=1870, 1990 pos=90;
id name;
format start end 4.;
label name = 'Player Name'
      start = 'Career Start'
      end = 'Career End'
      ;
title 'Random Sample of Players in the Baseball Hall of Fame';
run;

---------- Output listing begins ----------

Random Sample of Players in the Baseball Hall of Fame
```

```
                                                                                        max
Player                  Career Career  min                                              1990
Name                    Start   End   1870
                                        *-------------------------------------------------*
Aaron, Hank             1954    1976   |                                   b------------f  |
Alexander, Grover       1911    1930   |               b------------f                      |
Appling, Luke           1930    1950   |                        b------------f             |
Banks, Ernie            1953    1971   |                                   b------------f   |
Bottomley, Jamis        1922    1937   |                    b------f                       |
Brouthers, Dennis       1879    1896   |b------------f                                      |
Chance, Frank           1889    1914   |    b--------------------f                          |
Clarkson, John          1882    1894   |    b------f                                        |
Cobb, Ty                1905    1928   |            b------------f                          |
Comiskey, Charles       1882    1894   |    b------f                                        |
Cronin, Joe             1926    1945   |                   b------------f                   |
Dickey, Bill            1928    1946   |                   b------------f                   |
Duffy, Hugh             1888    1906   |       b------------f                               |
Faber, Red              1914    1933   |              b------------f                        |
Foster, Rube            1897    1926   |         b--------------------f                     |
Foxx, Jimmy             1925    1945   |                   b------------f                   |
Gibson, Josh            1929    1946   |                   b------------f                   |
Griffith, Clark         1891    1914   |          b------------f                            |
Hamilton, William       1888    1901   |       b------------f                               |
Hooper, Harry           1909    1925   |           b------------f                           |
Hoyt, Waite             1918    1938   |              b------------f                        |
Jennings, Hugh          1891    1918   |          b------------f                            |
Joss, Adrian            1902    1910   |           b------f                                 |
Kell, George            1943    1957   |                          b------f                  |
Killebrew, Harmon       1954    1975   |                            b------------f          |
Leonard, Buck           1933    1955   |                     b------------f                 |
Lyons, Theodore         1923    1946   |                  b------------f                    |
Mantle, Mickey          1951    1968   |                           b------f                 |
Mays, Willie            1951    1973   |                           b------------f           |
McGinnity, Joe          1899    1908   |         b------f                                   |
Medwick, Ducky          1932    1948   |                         b------f                   |
Ott, Mel                1926    1947   |                   b------------f                   |
Plank, Edward           1901    1917   |          b------f                                  |
Robinson, Frank         1956    1976   |                                b------------f      |
Ruffing, Red            1924    1947   |                  b------------f                    |
Ruth, Babe              1914    1935   |              b------------f                        |
Sisler, George          1915    1930   |              b------------f                        |
Stargell, Willie        1962    1982   |                                       b------------f|
Traynor, Pie            1920    1937   |                 b------f                           |
Waddell, George         1897    1910   |         b------------f                             |
Waner, Lloyd            1927    1945   |                   b------------f                   |
Williams, Billy         1959    1976   |                            b------------f          |
Wynn, Early             1939    1963   |                       b--------------------f       |
Young, Cy               1890    1911   |          b------------f                            |
                                        *-------------------------------------------------*
```

For More Information.   The basics of the TABULATE procedure are described in the *SAS User's Guide: Basics* and *SAS Procedures Guide, Release 6.03 Edition*, both published by SAS Institute. A separate publication, *SAS Guide to Tabulate Processing*, gives a thorough description of the syntax and has many complete, annotated examples.

Using TABULATE.   This section presents a moderately complex use of TABULATE. The program summarizes the National Park dataset used in Chapter 20. We stratify by COAST and formatted values of YRESTAB, taking the mean, sum, count, and various percentages of ACRES. These variables are indicated in the VAR and CLASS statements.

The TABLE statement specifies a two-dimensional table, each dimension separated by a comma. The ALL keyword and parentheses determine the levels of nesting and subtotaling in the row dimension. The F keyword and text in single quotes specify table cell formats and annotating text. The options following the "/" control row header widths and text to print when a value is missing.

Exhibit 25.9 presents the TABULATE program and output. As was the case with PROC FORMS output, consider how difficult producing this output would

---

EXHIBIT 25.9   PROC TABULATE used for complex cross-classified reports

```
------------------- Program listing -------------------

proc format;
value yrest 1870-1900 = '1870-1900'
            1901-1920 = '1901-1920'
            1921-1946 = '1921-1945'
            1946-1968 = '1946-1970'
            1969-high = '1969+' ;
value $coast 'E'='East' 'W'='West';
run;


options ps=60 ls=132;

proc tabulate data=book.natlpark formchar(3 5 9 11)='****';
class coast yrestab;
var acres;
table (coast all='Both Coasts') * (yrestab all='All years'),
      acres*(n*f=5.
              pctn<yrestab all>='% of year'*f=8.2
              pctn<coast all>='% of coast'*f=8.2
              sum*f=comma12.
              mean*f=comma12. )
      / printmiss rts=35 box='Coast-Year Estab. Combinations'
        misstext='None' ;
keylabel sum='Sum' n='# parks' mean='Average' all='Total';
format yrestab yrest. coast $coast.;
label coast='Coast'
      acres='Acreage'
      yrestab='Yr. established' ;
title 'PROC TABULATE Example: National Park Data';
run;
```

EXHIBIT 25.9 *(continued)*

```
------------------- Output listing -------------------

PROC TABULATE Example: National Park Data
```

Coast-Year Estab. Combinations		Acreage				
		# parks	% of year	% of coast	Sum	Average
Coast	Yr. established					
East	1870-1900	None	None	None	None	None
	1901-1920	1	12.50	9.09	41,365	41,365
	1921-1945	5	62.50	41.67	2,738,763	547,753
	1946-1970	1	12.50	14.29	14,695	14,695
	1969+	1	12.50	6.67	173,039	173,039
	All years	8	100.00	16.33	2,967,862	370,983
West	Yr. established					
	1870-1900	4	9.76	100.00	3,618,840	904,710
	1901-1920	10	24.39	90.91	7,959,620	795,962
	1921-1945	7	17.07	58.33	2,511,085	358,726
	1946-1970	6	14.63	85.71	1,151,008	191,835
	1969+	14	34.15	93.33	29,027,557	2,073,397
	All years	41	100.00	83.67	44,268,110	1,079,710
Both Coasts	Yr. established					
	1870-1900	4	8.16	100.00	3,618,840	904,710
	1901-1920	11	22.45	100.00	8,000,985	727,362
	1921-1945	12	24.49	100.00	5,249,848	437,487
	1946-1970	7	14.29	100.00	1,165,703	166,529
	1969+	15	30.61	100.00	29,200,596	1,946,706
	All years	49	100.00	100.00	47,235,972	963,999

be if your only tools were a MEANS output dataset and a DATA step. Even a cursory glance at the program suggests that the syntax is unintuitive. Here, more so than in most PROCs, a misplaced parenthesis or comma often renders results useless or comical.

Macros: Building Reusable Programs

Many applications require repeated execution of a program, each time making minor changes. For example, you might run a series of PRINTs, FREQs, and CHARTs on dataset FISCAL90, then on FISCAL91. The obvious way to accomplish this is by changing all references in the program of FISCAL90 to FISCAL91.

Another common feature of program development is the need to conditionally execute DATA step statements or PROCs. You may, for example, want to PRINT all or part of a dataset while debugging a program. A simple way to deactivate the PRINT procedure is to enclose it in a comment (a technique discussed in Chapter 3) or simply delete it from the program. Both solutions are unsatisfactory: commenting is tedious, particularly if several PROCs are affected and if they in turn use comments. An axiom of the programming profession suggests that deleting code guarantees that it will be needed (usually urgently) later.

SAS provides a tidy, powerful solution to these and other programming problems. Symbolic variables and macros (unavailable in release 5 of SAS for minicomputers) allow substitution of values throughout a program; permit conditional inclusion of part, all of, or several statements; and use a high-level programming language resembling that found in the DATA step.

Symbolic variables, also known as macro variables, are a familiar concept implemented with an unfamiliar syntax. The variable may be defined anywhere in a SAS program by a %LET statement. The value of the macro variable may be part, all, or several SAS statements. It may be used as often as necessary in subsequent program statements. It may also be changed later in the program if necessary. The macro variable is identified by a preceding ampersand (&) and a trailing period (.). When SAS sees the macro variable it substitutes the variable's value before executing the SAS statement. If a macro variable is defined with the statement

```
%let keepers = name id score1-score15 ;
```

and is used in a DATA step such as

```
data subset;
set  in.roster(keep=&keepers.);
```

SAS would execute the following statements:

```
data subset;
set  in.roster(keep=name id score1-score15);
```

The power and value of symbolic substitution becomes more obvious as programs increase in size. The substitution capability means you can define macro variables at the beginning of the program and be confident that every reference to that variable will be exactly the same throughout the program.

Macros are a very complex extension of the principles underlying symbolic variables. Like symbolic variables, they enable insertion of code into the SAS program before it begins to execute. They also have windows, functions, operators, and flow-of-control structures much like those in the DATA step. These tools permit construction of macros that can prompt for values, conditionally execute program segments, and even write other macros and programs. The key to their effective use is that the macros execute *before* SAS executes the programs they create—the macro is, in effect, a preprocessor, building the program according to the parameters and execution logic you give it.

For More Information. Macros are described in the *SAS User's Guide: Basics*. The *SAS Guide to Macro Processing* contains a thorough description of the syntax and has complete, annotated examples. Both are published by SAS Institute.

Using Macros. The examples in this section demonstrate the use of symbolic variables, then complicate and enhance the scenario by using a macro. The dataset's observations contain information on the 34 bridges in the world whose main span is greater than 500 meters.

Exhibit 25.10 (see p. 564) defines three macro variables prior to any PROCs or DATA steps. FMTTYPE refers to a user-written format used to group span lengths. Values of 100 and 200 are permitted. Variable NCASES controls how many observations will be processed by PRINT. SORTVAR identifies BY variables for the SORT procedure. Notice that some of these variables are used more than once in the program. A fourth macro variable, TOTAL, is created in the DATA step with the SYMPUT function. SYMPUT creates or replaces macro variables using the value of the second parameter. In this program, TOTAL is created at the end of the DATA step and contains the number of observations (_N_) in the dataset.

The output from the program is shown in Exhibit 25.11 (see p. 565). Notice how the use of macro variable values in the TITLE statements improves communication: we know that the PRINT procedure's output is sorted by variable COMPLETE, that there are 34 observations, and that only 10 are printed. The FREQ procedure's output is enhanced by its explicit reference to the grouping format used.

Consider how many changes to the program would be required if macro variables had not been used. Also think about the possibility for error: it is very easy to change the format in FREQ to 100 but leave the TITLE unchanged at 200. Effective use of macro variables ensures that all references will be changed correctly.

Now suppose that we want to get a bit more elaborate. We want to define the same variables as in Exhibit 25.11 plus have a cutoff for minimum span length. Since we might not always want to run both PRINT and FREQ, we create two variables, DOPRINT and DOFREQ, that control insertion of the PROCs into the program. Passing values of PRINTIT and FREQIT to the macro (in the parenthetical expression in the %MACRO statement) will include the appropriate procedure statements in the stream of instructions actually executed by SAS. The FORMAT procedure is exactly the same as in Exhibit 25.10. The macro is presented in Exhibit 25.12 (see p. 566).

The macro is not actually executed until it is "called" at the bottom of the program. Parameters are passed just as they are to SAS functions that are sensitive to parameter order. Some features of the macro include use of automatic macro variables in the FOOTNOTE: SYSTIME, SYSDAY, and SYSDATE indicate time and date values. IF-THEN structures control insertion of code: the value *printit* was passed to variable DOPRINT, so the %DO group containing the PRINT procedure was executed (the expression *&doprint. = printit* was true). The *nofreqs* was passed to variable DOFREQ, so the FREQ sequence was ignored (the expression *&dofreq. = freqit* was false). Also notice the creation of the TITLE2 statement: the macro substitutes part of the title, depending on the values of NCASES and TOTAL.

Exhibit 25.13 (see p. 567) shows the SAS Log and procedure output. The SAS System option MPRINT was in effect. This prints the code that SAS actually sees once the macro executes. The macro's variable substitutions and the IF-THEN logic produce the program that SAS actually runs. There is no reference to the FREQ procedure in the SAS Log since the call to macro BRIDGE "turned off" that portion of the macro. The FREQ code was never passed to SAS for execution.

EXHIBIT 25.10   Macro variables used to perform OBS, sort key, format name substitution

```
%let fmttype = 200;
%let ncases = 10;                               ◄──────────  Define the macro variables.
%let sortvar = complete;

proc format;
value $style 's' = 'Suspension'  'c' = 'Cantilever'
             'a' = 'Steel arch'  ;
value span100x 500-600   = '500-600'      601-700   = '601-700'
               701-800   = '701-800'      801-900   = '801-900'
               901-1000  = '901-1,000'    1001-1100 = '1,001-1,100'
               1101-1200 = '1,101-1,200'  1201-high = '1201+' ;
value span200x 500-700   = '500-700'      701-900   = '701-900'
               901-1100  = '901-1,100'    1101-1300 = '1,101-1,300'
               1301-high = '1,301+' ;
run;

data bridges;
infile '\data\raw\bridges' end=eof;
length type $1 name $25 country $15 span_len 4 complete 3;
input type $ name $ & country $ & span_len complete;
if eof = 1 then call symput('total',put(_n_,2.));   ──────  Create macro variable
format type $style. span_len comma5.;                        TOTAL at end of dataset.
run;

proc sort data=bridges;
by &sortvar.;               [10]
run;

proc print data=bridges(obs=&ncases.);                    [COMPLETE]
id &sortvar.; ◄──── [COMPLETE]

title1 "Bridge Span Length Dataset, Sorted by Variable '&sortvar.'";

title2 "First &ncases. records out of a total of &total."; ◄──── 34 (number of observations
                                                                  in dataset BRIDGES)
title3 "Only Spans 500+ Meters Long Are Included";
run;                    [10]

proc freq data=bridges;
tables span_len;

format span_len span&fmttype.x11.;  [200]

title1 "Bridge Span Length Dataset";

title2 "Group in lengths of &fmttype. meters";
run;
```

EXHIBIT 25.11   Output from program in Exhibit 25.10

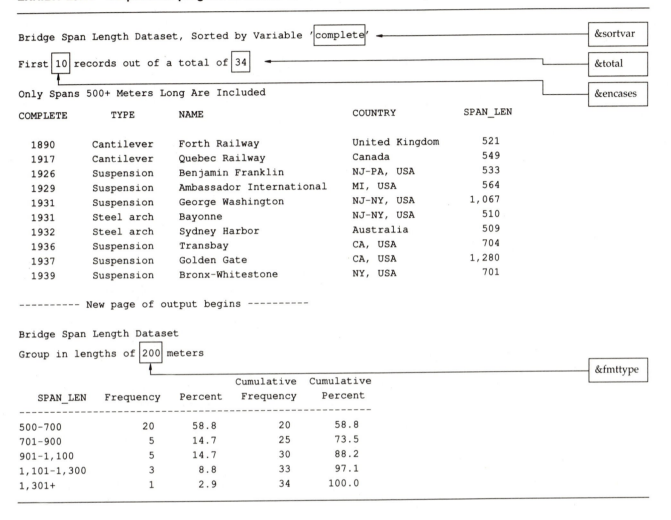

```
Bridge Span Length Dataset, Sorted by Variable 'complete'  ◄─────────────────    &sortvar

First 10 records out of a total of 34  ◄─────────────────────                     &total

Only Spans 500+ Meters Long Are Included                                          &encases

COMPLETE        TYPE        NAME                      COUNTRY          SPAN_LEN

    1890        Cantilever  Forth Railway             United Kingdom      521
    1917        Cantilever  Quebec Railway            Canada             549
    1926        Suspension  Benjamin Franklin         NJ-PA, USA         533
    1929        Suspension  Ambassador International   MI, USA            564
    1931        Suspension  George Washington         NJ-NY, USA       1,067
    1931        Steel arch  Bayonne                   NJ-NY, USA         510
    1932        Steel arch  Sydney Harbor             Australia          509
    1936        Suspension  Transbay                  CA, USA            704
    1937        Suspension  Golden Gate               CA, USA          1,280
    1939        Suspension  Bronx-Whitestone          NY, USA            701

---------- New page of output begins ----------

Bridge Span Length Dataset
Group in lengths of 200 meters
                                                                                 &fmttype
                                       Cumulative   Cumulative
                                       Frequency    Percent
    SPAN_LEN   Frequency   Percent
    ----------------------------------------------------------------
    500-700        20        58.8         20          58.8
    701-900         5        14.7         25          73.5
    901-1,100       5        14.7         30          88.2
    1,101-1,300     3         8.8         33          97.1
    1,301+          1         2.9         34         100.0
```

25.2   Specialized Products

Section 25.1 described some of the features available in the core of the SAS System. Over time, as the SAS user base expanded into diverse application areas, SAS Institute began developing specialized products. Each is licensed separately and requires the Base product to run. Dividing the product base in this manner allows customers to lease only the products they want.

Among these add-on products are AF, an interactive application development tool; IML, a high-level language for matrix manipulation; GRAPH, routines for producing presentation-quality charts, plots, maps, and other specialized graphic output; OR, QC, and ETS, collections of operations research routines, quality control, and econometric routines; and FSP, a collection of development tools for "full screen" data entry applications. Most of the add-on products are available for all environments supported by Base SAS.

This section gives a quick overview of FSP, GRAPH, and IML. As in Section 25.1, the emphasis is on product capability rather than syntax. Each product is briefly described and then followed by a common, realistic example of its use.

EXHIBIT 25.12   Program with macro to conditionally execute PROCs

```
%macro bridge(group,ncases,cutoff,sortvar,print,freq);
data bridges;
infile '\data\raw\bridges' end=eof;
length type $1 name $20 country $15 span_len 4 complete 3;
input type $ name $ & country $ & span_len complete;
if span_len >= &cutoff. then do;
   count + 1;
   output;
   end;
if eof = 1 then call symput('total',put(count,2.));
format type $style. span_len comma5.;
drop count;
run;

proc sort data=bridges;
by &sortvar.;
run;

footnote1 "Listing produced at &systime. on &sysday., &sysdate.";

%if &doprint. = printit %then %do;
   proc print data=bridges(obs=&ncases.);
   id &sortvar.;
   title1 "Bridge Span Length Dataset, Sorted by Variable '&sortvar.'";
   title2 %if &ncases. <= &total. %then
              "First &ncases. Observations Out of a Total of &total." ;
              %else "All &total. Observations Are Printed" ;
          ;
   title3 "Only Spans &cutoff.+ Meters Long Are Included";
   run;
   %end;

%if &dofreq. = freqit %then %do;
   proc freq data=bridges;
   tables span_len;
   format span_len span&fmttype.x11.;
   title1 "Bridge Span Length Dataset";
   title2 "Group in lengths of &group. meters";
   run;
   %end;
%mend  bridge;

%bridge(100,20,700,complete,printit,nofreqs)
```

FSP: Full-Screen Editing

SAS/FSP is a collection of procedures that facilitates SAS dataset editing, simplifies generation of form letters, and provides a powerful spreadsheet that can read and write SAS datasets. The user has extensive control over screen design. Text attributes (color, intensity, blinking, and so on) and location are definable by the user if the default layout is not satisfactory.

This section uses FSEDIT, the dataset editor, as an example since it is probably the most popular FSP procedure. Keep in mind that FSEDIT's "look and feel" are characteristic of the other FSP products.

FSEDIT is a user-friendly means to enter and update data (another, less powerful technique, DATA step windows, was discussed in Section 25.1). Variable values for an observation can be entered, validated, and formatted automatically. Using SAS Screen Control Language (SCL), the programmer can enter elaborate calculations and cross-field validation checks for each observation. Options include definition of required and protected fields, control of field color, video attributes, location, and annotation with user-supplied text. The user can move forward or backward in the dataset and can locate records based on complex search criteria. Screen layout is done interactively and has ample on-line help available.

EXHIBIT 25.13 Log and output when macro in Exhibit 25.12 is executed

```
------------------- SAS Log ---------------------------

  60      %bridge(100,20,700,complete,printit,nofreqs)
MPRINT(BRIDGE):    DATA BRIDGES;
MPRINT(BRIDGE):    INFILE '\data\raw\bridges' END=EOF;
MPRINT(BRIDGE):    LENGTH TYPE $1 NAME $20 COUNTRY $15 SPAN_LEN 4 COMPLETE 3;
MPRINT(BRIDGE):    INPUT TYPE $ NAME $ & COUNTRY $ & SPAN_LEN COMPLETE;
MPRINT(BRIDGE):    IF SPAN_LEN >= 700 THEN DO;
MPRINT(BRIDGE):    COUNT + 1;
MPRINT(BRIDGE):    OUTPUT;
MPRINT(BRIDGE):    END;
MPRINT(BRIDGE):    IF EOF = 1 THEN CALL SYMPUT('total',PUT(COUNT,2.));
MPRINT(BRIDGE):    FORMAT TYPE $STYLE. SPAN_LEN COMMA5.;
MPRINT(BRIDGE):    DROP COUNT;
MPRINT(BRIDGE):    RUN;
NOTE: The infile '\data\raw\bridges' is file C:\DATA\RAW\BRIDGES.
NOTE: 34 records were read from the infile C:\DATA\RAW\BRIDGES.
      The minimum record length was 29.
      The maximum record length was 46.
NOTE: The data set WORK.BRIDGES has 14 observations and 5 variables.
NOTE: The DATA statement used 11.00 seconds.
MPRINT(BRIDGE):    PROC SORT DATA=BRIDGES;
MPRINT(BRIDGE):    BY COMPLETE;
MPRINT(BRIDGE):    RUN;
NOTE: The data set WORK.BRIDGES has 14 observations and 5 variables.
NOTE: The PROCEDURE SORT used 4.00 seconds.
MPRINT(BRIDGE):    FOOTNOTE1 "Listing produced at 19:30 on Sunday, 26NOV89";
MPRINT(BRIDGE):    PROC PRINT DATA=BRIDGES(OBS=20);
MPRINT(BRIDGE):    ID COMPLETE;
MPRINT(BRIDGE):    TITLE1 "Bridge Span Length Dataset, Sorted by Variable 'complete'";
MPRINT(BRIDGE):    TITLE2 "All 14 Observations Are Printed" ;
MPRINT(BRIDGE):    TITLE3 "Only Spans 700+ Meters Long Are Included";
MPRINT(BRIDGE):    RUN;
NOTE: The PROCEDURE PRINT used 10.00 seconds.
NOTE: SAS Institute Inc., SAS Circle, PO Box 8000, Cary, NC 27512-8000
```

Annotations (right margin callouts):
- &cutoff
- Indentation in DO loop is lost when printed.
- Statements included by entering "PRINTIT" in call to BRIDGE macro
- &sortvar
- &systime
- &sysday
- &sysdate
- &ncases
- &sortvar
- &total
- &cutoff

EXHIBIT 25.13   *(continued)*

```
------------------- Output listing --------------------

Bridge Span Length Dataset, Sorted by Variable 'complete'
All 14 Observations Are Printed
Only Spans 700+ Meters Long Are Included


COMPLETE      TYPE        NAME             COUNTRY         SPAN_LEN

   1931     Suspension   George Washington  NJ-NY, USA      1,067
   1936     Suspension   Transbay           CA, USA           704
   1937     Suspension   Golden Gate        CA, USA         1,280
   1939     Suspension   Bronx-Whitestone   NY, USA           701
   1950     Suspension   Tacoma Narrows     WA, USA           853
   1957     Suspension   Mackinac Straits   MI, USA         1,158
   1964     Suspension   Verrazano-Narrows  NY, USA         1,298
   1964     Suspension   Forth Road         United Kingdom  1,006
   1966     Suspension   Ponte 25 de Abril  Portugal        1,013
   1966     Suspension   Severn             United Kingdom    988
   1967     Suspension   Angostura          Venezuela         712
   1973     Suspension   Bosporus           Turkey          1,074
   1973     Suspension   Kanmon Strait      Japan             712
   1981     Suspension   Humber             United Kingdom  1,410
```

	&systime
	&sysday

Listing produced at 19:30 on Sunday, 26NOV89 ← &sysdate

For More Information.   The *SAS/FSP Guide* for versions 5 and 6 contains a thorough description of FSP syntax and many complete, annotated examples. The guides are published by SAS Institute.

Using FSEDIT.   The FSEDIT examples presented in this section illustrate default and customized screens. In both cases SAS checks entries for proper data type (character or numeric) and will not allow entry to go beyond a field boundary (defined by the row of underscores). The default data entry screen for an existing mailing list dataset is contained in Exhibit 25.14. All the user had to enter to see this screen was

```
proc fsedit data=in.maillist;
```

Although it is conveniently produced, this screen lacks organization and clarity. The situation is improved by entering MODIFY on the COMMAND line. FSEDIT presents menus that let the user arrange the fields more logically and replace variable names with explanatory text. The customization process can also include specification of user-written formats for use in data validation, identification of protected and required fields, and specification of initial values for a field. The customized screen may be saved as a permanent dataset for use in other sessions and with other datasets. FSEDIT's features enhance the screen's visual appeal and help reduce data entry error rates. The customized screen is shown in Exhibit 25.15.

EXHIBIT 25.14 Default FSEDIT data entry screen

```
                          Edit SAS data set: IN.MAILLIST      Screen   1
Command ===>                                                  Obs      1


                     NAME:        _____
                     ORG:         _____
                     ADD1:        _____
                     ADD2:        _____
                     CITY:        _____
                     ST:          __
                     ZIP:         _____
                     NEWSLETR:    _
                     JOURNAL:     _
                     TECHNOTE:    _
                     SEMINARS:    _
                     XMASCARD:    _
                     ANNOUNCE:    _
                     SECTOR:      _
                     FCONT_M:     ___
                     FCONT_Y:     __
```

EXHIBIT 25.15 Customized FSEDIT data entry screen

```
                          Edit SAS data set: IN.MAILLIST      Screen   1
Command ===>                                                  Obs      1

Name              _____
Organization      _____
Address           _____
                  _____
                  _____  __   _____
                  City              State ZIP code

Publications (mark those received with a nonblank character):
_ Newsletter
_ "GigaTech Times" quarterly journal
_ "GigaTech Notes" periodic technical notes
_ Seminar notices
_ Christmas card
_ Product announcements


Sector (p=private, g=govt, n=nonprofit, ?=unknown) _


Date of first contact: ___   __
                       month year
----------------------------------------------------------------------
Press F1 for HELP, F3 to EXIT, F7 to go back, F8 to go forward
```

**GRAPH:
Presentation-Quality
Graphics**

The graphics procedures described in earlier chapters (PLOT, Chapter 21, and CHART, Chapter 6) produced quick, useful, and somewhat crude displays. SAS/GRAPH not only improves the quality of these procedures but extends their capability as well. GPLOT, for example, can draw confidence intervals and regression lines, while PLOT can only display individual data points.

GRAPH also offers a variety of mapping, three-dimensional and contour plotting, and utility procedures. A special facility, ANNOTATE, allows you to place special text and graphic symbols on the output. Output can be sent to virtually any device: hundreds of terminals, plotters, film recorders, and printers are supported by GRAPH. The output may be saved as a specially formatted SAS dataset and "replayed" in later sessions.

In its simplest form, by taking default options, GRAPH is used simply by specifying GPLOT or GCHART and using program statements unaltered from their Base SAS counterparts. Typically, however, the reason for using GRAPH is to produce high-quality output. This requires modification of items such as fonts, graphic image size, page layout, text height and rotation, colors, shading, and axis definitions. GRAPH allows a near-bewildering flexibility in handling these options, enabling the adventurous user to produce extremely complex and information-filled displays.

For More Information.   The *SAS/GRAPH User's Guide* for versions 5 and 6 contains a thorough description of the syntax and many complete, annotated examples. *SAS VIEWS: SAS Color Graphics* contains the notes from the SAS Institute's color graphics course. All publications are published by SAS Institute.

Using GRAPH.   SAS/GRAPH examples often fall prey to the programmer's urge to sacrifice clarity for font and shading variety. Exhibits 25.16 and 25.17 (see p. 572) demonstrate some of the simpler and more useful options available in the GLPOT procedure. Stomach cancer incidence rates for males in selected U.S. study areas are plotted.

Several features of GRAPH that make GPLOT's presentation more effective than its PLOT counterpart: data points for a pair of variables can be connected; and legends can be placed anywhere on the page and can contain more extensive annotation than is possible with PLOT's defaults.

Part of the program to generate the graph is presented in Exhibit 25.16. Notice that the enhanced output is not without extra effort: line styles, heights, fonts, label rotation, axis order, and the like must all be specified. The output from the program is presented in Exhibit 25.17.

**IML: High-Level Matrix
Handling**

Data management and analysis tasks often require that you work within an observation, then across observations. For example, you may want to identify the test subject with the smallest average score for a set of measures. This requires computing averages for each observation, followed by a MEANS or UNIVARIATE. This is one type of scenario addressed by the Interactive Matrix Language (IML).

IML, an enhanced descendant of the MATRIX procedure, also addresses statistically oriented users. Most statistical formulas may be expressed in compact, matrix algebra notation. Simple measures like means as well as the exotic multivariate techniques such as factor analysis are all succinctly expressed in matrix notation. Row and column operations, eigenvalues, inverses, transposes, and other operations are fundamental to such calculations.

IML is a procedure that provides an elegant data manipulation language based on matrix algebra notation. Unlike the DATA step, which processes data

EXHIBIT 25.16   SAS/GRAPH program: DATA step and call to GPLOT

```
title1     h=1.0 f=none "Incidence of Stomach Cancer (ICD 151)";
title2     h=1.0 f=none "US SEER Areas 1973-1986, Males, All Races";
footnote1  h=0.7 f=none j=1 "Source: Cancer Statistics Review 1973-1986";

* Define line join, identification symbols ;
symbol1 i=join h=.6 l=1 v=2;
symbol2 i=join h=.6 l=1 v=4;
symbol3 i=join h=.6 l=1 v=5;
symbol4 i=join h=.6 l=1 v=6;
symbol5 i=join h=.6 l=1 v=7;
symbol6 i=join h=.6 l=1 v=8;

* Create ANNO1, an "annotate" dataset which will be displayed on top
  of the PLOT produced by PROC GPLOT.  It is the programmer's
  responsibility to position the box text with the X and Y variables. ;
data anno1;
length function style $8 text $16;
style = 'NONE';  xsys='1';  ysys='1';  size =.8;
function = 'LABEL';  position = '6';  x=77;
y =97;     text = 'Age at diagnosis';  output;
y =y-3.5; text = '2=25-44  4=45-54';  output;
y =y-3;   text = '5=55-64  6=65-74';  output;
y =y-3;   text = '7=75-84  8=85+'  ;  output;
run;

proc plot data=cancers anno=anno1;
axis1 value=(height=.8) label=(f=none height=.99) order=1973 to 1987 by 2;
      minor=none origin=(6);
axis2 value=(height=.8) label=(f=none a=90 r=0 height=1.10)
      order=0 to 20 by 2 minor=none;
plot rate * year = agediag
     / vzero frame nolegend vaxis=axis2 haxis=axis1;
label agediag  = 'Age Group'
      rate     = 'Incidence per 100,000'
      year     = 'Year of Diagnosis';
run;
```

one observation and variable at a time, IML's expressions work on matrices. This makes translating formulas into programs a relatively straightforward task.

SAS datasets may be read and modified in IML, and windows similar to those described in Section 25.1 may be used to facilitate program input and output. IML is a truly interactive procedure. A RUN statement is not required to initiate execution of a program since IML knows what activity (data movement, calculation, printing) each statement initiates. Programs may be compiled and stored in an executable form in later sessions.

For More Information. The *SAS/IML Guide for Personal Computers, Version 6 Edition* and *SAS/IML User's Guide, Version 5 Edition* describe the language and provide complete, annotated examples. Both are available from SAS Institute.

Using IML. Rather than develop a single example program, this section presents some of the more interesting features of IML's syntax and capability. First we

EXHIBIT 25.17   Output from program in Exhibit 25.16

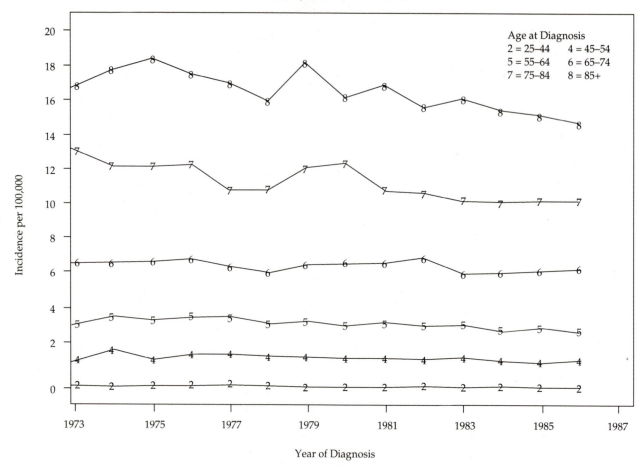

Incidence of Stomach Cancer (ICD 151)
US SEER Areas 1973-1986, Males, All Races

Age at Diagnosis
2 = 25–44   4 = 45–54
5 = 55–64   6 = 65–74
7 = 75–84   8 = 85+

Incidence per 100,000

Year of Diagnosis

Source: *Cancer Statistics Review 1973–1986*

see some compact and powerful operators. SUM and MEAN are scalars that are the sum and mean, respectively, of all elements in matrix GRADES:

```
sum = grades[+];
mean = grades[:];
```

To identify the name of the student with the lowest grade, use the >:< reduction operator to point to the corresponding element in the NAMES array:

```
lowest = names[grades[,>:<]];
```

The next example will read SAS dataset PFOLIO, selecting only those stocks with yields over 10%. Lower values are zeroed out.

```
use pfolio var(yield y87q1--y91q3);
find all where(yield > .10) into pf;
zap = loc(yield <= .10);
pfolio[zap] = 0;
```

In this example, the vector USETHESE is used to identify company id's for listing:

```
usethese = {"a128","a010","w101","w500","b003"};
list var{id,loc,perf,ratio,asset}
     where{id=usethese};
```

In this next example, all numeric variables in a matrix are summarized. The syntax and output are similar to the MEANS procedure:

```
use termdep where(due >= 24);
summary class(due_yr due_mth) var(branch1-branch5)
        stat(min man sum mean);
```

In the last example, regression parameter estimates (betas) are printed. Notice that IML creates the matrices with the correct row and column dimensions; no prior setups or declarations are required. Sophisticated functions and operators are available for more complex analyses.

```
xprodx   = indep ` * indep ;
xprody   = indep ` * dep ;
invxprod = inv(xprodx) ;
beta     = invxprod * xprody ;
```

Summary

This chapter broadly outlines some of the capabilities in SAS not covered in the rest of the book. Although this introduction is brief, it should reinforce what has been demonstrated throughout the book: that SAS is a diverse, powerful, and capable language. The chapter is divided into two main areas: products and features that are part of the standard, "base" SAS product and specialized add-on products.

Base SAS products and features outlined here include windows, customized displays for informative messages, and data input and output. The FORMS procedure facilitates compact arrangement of data with a minimum of programming effort. The TIMEPLOT procedure enhances printing of temporally oriented data by adding graphical output. The TABULATE procedure uses a complex, powerful syntax to produce attractively formatted tables containing descriptive statistics. The chapter also presents an overview of the SAS macro language. Macros and macro-variables allow high-level substitution of values in programs and conditional inclusion of statements to be executed. Macros encourage programming efficiency and help reduce errors in programs in which simple changes are often required.

Among the add-on products described in the chapter are FSP and its full-screen data entry procedure FSEDIT, which allow the programmer to design customized data entry screens with underlying range, consistency, and cross-form checks. The apparent similarity to the DATA step windowing capability quickly ends when you begin to investigate the wealth of screen management and data entry features built into FSEDIT. The GRAPH add-on takes the simple graphics found in the base product and adds color and shading, extended labeling and overlaying features, and three-dimensional renderings. Also described is the interactive matrix language, IML, a tool for problems that lend themselves to matrix manipulation. IML has an easy-to-learn syntax that can implement virtually any problem expressable in matrix notation. It is also an ideal tool for users who want to explore thoroughly the underlying principles behind the statistics other SAS procedures so easily generate.

Some Conclusions About SAS

Finally, there is the appeal of the SAS System to those who enjoy tinkering. SAS software offers a near limitless venue to explore. The DATA step has many features not discussed in this book. The macro and matrix languages are small self-contained worlds. The graphics and full-screen procedures have rich, specialized vocabularies and capabilities. For those with the urge to exlore, there is enough to satisfy appetites for months, years, or even careers. So, the SAS System not only aids the solution of *existing* problems, it also helps inquisitive minds *create* new problems. Seen in this light, SAS is truly a comprehensive system for your computing needs!

A APPENDIX: The ASCII and EBCDIC Character Sets

When SAS sorts a dataset, uses character variables in IF-THEN statements, or arranges groups in procedures such as FREQ or MEANS, it uses a character set to perform the comparisons. A *character set* is simply a standard arrangement of commonly used letters, numbers, and special symbols. The set designates some characters with "smaller" values than others. This designation is intuitive for letters and numbers (e.g., A is less than Z) but is not as obvious for special characters such as *, (, and \.

SAS operates in environments using the two most popular character sets: ASCII and EBCDIC. EBCDIC is used on IBM-compatible mainframe computers, and ASCII is used on virtually all other machines. It is good practice to know which character set is used on your computer and to be aware of some of its subtleties. If, for example, you are using the EBCDIC character set and are testing to see that an input character is a capital letter, the IF statement

```
if 'A' <= letter <= 'Z' then ...
```

would, strictly speaking, not be correct. EBCDIC contains the character "{" between I and J and "\" between R and S. If the input data contained these symbols, the IF statement would not yield a "false" value and the THEN clause would execute. Similar apparent anomolies exist elsewhere in both character sets.

Another reason to be aware of each set's character magnitude is cross-system development. Comparisons made in an ASCII environment may not perform as anticipated when moved to an EBCDIC computer. For example, consider the following statement:

```
if style >= 'A' then do;
```

The intent of the comparison is to identify values of STYLE that are alphabetic. In ASCII environments this will work. If run on an EBCDIC-based machine, however, this open-ended comparison will also be true for values of STYLE beginning with numbers. A better phrasing would be

```
if 'a' <= style <= 'z' | 'A' <= style <= 'Z' then do;
```

Knowledge of these and other peculiarities will prevent such logic errors or at least make their diagnosis a bit quicker.

A.1   Character Set Listings

Exhibit A.1 presents the character sets from their low to high values. Each character's representation can be expressed as a decimal or hexidecimal (base 16) number. These numbers are listed along with the characters themselves (if only to make the programmers reading this book feel more at home!). Note that ♭ indicates a blank character, while a blank in a position simply means that no character is defined for the character set or that the value is not one commonly used in most programs and data.

EXHIBIT A.1   Comparison of ASCII and EBCDIC character sets

Column Headings:
. Dec is the character's decimal value.
. Hex is the character's hexadecimal (base 16) value.
. A is the ASCII character associated with this decimal/hexidecimal value.
. E is the EBCDIC character associated with this decimal/hexidecimal value.

Dec	Hex	A	E	Dec	Hex	A	E	Dec	Hex	A	E	Dec	Hex	A	E	Dec	Hex	A	E
032	20		♭	065	41		A	098	62		b	132	84		d	201	C9		I
033	21	!		066	42		B	099	63		c	133	85		e	208	D0		}
034	22	"		067	43		C	100	64		d	134	86		f	209	D1		J
035	23	#		068	44		D	101	65		e	135	87		g	210	D2		K
036	24	$		069	45		E	102	66		f	136	88		h	211	D3		L
037	25	%		070	46		F	103	67		g	137	89		i	212	D4		M
038	26	&		071	47		G	104	68		h	145	91		j	213	D5		N
039	27	'		072	48		H	105	69		i	146	92		k	214	D6		O
040	28	♭	(	073	49		I	106	6A		j	147	93		l	215	D7		P
041	29	)		074	4A	¢	J	107	6B	,	k	148	94		m	216	D8		Q
042	2A	*		075	4B	.	K	108	6C	%	l	149	95		n	217	D9		R
043	2B	+		076	4C	<	L	109	6D	_	m	150	96		o	224	E0		\
044	2C	,		077	4D	(	M	110	6E	>	n	151	97		p	226	E2		S
045	2D	-		078	4E	+	N	111	6F	?	o	152	98		q	227	E3		T
046	2E	.		079	4F	|	O	112	70		p	153	99		r	228	E4		U
047	2F	/		080	50	&	P	113	71		q	161	A1	~		229	E5		V
048	30	0		081	51		Q	114	72		r	162	A2		s	230	E6		W
049	31	1		082	52		R	115	73		s	163	A3		t	231	E7		X
050	32	2		083	53		S	116	74		t	164	A4		u	232	E8		Y
051	33	3		084	54		T	117	75		u	165	A5		v	233	E9		Z
052	34	4		085	55		U	118	76		v	166	A6		w	240	F0		0
053	35	5		086	56		V	119	77		w	167	A7		x	241	F1		1
054	36	6		087	57		W	120	78		x	168	A8		y	242	F2		2
055	37	7		088	58		X	121	79		y	169	A9		z	243	F3		3
056	38	8		089	59		Y	122	7A	:	z	192	C0	{		244	F4		4
057	39	9		090	5A	!	Z	123	7B	#	{	193	C1		A	245	F5		5
058	3A	:		091	5B	$	[	124	7C	@	|	194	C2		B	246	F6		6
059	3B	;		092	5C	*	\	125	7D	'	}	195	C3		C	247	F7		7
060	3C	<		093	5D	)	]	126	7E	=	~	196	C4		D	248	F8		8
061	3D	=		094	5E	;	^	127	7F	"		197	C5		E	249	F9		9
062	3E	>		095	5F	^	_	129	81		a	198	C6		F				
063	3F	?		096	60	-	`	130	82		b	199	C7		G				
064	40	@		097	61	/	a	131	83		c	200	C8		H				

The EBCDIC character set is shown in Exhibit A.2. The "smallest" values are listed first. Thus EBCDIC considers the magnitude of "#" to be less than "a", and "a" in turn is less than "0". A blank (" ") is the smallest value.

EXHIBIT A.2   EBCDIC character set

```
b¢.<(+|&!$*);^-/|,%_>?:#@="abcdefghijklm
nopqr˜stuvwxyz{ABCDEFGHI}JKLMNOPQR\STUVW
XYZ0123456789
```

The ASCII character set is displayed in Exhibit A.3. Like the EBCDIC set, a blank is the smallest value.

EXHIBIT A.3   ASCII character set

```
b!"#$%&'()*+,-/0123456789:;<=>?@ABCDEFGH
IJKLMNOPQRSTUVWXYZ[\]^_`abcdefghijklmnop
qrstuvwxyz{|}˜
```

B APPENDIX: The SAS Style Sheet

Most SAS users, whether by accident or design, adhere to self-imposed rules even when writing trivial applications. This appendix discusses a set of SAS programming style guidelines. It is taken from the paper, "Good Code, Bad Code: Strategies for Program Design," presented at the Second Northeast SAS User's Group Conference, 1989.

The guidelines are deliberately general, making them appropriate for adoption or consideration in a variety of programming environments. Processes of program development and suggestions for program appearance are intermingled: no overriding model of program development is being built. The appendix's sole objective is to create a "style sheet" that may be helpful to novices and experts alike. Remember that even though the items are presented as absolutes, you should consider their appropriateness for your computing environment.

B.1 Program Design

This section discusses issues of program design and presents guidelines common to DATA steps and procedures.

Presentation

Use Consistent Case. Enter all program text except comments and character constants in either upper- or lowercase. This improves readability and makes most program editors' search functions perform more reliably.

Use Blank Space Liberally. Highlight subordinate code by using indentation. This is particularly useful when coding DO-loops. Split long or complex statements across generously spaced multiple lines.

Clearly Separate Units of Work. Use RUN statements to emphasize DATA step and PROC boundaries. Blank lines clarify the structure of long DATA steps by providing a visual break between logically related groups of statements.

The program segments below illustrate a simple application of these guidelines. The programs are functionally equivalent but are vastly different in their readability and the effort required to modify them. The first example is in "bad" code:

```
data out.test;
input id v1-v50;
length nbad 3; retain nbad;
if sum(of v1-v50 < 25 | nmiss(of v1-v50)
 > 10 then do;
nbad = nbad + 1; put 'bad obs, id: ' id;
if nbad > 20 then stop; else return;
end;
output;


proc means; var v1-v20;
```

The next example is a "good" (or at least more readable) program. The indentation clearly shows which statements are within the range of the DO, the alignment of the conditions in the IF make the statement's intent clear, and the asterisks accompanying the RUN statement highlight the separation of DATA step and PROC:

```
data out.test;
input id v1-v50;
length nbad 3;
retain nbad;
if sum  (of v1-v50) < 25 |
   nmiss(of v1-v50) > 10 then do;
   nbad = nbad + 1;
   put 'bad obs, id: ' id;
   if nbad > 20 then stop;
      else              return;
   end;
output;
run; ****************************;

proc means;
var v1-v20;
run; ****************************;
```

Naming

Choose Meaningful Names.   Datasets, variables, labels, formats, and other entities named by the user should communicate their content or activity. The eight-character limit for entity names sometimes thwarts effective identification: the variable name MGR2SBFP is forgettable and does not imply content. In these situations it may be more effective to name the variable via a cross reference to a data dictionary or data collection form such as a questionnaire. The seemingly obtuse Q12A may in fact provide more cues to meaning than the original MGR2SBFP, for example.

Avoid Default Names.   Be sure of dataset names: avoid defaults in the DATA step and the DATA= option and output datasets in PROCs. Many inexplicable analytic results can be traced to a nonexistent or vaguely specified dataset name.

Documentation

Use Labels.   Variable labels and their often overlooked cousins, dataset labels, enhance the description of entities.

Establish an Audit Trail.   Leave identifying information on all but the most formal output. This greatly facilitates reconciliation of program and hard copy output days or weeks after a run.

The DATE and NUMBER system options should be turned on. TITLEs and FOOT-NOTEs should be used to indicate the file name of the program and other relevant administrative details.

Explain with Comments. Use a comment block at the start of the program to explain the program's function, then use smaller blocks as necessary before and within DATA steps and PROCs. Insert comments at critical or potentially confusing sections of DATA steps.

Do not overcomment. Too much verbiage is as bad as too little. Increased text also increases the likelihood that updates to the program proper will not be reflected in the comments.

The following program fragments illustrate good use of comments. The first example is a program header:

```
*--------------------------------------------------
|
| NJOHNSON.TESTLIB.SAS(BUILD1)
|
| Read test data, use SUMMARY to collapse across
| id's, then CHART results.
|
| Calc's in DATA result of committee meeting
| 9/25/90.
|
| Macro COLLAPSE is in the AUTOCALL library.
|
| Next program can be either STAT1 or GRAPH1, same
| library.
|
*-----------------------------------------------;
```

The next example effectively uses comments in a DATA step to explain a valid but tricky section of code:

```
* NB: order of declaration is important. See defi-
      nition of Q array below. ;
length q1-q20 q21a q21b q21c q22-q25 $1 ;
* If question order above is changed, maybe change the
  ARRAY order below. ;
array q{27} q1--q25;
```

Environment

Choose Data Types Carefully. Do not assume that if a field looks numeric that it should be stored as such. Ask how the variable will be used: will it be a CLASS, STRATA, or other classification or grouping variable? Or will it be used in arithmetic operations? Most nominal-scale variables (e.g., REGION values of 10 and 15) are candidates for character data, as are items that will not be used arithmetically.

Careful selection of data types ensures a bit more "up front" thinking and careful planning. It also tends to move data into the more efficiently handled character representation, clearly a case of forethought being rewarded.

Document Use of Obscure Options. Some options make programs difficult to read and obscure their function. Highlight their use with an explanatory comment. Examples of such options are S= and NOSOURCE.

Strategy

Consider Cross-System Incompatibilities. If developing a program on one platform for eventual use on another, be sure that you use features of the SAS language common to both. A user developing an exotic PC-based program will be very disappointed when trying to execute the program in VAX version 5.18!

ASCII-EBCDIC character representations can also be problematic. A dataset in ASCII sort order is not in EBCDIC sort order. If you are developing systems in different environments, examine the collating sequences carefully, rewrite open-ended comparisons, and add SORT procedures as necessary. (See Appendix A for collating sequence details.)

Don't "Clever Code." One person's efficient, compact code is another's frustration. Write code to the level of its consumer, excising valid but "not obvious" coding conventions. A verbose series of statements may be less elegant than a single statement that is clever and compact, but it usually has the advantage of being easily grasped and modified. Which of these two equivalent program fragments is easier to read?

```
x = a>b * c>d;

if a > b & c > d then x = 1;
   else                 x = 0;
```

Do Not Overcode. Exploit SAS's powerful mixture of third and fourth generation programming languages. Use the PROCs to calculate statistics and perform other high-level activities. Use DATA steps to implement complex logic and carry out tasks not addressed by the PROCs.

The freeform structure of SAS programs complements this PROC–DATA step use. Rather than write a single long DATA step, consider whether the task could be accomplished by a series of DATA steps and PROCs gradually massaging data into the desired format. This flexibility is taken for granted by those of us raised on SAS. However, it is a continual source of confusion and apprehension for lower-level language programmers just learning SAS.

Develop Incrementally. Do not code the entire problem at once. Address the most important features first and deal with display and fine-tuning issues later. Prototyping in interactive SAS with a subset of the data or a "synthetic" dataset can cut development time significantly.

Assume Growth and Change. Even the most innocent-looking applications tend to grow in size and complexity. Avoid restrictive variable naming conventions and calculation methods. Programs, raw data, SAS datasets, and formats should be placed in separate libraries even if the initial statement of the project does not warrant such elaborate organization.

Use Libraries. Use your operating system's file management facilities (partitioned datasets, text libraries, directories, and so on) to your advantage. Programs can be logically separate from raw and SAS datasets.

Investigate the system's ability to annotate file contents. In IBM microcomputer environments, for example, the Norton Utilities FI and 4DOS DESCRIBE commands attach file descriptions much like SAS uses variable and dataset labels.

B.2 The DATA Step

This section discusses stylistic considerations unique to the DATA step. Many coding conventions useful in DATA have already been covered. Issues such as controlling the flow of execution, performing calculations, and designing effective, readable programs are addressed here.

Design

Group Unexecutable Statements. Statements that declare variables, set up arrays, and perform other housekeeping chores should be grouped together. LENGTH, RETAIN, and ARRAY statements should precede references to their entities' use in executable statements and so should be placed at the beginning of the DATA step. DROP, KEEP, ATTRIB, FORMAT, and LABEL statements can be grouped at the bottom. Consistent placement enables quick location of a particular statement. These conventions are illustrated below:

```
data temp;
length idflag $1;
retain idflag ;
array sums{20} sum1-sum20;
[ intervening statements ]
drop idflag;
label region = 'Sales region'
      area   = 'Region subdistrict';
format _numeric_ comma10.;
```

Declare Character Variables. Use LENGTH statements to declare lengths for character variables. This is particularly useful when creating variables via character-handling functions: although SAS follows rules for the lengths of new variables, it is far easier to state the lengths rather than try to fathom the rules.

Identify RETAINed Variables. SAS treats retained variables differently than others. You should do the same. Be explicit about them: use a RETAIN statement and begin their names with a readily identifiable character such as the underscore (_). Coding references to retained variables as $x + 1$ rather than $x = x + 1$ further emphasizes the retained nature of the variable.

Flow of Control

Ensure Closure of IF-THEN Sequences. When computing values via a series of IF-THEN statements, test *all* logical conditions and use an unconditional ELSE to catch untested categories. This practice is particularly important when the body of the DATA step is in a DO-loop and the DATA statement's normal data management facilities are bypassed.

Consider the example below. The DATA step, written without IF-THEN closure, and the affected variables are shown:

```
data badcode;
do until (eof=1);
   set x end=eof;
   if        1 <= trial <=  5 then t = '1';
     else if 6 <= trial <= 10 then t = '2';
      /* NOTE: no unconditional ELSE to end the
               IF-THEN-ELSE sequence */
   output;
end;

trial  t
   1    1
   8    2
   6    2
  12    2
   .    2
```

TRIAL values of 12 and . have a T value of 2 because the 2 was left over from the third observation. The DATA statement's resetting to missing values is bypassed since the SET statement is in a loop: the DATA statement is executed only once during the entire DATA step. The following example fixes this problem by adding an unconditional ELSE:

```
data goodcode;
do until (eof=1);
   set x end=eof;
```

```
     if          1 <= trial <=  5 then t = '1';
        else if 6 <= trial <= 10 then t = '2';
        else                          t = '?';
     output;
end;

trial  t
   1   1
   8   2
   6   2
  12   ?
   .   ?
```

Simplify Logical Expressions.   Write logical conditions and compound expressions as you would speak them: avoid excessively long statements, and reword "not"-filled conditions to their easier-to-grasp, positively stated equivalents. Use parentheses to highlight the order of evaluation of compound conditions.

Calculations

Simplify Complex Calculations.   Break complex calculations into several statements. The intermediate variables created by this technique are useful during the debugging process. At a minimum, split the calculation across several lines to improve clarity and highlight repeated or parallel operations.

Compare the two functionally equivalent assignment statements below. Which is easier to read? easier to modify?

```
state = put(input(substr(id,1,2),2.),stname.);

st_part = substr(id,1,2);
num_st  = input(st_part,2.);
state   = put(num_st,stname.);
```

Use Functions.   It is hard to think of a valid reason to hand-code a square root function or loop through an array to compute a univariate statistic. Use functions to simplify the program and make its output more reliable.

Use Arrays.   Just as functions make individual calculations more compact, arrays reduce code volume for groups of calculations. Statements that fall into a predictable pattern are usually suitable for array processing. Array coding makes the nature of the calculation apparent and simplifies the change process.

Compare the following sets of statements for clarity and compact expression. Also consider how the coding requirements would differ if 50 pairs of variables, rather than 5, had to be added.

```
x1 = y1 + y2;
x2 = y2 + y3;
x3 = y3 + y4;
x4 = y4 + y5;
x5 = y5 + y6;

array x{5} x1-x5;
array y{6} y1-y6;
do i = 1 to 5;
   x{i} = y{i} + y{i+1};
end;
```

Avoid Mixed Data Type Calculations.   SAS does its best to use character variables arithmetically. It usually, but not always, performs as expected, and always takes a surprising amount of CPU time to do so. Use the INPUT function to convert "character numbers" prior to the arithmetic operation.

Use Formats and IN to Recode. The IN operator and user-written formats are extremely effective and efficient means to group character and numeric data. Unless the number of objects and target groups is small (fewer than, say, 10), avoid using IF-THEN sequences for recoding.

B.3   Procedures

Although most procedures do not have as great a potential for improvement from good coding practices as DATA steps, there are some steps to take. These are mostly high-level issues of selection and use rather than the finer-grained DATA issues of display and organization.

Review Periodically. Most users tend to use PROCs for particular tasks. Periodically reread the documentation for "familiar" procedures. Being reminded of options and capabilities can help avoid unnecessary work. Review is especially important when starting to use a new release of SAS or switching operating systems.

In a similar vein, review the default assumptions of often-used PROCs. Although they are usually reasonable, your needs may change and be incompatible with the initial settings. Use the devil's definition of default: "the vain attempt to avoid errors by inactivity."

Use New PROCs Gradually. When using a new or "long lost" procedure, do not start with a complex request unless you are sure of the impact of the options and parameters. Review default options and start with as many as possible, gradually developing the more complex analysis. This is especially appropriate for multivariate, exploratory statistics and graphics. (Who, after all, ever got a complex CHART request correct the first time?)

Reconcile Log to Output. Suppress the "I finally got some output!" response. Output may appear to be reasonable but there may be Log messages indicating problems with numbers of observations, missing values, and so on. Not all procedures print such diagnostics on the output.

Use DATASETS. Use the DATASETS procedure when the only activity you need to perform on a disk dataset involves directory maintenance (rename entities, add formats and labels, delete datasets). Such activities do not require a pass through the data.

Exploit Output Datasets. Output datasets from PROCs can save huge amounts of work. Familiarize yourself with the contents of procedure output datasets. The easiest way to understand them is to print them and run a CONTENTS procedure. PROCs that utilize CLASS or BY statements or that have many computational options often reflect this complexity in the output datasets as well.

Let the PROCs Do Your Work. Unless computer time is at a premium or there is a "hand coding" ethic at work, there is no reason to compute statistics or design routine graphics within a DATA step. Let PROCs do the generalized tasks they were designed for and do so efficiently.

Choose among Overlapping PROCs. Many procedures, particularly for univariate statistics and regression, have overlapping capabilities. You can often save a lot of time if you know which PROC is best suited for a task. For example, if you simply need a listing of means, MEANS is more appropriate than UNIVARIATE. If the work is more exploratory, UNIVARIATE's comprehensive default statistics is better suited.

B.4 A Word About Style

These programming guidelines reflect the author's style and biases. Are there other forms that might be considered good style? Certainly. "Good" is in large part in the eye of the beholder: some like generous indentation and uppercase, others minimal blank space and capital letters. Users in both camps can also produce accurate, correct programs. The real determination of "good style" is not in the appearance of the program and the design of the system but in the effectiveness of the product.

This effectiveness is measured by the criteria of accuracy, consistency, and efficiency. The style can rightfully be assessed "good" if the coding style results in a product that meets specifications, consistently employs techniques that highlight the program's logic, and creates a better balance of person-machine resources than would have existed in its absence.

C APPENDIX: Getting Started at Your Installation

No SAS book can be simultaneously exhaustive, readable, and physically portable. Because of the language's broad scope, many decisions about what topics to cover and what level of detail to include had to be made during the writing of this book. The breadth-depth trade-off is especially evident when you actually begin using SAS at your site. There are simply too many system- and site-dependent features to cover thoroughly in a general-purpose book.

What *can* be anticipated, however, are some of the questions a newcomer to SAS typically asks. This appendix includes both SAS- and system-specific questions you should ask as a new SAS user at your installation. The questions are particularly appropriate for an academic setting but also apply to any environment. Many resources that will help you run SAS programs and learn more about the product are also discussed.

C.1 SAS-Specific Questions

The first thing you should do is ask some questions about the SAS products available at your installation. SAS at one site is not necessarily the same as SAS at another; it often has different meanings and connotations, depending on the environment. Some of the things you should find out include what version of SAS is used, what additional products are available, how SAS is run at the site, what user-written PROCs are available locally, what the supplemental library includes, and if a micro-host link exists.

Version

What release of SAS is used at your site? For fairly straightforward applications like those described in this book, the differences between versions are slight. More advanced applications, however, may rely on the most current release. By the middle of 1991, version 6.06 should be available at most minicomputer and IBM mainframe sites. Two of its major enhancements to earlier releases include an improved user interface and better dataset access methods. These features are summarized in Appendix F. Your successful use of SAS as described in this book does not depend on using the most current release: the basic data manipulation, analysis, and reporting capabilities you need to get started will be found in *all* versions of the language. Version 6.06 has nicer packaging and has some features of interest to "power users."

Additional Products

SAS Institute produces a wide variety of additional products that run under the SAS System "umbrella" (some of these were discussed briefly in Chapter 25). These products address specialized topics such as econometrics, quality control, and operations research. Powerful graphics and matrix manipulation products are also available. Find out what is available at your site. This may save you either doing it the hard way in SAS or having to learn SAS and another, more specialized package to satisfy your computational needs.

Running SAS

Each installation has a slightly different way of running SAS. Find out the commands to run SAS in both interactive and batch modes. If you are familiar with writing command procedures (DOS BAT files, VAX COM files, TSO CLISTs, and so on), you may want to customize the system commands to preallocate files, use nonstandard options, and so forth. An alternative to the custom command procedure is a start-up file, discussed in Appendix D.

Local PROCs and Functions

A special library of SAS Institute routines helps programmers develop their own procedures and functions and run them just like any other SAS product. Find out if any of these user-written routines are available at your site. These programs may save you a great deal of time and effort, and not force you to reinvent the proverbial wheel.

Supplemental Library

Version 5 of the SAS System sometimes has an accompanying "supplemental library" of user-written procedures. Much like locally written procedures and functions, these perform useful tasks not covered by the SAS System itself. Although the Institute distributes these procedures, it does not usually support them, so you run the (slight) risk of having to contact the author in case something goes wrong or needs explanation.

Micro-Host Link

Many sites support SAS products that allow a microcomputer to behave both as a stand-alone micro and as a terminal connected to a minicomputer or mainframe. SAS datasets on the mainframe can be read or written by the micro or downloaded to it. Find out whether such a link exists and what software and hardware are required. Locally written documentation detailing the steps required to perform the file transfer is almost always available.

C.2 System-Specific Questions

No matter how much you rely on SAS to fill your data processing and analysis needs, you still need to interact with the operating system. A few issues to address include what operating system is used, what usage restrictions are enforced, how jobs are executed, and what program editors are available.

System Choice

Is more than one operating system available? If so, which is the system of choice for running batch SAS? interactive? The choice between systems may already be made for you. If not, solicit the opinions of fellow students or coworkers. Such a seemingly innocent query often reveals near-religious "belief" in a particular system. Be prepared for a variety of emphatic, and differing, opinions.

Restrictions

When running on a minicomputer or a mainframe, you often have to deal with usage restrictions. Some systems (e.g., many IBM WYLBUR sites) do not allow access to interactive SAS. Other sites may restrict larger jobs to run during non-"prime time" shifts. Such restrictions may encourage microcomputer usage for entire projects or, at a minimum, for program development and testing. SAS's similar capabilities across operating environments make the cross-system development approach an appealing one: with a little care and planning, you can develop a fully functional program or system on your microcomputer, then run the program on the mainframe or minicomputer.

Job Execution

Methods for job submission and retrieval vary greatly from site to site. For batch programs, find out how to submit a job, know when it has finished executing, cancel it before it begins to execute, display output at your terminal, and print the output. Interactive execution is more straightforward but may still have some quirks that are documented locally.

Interactive sessions in some sites may take advantage of the SAS Display Manager System (DMS). This is a session controller with very powerful but resource-hungry features. With Display Manager you have the ability to read and modify a program, run it, view the output, save and print the output, and perform many of the dataset management tasks described in Chapter 10 via special windows. Window color and size may be configured by the user. Function key settings may also be changed and reused in later sessions.

Program Editors

Minicomputer and mainframe sites usually have several program editors (also known as text editors) available. Some are easier to use and more powerful than others. Find out which editor is usually used with SAS and try it out. The Display Manager has a very powerful and intuitive editor and is a good alternative to many stand-alone editors.

C.3 Resources: Where to Go for Help

Knowing which questions to ask is the first step. Knowing where to ask them is the next. In addition to identifying sources of assistance, this section also discusses where to go for further instruction and lists references for SAS products.

Troubleshooting

A few people-oriented sources of help are usually available to answer questions of program design and error correction.

Help Desks. Also referred to as *user service* or *information centers*, these groups are organized to provide walk-in and telephone assistance to computer users in an organization. They also usually conduct seminars for the user community (discussed in the Training section below). Find out their location and phone number and do not be shy about asking what may seem to be silly questions. *Everyone* was, after all, a beginner at some point in his or her career.

Gurus. These are the people in your organization who appear to know everything, are able to explain what they know, and do not mind giving advice even if it is not an official part of their job. Word of mouth is the best way to identify these people. They are usually more willing to help if the problem is complex and challenging. These people can be an invaluable resource. Do not abuse their good will: approach them only if the help desk is unable to solve your problem.

References

Many illustrations of SAS syntax and usage exist in many forms.

SAS Sample Library.   SAS Institute distributes a sample library with each of its products. These libraries contain data and programs that illustrate effective use of the DATA step and PROCs. Ask someone at the help desk for the location of this library, then look through it to get a feel for its contents.

Usage Notes.   Another product shipped by SAS Institute is its Usage Notes. This is a SAS dataset indexed by application area (DATA step in VMS systems, PROC CHART, and so on) describing known problems in the SAS System. Sometimes (rarely) you follow the syntax instructions to the letter and the program still does not behave as expected. Such behavior suggests referral to the Usage Notes. Ask someone at the help desk for their location. Some sites restrict access to these notes to systems personnel.

On-Line Help.   A useful feature of most SAS environments, including batch mode, is the help facility. Entering *HELP*; as a program statement gives a list of help topics, while *HELP topic*; lists help on a particular topic. The help command issued from the command line of a Display Manager window opens the HELP window. The window can remain open throughout the SAS session, making help almost instantly accessible. Be warned that on-line help is always more terse and contains fewer examples than printed documentation. It is used to best advantage when you know the PROC or DATA step feature you want but cannot remember the exact syntax.

Menus.   If you use the SAS Display Manager during interactive sessions, you may also take advantage of a series of menus supplied with the SAS System. These are fill-in-the-blank screens that guide you through using most of the procedures available on your system. They do *not* address DATA step programming. To use the menu system, enter MENU on the command line of any screen, then press the ENTER key.

The menus are helpful if your needs are fairly simple and you do not mind the relatively sluggish response times created by the menu system's overhead. Some users find that for more complex analyses it is faster to simply look up the procedure's syntax and write the PROC statements.

SAS Institute Documentation.   SAS Institute publishes a wide variety of documentation for its products. Its guide to the DATA step and basic statistical procedures, *SAS User's Guide: Basics*, is the definitive reference for anyone who wants to seriously explore the language's possibilities. Other documentation includes the operating system-specific *Companion* series and a *SAS Views* series (the bound notes for the Institute's courses). The Institute also publishes a quarterly magazine, *SAS Communications*, distributed free of charge. For information about any of these products, call SAS Institute at (919) 677-8000 and ask for the publications division, or write for information at P.O. Box 8000, Cary, North Carolina 27511-8000.

Local Documentation.   Most organizations produce documents that describe how to run SAS at the site, presenting examples of typical tasks and sometimes even teaching some of the language. Ask your help desk personnel about these publications.

Training

A wide variety of classroom and other forms of training is available.

Short Courses.   Many sites offer one- or two-session seminars that cover aspects of SAS usage. These are typically noncredit informal sessions, are usually free of charge, and are particularly good for learning about quirks or features of SAS usage at your organization. Help desk personnel, electronic bulletin boards, and computation center newsletters usually contain course schedules.

Third-Party Courses. SAS Institute, other training organizations, and consultants also offer courses. Their completion sometimes earns continuing education credits. In some cases an organization can have the course taught on-site and tailored to its particular requirements.

For-Credit Courses. Some universities and community colleges offer SAS instruction either as a course in itself or as an adjunct to a data management course. Look for these courses as part of data processing, biostatistics, or, less frequently, computer science department offerings.

Computer-Based Training. SAS Institute and several other vendors offer computer-based training (CBT) courses aimed at a variety of levels of user expertise. These courses are self-paced and usually allow direct interaction with the SAS System itself. Help desk personnel can advise on their availability at your site.

Video Courses. SAS Institute and other training groups sell or rent video courses on several beginning- to intermediate-level topics. These are especially useful learning tools when used in live, classroom training settings.

Other Sources

And if all the sources above were not enough, here are two more. These are less organization-bound and more free-form than the others.

User Groups. SAS user groups have formed all over the world. They usually have periodic meetings where people present papers or tutorials, discuss common problems, and "network." The groups often maintain electronic bulletin boards, publish a newsletter, and offer a job referral service. The largest of these groups are the SAS User's Group International (SUGI) and the Northeast [United States] SAS User's Group (NESUG). SUGI's annual conference draws more than 4,000 people, NESUG's about 1,000. Hundreds of papers are presented and later published in a Proceedings. The Proceedings are a valuable source of hints, tips, industry-specific usage, and tutorials. For information about SUGI and NESUG and assistance in locating local groups, call SAS Institute at (919) 677-8000.

List Servers. SAS discussion groups are found on several worldwide computer networks. The most active of these is the SAS-L list server on BITNET (or CREN), with over 850 regular discussants. Topics include routine problem solving, requests for information about competing or complementary products, and the occasional "flame war" (discussions of the "I can't believe anyone could be so wrong" variety). The servers are an invaluable gateway to the outside world, particularly in sites where local expertise is lacking. If you are already on a network such as USENET, BITNET, or CompuServe, try to obtain a list of discussion groups, then read the system help files for information on how to sign on to them.

D

APPENDIX: System-Specific

Illustrations

No matter how well a SAS program is written, it is useless if it cannot interact with your computer's operating system. The operating system is the software that controls how commands are processed, datasets are stored, and so on. When an INFILE statement specifies a dataset name, for example, SAS uses operating system routines to verify that the dataset exists. The operating system is extremely complex but the user usually requires only a cursory knowledge of its commands to run SAS effectively.

SAS runs on many brands of computer hardware and operating systems. This appendix describes some typical tasks in four of the major systems: IBM's MVS and CMS, Digital Equipment Corporation's VMS, and the Microsoft and IBM implementations of DOS. Each section describes how to allocate and use raw datasets, SAS datasets, and user-written format libraries. Unique and noteworthy features of the systems are also presented when appropriate. When reviewing the examples, keep one overriding consideration in mind: most sites implement slight variations of the generic examples presented. The examples are good templates for communicating your needs to the system, but should not be taken as is. Consult your help desk personnel for site-specific information (see Appendix C for more details).

Despite often profound differences in syntax and philosophy, the four systems share some features. Section D.1 discusses common features of SAS's operating environments in the four systems. The remaining sections demonstrate a series of typical tasks in each system.

D.1   Operating Environments

The SAS System may be used in a variety of operating environments. You may use a "full screen" editor, send the SAS job to another microcomputer on a local area network (LAN), or specify a program and let SAS take over your terminal while it runs. The way SAS runs is determined by the operating environment you select. SAS supports two major modes, or environments: batch and interactive.

While reviewing the environment descriptions below, bear in mind that during the development of a program you are not restricted to only one environment. SAS statements described in this book run identically in both batch and interactive modes. Very often a simple, stripped-down version of a program with test data will first be run interactively. Once the basic logic is verified and the fine-tuning completed, the program may run against the entire dataset in batch. Be aware that you have the flexibility to switch between environments to take advantage of each one's strengths.

Batch Environments

In a batch environment the SAS program is processed without any direct control by the user. The program is contained in a file submitted to the system for processing. Once SAS starts processing the program, it assumes that the syntax is correct, the required files are available, and so on: there is no opportunity for the user to alter the program once it starts. The results of the program are either routed to a printer or sent back to the user's terminal for examination.

Programs run in batch mode are usually less expensive than their interactive equivalents. They also have the advantage of freeing the user to do other tasks while the program is running. A drawback of using batch environments is the delay between the time you send the job to the system and when it finishes processing. If many people are contending for system resources, this turnaround time may become unacceptably long. A more serious drawback is the inability to change the program once it starts. Sometimes a static, predefined description of your analysis is adequate, but usually it is better to be able to alter analyses and perform new ones on an ad hoc basis. These needs are addressed in interactive processing.

Interactive Environments

The interactive environment is characterized by a greater degree of control and flexibility than batch environments afford. In two of its three variations the user enters or edits SAS statements, runs them, and sees results immediately. The user then has the option of changing the statements or writing new ones, running these, and again seeing immediate feedback. This cycle can be repeated many times without ever leaving SAS.

The benefits of the interactive SAS environment are its immediacy and flexibility. If you have syntax or logic errors, you can correct them immediately. If an analysis has anomalous or intriguing results, you can immediately run other DATA steps and procedures to explore the problem further. This ability to correct errors immediately is a powerful incentive to use interactive SAS. In environments that charge for computer usage, however, the flexibility of the interactive environment often carries a heavy price. Because the interactive use of computer resources takes priority over other (batch) users, the charging rate for interactive SAS is often much higher than for batch. There are three varieties of interactive modes: command line, line, and Display Manager.

Command-Line Mode. Sometimes referred to as noninteractive mode, this method of running SAS has the immediacy of interactive mode but the inflexibility of batch. It is usually used when a program is fully debugged and when there is no need to perform the ad hoc analysis common in the other forms of interactive SAS. This is the least resource-consumptive method of running SAS interactively. Details on running it in each operating system are presented later in this appendix. The general form of the command entered at the system prompt is

```
SAS pgmname [options]
```

where *pgmname* is the name of the file containing the SAS statements to execute and *options* identifies one or more SAS System options.

Line Mode. The second form of interactive SAS is line mode, so called because statements are sent to SAS one line at a time. If the statement is syntactically correct, SAS prompts for another line. If not, you may have the chance to correct the line. A RUN statement actually sends the statements to SAS for processing. Line mode is useful when

DATA steps or PROCs are short and not much typing needs to be done. The lack of good error recovery diminishes its appeal.

Details on running line mode in each operating system are presented later in this appendix. The general form of the command entered at the system prompt is

```
SAS [nodms other_options]
```

where *nodms* is a SAS option to suppress the SAS Display Manager (the third interactive mode), and *other_options* identifies additional SAS System options. The *nodms* may be the default in some installations.

To end a line mode SAS session, enter the ENDSAS command (discussed in Chapter 3), as shown in the example below:

```
proc print data=x.app01;
endsas;
```

In this example, the PRINT procedure will print dataset *x.app01* before ending the SAS session.

Display Manager Mode. Also referred to as full-screen mode, this form of interactive SAS is the most powerful and the most interactive. The complexity and scope of the Display Manager System (DMS) put it beyond the scope of this book. A brief description, however, is worthwhile.

The DMS is a collection of screen partitions, or windows, corresponding to the units of work in a SAS session. Among these are the "basics": SAS program statements, the SAS Log, and output. More sophisticated windows are also available: extensive command and procedure help, user-defined data entry windows, function key and system option redefinition, and SAS library browsing.

The program window puts a powerful program editor at your disposal. Programs can be included from external files, modified with the DMS editor, executed, and saved back to the external file. Log and procedure output can be examined, then deleted, printed, and saved to an external file. Even if only a portion of its capabilities are used, the DMS usually increases productivity and makes interactive SAS a more pleasant experience.

The downside to using DMS is its memory requirements. In microcomputer environments it usually requires about 110,000 bytes more than line mode or batch. In some cases the "cost" of extra memory is acceptable. Sometimes, however, DATA steps or PROCs might require so many resources that execution within DMS is impossible. Since DMS and line mode SAS are virtually identical it is usually a simple matter to save the program, exit DMS, and rerun the program in batch or line mode.

Details on running DMS in each operating system are presented later in this appendix. The general form of the command entered at the system prompt is

```
SAS [dms other_options]
```

where *dms* requests use of the Display Manager and *other_options* identifies additional SAS System options. The *dms* may be the default at some mainframe computer centers and is usually the default in microcomputer environments. To end a Display Manager session, execute the ENDSAS command or enter BYE on the command line of any window, then press the ENTER key.

Running Host Commands in SAS

Host operating system commands may be executed from an interactive SAS session or a batch job. The principle is the same among the four operating systems discussed in this appendix: you indicate to SAS that you want to execute a host command or, for interactive sessions, that you want to leave SAS temporarily and enter a series of host system commands. SAS responds to your request either by executing the command or by placing you in the host system. In the latter case you return to the SAS session by entering a special command (EXIT, RETURN, or END).

Host command execution is handy when you are in the middle of a session and realize you need to allocate another dataset, want to see one of the system help files, respond to electronic mail, and so on. Rather than exit SAS to execute the commands, you can stay in SAS and run them (provided your session or computer is equipped with sufficient memory).

Host commands may be placed anywhere in the SAS program. They may also be executed (minus the semicolon) on the command line of a Display Manager window. The command is executed immediately: SAS does not wait for a RUN; statement or a new PROC or DATA step to begin processing. Execute the command by entering one of the following commands:

```
X ['command'] ;      (PS/MS-DOS, VMS, MVS, TSO)
TSO [command] ;      (TSO)
CMS [[CP] command];  (CMS)
```

TSO, CMS, and *CP* are special formats for TSO and CMS/CP environments. In all cases, if *command* is omitted or blank, SAS passes control of the session to the operating system. You may enter as many host (TSO, CMS, CP, DOS, VMS, and so on) commands as required, then return to the SAS session by entering one of the following commands:

```
end       (TSO, CMS)
return    (CMS)
exit      (PC/MS-DOS)
logoff    (VMS)
```

The following statements either execute host system commands or pass control of the SAS session to the operating system.

```
tso;
tso help allocate operands;
x "alloc da('toxic.masters.sas') fi(sasdata)";
cms filedef input disk part1 data ;
cms cp link sys01 191 193 rr;
cms access 193 d;
cms filedef library disk x x b;
x '';
x 'dir \clintri\laetrile\*.ssd';
x 'dir [mortality.data]';
```

D.2 IBM MVS/TSO

IBM's MVS and TSO environments are not the easiest for beginners to learn, especially for those coming from microcomputer or minicomputer systems. There are two reasons for this. First, the batch and interactive data management languages (JCL and TSO, respectively) are different: using SAS in both requires learning two fundamentally different languages. Second, the level of detail required to make datasets available to SAS is much greater than experienced users of other systems are used to and more than novices would usually like.

The good news is that once you get past the JCL and TSO, SAS is the same in both environments. Only the "hooks" into the system differ. A program tested in interactive SAS usually requires modification only of the control language to work properly in batch. Raw and SAS datasets and format libraries created in batch SAS programs may be used with interactive SAS, and vice versa.

> **VERSION 6.06 UPDATE**
>
> The differences between SAS programs using TSO and JCL decreases dramatically in version 6.06 of the SAS System. FILENAME was not implemented and LIBNAME only partially implemented in version 5. In version 6, however, LIBNAME and FILENAME statements will be fully supported and the creation of format libraries will be simplified. Since FILENAME and LIBNAME are discussed elsewhere in this book, the examples in this section emphasize release 5 of the SAS System.

When using this section, keep in mind the following points about its presentation:

- All JCL text must be entered in uppercase. The single exception is specification of SAS options, discussed in the following "Running SAS" section.

- Although JCL and TSO commands are fairly standardized from site to site, there are local variations in how some features are specified. Items that often vary (other than, of course, dataset names) are presented in italics in the text. Keep in mind, though, that you may have even more site-specific idiosyncracies. Refer to Appendix C for information on who to contact for a definitive explanation of running SAS at your site.

- The TSO ALLOCATE command is usually enough to meet file creation, use, and allocation requirements. Like any TSO command, it may be continued on another line of the terminal by entering a hyphen (-) as the last character in a line. This continuation feature is demonstrated in several examples (D.1, D.4, D.7, D.8, D.14). Another useful feature of ALLOCATE is the REUSE option. If you want to use a FILE name that has already been used, specify REUSE. This releases the previous dataset allocation and makes the name available for use again. Finally, if you do not specify required parameters, ALLOCATE will either prompt you for them or assign default values.

- Specification of names in interactive mode (i.e., TSO) may take two forms. The first is the "fully qualified" name: user identification followed by one or more dataset name "levels." These names are always specified in single quotes (e.g., 'MIS098.FISCAL .DATA'). The second form assumes that the user identification used to log into the TSO session is the same as the first node of the dataset name. If this is the case, the single quotes and the first node of the dataset name are not required. If user MIS098 logged into TSO, the fully qualified dataset name used above could be entered as FISCAL.DATA.

- In the JCL examples, only the relevant JCL statements and SAS program fragments are presented. JOB and EXEC statements are usually excluded to prevent clutter.

Running SAS

Batch. The SUBMIT command is usually used to submit a file for batch processing. Suppose a SAS program was contained in the file FIN201.SAS. While in an interactive environment (TSO, WYLBUR, and so on), enter the command

```
SUBMIT FIN201.SAS
```

to begin execution of the program. Consult your local documentation for details about job queues and other options for the SUBMIT command.

Usually all the JCL required to run batch SAS is an EXEC statement of the form

```
//    EXEC SAS
```

To specify SAS system options (OBS, CENTER, and so on) modify the EXEC as follows:

```
//    EXEC SAS,OPTIONS='NODATE,NONUMBER,OBS=1000'
```

The text enclosed in the single quotes may be entered in either upper- or lowercase.

Some sites do not simply call the SAS procedure "SAS" but also require some indication of which version of the SAS System you want to run. Thus the name may be SAS518, SASNEW, and so on. Another variation is the requirement of a special PROCLIB statement pointing to the location of the system procedure required to run SAS. Ask your help desk personnel for the exact syntax appropriate at your installation.

Interactive.   In most cases all that is required to begin an interactive SAS session is a command of the form

```
sas
```

To run SAS and specify SAS system options, use the OPTIONS parameter, illustrated below:

```
sas options('nocenter nodate nonumber')
```

To run SAS in foreground mode (i.e., let SAS take control of your session but not prompt you for statements to execute), use the INPUT option of the SAS command:

```
sas input('sysadmin.monthly01.sas')
```

The value associated with INPUT is the name of the SAS program to execute. See the "Program File Format" section on page 604, for notes about this dataset's physical characteristics.

If you will be sorting data during the SAS session, use the SORT parameter:

```
sas sort
```

If *sort* is not specified, SAS will not allocate the special files required for PROC SORT's temporary files. This default option may be explicitly specified by entering *nosort*.

The cautions about site-specific changes to batch SAS also apply to interactive SAS: the command might not be simply "SAS", the method for invoking options may be different, and so on. This section simply illustrates some of the key options and how they are *usually* specified. To get the exact specifications for your site, enter HELP SAS from the TSO READY prompt or contact your help desk personnel (see Appendix C for more details).

Handling Raw Data

This section presents examples of reading and writing raw data in IBM MVS and TSO environments. For each task a batch (JCL) example is given, followed, when appropriate, by its equivalent interactive (TSO) example.

Example D.1. Read a disk dataset. Since the dataset is not cataloged, specify the volume name and type on which it resides. SHR indicates that the dataset may only be read, a good tool to use to prevent accidental overwriting. The JCL example follows:

```
//IN DD VOL=SER=OPEN20,UNIT=DISK,DISP=SHR,
//      DSN=MIS098.FISCAL89.DATA
```

Its equivalent in TSO looks like this:

```
alloc file(in) dataset('mis098.fiscal89.data') -
      shr volume(open20) unit(disk)
```

In both cases the SAS program could use an INFILE specification of INFILE IN;.

Example D.2. This example is the same as Example D.1 except that the dataset is cataloged. The operating system knows where to look for the dataset, so all you have to supply is DSN and DISP. This is the most common way to use raw datasets. The JCL example follows:

```
//IN DD DISP=SHR,DSN=MIS098.FISCAL89.DATA
```

Its equivalent in TSO looks like this:

```
alloc file(in) shr dataset('mis098.fiscal89.data')
```

Example D.3. Use a member of a cataloged partitioned dataset (commonly referred to as a PDS). A PDS is a single dataset that is subdivided into smaller entities called members. The desired member name is indicated in parentheses at the end of the dataset name. The JCL example follows:

```
//IN DD DISP=SHR,DSN=MIS098.FISCAL.DATA(FY1989)
```

In the equivalent TSO command, we assume that the user is logged on with the user id "MIS098." This relaxes the requirement of entry of the fully qualified dataset name.

```
alloc shr dataset(fiscal.data(fy1989)) file(in)
```

A special feature of SAS in IBM environments allows the user to specify the member name in the INFILE statement rather than in the JCL or TSO commands. This allows you to refer to different members in several DATA steps but code only one DD or ALLOC statement. If we omitted the (FY1989) member references above, we could write the following IN-FILE statements:

```
infile in(fy1989);
infile in(fy1989) firstobs=300 obs=400;
```

Example D.4. Concatenate several raw datasets. Notice that they can be any combination of PDS's and "regular" datasets. They are read by the DATA step in the order in which they are defined in the TSO command or JCL. The JCL example follows:

```
//DATAIN DD DISP=SHR,DSN=MIS098.FISCAL87.DATA
//       DD DISP=SHR,DSN=MIS098.FISCAL88.DATA
//       DD DISP=SHR,DSN=MIS098.FISCAL(FY1989)
//       DD DISP=SHR,DSN=MIS098.FISCAL(FY1990)
```

In the equivalent TSO command, we once again assume that MIS098 was used to log into the session:

```
alloc file(datain) shr dataset(fiscal87.data -
      fiscal88.data fiscal(fy1989) fiscal(fy1990))
```

Example D.5. Read a dataset from a standard label tape. Tape access is typically available only in batch mode, so only a JCL example is presented here:

```
//RAWDAT DD DISP=SHR,UNIT=TAPE,VOL=SER=INDEX02,
//         LABEL=(4,SL),DSN=SIC.REGION2
```

If the data were from a nonlabelled tape, the system would need to know how the data were arranged and formatted. This information is supplied by data control block (DCB) parameters. An example is shown below:

```
//RAWDAT DD UNIT=TAPE,VOL=SER=INDEX02,LABEL=(3,NL),
//         LABEL=(1,NL),DISP=SHR,
//         DCB=(RECFM=FB,LRECL=80,BLKSIZE=6400)
```

Example D.6. Define an in-stream dataset: the raw data is contained in the SAS program submitted for batch processing but is located in the JCL rather than in the program itself. This makes reading the program easier, since the data lines do not interrupt the program text.

```
//RAWDATA DD *
.
. data lines go here
.
/*
```

The DD statement's "*" signals the beginning of in-stream data, while "/*" marks its end. In-stream data cannot be concatenated with other (tape, disk) datasets.

There is no easy way to define an in-stream dataset in TSO, so no example is presented. PDF edit macros, ISPF skeletons, CLISTs, and REXX command files are capable of simulating in-stream data. The techniques are not for the faint of heart and are beyond the scope of this overview.

Example D.7. Write raw data to a new disk dataset. The JCL example shows both the DD statement and the SAS program. As always, the key to the program's success is using the same file reference in both the JCL and the FILE statement (in this case, DATAOUT).

```
//DATAOUT DD DISP=(NEW,CATLG),UNIT=FILE,SPACE=(TRK,5),
//              VOL=SER=FILE20,DSN=MIS098.SUBSET.FY89,
//              DCB=(RECFM=FB,LRECL=32,BLKSIZE=3200)
data _null_;
file dataout notitles;
put (area sales pctchg)($s2. +1 20.2 +1 6.2 +1 '89');
```

The TSO statement for the dataset allocation follows:

```
alloc file(dataout) dataset('mis098.subset.fy89') -
      unit(file) space(5) tracks vol(file20) -
      recfm(f) lrecl(32) blksize(3200) catlg
```

Example D.8. Write raw data to a new partitioned dataset. Note that the member name (in this case, EST93) may be specified either in the JCL or as an option in the FILE statement (the usage is identical to that of the INFILE example in Example D.1). The "10" operand of the SPACE parameter allocates directory blocks. These blocks store information about the members in the PDS directory. Usually one directory block contains information about five members.

```
//DATAOUT DD DISP=(NEW,CATLG),UNIT=FILE,
//              VOL=SER=FILE20,DSN=MIS098.SUBSETS(EST93),
//              DCB=(RECFM=FB,LRECL=100,BLKSIZE=10000),
//              SPACE=(TRK,(5,2,10))
data _null_;
* If (EST93) was not specified in the JCL, you could
  write the following statement as:
      file dataout(est93) notitles ;
file dataout notitles;
put area $s4. (est1-est8)(12.);
```

The equivalent operation in TSO follows:

```
alloc file(dataout) dataset('mis098.subsets(est93)') -
      unit(file) space(5 2) dir(10) tracks catlg -
      vol(file20) recfm(f) lrecl(100) blksize(10000)
```

Example D.9. Overwrite, or replace, an existing cataloged dataset. Note the change of DISP. OLD allows you to write to the dataset. The batch example follows:

```
//REPL DD DISP=OLD,DSN=MIS098.FISCAL89.DATA
```

The TSO equivalent is illustrated below:

```
alloc old file(repl) dataset('mis098.fiscal89.data')
```

Example D.10. Append data to a cataloged dataset. Again, the DISP value controls what happens: MOD points SAS to the end of the dataset. The original data remain intact, and the new data are added to the end of the file.

```
//REPL DD DISP=MOD,DSN=MIS098.FISCAL89.DATA
```

The TSO equivalent follows:

```
alloc mod dataset('mis098.fiscal89.data) file(repl)
```

Notice that in both the JCL and TSO cases the dataset disposition MOD could have been replaced with OLD if the FILE statement included the MOD parameter (see Chapter 19 for details). It is usually best to specify MOD or OLD directly in the JCL and leave the writing and formatting of the data to SAS.

Example D.11. Write a dataset to an IBM standard-label tape. Notice that the SPACE parameters used with the earlier, disk-oriented examples (D.1–D.10) are not relevant here.

Many sites require the use of special JCL statements that request the tape to be mounted and enabled for both reading and writing. Contact your help desk or local user's manual for details. Since tape access is usually unavailable in interactive mode, only the JCL example is presented here:

```
//TAPEOUT DD LABEL=(2,SL),UNIT=TAPE4,DISP=(NEW,KEEP),
//              DCB=(LRECL=80,BLKSIZE=12000,RECFM=FB),
//              DSN=DIST0120.FISCAL80,VOL=SER=ARC0012
```

Using SAS Datasets

The meaning of JCL and TSO space, disposition (DISP), and other parameters is much the same when using SAS datasets as with raw datasets. There are, inevitably, some differences. Keep in mind the following points when using SAS datasets in IBM MVS and TSO systems:

- When you write a permanent SAS dataset to disk for the first time, you must supply an estimate of how much disk space it will use. A simple way to do this is to run the program on just a few observations (e.g., OPTIONS OBS=100), not saving the dataset. Look at the SAS Log for a message like "the dataset TEMP has 100 observations, 508 observations/track." Divide the number of expected observations by 508, then add 10% to 20% just to be safe. This is the number of disk SPACE units, or tracks, to allocate when you create the dataset.

- Unlike raw datasets, SAS datasets do not require DCB parameters either on tape or disk. SAS supplies these to the operating system automatically.

- You can store more than one SAS dataset in a single OS dataset (identified by the DSN parameter). Many people find this lack of one-to-one correspondence confusing and prefer to stick to one OS dataset for each SAS dataset. You *cannot* mix SAS datasets, raw data, and user-written formats in the same dataset.

- When replacing a permanent SAS dataset, you do not reuse the old space. Instead, the new data are usually written at the end of the dataset. This frequently causes the operating system to abort your job or session due to lack of space. If this happens, allocate a new dataset, copy the old one to it, and rerun the job.

VERSION 6.06 UPDATE

Prior to version 6.06, SAS datasets were stored using JCL parameters that made it difficult, if not impossible, to move or copy the datasets with any software other than SAS (for the technically minded: DSORG=DA, RECFM=U). These parameters are different in version 6.06 (DSORG=PS, RECFM=FS) and make it possible to copy SAS datasets with software other than SAS (e.g., IBM utilities, TSO COPY command, ISPF MOVE/COPY, and so on). Ask the help desk staff for the preferred method of copying or moving at your installation.

Example D.12. The SAS datasets PANEL01 and PANEL02 are stored in a cataloged disk dataset. Use them in two PROCs. The DISP parameter SHR has the same effect as with raw data: it permits reading from, but not writing to, the dataset.

```
//SASIN  DD DSN=TOXIC.MASTERS.SAS,DISP=SHR
```

```
proc print data=sasin.panel01;
proc freq data=sasin.panel02(keep=rank01-rank05);
```

The TSO statement to perform the equivalent JCL follows:

```
alloc da('toxic.masters.sas') fi(sasin) shr
```

Note that the LIBNAME statement is supported for existing datasets in version 5 of the SAS System. Thus you could avoid the TSO and JCL by entering the following SAS statement prior to using PANEL01 and PANEL02:

```
libname sasin 'toxic.masters.sas';
```

Example D.13. Write to an "old," preexisting dataset. Notice the DISP parameter OLD: this allows you to both read from and write to the dataset. The LIBNAME statement in Example D.12 could have been used here as well, since the dataset TOXIC.MASTERS.SAS existed prior to the SAS session/job.

```
//MASTER DD DISP=OLD,DSN=TOXIC.MASTERS.SAS

proc means data=master.panel01 noprint;
class group;
var lipid00-lipid60;
output out=master.summ01 mean=;
```

The TSO statements to perform the equivalent JCL follow:

```
alloc da('toxic.panel.masters.sas') fi(sasin) shr
```

Example D.14. Read from an existing SAS dataset, and write to a new one. Notice the absence of the DCB parameters required for raw datasets.

```
//MASTER DD DISP=OLD,DSN=TOXIC.MASTERS.SAS
//SUMMS  DD DISP=(NEW,CATLG),DSN=TOXIC.SUMMARY.SAS,
//          UNIT=3380,VOL=SER=TOXD20,
//          SPACE=(TRK,(20,5))

proc means data=master.panel01 noprint;
class group;
var lipid00-lipid60;
output out=summs.summ01 mean=;
```

The TSO equivalent follows:

```
alloc data('toxic.summary.sas') catlg fi(summs) -
volume(toxd20) unit(3380) space(20 5) tracks
```

Example D.15. Read from an existing cataloged disk dataset, and write an output dataset to tape. See Example D.11 about tape usage practices at your site. Since it is an IBM standard-label tape, the SL notation used in earlier examples (D.5, D.11) is not necessary. As with raw datasets, SPACE parameters have no meaning when using tapes. Since interactive tape handling is usually not permitted at most installations, only the JCL example is shown.

```
//MASTER DD DISP=OLD,DSN=TOXIC.MASTERS.SAS
//SUMMS  DD DISP=(NEW,KEEP),DSN=TOXSUMM.SAS,
//          UNIT=TAPE4,VOL=SER=TOXT85,LABEL=4

proc means data=master.panel01 noprint;
class group;
var lipid00-lipid60;
output out=summs.summ01 mean=;
```

Format Libraries

User-written format libraries in both JCL and TSO applications follows two rules:

1. Libraries may only be read from and written to disk datasets.
2. When writing or updating a library, keep in mind that it will hold a limited number of formats. This limit is established by the number of directory blocks (the third parameter of the SPACE statement) used when the library was created. Approximately five formats per block can be stored.

VERSION 6.06 UPDATE

Format libraries are stored differently in version 6.06. Rather than create a partitioned dataset, you must create a SAS "catalog." Catalogs are created in exactly the same manner as SAS datasets.

Example D.16. Create a format library. Although only one format is currently stored in it, there will eventually be 20. Thus the SPACE directory block value is set to 5 (20 divided by 4 plus one to be safe).

```
//SASLIB DD DSN=SYSADMIN.FORMAT.LIB,DISP=(NEW,CATLG),
//         SPACE=(TRK,(10,5,5)),UNIT=3380,
//         VOL=SER=SYS030

proc format ddname=saslib;
value $sregion '01'-'20' = 'North' other = 'South' ;
```

The TSO equivalent of the JCL follows:

```
alloc fi(saslib) dataset('sysadmin.format.lib') -
catlg space(10 5) dir(5) unit(3380) -
volume(sys030)
```

Example D.17. Add a format to the library created in Example D.16. Like-named formats (in this case, $SREGION) replace existing ones, while new names are added to the format library's directory. The JCL example follows:

```
//SASLIB DD DSN=SYSADMIN.FORMAT.LIB,DISP=OLD

proc format ddname=saslib;
value fence1x '01oct88'd - '30sep89'd = 'fy 88-89'
              '01oct89'd - '30sep90'd = 'fy 89-90'
              other                   = 'other'    ;
```

The library now contains formats $SREGION and FENCE1X. The TSO command to perform the equivalent JCL follows:

```
alloc fi(saslib) old dataset('sysadmin.format.lib')
```

Example D.18. Use the format library constructed in Examples D.16 and D.17.

```
//SASLIB DD DISP=SHR,DSN=SYSADMIN.FORMAT.LIB

proc print data=temp;
var branch expire01-expire20 revcurr revprev;
format branch $region. expire01-expire20 fence1x.
       revcurr revprev dollar20.;
```

The TSO statements to perform the equivalent JCL follow:

```
alloc fi(saslib) shr dataset('sysadmin.format.lib')
```

Example D.19. Concatenate several format libraries. Once SAS cannot locate a format name in its own libraries, it searches the SASLIB datasets for the format. If there are multiple occurrences of a name in the libraries, SAS uses the first one encountered. Thus if the datasets pointed to by the SYSADMIN and PROD0001 libraries each had a format named TIME0X, the one in SYSADMIN would be used since its library was defined prior to PROD0001's in the JCL.

```
//SASLIB DD DISP=SHR,DSN=SYSADMIN.FORMAT.LIB
//       DD DISP=SHR,DSN=PROD0124.SCHEDULE.FORMAT.LIB
//       DD DISP=SHR,DSN=PROD0001.MASTER.LIB
```

The TSO equivalent, formatted to enhance legibility, follows:

```
alloc shr file(saslib) -
      dataset('sysadmin.format.lib' -
              'prod0124.schedule.format.lib' -
              'prod0001.master.lib')
```

Miscellaneous Topics   **The Sometimes Mysterious /*.**   The beginning of embedded comments is indicated by a /*. Unfortunately, in IBM installations this character string also indicates the end of an in-stream file if it begins in column 1 of a line. Thus a seemingly innocent program such as

```
//        EXEC SAS
//SYSIN   DD  *
/* This program completes the correlation analysis
sequence suggested in the April 20th meeting. */
data part3;
.
. other statements
.
```

will not only fail to produce the expected output files but will not even display the program statements in the SAS Log! The /* in the first column of the first line terminated the input to SAS—that is, it ended the stream of instructions to the SAS System. This is a simple, common, and endlessly befuddling error.

Program File Format.   The batch mode SYSIN dataset, the SAS program, should be fixed length, card-image 80-byte records. It may not be compressed. Some systems, such as WYLBUR, employ space-saving compression routines. These save disk space but cannot be read by SAS. Be sure to save SAS programs in these systems using options that avoid compression and store card images.

D.3   IBM VM/CMS

Since most novice SAS users do not have to deal with handling tape datasets in the CMS environment, that topic is omitted in this section. Tape use is often restricted during working hours and often has site-dependent features that cannot be covered in the brief treatment allowed here. See your help desk personnel for assistance in using tapes (refer to Appendix C for ideas about where to go for help at your installation).

Running SAS

Batch.   Not all CMS installations have batch capabilities. Typically a batch SAS job requires the addition to the program of some accounting and routing information. The file is then spooled to a special user identification, often CMSBATCH, for processing. Once SAS completes execution, the output is spooled back to your user identification. See your help desk personnel about batch processing availability and the procedures required to use it.

Interactive.   The three forms of interactive SAS are supported in the interactive environment. Command-line SAS is executed as shown below:

```
sas filename ... [(options]
```

where *filename* is the file name of a program containing SAS statements. It must have a file type of SAS and may be present on any accessed minidisk. Up to 12 *filename*s may be specified. The *options* is a list of SAS options. In the CMS environment, a keyword and its value are separated with a blank rather than an equals sign (=). The following are valid executions of command-line SAS:

```
sas test
sas test1 test2
sas test1 test2 (nocenter nodate
```

SAS produces Log and procedure output files with a file name of the first file in the list of *filenames* and file types of SASLOG and LISTING, respectively. Thus in the third execution above, SAS writes the files TEST1.SASLOG and TEST1.LISTING.

Line mode SAS is entered by specifying the SAS command and, if necessary, the NODMS option:

```
sas (nodms [other_options]
```

The *nodms* suppresses the loading of the Display Manager. This may be the default at your site, in which case its specification is redundant but not harmful.

The SAS Display Manager is entered by specifying the SAS command and, if necessary, the DMS option:

```
sas (dms [other_options]
```

The *dms* explicitly requests the loading of the Display Manager. If it is the default at your site the specification is redundant but not harmful.

Handling Raw Data

The CMS FILEDEF command establishes the link between a raw dataset on an accessed minidisk and the SAS program's INFILE statement. The FILEDEF may be run either prior to the running of SAS or within SAS, using the CMS statement (see the "Miscellaneous Topics" section on page 607 for a description of the CMS statement). The general forms of the FILEDEF statement follow:

```
FILEDEF refname DISK file_name file_type file_mode [(options
FILEDEF refname TERM
FILEDEF refname CLEAR
```

In the first form, *refname* is the name used to link the dataset to the INFILE statement. It must begin with an alphabetic character, contain only letters and numbers, and not exceed eight characters in length. The *file_name*, *file_type*, and *file_mode* identify the file. These conform to the usual CMS naming conventions and must refer to an accessed disk (i.e., have a valid file mode). Use the CMS LINK and ACCESS commands to define a minidisk not currently in your search path. The *options* specify file format features and are not usually required for disk datasets since SAS automatically determines file characteristics.

The second and third forms are handy for more specialized tasks. The *term* indicates input from or output to the terminal (see Example D.23). The *CLEAR* wipes out any existing definitions for *refname*, thus avoiding naming conflicts when programs are run repeatedly in the same session. A safe way to avoid CMS errors and messages about the potential naming conflict is to use a FILEDEF that clears a refname, then one that defines it.

Example D.20. Make a dataset on the A disk available for reading.

```
filedef input disk part1 data a
```

Within SAS, link to this raw data file by a statement such as this:

```
infile input;
```

Example D.21. The dataset of interest is on another user's minidisk. LINK and ACCESS the disk, then issue the FILEDEF.

```
cp link sys01 191 193 rr
access 193 d
filedef datahere disk invntry feb91 d
```

Once in SAS an INFILE statement may refer to *filedef* DATAHERE.

Example D.22. Append data to a raw dataset. Assume that the user is allowed to write to the C minidisk. Notice that the MOD option in the FILEDEF could have been specified as an option in the FILE statement.

```
filedef input  disk  master invntry a
filedef output disk  master subset  c (disp mod
```

The SAS program's DATA step reads, in part, like the following:

```
data _null_;
infile input;
file output notitles;
input code $12. quant 10.;
if substr(code,8,3) = 'ACD' then
   put code $12. +5 quant 12.;
```

Example D.23. While processing a large dataset, write a message to the terminal to keep the user appraised of the program's progress. The message is written every 5,000th input record.

```
filedef msg term
```

The SAS program may look like this:

```
data partial;
set in.complete;
file msg;
if mod(_n_,5000) = 0 then
   put 'Processed obs # ' _n_ comma9.;
```

Using SAS Datasets

Naming Conventions.   SAS datasets in CMS are organized using simple yet unintuitive rules. The FILEDEF merely has to indicate the correct *refname*, with file name and type given "dummy" values. The file type of the SAS dataset matches the *refname*, while the file name is the second-level name used in the SAS program. For example, the SAS statement

```
data out.males out.females;
```

would create two files:

```
males     out  a
females   out  a
```

The FILEDEF to associate the minidisk to the SAS session might be

```
filedef out disk x x a
```

Note the use of the dummy values for file name and type. These are placeholders, used so that CMS can identify the sixth value in the command (the "a") as the file mode, or minidisk, to use.

When to Use FILEDEF.   Normally you do not have to issue a FILEDEF to use a SAS dataset. Use FILEDEF only if you want to override the rules used in the default search order. These rules are summarized below:

- SAS first checks all existing FILEDEFs to see if the *refname*, or first level of the SAS dataset name, has been defined. If so, SAS uses that file.
- If no FILEDEF for the *refname* exists, SAS searches all accessed mindisks for the *refname* and attempts to use the first disk with a matching *refname*. The user must have the minidisk accessed correctly (read/write access for writing, read access for reading).
- If no minidisk has a matching *refname*, SAS writes the file to the user's A disk. The A-disk assumption is not made when reading SAS datasets.

Dataset Maintenance.   PROC DATASETS and CONTENTS (see Chapter 10) are sufficient to perform all data management functions (copying, renaming, deleting datasets).

The CMS ERASE and RENAME commands may also be used for deleting and renaming datasets, since each SAS dataset is stored as an individual CMS file. The COPYFILE and MOVEFILE commands are usually able to move SAS datasets between minidisks. Consult your help desk personnel before using these or other file-movement commands.

Format Libraries

Creating and using user-written formats is usually simple. The FORMAT procedure statement should indicate a LIBRARY value of LIBRARY, thus telling the procedure that the formats will be stored permanently. By default, the formats are written to the write-accessed disk with the most available space. Alternately, the formats may be directed to a specific disk via a FILEDEF command. Both methods are illustrated below (Examples D.24 and D.25).

SAS creates format files whose *filename* corresponds to the format's name, with a *filetype* of TEXT. It may be more convenient to group these TEXT files into a single file called a TXTLIB (the CMS TXTLIB command performs this grouping). Using a TXTLIB requires use of the SASLIB option when starting SAS. Its use is demonstrated below:

```
sas (saslib master
```

This SASLIB option instructs SAS to search for a file on any accessed disk with a *filename* "master" and a *filetype* "txtlib". Entry names in this TXTLIB correspond to formats that may be used during the SAS session.

Example D.24. Create two formats, letting SAS determine their location.

```
proc format library=library;
value reg2x 010-020='Zone 1' 021-099='Zone 2';
value reg3x 010-050='Zone 1a' 051-020='Zone 1b'
            021-099='Zone 2';
```

The files *reg2x text a* and *reg3x text a* are written.

Example D.25. Override the default selection of minidisks. Direct the format TEXT files to the minidisk B via a FILEDEF.

```
filedef library disk x x b
```

The SAS program in Example D.24 would remain unchanged.

Miscellaneous Topics

Special Keys. Interactive SAS sessions can be interrupted by using the attention or break keys (line mode) or ENTER key (full-screen mode). Once the program is no longer in its RUNNING state, you can either stop execution of the entire SAS session (entering the HX command) or terminate the display of output at the terminal (HT command). Entering the RT command resumes display to the terminal. If HX is entered, output from subsequent PROCs and DATA steps will be displayed: HT and RT only affect the program segment currently executing.

Special Routing Options. Display Manager sessions have many options available for saving and printing Log and procedure output. These techniques are beyond the scope of this book but can be gleaned in part from the on-line help available within the Display Manager.

Line mode and command-line (noninteractive) SAS have several defaults and options for routing (some of these were reviewed in Chapter 12). The LDISK, LTYPE, LPRINT, PDISK, PTYPE, and PPRINT options control the routing of Log and output files to any combination of your A-disk, terminal, and printer, respectively. These options must be specified when SAS is first called. They cannot be specified in an OPTIONS statement within the program.

The NAME option may be used together with the file redirection features noted above. NAME indicates the *filename* of the LISTING and/or SASLOG files created by LDISK and PDISK. Its format in a typical execution of SAS is

```
SAS (NODMS LDISK PDISK NAME ADHOC
```

In this example, the Log and output files from this line mode SAS session will be contained in the files "ADHOC SASLOG A" and "ADHOC LISTING A," respectively. NAME, like the routing options it complements, must be specified as an option when SAS is being called.

"Disk Full" Problems.   If the SAS program cannot find enough room to write a temporary or permanent dataset, the message

```
DISK x(add) IS FULL
```

is displayed, where x is the name of the filled disk and *add* is its address. The program will not continue until you make more disk space available to it. Once the appropriate CMS file-management commands (COPYFILE, ERASE, and so on) make space available, return to the SAS session by entering

```
RETURN
```

If the program is executed in batch or command-line (noninteractive) mode, you will not have a chance to clear space from the disk. The program will abnormally terminate.

D.4   DEC VMS

Digital Equipment Corporation's (DEC) VMS is the most popular operating system in its wide range of minicomputer products. It is characterized by a relatively straightforward and consistent syntax. Users familiar with the DOS microcomputer operating system will recognize many stylistic similarities to VMS.

Running SAS

Batch.   While most installations support SAS batch processing, there is not a standard command to submit a SAS job. There are often system commands with names such as BATCHSAS or SASSUBMIT. Ask your help desk personnel for details.

An alternative is to write your own customized COM file. One approach to this do-it-yourself approach is described here. (*Note*: This is simply a model and may not work at all installations if copied verbatim.) First, create a COM file, say MYBATCHSAS.COM, with your editor. This file contains the name of the program containing the SAS statements and makes the necessary SAS System setups. An example is shown below:

```
$ set default sys$login:
$ @setup_sas
$ sas 'p1'.sas
```

The first line saves the name of the directory in which the SAS program is located. The second line performs the necessary SAS file allocations. The last line actually runs SAS: "p1" is the file name of the program and is passed to the COM file when the SUBMIT takes place.

To use MYBATCHSAS to submit a SAS program named FINAL-03.SAS, enter the following at the DCL prompt:

```
submit/notify/param=final03/noprint mybatchsas
```

SAS will place files FINAL03.LOG and, possibly, FINAL03.-LIS in the current directory.

Interactive.   In some installations it may be necessary to run a setup procedure prior to actually running SAS. It may be called @SETUP_SAS (contact your help desk staff for particulars). Once SAS is made known to the operating system you can run it by entering the following:

```
SAS [/option] [program [...]] [/option]
```

The /option indicates a SAS System option (see Chapter 12 for examples) and *program* specifies one or more program names to execute. If *program* uses just a file name (no "."), a file type of *SAS* is assumed.

Using any form of the program name parameter simply executes the programs. The Log and output files are found in *program.LOG* and *program.LIS*, respectively. If the program name is omitted, SAS enters interactive mode.

The following are all valid ways to start interactive SAS:

```
$ sas
$ sas /nodms
$ sas/ld
$ sas test1,test2
$ sas/lt/nonumber try1.pgm
$ sas/nodate/nocenter  report_1  /lt
```

Handling Raw Data

You can allocate raw datasets either directly in the INFILE or FILE statements or with a FILENAME statement. Both methods support the VMS % and * wildcards for input datasets. File names can be as complete or abbreviated as necessary. If no directory information is provided, SAS assumes the dataset is located in the current directory (the one that was active when the SAS command was issued). Examples D.26 through D.28 illustrate this distinction.

Example D.26. Use FILENAME to identify a file to be read by a DATA step. Since no node and directory information is provided, SAS assumes that the dataset is in the current directory. The −1 specification in the file name indicates the next-to-most-recent version of the dataset will be used. The default is the most recent.

```
filename input 'test_scores.data;-1';

data class1;
infile input;
```

Example D.27. Use FILENAME to identify a series of datasets to be used as input. The %%, or "placeholder wildcard," includes any dataset beginning with YEAR and having two additional letters in the dataset name. The file specification is complete: the program can be run from any directory and SAS will still know where to look for the data (contrast this to the directory-dependent specification of Example D.26).

```
filename allyears 'eve::disk38:[anderson.data]year%%';

data alldata;
infile allyears;
```

Example D.28. Specify a file name directly when writing a raw dataset. The [] in the file name has the same effect as no directory specification: the file is written to the current directory.

```
data _null_;
set final;
file '[]version_1.rept';
```

Using SAS Datasets

An advantage of handling SAS datasets in the VMS environment is the ability to recover fairly painlessly from your mistakes by using previous versions of a dataset. This is illustrated in Example D.29 and D.30.

Example D.29. Access a SAS dataset library in the current directory. Write a new version of it to a different directory.

```
libname old '[]';
libname new '[tyson.inventory]';

data new.widgets;
set  old.master;
if upcase(type) = 'WIDGET' then output;
```

SAS looks for file MASTER.SSD in the current directory and writes WIDGETS.SSD to the INVENTORY directory.

Example D.30. You discover that the new dataset just written does not contain the appropriate information and that the previous version was in fact correct. You want to restore the old version. First, see if it exists:

```
$ set default [tyson.inventory]
$ dir widgets.*
    widgets.ssd;12    widgets.ssd;11    widgets.ssd;10
```

There are three versions of the dataset in the directory. Delete the highest numbered version (12, the most current), thus making version 11 the most current.

```
$ del widgets.ssd;12
$ dir widgets.*
    widgets.ssd;11    widgets.ssd;10
```

See the "Version Control" section on page 611 for details about controlling the number of file versions retained in a directory.

Format Libraries

Accessing user-written format libraries requires using a new statement. LIBSEARCH identifies a list of LIBNAME file references. When SAS encounters a format name that is not part of the SAS System format library, it looks for the format in the directories identified by LIBSEARCH. Just as SAS datasets are automatically given a file type of SSD, numeric formats use SFN and character formats SFC. The syntax of LIBSEARCH is shown below:

```
LIBSEARCH ref1 [ref2...refn];
```

The *ref1*, *ref2*, and so on through *refn* identify *filerefs* used in preceding LIBNAME statements. The *refs* may also contain nonformat files (SAS datasets, programs, and so on).

VERSION 6.06 UPDATE

Format libraries are handled differently in version 6.06 of the SAS System. Formats are stored in a SAS "catalog." These specially formatted libraries are created exactly as you would create a SAS dataset. When using the library, simply enter a LIBNAME pointing to the library holding the formats. Use the special *ref* of LIBRARY; otherwise SAS will not be able to locate the formats.

Example D.31. Create a format library.

```
libname areafmt '[mis.control1]';

proc format library=areafmt;
value $region '1'-'5' = 'E'   other='W';
```

Example D.32. Use the format library created in Example D.31 as well as other user-written libraries. Since LIBSEARCH reads the directories in the order in which they were defined, a numeric format REGION in the second library will be used instead of a like-named format in later libraries. If a LIBNAME is specified but its *fileref* is not included in the LIBSEARCH, SAS will have no way of knowing to use the library and will produce an error if you try to use a format contained in the unspecified library.

```
libname cont1 '[mis.control1]';
libname cont2 '[mis.control2]';
libname cont3 '[mis.control3]';

libsearch cont1 cont2 cont3;

proc print data=cont3.trial2;
format region $area.;
```

Miscellaneous Topics

Using a Start-up File. A handy feature of SAS in the VMS environment is the ability to identify a start-up file. This file contains SAS statements that will be executed before your program's or session's commands are executed. For example, you can use the start-up file to set system options or define commonly used LIBNAMES and FILENAMES. A sample file follows:

```
options nodate nonumber nocenter;
libname prior 'dsk_sys:[sys.master]';
libname prev  'dsk_sys:[sys.master]';
```

Start-up files not only save you some programming effort but also help standardize file definitions and option selections, as shown in Examples D.26 through D.32. Programs using the start-up file are assured of using the same settings and file names.

Use the start-up file facility by identifying the file name with a DCL logically named SAS$INIT:

```
$ define sas$init [fcd.misc]startup.sas
```

When SAS begins execution it will attempt to execute the SAS statements in the file *startup.sas*. SAS$INIT may be defined as part of the LOGIN.COM file. The file name does not need to have a file type of SAS.

VERSION 6.06 UPDATE

Several system options in version 6.06 improve the AUTOEXEC file capability. AUTOEXEC allows direct specification of the start-up file, while ECHOAUTO forces display of the statements in the start-up file. The following DCL command invokes the SAS System with an AUTOEXEC file and will display the statements in the file:

```
sas/autoexec=[fcd.misc]startup.sas/echoauto
```

Running VMS Commands within SAS. VMS DCL commands may be executed within a batch or interactive SAS session. The syntax of the command is

```
x ['[command]'];
```

where *command* is any DCL command or COM file name. The *x* command may be entered as a SAS program statement, in which case the semicolon (;) is required. The *x* may also be used as a command on the command line of a DMS window.

If *command* is omitted (i.e., just *x*; or *x* "; was entered) in interactive SAS, SAS spawns a DCL process. Enter DCL commands as needed, then enter *lo* or *logoff* to return to the SAS session.

Version Control. The number of versions of SAS datasets, raw datasets, programs, and formats retained in each directory is controlled by the SET DIR command, shown below:

```
$ SET DIR [dirname] /VER=n
```

The *dirname* is the name of the directory, and *n* is the number of versions of new datasets to retain. Existing datasets are not affected by this command. To keep five versions of datasets in the directory PEROT.HOLDINGS, enter the following at the DCL prompt:

```
set dir [perot.holdings] /ver=5
```

Use the PURGE command to erase all but the most recent versions of all or selected datasets.

Special Keys. Pressing *control-C* while SAS is executing an interactive program will interrupt the DATA step or PROC. The *control-Y* prompts you for termination of the entire program. The *control-O* (letter "O") suppresses display of output during an interactive session. Pressing it the first time turns the output off; pressing it again resumes listing of the output. Display since the first *control-O* is lost (but will be written to the disk if the LD option was in effect).

D.5   PC/MS-DOS

PC/ and MS/DOS are the most popular operating systems for IBM and compatible micro-computers. The syntax of the commands is usually straightforward. Many DOS "shell" programs are also available to facilitate the system's use. Fortunately, entering SAS is simple and contact with DOS is minimized once you begin running programs.

Running SAS

Batch. Batch processing is available for microcomputers attached to LANs. LANs offer a considerable flexibility and power, but their site-to-site variation puts them beyond the scope of this book. See your help desk personnel about LAN availability at your site.

Interactive. Before you can execute SAS, you must tell DOS where the SAS System library is located. This is usually done using two methods. The first is inclusion of the SAS subdirectory in the system's PATH. This specification is usually included in the AUTOEXEC.BAT file. It is a list of directories that DOS searches when a command is issued. Assuming the SAS System library is stored in the directory C:SAS, a PATH assignment may look like this:

```
path=c:\;c:\dos;c:\wp5;c:\sas
```

When DOS is given the command SAS, it will search the root (*c:*), *dos*, *wp5*, and *sas* directories for a program or BAT file named SAS. Once it locates SAS in the path, DOS starts executing the program.

The second method of executing SAS is by using a BAT file. In this case, the SAS directory need not be part of the path, since the BAT file will direct DOS to the proper location. A simple BAT file for DOS 3.1 or above follows:

```
@echo off
echo Begin running SAS Version 6.03 ...
c:\sas\sas %1 %2 %3 %4 %5 %6 %7 %8 %9
```

The BAT file displays a message and runs the SAS program, allowing up to nine parameters to be passed (%1, %2, ..., %9).

SAS has three interactive modes in DOS environments. The first interactive environment is command-line, demonstrated below:

```
sas pgmname [-opts ...]
```

The *pgmname* is the name of the file containing SAS program statements, *-opts* are one or more SAS System options. If *pgmname* specifies only a file name, an extension of SAS is assumed:

```
sas \bios163\hw1
```

This implies that file HW1.SAS is in directory BIOS163. SAS executes the program, creating the file HW1.LOG (the SAS Log) and possibly HW1.LST (if there was any procedure or PUT statement output).

The second SAS environment is line mode. This may be the default on your computer. If not, or if you are not sure, enter the *nodms* option:

```
sas -nodms
```

The third SAS environment is full screen, usually referred to as Display Manager mode. This is often the system default but may be specified explicitly:

```
sas -dms
```

The *-dms* option requests use of the SAS Display Manager, described in Section D.1.

Note that in all three modes you may specify one or more SAS System options (see Chapter 12). Enter a hyphen (-) followed by the option. If an option requires a value, separate the option name and the value with a blank and do not precede the value with a hyphen. Examples are listed below:

```
sas c:\budget\balance -nocenter -obs 100
sas -nodate -nonumber -nodms
sas -dms -nofmterr
sas v1 -firstobs 100 -obs 200
```

Handling Raw Data

You can define raw datasets either directly in the INFILE or FILE statements or with a FILENAME statement. You cannot use wildcards (*) in file names nor can you concatenate input files. File names may be as complete or abbreviated as necessary. Just remember that if incomplete path names are given, SAS assumes the dataset is located in the current directory. Example D.34 illustrates this distinction.

The syntax for the FILENAME statement is described in Chapter 5. An enhancement to the statement is described below:

```
filename ref [device] 'pathname';
```

The *ref* and *pathname* follow the rules outlined in Chapter 5 and used throughout the book. The *device* is an optional specification for the routing of the output. It can be either *printer* or *terminal*. Using this option and specifying a *pathname* writes a disk file containing carriage control and other special printer characters. Later on you can use a print command to route the output to a printer. The *device* can be very handy for testing displays: initial runs of the program may be written to the screen (*terminal*), then written to disk once the program is debugged. If output will be directed to the *printer* or *terminal*, you must specify an empty *pathname* (' '). The *device* may be used in both batch and interactive SAS. Using it implies use of the NOTITLES option of the FILE statement (see Chapter 19 for details).

Example D.33. Use FILENAME to identify a file to be read by a DATA step. Since a complete path name (drive, directory, file name) is not provided, SAS assumes that the dataset is located in the current directory.

```
filename input 'scores.dat';

data class1;
infile input;
```

Example D.34. Specify a file name directly when writing a raw dataset. The [] in the file name has the same effect as no directory specification: the file is written to the current directory.

```
data _null_;
set final;
file '[]ver1.rpt';
```

Example D.35. Send output from the SAS program directly to the default printer. This is usually defined to the system as COM1 or PRT.

```
filename pdirect printer 'com1';

data _null_;
set pmast1;
file pdirect;
put name $20. +1 (est1-est20)(2.);
```

Using SAS Datasets

The LIBNAME statement identifies directories containing permanent SAS datasets. Its syntax resembles FILENAME but does not include specification of file names and extensions (this is handled automatically by SAS). The DOS dataset uses the SAS dataset name for the file name and SSD as the extension. Thus a SAS dataset named CURRENT has a DOS dataset name of CURRENT.SSD.

Example D.36. Access a SAS dataset library in the current directory. Write a new version of it to a different directory.

```
libname old '';
libname new 'd:\tyson\inventory';

data new.widgets;
set  old.master;
if upcase(type) = 'WIDGET' then output;
```

SAS looks for file MASTER.SSD in the current directory and writes WIDGETS.SSD to the INVENTORY directory.

Format Libraries

Libraries of permanent, user-defined formats are made available to the SAS program via the LIBNAME statement. Its syntax is identical to its use with SAS libraries, described earlier. By default, the format library is named FORMATS.SCT.

Example D.37. Create a format library.

```
libname library '\mis\control1';

proc format library=library;
value $region '1'-'5' = 'E'    other='W';
```

Example D.38. Use the format library created in Example D.37. Assume that when SAS began execution *mis**control* was the current directory. It is usually safer to take the time to be explicit about the path name rather than rely on your memory to run the program from a particular directory each time: a fully specified path ensures that the program will run correctly regardless of which directory was current when SAS was executed.

```
libname library '';
libname data 'c:\mis\sasdata';

proc print data=data.may93;
format dest1 dest2 $region.;
```

Miscellaneous Topics

Interrupting the Session. At some point during your use of interactive PC SAS, it is likely that you will accidentally begin execution of a DATA step you did not mean to run or start a PRINT on a huge dataset. Rather than yield to impulse and reboot your computer or accept your fate and let the program run, PC SAS provides an alternative.

To interrupt work being performed in your session, simultaneously press the Control and Break keys (Ctrl-Break). SAS will usually respond with a message resembling

```
Press Y to cancel submitted statements, T to halt
data/proc, N to continue.
```

At this point you can enter *Y* to cancel the statements submitted (via a DMS SUBMIT command or a RUN statement in line mode) or *T* to stop execution of the current unit of work (DATA step or proc). Execution continues with the next unit of work in the group of submitted statements. For example, suppose you submit the following statements and interrupt execution during PRINT:

```
proc print data=mast.repeat1;
proc means data=second;
```

A response of *T* to the SAS prompt will stop the PRINT procedure and resume execution at MEANS. *N* resumes execution without interrupting the current or subsequent units of work submitted for processing.

AUTOEXEC: Using a Start-up File.

A handy feature of SAS in the PC/MS-DOS environment is the AUTOEXEC facility. Frequently you find yourself using the same options and making the same LIBNAME assignments in many programs. AUTOEXEC points SAS to a file that does these tasks for you automatically, prior to execution of any batch or interactive SAS statements. Any valid SAS statement may be included in the AUTOEXEC file. A sample file follows:

```
options nodate nonumber nocenter;
libname prior 'c:\analysis\prior';
libname curr  'c:\analysis\curr';
```

When this AUTOEXEC is used, the *nodate*, *nonumber*, and *nocenter* system options are turned on and LIBNAMEs PRIOR and CURR defined before your SAS program is executed (in batch) or SAS prompts you to enter statements (in interactive). This facility can standardize your file definitions and option selections, and it reduces the volume of code in your programs.

The rules for using AUTOEXEC are straightforward. SAS looks in the current directory, system PATH, and root directory for the first occurrence of a program named AUTOEXEC.SAS. It uses this file as the start-up program. If the start-up program does not use this name or is not in any of the default directories use the AUTOEXEC system option when running SAS:

```
sas hw1part2 -autoexec \userlib\fcdauto.sas
```

In this case, SAS will execute statements in FCDAUTO.SAS prior to its execution of HW1PART2. Using the AUTOEXEC option when calling SAS overrides the AUTOEXEC .SAS file in the default directories.

E APPENDIX: Sample Datasets

This appendix presents the sample datasets used throughout the book. The following items are discussed for every dataset:

- *SAS dataset name.* The SAS dataset name used when writing the examples. The first-level name (that used in the LIBNAME statement) may vary from session to session.

- *Description.* A brief description of the dataset's contents.

- *Size.* Number of observations and variables.

- *Source.* Source of the data. This is usually a reference to a government agency.

- *Chapter references.* Chapters using the dataset.

- *Variables.* Several items are presented for each variable. The name and label of each variable are listed, along with its type (character or numeric) and location in the raw data records. Data-line/column locations are shown, or, if list input is appropriate, the order of the variables is indicated. The length of character variables is in the next column, followed by the minimum and maximum values of numeric variables.

- *Footnotes.* Sometimes the dataset will have unique features that require explanation. Footnotes aid interpretation of the variable tables and sometimes have sample DATA steps and other hints for reading the data.

- *Listing of raw data.* Finally, a listing of the raw data is displayed. A column ruler begins the listing to make the location of fixed-field (column input) data easier.

```
--------- BRIDGES -------------------------------------------------------
SAS Dataset Name: BRIDGES

Description: Data for bridges with a main span of at least 500 meters.

Size: 34 observations, 5 variables

Sources: U.S. Department of Transportation, Federal Highway Administration

Chapter references: 15, 25

Variables
```

Name	Label	Type	Raw Data Cols (1)	Length	Minimum	Maximum
TYPE	Bridge style: Suspension/ Cantilever/Arch (coded "s", "c" and "a", respectively)	char	1	1		
BR_NAME	Bridge name	char	2	25		
LOCATION	Location: state (if US) and country	char	3	15		
LENGTH	Main span length, in meters	num	4		500	1410
YEAR	Year completed	num	5		1890	1981

```
(1) Column locations refer to the variable's position in the raw data record.
    Use list-style input to read this dataset.

Listing of Raw Data

----+----1----+----2----+----3----+----4----+----5----+----6----+----7----+----8

s  Humber  United Kingdom  1410  1981
s  Verrazano-Narrows  NY, US  1298  1964
s  Golden Gate    CA, US  1280   1937
s  Mackinac Straits  MI, US  1158  1957
s  Bosporus  Turkey  1074  1973
s  George Washington  NJ-NY, US  1067  1931
s  Ponte 25 de Abril  Portugal  1013 1966
s  Forth Road  United Kingdom  1006  1964
s  Severn  United Kingdom  988  1966
s  Tacoma Narrows  WA, US  853 1950
s  Kanmon Strait  Japan  712  1973
s  Angostura  Venezuela  712 1967
s  Transbay  CA, USA  704 1936
s  Bronx-Whitestone  NY, USA  701 1939
s  Pierre Laporte  Canada  668 1970
s  Delaware Memorial  DE-NJ, US  655  1968
s  Seaway Skyway  Canada  655 1960
s  Gas Pipe Line  LA, USA    610 1951
s  Walt Whitman  NJ-PA, USA  610  1957
s  Tancarville  France  608  1959
s  Lillebaelt  Denmark  600 1970
s  Ambassador International  MI, USA  564 1929
s  Throgs Neck  NY, USA  549  1961
s  Benjamin Franklin  NJ-PA, USA    533  1926
s  Skjomen  Norway  525 1972
s  Kvalsund  Norway  525 1977
s  Kleve-Emmerich  W. Germany  500 1965
c  Quebec Railway  Canada  549 1917
c  Forth Railway  United Kingdom  521  1890
```

```
c  Minato Ohashi  Japan  510  1974
c  Commodore John Barry  PA, USA   501  1974
a  New River Gorge  WV, USA  518  1977
a  Bayonne  NJ-NY, USA  510 1931
a  Sydney Harbor  Australia  509  1932
```

---------- **CMI** ---

SAS Dataset Name: CMI

Description: Study of the effect of neighborhood racial characteristics on the
            health of low-income housing residents. Key variable is a self-
            perceived measure of physical and mental health, the Cornell
            Medical Index (CMI). In theory, as neighborhood racial homogeneity
            increases, so should CMI.

Size: 100 observations, 6 variables

Source: Data taken from Kleinbaum and Kupper, Applied Regression Analysis
        and Other Multivariate Methods, Duxbury Press, Boston, MA, 1978.

Chapter reference: 24

Variables

Name	Label	Type	Raw Data Cols [1]	Length	Minimum	Maximum
AREA	Neighborhood code	char	1 / (1)	1		
HHOLDS	# hholds in neighborhood	num	1 / (2)		40	212
PCTBLACK	% black in surrounding neighborhood	num	1 / (3)		17	100
RACIAL	> 50% black = H, others = L	char	[2]	1		
HOUSING	> 100 hholds in area = H, others = L	char	[2]	1		
CMI	CMI (well-being) score	num	2/(1-25)		1	83

[1] Column locations refer to the variable's position in the raw data record.
    Use list-style input to read these variables. There are 25 CMI scores per
    AREA. All are entered in list input on each AREA's second data line. Use the
    following DATA step as a model for reading the data:

```
    data cmi;
    infile [location of data];
    input area $ hholds pctblack;
    do i = 1 to 25;
       input cmi @@ ;  * @@ is discussed in Chapter 11 ;
       output;
    end;
```

[2] These variables are calculated on the basis of HHOLDS and PCTBLACK. They
    are not present in the raw data.

Listing of Raw Data

----+----1----+----2----+----3----+----4----+----5----+----6----+----7----+----8

```
C 98 17
49 12 28 24 16 28 21 48 30 18 10 10 15 7 6 11 13 17 43 18 6 10 9 12 12
M 211 100
5 1 44 11 4 3 14 2 13 68 34 40 36 40 22 25 14 23 26 11 20 4 16 25 17
N 212 36
20 31 19 9 7 16 11 17 9 14 10 5 15 19 29 23 70 25 6 62 2 14 26 7 55
E 40 65
13 10 20 20 22 14 10 8 21 35 17 23 17 23 83 21 17 41 20 25 49 41 27 37 57
```

---------- **COASTAL** ---

SAS Dataset Name: COASTAL

Description: Data for states with tidal shorelines. "General outline"
            refers to the coarsely measured portion of the state directly
            bordering on the ocean. "Detailed shoreline" is any land that
            comes in direct contact with the ocean, and includes more precise
            measurements of inlets, islands, and so on.

Size: 26 observations, 5 variables

Sources: U.S. Department of Commerce, National Oceanic and Atmospheric Admin-
          istration

Chapter references: 4, 5, 17

Variables

			Raw Data			
Name	Label	Type	Cols (1)	Length	Minimum	Maximum
========	==============================	====	========	======	=======	=======
COAST	East or West coast (coded	char	1 - 2	2		
	"ec" and "wc")					
OCEAN	Body of water bordered on	char	3 - 4	2		
	("at" = Atlantic, "gu" =					
	Gulf of Mexico, "pa" =					
	Pacific, "ar" = Arctic)					
STATE	State postal code (lowercase)	char	7 - 8	2		
GENCST	General outline	num	12 - 14		13	5580
TIDALCST	Detailed shoreline	num	18 - 21		89	31383
========	==============================	====	========	======	=======	=======

(1) Locations refer to the variable's column locations in the input record. Use
    formatted input to read this dataset. Note that this dataset may be read
    with a combination of list and formatted input.

Listing of Raw Data

```
----+----1----+----2----+----3----+----4----+----5----+----6----+----7----+----8

ecat  me   228   3478
ecat  nh    13    131
ecat  ma   192   1519
ecat  ri    40    384
ecat  ct     .    618
ecat  ny   127   1850
ecat  nj   130   1792
ecat  pa     .     89
```

```
ecat  de     28    381
ecat  md     31   3190
ecat  va    112   3315
ecat  nc    301   3375
ecat  sc    187   2876
ecat  ga    100   2344
ecat  fl    580   3331
ecgu  fl    770   5095
ecgu  al     53    607
ecgu  ms     44    359
ecgu  la    397   7721
ecgu  tx    367   3359
wcpa  ca    840   3427
wcpa  or    296   1410
wcpa  wa    157   3026
wcpa  hi    750   1052
wcpa  ak   5580  31383
wcar  ak   1060   2521
```

---------- **COUNTRYS** --

SAS Dataset Name: COUNTRYS

Description: Natural resource, other measures for the world's 15 most populous
            countries (as of late 1980s).

Size: 15 observations, 8 variables

Source: United Nations

Chapter reference: 15

Variables

Name	Label	Type	Raw Data Cols (1)	Format	Minimum	Maximum
ID	Country identification #	num	1 – 3	3.	101	615
COUNTRY	Country name	char	4 – 18	$15.		
POP	1987 population estimate, in millions	num	19 – 22	4.	57	1062
PCTURB	% living in urban areas	num	23 – 24	2.	19	85
ENERGY	Energy consumption per capita, 1984	num	25 – 27	3.	4	280
CULT	% land cultivated	num	28 – 29	2.	9	51
PASTURE	% land in pasture	num	30 – 31	2.	1	38
FOREST	% land forested	num	32 – 33	2.	4	68

(1) Locations refer to the variable's column locations in the input record. Use
    formatted input to read this dataset.

Listing of Raw Data

```
----+----1----+----2----+----3----+----4----+----5----+----6----+----7----+----8
101China          106232 19113014
107Japan          1227611113 268
113Philippines      6240  938 440
102India           80025  751 421
```

```
112Vietnam          6219  420 140
109Bangladesh       107 . . . . .
105Indonesia        175 .  . . . .
110Pakistan         10528  625 6 4
304United States    24374280202628
311Mexico           8270 50133823
406Brazil           14171 19 91966
508Nigeria          10928  7342316
603USSR             28465176101742
614West Germany     6185163301929
615Italy            5772 90411621
```

```
---------- LIFESPAN ---------------------------------------------------------
SAS Dataset Name: LIFESPAN
```

Description: Life expectancy, United States sex-race groups, 1985.

Size: 18 observations, 6 variables

Sources: U.S. Department of Health and Human Services, National Center for
          Health Statistics

Variables

Name	Label	Type	Raw Data Cols (1)	Format	Minimum	Maximum
AGE	Starting value, 5-yr age interval (0, 5, 10, ..., 85)	num	1 - 2	2.	0	85
TOTAL	Expectancy: total population	num	4 - 7	4.1	16.7	74.7
WM	Expectancy: white males	num	9 - 12	4.1	14.6	71.9
WF	Expectancy: white females	num	14 - 17	4.1	18.7	78.7
OM	Expectancy: non-white males	num	19 - 22	4.1	14.0	67.2
OF	Expectancy: non-white females	num	24 - 27	4.1	17.6	75.0

(1) Locations refer to the column locations in the input record. Use formatted
    input to read this dataset. Note that this dataset may be read with EITHER
    list or formatted input or a combination of the two.

Listing of Raw Data

```
----+----1----+----2----+----3----+----4----+----5----+----6----+----7----+----8
 0 74.7 71.9 78.7 67.2 75.0
 5 70.7 67.8 74.5 63.6 71.2
10 65.8 62.9 69.6 58.7 66.3
15 60.9 58.0 64.6 53.8 61.4
20 56.1 53.3 59.8 49.1 56.5
25 51.4 48.7 54.9 44.6 51.7
30 46.7 44.0 50.1 40.2 47.0
35 42.0 39.4 45.2 35.8 42.3
40 37.3 34.7 40.4 31.6 37.7
45 32.7 30.2 35.7 27.5 33.2
50 28.3 25.8 31.1 23.7 28.9
55 24.2 21.7 26.7 20.1 24.9
60 20.3 18.0 22.6 16.9 21.1
65 16.7 14.6 18.7 14.0 17.6
```

```
70 13.5 11.6 15.0 11.4 14.4
75 10.6  9.0 11.7  9.2 11.5
80  8.1  6.8  8.7  7.2  8.9
85  6.0  5.1  6.4  5.9  7.0
```

---------- **NATLPARK** --

SAS Dataset Name: NATLPARK

Description: Size and history of U.S. National Parks.

Size: 49 observations, 5 variables

Sources: U.S. Department of the Interior, National Park Service

Chapter references: 20, 25

Variables

Name	Label	Type	Raw Data Cols (1)	Format	Minimum	Maximum
======	=======	======	========	========	========	========
PARK	Park name	char	1 - 20	$20.		
ST	Principal state (uppercase postal code: UT, CA, and so on)	char	22 - 23	$2.		
COAST	East/west of Mississippi River ("E" or "W")	char	25 - 25	$1.		
YRESTAB	Year established	num	27 - 30	4.	1872	1986
ACRES	Acres in park	num	31 - 39	9.	5839	8331604

(1) Locations refer to the variable's column locations in the input record. Use
    formatted input to read this dataset. Note that this dataset may be read
    with either list or formatted input or a combination of the two.

Listing of Raw Data

```
----+----1----+----2----+----3----+----4----+----5----+----6----+----7----+----8
Acadia              ME E 1919    41365
Arches              UT W 1971    73378
Badlands            SD W 1978   243302
Big Bend            TX W 1935   735416
Biscayne            FL E 1980   173039
Canyonlands         UT W 1964   337570
Capitol Reef        UT W 1971   241904
Carlsbad Caverns    NM W 1930    46755
Channel Islands     CA W 1980   249353
Crater Lake         OR W 1902   183224
Denali              AK W 1917  4716726
Everglades          FL E 1934  1398938
Gates of the Arctic AK W 1980  7523888
Glacier             MT W 1910  1013572
Glacier Bay         AK W 1980  3225284
Grand Canyon        AR W 1919  1218375
Grand Teton         WY W 1929   310521
Great Basin         NV W 1986    77109
Great Smoky Mts.    NC E 1926   520269
Guadalupe Mts.      TX W 1966    76293
```

```
Haleakala              HI W 1960     28655
Hawaii Volcanoes       HI W 1916    229177
Hot Springs            AK W 1921      5839
Isle Royale            MI E 1931    571790
Katmai                 AK W 1980   3716000
Kenai Fjords           AK W 1980    670000
Kings Canyon           CA W 1940    461901
Kobuk Valley           AK W 1980   1750421
Lake Clark             AK W 1980   2636839
Lassen Volcanic        CA W 1916    106372
Mammoth Cave           KY E 1926     52420
Mesa Verde             CO W 1906     52085
Mount Rainier          WA W 1899    235404
North Cascades         WA W 1968    504780
Olympic                WA W 1938    914818
Petrified Forest       AR W 1962     93532
Redwood                CA W 1968    110178
Rocky Mountain         CO W 1915    265200
Sequoia                CA W 1890    402482
Shenandoah             VA E 1926    195346
Theodore Roosevelt     ND W 1978     70416
Virgin Islands         VI E 1956     14695
Voyageurs              MN W 1971    218059
Wind Cave              SD W 1903     28292
Wrangell-St. Elias     AK W 1980   8331604
Yellowstone            WY W 1872   2219784
Yosemite               CA W 1890    761170
Zion                   UT W 1919    146597
Bryce Canyon           UT W 1924     35835
```

---------- **PARKPRES** ---

SAS Dataset Name: PARKPRES

Description: U.S. National Park data summarized by presidential terms of office.

Size: 16 observations, 4 variables

Sources: U.S. Department of the Interior, National Park Service

Chapter reference: 20

Variables

Name	Label	Type	Raw Data Cols (1)	Format	Minimum	Maximum
PARTY	Political party	char	1 - 3	$3.		
PRESNAME	President's name	char	5 - 16	$12.		
_PRESPRK	# parks added during presidency	num	19 - 20	2.	1	11
_PRESACR	Acreage added during presidency	num	22 - 29	8.	5839	28590146

(1) Locations refer to the variable's column locations in the input record. Use
    formatted input to read this dataset. Note that this dataset may be read
    with either list or formatted input or a combination of the two.

Listing of Raw Data

```
----+----1----+----2----+----3----+----4----+----5----+----6----+----7----+----8
Rep Grant          1  2219784
Rep Harrison, B.   2  1163652
Rep McKinley       1   235404
Rep Roosevelt, T.  3   263601
Rep Taft           1  1013572
Dem Wilson         7  6723812
Rep Harding        1     5839
Rep Coolidge       4   803870
Rep Hoover         3   929066
Dem Roosevelt, F.  4  3511073
Rep Eisenhower     2    43350
Dem Kennedy        1    93532
Dem Johnson, L.    4  1028821
Rep Nixon          3   533341
Dem Carter        11 28590146
Rep Reagan         1    77109
```

---------- **PRES** ---

SAS Dataset Name: PRES

Description: Political and demographic data for U.S. presidents.

Size: 40 observations, 7 variables

Chapter references: 4, 9, 10

Variables

Name	Label	Type	Raw Data Cols (1)	Format	Minimum	Maximum
NAME	President's name	char	1 - 20	$20.		
PARTY	Political party ("fed," "dem," and so on)	char	21 - 29	$9.		
BORN	Year born (4 digits)	num	31 - 34	4.	1732	1924
BORNST	State born (postal code, lowercase)	char	37 - 38	$2.		
INAUG	Year inaugurated	num	43 - 46	4.	1789	1981
AGEINAUG	Age at inauguration	num	51 - 52	2.	42	69
AGEDEATH	Age at death	num	56 - 57	2.	46	90

(1) Locations refer to the variable's column locations in the input record. Use
    formatted input to read this dataset. Note that this dataset may be read
    with either list or formatted input or a combination of the two.

Listing of Raw Data

```
----+----1----+----2----+----3----+----4----+----5----+----6----+----7----+----8
Washington          fed      1732  va    1789    57   67
Adams, J.           fed      1735  ma    1797    61   90
Jefferson           dem/rep  1743  va    1801    57   83
Madison             dem/rep  1751  va    1809    57   85
Monroe              dem/rep  1758  va    1817    58   73
Adams, J.Q.         dem/rep  1767  ma    1825    57   80
```

Jackson	dem	1767	sc	1829	61	78
Van Buren	dem	1782	ny	1837	54	79
Harrison, W.	whig	1773	va	1841	68	68
Tyler	whig	1790	va	1841	51	71
Polk	dem	1795	nc	1845	49	53
Taylor	whig	1784	va	1849	64	65
Fillmore	whig	1800	ny	1850	50	74
Pierce	dem	1804	nh	1853	48	64
Buchanan	dem	1791	pa	1857	65	77
Lincoln	rep	1809	ky	1861	52	56
Johnson, A.	union	1808	nc	1865	56	66
Grant	rep	1822	oh	1869	46	63
Hayes	rep	1822	oh	1877	54	70
Garfield	rep	1831	oh	1881	49	49
Arthur	rep	1829	vt	1881	50	57
Cleveland	dem	1837	nj	1885	47	71
Harrison, B.	rep	1833	oh	1889	55	67
Cleveland	dem	1837	nj	1893	55	71
McKinley	rep	1843	oh	1897	54	58
Roosevelt, T.	rep	1858	ny	1901	42	60
Taft	rep	1857	oh	1909	51	72
Wilson	dem	1856	va	1913	56	67
Harding	rep	1865	oh	1921	55	57
Coolidge	rep	1872	vt	1923	51	60
Hoover	rep	1874	ia	1929	54	90
Roosevelt, F.	dem	1882	ny	1933	51	63
Truman	dem	1884	mo	1945	60	88
Eisenhower	rep	1890	tx	1953	62	78
Kennedy	dem	1917	ma	1961	43	46
Johnson, L.	dem	1908	tx	1963	55	64
Nixon	rep	1913	ca	1969	56	.
Ford	rep	1913	ne	1974	61	.
Carter	dem	1924	ga	1977	52	.
Reagan	rep	1911	il	1981	69	.
Bush	rep	1924	ma	1989	64	.

---------- **STATES** --

SAS Dataset Name: STATES

Description: Demographic, geographic, social, economic, and other data for U.S.
            States and the District of Columbia.

Size: 51 observations, 62 variables

Sources: U.S. Department of Commerce, Bureau of the Census; U.S. Department of
         Labor

Chapter references: 2, 4, 6, 21, 22, 23, 24

Variables

Name	Label	Type	Raw Data Cols (1)	Format	Minimum	Maximum
STATE	State FIPS code	char	1/ 1- 2	$2.		
NAME	Name of state	char	1/ 3- 22	$20.		
REGION	Region of country	char	1/23- 29	$7.		
EASTWEST	East or west of Mississippi River ("e" or "w")	char	1/30- 30	$1.		

CENREG	U.S. Census Bureau district code	char	1/31- 31	$1.		
CENST	U.S. Census Bureau state code	char	1/32- 33	$2.		
SQMILES	# square miles	num	1/34- 39	6.	63	570833
HIGHTEMP	Highest recorded temperature	num	1/40- 42	3.	100	134
LOWTEMP	Lowest recorded temperature	num	1/43- 45	3.	-80	14
BTUS	BTU's used per capita, 1985	num	1/46- 48	3.	190	858
POP	Population, 1980	num	1/49- 56	8.	401851	23667902
DEN_GRP	Population density quartile, 1980	num	1/57- 58	2.	1	4
POPSQMI	1980 population/square mile	num	1/59- 65	7.1	0.7	10180.7
URBPOP	# persons in urban areas, 1980	num	1/66- 73	8.	172735	21607606
RURFARM	Rural farm population, 1980	num	2/ 1- 6	6.	0	391070
PCTFARM	% population on farms, 1980	num	2/ 7- 11	5.2	0	16.337468
PCTURB	% population in urban areas, 1980	num	2/12- 17	6.2	33.77	100
_5YRPLUS	# persons > 4 years old, 1980	num	2/18- 25	8.	362922	21969725
_65PLUS	# persons > 64 years old, 1980	num	2/26- 32	7.	11547	2414250
DTH1K	Deaths per 1,000 pop., 1980	num	2/33- 36	4.1	4	10.9
DIV1K	Divorces per 1,000 pop., 1980	num	2/37- 40	4.1	2.9	17.3
BRTH1K	Births per 1,000 pop., 1980	num	2/41- 44	4.1	12.5	28.6
DIFFST	# living in different state, 1975 v 1980	num	2/45- 51	7.	72240	2898992
PCTIN	% 5+ yrs 1980 living diff. state 1975	num	2/52- 56	5.2	5.93	34.25
PERHH	# persons/household, 1980	num	2/57- 60	4.1	2.4	3.2
OCCUNIT	# occupied housing units, 1980	num	2/61- 67	7.	131463	8629866
CROWD	# housing units > 1 person/ room, 1980	num	2/68- 73	6.	4534	638333
PCTCROWD	% occupied units, > 1 person/ room, 1980	num	2/74- 78	5.3	2.050	15.271
SSECRET	Soc. sec. retiree payments/mth ($k) 1980	num	3/ 1- 6	6.	3347	636720
UNEMPBEN	Max. weekly unemp. benefit, 1988	num	3/ 7- 9	3.	120	354
PCAPINC	Per capita income, 1979	num	3/10- 14	5.	5183	10193
FAMILIES	# families, 1980	num	3/15- 21	7.	96840	5978084
FEMHHOLD	# families with female head, 1980	num	3/22- 27	6.	9288	883177
PCTFEM	% female-headed families, 1980	num	3/28- 32	5.2	7.52	35.74
FEMPOV	# female-headed families below pov, 1979	num	3/33- 38	6.	2196	273416
PCTFPOV	% female-headed fam's below pov., 1979	num	3/39- 43	5.2	18.57	43.55
EDBASE	# persons 25 years or older, 1980	num	3/44- 51	8.	211397	14043986
COLLEGE	# 25+ yrs old with college ed, 1980	num	3/52- 58	7.	43767	2752865
PCTCOLL	% 25+ years with college education, 1980	num	3/59- 63	5.2	10.41	27.46
EDEX_CUR	Ed. expend./pupil, current op's, 1986	num	3/64- 67	4.	2096	7365
EDEX_SAL	Ed. expend./pupil, salaries, 1986	num	3/68- 71	4.	1555	4892
TRANSWRK	# reporting means transit to work, '80	num	4/ 1- 8	8.	180553	10585675
CARPOOL	# workers carpooling, 1980	num	4/ 9- 15	7.	38777	1788527
PUBTRANS	# workers using public transit, 1980	num	4/16- 22	7.	1230	1924027
PCTSHARE	% commuters sharing transit, 1980	num	4/23- 27	5.2	14.71	53.65
EMPRES	Employment/residence ration, 1980	num	4/28- 32	5.3	0.887	2.334
LABFORCE	# persons age 16+ employed, 1980	num	4/33- 40	8.	164874	10640405

MFGEMP	# employed in manufacturing, 1980	num 4/41- 47	7.	10349	2159838
PCTMFG	% labor force in manufacturing, 1980	num 4/48- 52	5.2	4.51	32.76
MFGESTAB	# manufacturing establishments, 1977	num 4/53- 57	5.	429	45289
MFGVAL	Value ($million) manufacturing, 1977	num 4/58- 63	6.	942	120895
NFARMS	# of farms, 1978	num 4/64- 69	6.	0	194253
AVACRES	Avg. # acres/farm, 1978	num 4/70- 74	5.	0	5047
FARMPCT	% total land used by farming, 1978	num 4/75- 79	5.1	0	94.8
RESOPER	# farms with resident operator, 1978	num 5/ 1- 6	6.	0	106462
PCTRESFM	% farms operated by resident, 1978	num 5/ 7- 11	5.2	24.44	62.95
FAMFARMS	# family farms, 1978	num 5/12- 17	6.	0	172979
PCTFAMFM	family farms as % total # farms, 1978	num 5/18- 22	5.2	77.24	92.56
CRIMERT	Reported crimes per 100,000 pop., 1981	num 5/23- 27	5.	2580	10692
MIL_CONT	Per capita military·contracts, 1985	num 5/28- 32	5.	39	26387
MV_DTHS	Motor vehicle deaths/100,000 pop., 1985	num 5/33- 37	5.1	1.4	4.5
BUSHVOTE	% voting for Bush, 1988	num 5/38- 39	2.	14	67

======== ============================== ==== ======== ====== ========= =========

(1) Locations refer to the variable's "card" number and column locations in the
    input record. Use formatted input to read this dataset.

Listing of Raw Data

```
----+----1----+----2----+----3----+----4----+----5----+----6----+----7----+----8
01Alabama          South  e663 50767112-27346 3893888 2   76.7 2337713
 87757 2.25 60.04 3598388 4400159.106.9016.3 348660 9.69 2.81341856 726685.415
 91991120 58941042571157355155.09 6120438.90 2217315 27006312.1821861763
 1506232 351932  2051924.730.984 1511928 39491326.12 5863 21010 57503  201 35.6
 3984545.40 5246991.25 4801 1127 2.9060
02Alaska           Pacificw994570833100-80858  401851 1   0.7  258567
   822 0.20 64.34  362922  115474.008.8023.7 11420731.47 2.9 131463 1322510.06
 334726010193  96840 1005510.38  266926.54  211397  4455421.0873654892
  180553  38777  574324.661.013  164874  10349 6.28  429  1250   383 3359  0.4
  31938.81   32785.38 6410  386 2.5062
04Arizona          West   w886113508127-40257 2718215 1   23.9 2278728
 13770 0.51 83.83 2505455 3073627.807.3018.4 65006325.95 2.8 957032 693077.242
 85641135 7041 709912 8294911.68 2297427.70 1558891 27146017.4131052063
 1112482 220538  2179421.781.000 1113270 16130214.49 2892  7022  7660 5047 53.3
  506236.76  591777.25 7614 1360 4.5061
05Arkansas         West   w771 52078120-29323 2286435 2   43.9 1179556
107648 4.71 51.59 2111224 3124779.906.9016.3 27572713.06 2.7 816065 426505.226
 65564204 5614 628006 7663812.20 2920438.11 1337118 14467710.8223991725
  864888 195045  729023.391.003  875733 21972025.09 3595 12276 58766  265 46.9
 4144738.50 5300690.20 3743  629 3.4057
06California        Pacificw993156299134-4524123667902 3  151.421607606
176460 0.75 91.292196972524142507.905.6017.0289899213.20 2.786298666383337.397
 636720166 8295597808488317714.7723048626.1014043986275286519.6035272432
 105856751788527 61574322.710.99610640405215983820.3045289120896 81706  405 33.1
 5633931.93 6589480.65 759226387 2.4052
08Colorado         West   w884103595118-60273 2889964 2   27.9 2329869
 59152 2.05 80.62 2673872 2473256.506.4017.2 60665922.69 2.61061249 302162.847
 61025213 7998 744228 8625811.59 2245226.03 1663891 38201322.9635462397
```

```
        1360923 275217   5783524.470.999 1362017 19230514.12 3948 10018 29633 1197 53.4
    2194337.10 2493584.15 7189 1007 2.2054
09Connecticut           N'east e116 4872105-32218 3107576 4  637.9 2449774
        7383 0.24 78.83 2922810 3648648.404.3012.5 32521311.13 2.81093678 285952.615
113818266 8511 81818711501814.06 3017226.23 1900164 39256420.6640542849
    1461374 285878  7433024.650.992 1482309 45881630.95 6485 19842  4560  110 16.1
        380751.56  395186.64 6552 5132 1.9053
10Delaware         South  e551 1932110-17330  594338 3  307.6  419819
    10246 1.72 70.64  553319  591798.603.9015.8  8117414.67 2.8 207081  53512.584
17486205 7449 155073 2245914.48  675430.07  344657  6020617.4742492604
     262003  55748 1065525.341.023  262809 6207723.62  619  5209  3632  187 53.5
        275626.90  317887.50 6644  220 2.7057
11DC               South  e553    63  .  .  .  638333 410180.7  638333
         0 0.00100.00  604289  7428710.97.3014.7 11521419.07 2.4 253143 205188.105
15071250 8960 135569 4846135.75 1378928.45  398653 10947827.4646323684
     295131  46254 11210053.662.334  298107  13474 4.52  555   984   0    0 0.0
         0  .    0  . 10692   . 1.4014
12Florida          South  e559 54153109 -2225 9746324 3  180.0 8212385
    58679 0.60 84.26 9180221168757310.77.3013.5204019322.22 2.537442541995295.329
441900200 7270270648536326013.4211027830.36 6250125 93013414.8835532258
    3978407 806416 10654622.950.997 4002330 50456012.6112399 20981 44068  302 38.4
    2835048.31 3673483.36 8048 4650 3.1061
13Georgia          South  e558 58056113-17296 5463105 2  94.1 3409081
121089 2.22 62.40 5049559 5167318.106.4016.9 63615512.60 2.81871652 994235.312
114007155 6402143233123539316.43 8359835.51 3085528 45026714.5927101957
    2350978 519949  9252826.051.001 2335835 56202324.06 8623 32856 58648  234 37.0
    4166434.41 5231089.19 5536 2449 3.1060
15Hawaii         Pacificw995  6425100 14221  964691 3  150.1  834592
     4523 0.47 86.51  761505.004.6018.8 20217422.77 3.1 294052 4490515.27
22127223 7740 227974 2851412.51  798528.00  547608 11141520.3535732165
     457717 106039  3810031.490.982  415181  32914 7.93  949  1974  4310  461 48.3
     217748.13  361483.85 6543  639 1.7045
16Idaho          West   w882 82412118-60346  943935 1  11.5  509702
    69129 7.32 54.00  850427  936807.207.0021.4 18175621.37 2.8 324107 144624.462
25510188 6248 248258 22487 9.06  744233.09  514365  8141315.8320961587
     378443  65997  933719.910.992  383652  5345513.93 1495  3658 26478  562 28.1
    2052529.69 2266585.60 4525   49 3.5063
17Illinois         Midweste333 55645117-3530111426518 3  205.3 9518039
313978 2.75 83.301058635612618858.904.5016.6 829881 7.84 2.840453741690564.179
342537230 8066294510842914314.5713207730.78 6678759108228516.2032702244
    4956278 887603 59651529.940.997 5068428130799925.8119517 93081109924  270 83.3
    8084725.75 9479786.24 4906 1535 2.3051
18Indiana         Midweste332 35932116-35416 5490224 3  152.8 3525298
276154 5.03 64.21 5071880 5853848.607.3016.1 415859 8.20 2.81927050 600093.114
164713161 7142146164517510911.98 4691126.79 3135772 39259012.5227822059
    2296432 459896  3980021.760.990 2366263 73180030.93 8061 52172 88427  193 73.7
    6749524.44 7639886.40 4726 2117 2.6060
19Iowa            Midwestw442 55965118-47319 2913808 2  52.1 1708232
39107013.42 58.63 2692363 3875849.004.1016.4 231394 8.59 2.71053033 215902.050
97715205 7136 773311 70509 9.12 1818625.79 1700102 23705513.9431462074
    1275576 234585  2449120.310.997 1304638 26411920.24 3783 23515126456  266 93.8
    9784625.0210872385.98 4657  407 2.3045
20Kansas          Midwestw447 81778121-40411 2363679 2  28.9 1575899
172901 7.31 66.67 2183102 3062639.305.7017.2 30596814.02 2.6 872239 206792.371
75865204 7350 638387 63271 9.91 1619425.59 1388102 23642717.0334202358
    1080211 207666  1121020.260.984 1078741 20747419.23 3270 15987 77129  619 91.2
    5172729.92 6814088.35 5394 1575 2.6057
21Kentucky        South  e661 39669114-34320 3660777 2  92.3 1862183
244589 6.68 50.87 3378317 4098289.204.6016.3 328063 9.71 2.81263355 587884.653
85642151 5978 98649512587512.76 4410435.04 2086692 23113911.0823491555
    1385747 316148  3437525.290.997 1388046 31202222.48 3548 22875109980  137 59.3
    7240329.60 9415585.61 3286  420 2.7056
22Louisiana        West   w772 44521114-16710 4205900 2  94.5 2887309
    58945 1.40 68.65 3845505 4042798.404.3019.5 368418 9.58 2.91411788 999657.081
```

```
       76558191 6430107447917475416.26 7244941.46 2281481 31722313.9027091849
       1621303 346386  6908625.631.003 1639394 23607414.40 4276 29493 38923   247 33.4
       2683045.52 3509690.17 5042 1484 2.6055
       23Maine              N'east e111 30995105-48261 1124660 2   36.3  534072
       13963 1.24 47.49 1046188 1409189.605.5014.6 12203311.66 2.7 395184 121373.071
       36411241 5768 295488 3507411.87 1079030.76  661840  9528414.4029542007
        455378 111881  691326.090.988  459522 12535827.28 2157  5145  8158  197  8.1
        692549.60  744691.27 4244  405 2.1056
       24Maryland           South e552  9837109-40257 4216975 3  428.7 3386555
        44934 1.07 80.31 3945488 3956098.104.1014.2 48542012.30 2.81460865 448243.068
       102705195 8293109438617677016.15 4493425.42 2499096 50869620.3635652638
        1942280 448257 17015131.840.887 1946612 27974014.37 3937 15930 18727  145 42.9
        1468132.67 1654488.34 6561 3540 2.3051
       25Massachusetts      N'east e114  7824107-34213 5737037 4  733.2 4808339
         9839 0.17 83.81 5400422 7265319.603.1012.7 470388 8.71 2.72032717 531922.617
       199342354 7458144498522513915.58 6181327.46 3463256 69363020.0339912505
        2623038 501912 24361128.421.006 2674275 69419225.9611133 30144  5891  115 13.6
         478048.58  501785.16 5495 6328 1.8046
       26Michigan           Midweste334 56954112-51287 9262078 3  162.6 6551551
       177591 1.92 70.74 8577844 9122588.104.9015.7 518044 6.04 2.83195213 999293.127
       266305229 7688240491035225614.6510809530.69 5254040 74918714.2636752437
        3630817 646710  8949120.280.993 3750732113627130.2915627 93757 68237  168 31.5
        5614531.61 6139989.98 6820 1782 2.3054
       27Minnesota          Midwestw441 79548114-59292 4075970 2   51.2 2725202
       315400 7.74 66.86 3768829 4795648.203.8016.6 311560 8.27 2.71445222 328322.272
       121003250 7451104353210463910.03 2385922.80 2345701 40728117.3639092765
        1837689 349209 10174924.541.001 1885521 38076320.19 6637 23021102963  279 56.5
        8234226.11 9194489.30 4777 1605 1.8046
       28Mississippi         South e664 47233115-19312 2520638 2   53.4 1192805
        84758 3.36 47.32 2305854 2893579.405.5019.0 228369 9.90 3.0 827169 663858.026
        55133130 5183 64545310342516.02 4504943.56 1367792 16808612.2923761631
         934732 236667  1112126.510.980  937206 23010824.55 3289 12766 54182  256 45.8
        3737844.10 4917190.75 3506 1840 4.0060
       29Missouri           Midwestw443 68945118-40293 4916686 2   71.3 3349588
       282074 5.74 68.13 4563086 64812610.05.6016.1 46215310.13 2.71793399 600973.351
       160497140 6917131695516132512.25 4359227.02 2918656 40551513.8930392100
        2078854 453789  7889125.621.024 2103907 46166221.94 7355 33163121955  253 69.9
        8889331.5110815188.68 5601 5638 2.7052
       30Montana            West  w881145388117-70421  786690 1    5.4  416402
        58396 7.42 52.93  722313  845598.506.3018.1 11496315.92 2.7 283742 107843.801
        22101181 6589 207525 19952 9.61  607230.43  450862  7874717.4738132327
         325079  56099  352918.341.000  328316  24286 7.40 1168  2670 24469 2545 66.8
        1803930.89 1990181.33 3733  117 2.9053
       31Nebraska           Midwestw446 76644118-47310 1569825 1   20.5  987859
        17811311.35 62.93 1447251 2056849.204.1017.4 16746211.57 2.7 571400 120522.109
        50240134 6936 414503 39451 9.52  992925.17  912153 14154215.5235232205
         716048 127697 1795520.341.016  716633  9904613.82 1965  8713 65916  702 94.5
        4669926.22 5687286.28 4252  163 2.3060
       32Nevada             West  w888109894122-50327  800493 1    7.3  682947
         5539 0.69 85.32  744698  657567.3017.316.6 25507634.25 2.6 304327 142554.684
        19394177 8453 208934 2739013.11  508818.58  479601  6924714.4432412224
         397368  77349  754921.371.027  398566  23353 5.86  729   942  2877 3641 14.9
         224040.44  223777.75 8414  159 3.0060
       33New Hampshire       N'east e112  8993106-46200  920610 3  102.4  480325
         6640 0.72 52.17  858108 1029678.205.7014.9 16651319.40 2.7 323493  78262.419
        30822156 6966 239647 2659411.10  653824.58  541953  9868418.2131491913
         425908 101011  546625.000.940  432622 13795231.89 1825  4032  3288  164  9.4
         291743.93  300791.45 4235  541 2.1063
       34New Jersey          N'east e222  7468110-34282 7364823 4  986.2 6557377
        18984 0.26 89.04 6903354 8597719.303.8013.2 681060 9.87 2.82548594 895643.514
       252505241 8127194210829727615.31 8684829.21 4504247 82604018.3446192997
        3222925 590638 29786027.570.939 3288302 82014424.9415696 51279  9895  106 21.8
         799442.11  845585.45 6183 2640 1.8057
       35New Mexico         West  w885121335116-50316 1302894 1   10.7  939963
```

```
 20087 1.54 72.14 1188276 1159066.908.0020.0 22885519.26 2.9 441466 380318.615
 27269158 6119 334917 4343812.97 1615937.20  707147 12434517.5830641823
  511084 103232   931522.021.000  508238  37737 7.43 1323  2009 14253 3389 62.2
 944747.03 1215785.29 5960  463 3.7052
36New York          N'east e221 47377108-5219017558072 3  370.614858068
123109 0.70 84.621642901121607679.803.5013.61072612 6.53 2.763404293118484.918
624709180 7498446803179978517.9027341634.1910721012192354717.9452523114
 72516031152045192402742.421.025 7440768155715020.9336578 86216 49273  201 32.4
 4158733.78 4353988.36 6691 9635 2.2048
37North Carolina      South  e556 48843109-29269 5881766 3  120.4 2822852
188437 3.20 47.99 5478813 6031818.204.8014.4 58230910.63 2.82043291 918544.495
144607204 6133158349023304814.72 7369731.62 3403219 45042313.2428961960
 2652593 653985  4010026.171.001 2607925 85454932.77 9954 40912 89367  127 36.3
 5951331.58 7858187.93 4509  786 3.1058
38North Dakota       Midwestw444 69300121-60439  652717 1   9.4  318310
10388115.92 48.77  598049  804458.603.3018.4  8236313.77 2.7 227664  61782.714
 18588179 6417 168418 13207 7.84  351926.64  364601  5404214.8233012114
  277849  45442   202317.081.028  272620  15877 5.82  571  1313 41169 1021 94.8
 2866927.60 3614087.78 2997  137 1.9057
39Ohio             Midweste331 41004113-3933410797630 3  263.3 7918259
271542 2.51 73.331001158011694609.105.4015.7 637094 6.36 2.83833828 956462.495
302428248 7285286394737801513.2011156929.51 6291667 85896313.6534612317
 4422842 764743 17704121.291.005 4558442137317230.1217354 95235 95937  168 61.3
 7357227.09 8364287.18 5385 3365 2.1055
40Oklahoma         West   w773 68655120-27405 3025290 2   44.1 2035082
129874 4.29 67.27 2792986 3761269.308.0017.2 42356515.17 2.61118561 410733.672
 82508197 6858 830508 9302711.20 2890631.07 1769761 26676215.0728372061
 1288460 261829  1331021.350.989 1287857 21477916.68 3818 12565 79388  433 78.0
 5185739.93 7249991.32 4793  612 2.4058
41Oregon          Pacificw992 96184119-54299 2633105 2   27.4 1788354
 78045 2.96 67.92 2435197 3033368.306.7016.4 45292918.60 2.6 991593 289672.921
 87600222 7557 703728 7945811.29 2226528.02 1579841 28203617.8538372307
 1111750 195558  5581622.611.017 1138425 22201719.50 5716 14370 34642  532 29.9
 2929737.54 3042687.83 7019  181 2.8047
42Pennsylvania       N'east e223 44888111-4227611863895 3  264.3 8220851
158183 1.33 69.2911117871153093310.42.9013.4 659972 5.94 2.742196061012902.400
420477260 7077314780943604713.8511954827.42 7240244 98141613.5639082523
 4842003 978735 39937528.461.003 4961501142083728.6418735 79845 59942  146 30.4
 4919831.10 5407690.21 3662 3329 2.6051
44Rhode Island       N'east e115  1055104-23202  947154 4  898.0  824004
  1115 0.12 87.00  890643 1269229.803.8012.9  9489510.65 2.7 338590  86002.540
 36615281 6897 246342 3692114.99 1081529.29  575243  8876815.4340532775
  418158  89522  1790225.690.983  426812 13858632.47 3107  5365   866   86 11.1
  70262.96   75687.30 5881  381 1.9044
45South Carolina      South  e557 30203111-20298 3121820 3  103.4 1689253
 53595 1.72 54.11 2884059 2873288.104.4016.6 35873912.44 2.91029981 615085.972
 67843132 5886 80997412712215.69 4546735.77 1733022 23262913.4225691930
 1349456 345252  1797826.920.984 1319970 43006532.58 4229 18882 33430  189 32.7
 2324543.37 3013590.14 5267  400 3.7062
46South Dakota       Midwestw445 75952120-58271  690768 1   9.1  320777
11285416.34 46.44  632385  910199.404.1019.2  7491311.85 2.7 242523  87393.603
 21079140 5697 178756 16626 9.30  572434.43  389991  5462414.0130591942
  297051  42470   123014.710.994  296679  28555 9.62  740  1794 39665 1123 91.6
 2969726.31 3504188.34 3089   42 2.0053
47Tennessee        South  e662 41155113-32344 4591120 3  111.6 2773573
175673 3.83 60.41 4265638 5175888.906.6015.1 48025211.26 2.81618505 707444.371
113150145 6213125222617846014.25 6036433.82 2692256 33917312.6021251588
 1891670 438316  4664925.641.010 1914920 51112926.69 6487 28752 97036  136 49.7
 6680138.03 8669689.34 4307  828 3.0058
48Texas           West   w774262017120-2354514229191 2   54.311333017
268893 1.89 79.651306459613711617.606.8019.2177878013.62 2.849292673565397.233
297878210 7205369665645668912.3514036430.74 7944161134098116.8829442258
 63105981329203 14619823.381.000 6311845112926717.8918107 92736194253  708 82.0
10646239.5917297989.05 5826 8229 2.4056
```

```
49Utah              West   w887 82073116-50322 1461037 1   17.8 1233060
 18372 1.26 84.40 1271285 1092205.505.3028.6 22897218.01 3.2 448603 258245.757
29518202 6305 354171 33422 9.44   937228.04   704790 14010219.8823601559
  581414 132741  2034426.330.998  585921  9255715.80 1748  5093 13833   760 20.0
 889848.43 1163384.10 5362  722 2.4067
50Vermont           N'east e113  9273105-50215  511456 2   55.2  172735
 18079 3.53 33.77  475452  581669.005.1015.4  7224015.19 2.7 178325  45342.543
15601160 6178 129036 1524311.81  430228.22  295051  5618419.0434752026
  221442  55187   314326.340.998  227195  5420123.86 1030  2189  7273   241 29.6
 637135.24  657290.36 3958  180 2.2051
51Virginia          South  e554 39704110-29280 5346818 3  134.7 3529423
113115 2.12 66.01 4986648 5053047.904.4014.7 80537416.15 2.81863073 640813.440
118931167 7478140474519709814.03 5808029.47 3132882 59806919.0934672250
 2438578 615663 12452030.350.936 2348401 44528018.96 5519 23989 56869   175 39.1
4052535.83 5030288.45 4626 7072 2.2060
53Washington        Pacificw991 66511118-48382 4132156 2   62.1 3037014
 81664 1.98 73.50 3826416 4315627.706.9016.4 70901418.53 2.61540510 443802.881
122928205 8073108631912278811.30 3465928.23 2439417 46231818.9533292273
 1799975 340670  9583424.250.985 1794354 34997719.50 6723 21747 37730   451 39.9
3056837.43 3260586.42 6579 3986 2.0049
54West Virginia     South  e555 24119112-37379 1949644 2   80.8  705319
 28730 1.47 36.18 1804171 2378689.805.3015.1 163940 9.09 2.8 686311 282324.114
52455225 6141 531248 6230811.73 1886630.28 1147042 11948810.4230132204
  668914 165042  1240026.531.012  689461 12658218.36 1857  8706 20532   188 25.1
1556854.19 1900592.56 2580  122 3.4048
55Wisconsin         Midweste335 54426114-54271 4705767 2   86.5 3020732
282722 6.01 64.19 4358993 5641978.303.7015.9 319421 7.33 2.81652261 402702.437
159254200 7243121502313331310.97 3342325.07 2705388 40107614.8338462408
 2065853 396485  8151023.140.988 2114473 60250728.49 8678 38725 89945   201 51.9
7603726.89 7943688.32 4520  775 2.0048
56Wyoming           West   w883 96989114-63688  469557 1    4.8  294639
 19407 4.13 62.75  424510  371756.808.5022.5 12524029.50 2.8 165624  69614.203
 9756198 7927 123420  9288 7.53  219623.64   255149  4376717.1551623114
  215534  49107   406324.671.016  217374  11821 5.44   505  1288  8495 3969 54.2
 659433.98  676479.62 5142   39 3.1061
```

F APPENDIX: What's New in Version 6.06

During the first quarter of 1990, SAS Institute began shipping version 6.06 of the SAS System. Typical of many products of the late 1980s and early 1990s, version 6.06 displayed vastly superior display and ease of use over previous versions. Happily, the "look and feel" of the system was not the only improvement. Many DATA step, procedure, and SAS environment features were added or enhanced. Much of version 6.06 will look familiar to users accustomed to the microcomputer release, version 6.03. Mainframe and minicomputer users who take advantage of version 6.06 will be greatly, and pleasantly, surprised at the change in the system.

This appendix outlines some of the more significant changes in the SAS System in version 6.06. It does not present syntax descriptions or usage notes. Instead it focuses on describing SAS Base product features that may be of interest to the novice to intermediate-level user. For complete documentation on version 6.06 contact the Publications Department at SAS Institute (919) 677-8000. Version 6.06 orientation courses may be offered by your computation center's training staff. Refer to Appendix C for suggestions about identifying these and other local resources.

This appendix is divided into two sections. The first describes new and improved features of the way the SAS System looks and handles in interactive environments. The second section moves beyond these mostly cosmetic features and outlines some of the major improvements in the SAS System's functionality. New DATA step features and procedures are outlined.

F.1   Improvements in Look and Feel

Many of the changes in version 6.06 address issues of how interactive SAS "looks and feels." Some of the changes are cosmetic; others, more substantive. This section describes changes to the Display Manager, SAS/ASSIST menu-driven procedures, and PMENUS, an alternative to the command line.

Display Manager

Users of the SAS Display Manager System (outlined in Appendix D) in both versions 5 and 6.03 will notice many changes to this powerful user interface. Many screens, or panels, have been added.

Library Management. The FILENAME, LIBNAME, DIR, VAR, and CATALOG windows perform many of the functions of the DATASETS procedure without DATASETS' sometimes confusing syntax. You can use these windows to determine the LIBNAMEs and FILENAMEs in effect, see which datasets are stored in a library, produce a modified CONTENTS listing of SAS datasets, and delete and rename datasets.

Output Manager. This window closely resembles the OUTPUT window of earlier versions. It adds two important features to OUTPUT. First, it has an index to different procedures' output. If you ran a MEANS, then a PRINT, and wanted to view only PRINT results, you could point directly to the portion of the window containing that output. Output Manager's second improvement over OUTPUT is its ability to edit rather than simply read output.

Calculator, Appointment, and Notepad. These windows are among the desktop accessories furnished as part of the SAS Base product.

Display Manager allows you to move windows within your screen, resize them, present them side by side ("tiled" windows), or piled on top of each other, one line visible on all but the top window ("cascading"). Cut-and-paste and text-file import/export features allow fairly easy movement of text from one window to another or into and out of the SAS System environment.

SAS/ASSIST

SAS/ASSIST software is provided with the Base SAS product. ASSIST is a series of checklist and fill-in-the-blanks menus and help screens that help new and occasional users access SAS without having to learn (or remember) the SAS language. All procedures have ASSIST panels to guide the user through simple use of most of their options. Advanced use of the procedures, however, may find ASSIST's options and capabilities too restrictive. Such sophisticated usage may require actual coding of SAS statements.

ASSIST does not help you construct DATA steps. This makes it an unsuitable choice for users who need to build programs or small systems from scratch: the dataset reading, combining, and formatting features that are part of such applications are not addressed by the ASSIST product.

Pull-Down Menus: The PMENU Facility

Pull-down menus may be used as a substitute for the command line in the Display Manager System. Entering PMENU ON on the command line will activate the facility. Groups of system commands and options are displayed in an "action bar" that replaces the command line. Rather than have to remember Display Manager commands, you simply move the cursor or mouse to the appropriate group of commands (HELP, EDIT, FILE, OPTIONS, and so on) and select an option. PMENU is ideal for occasional users of the SAS System who are prone to forgetting command names and syntax. It is also a good way for new users to become acquainted with SAS System capabilities.

You may write your own PMENU. The PMENU procedure enables you to define your action bar and the activities and dialogs associated with each option. These custom PMENUs are a very powerful and reasonably easy-to-learn means of customizing the SAS environment for specific end-user applications.

F.2   Improvements in Functionality

A number of changes have been made to the DATA step in version 6.06. Some of these will be familiar to users of version 6.03, the microcomputer release of the SAS System.

DATA Step Changes

Among the changes and enhancements to the DATA step in version 6.06 are those to INFILE and FILE, ARRAY, RETAIN, WHERE, and stored programs.

INFILE and FILE.
The file specification may be entered directly in the INFILE and FILE statements. Thus a statement such as

```
infile 'uabr.fiscal89(q1)';
```

is now valid in all operating systems.

ARRAY.
You can enter initial values for array elements in the ARRAY statement. Thus a statement such as

```
array q{5} q1-q5 ('f', 'f', 'f', 'f', 'f')
```

is now valid in all operating systems.

RETAIN.
You can specify an array name in a RETAIN statement. This will retain values of all elements in the array.

WHERE.
The WHERE statement selects observations from a SAS dataset and passes them along to the DATA step or PROC for processing. The following use of WHERE bypasses a preprocessing DATA step: only states with median incomes over $10,000 are printed.

```
proc print data=survey;
where med_inc > 10000;
```

WHERE can eliminate the need for DATA steps used only for observation selection. WHERE can also create subtle problems in DATA steps when used with conditional OUTPUT statements and retained variables. To be safe, consult the SAS Institute documentation before using WHERE in a DATA step.

Stored Programs.
DATA steps may now be compiled and stored in executable format. The advantage of this "Stored Program Facility" is faster execution speed, especially when the DATA step contains many statements or processes many observations. You still need the SAS System to run the program at a later point: the stored program is not a stand-alone module.

Dataset "Engines"

One consequence of the release of version 6.06 is the very real possibility that a single site may have three versions of the SAS System in use: versions 5, 6.03, and 6.06. Most SAS program statements are understood by the three releases. SAS datasets are another matter. A dataset written by version 6.03 cannot be directly read by version 6.06, nor can a version 5 dataset.

To simplify the transmission of data from one release to another, SAS Institute developed database "engines." These are instructions SAS uses when reading a particular kind of dataset. If you specified a version 5 engine while running version 6.06, SAS would read the dataset using version 5's dataset layout rules. Engines are specified in the LIBNAME statement:

```
LIBNAME ref engine 'path';
```

Many non-SAS databases may also be read by supplying an engine. Using these engines allows you to refer to the "foreign" file in exactly the same manner as you would a SAS dataset. There is no need for an intermediate conversion of the dataset into SAS format. SAS Institute provides engines for DB2, System 2000, SQL/DS, ORACLE, Rdb, dBASEII, dBASEIII, dBASEIV, SPSS, BMDP, and OSIRIS. These non-SAS database engines are not part of the Base SAS product and must be leased separately.

Fine-Tuning Features

Two new features in version 6.06—indexing and compression—give you the means to improve SAS's performance on your system. Full discussion of their use is not possible here, but an overview of their capabilities is useful.

Indexing. SAS datasets in version 6.06 may be indexed. The index is a separate file containing record location information on one or more dataset variables. If an indexed variable is used in, say, a WHERE statement, SAS automatically uses the index to locate observations with appropriate values. Indexing is extremely effective in situations where the "hit rate" is low: these are cases in which relatively few of the observations have appropriate index variable values. The extra processing time to use the index is, in these cases, more than offset by the reduction in input seek and transfer times required to read unneeded records. SAS Institute documentation has a complete discussion of index creation, maintenance, and benefits and drawbacks.

Compression. The typical computer user is probably no stranger to disk space shortages. It is especially frustrating for SAS users with long character variables. The variables *always* occupy enough space to allow for the "worst case" (longest name, longest address, longest survey comment or question) scenario. A new technique in version 6.06 allows the dataset to be written using only as much space as it needs. If, for example, variable NAME is read with a $30. format, the SAS data "compression" routines will store the value "Chad" in only six characters.

As with indexing, the benefits of compression are not without attendant costs. Extra processing time is required to expand the observation into a format usable by SAS. Consult SAS Institute documentation for syntax and usage notes.

Changes to PROCs

Many changes and enhancements have been made to procedures in version 6.06, including those to SORT, PRINT, and CONTENTS.

SORT. The NODUPKEY option has been added. This eliminates observations from the output dataset if there are duplicate BY variable values. The NODUP option eliminated only if all variables (BY and others) were identical.

PRINT. Column headings using the SPLIT option are no longer limited to three lines (three split characters in the variable label). Any number of lines may now be specified.

CONTENTS. Much more information is now printed by the procedure. Among the new items reported are index and compression status (described in the preceding section). Other information includes the engine used to write the dataset, creation and recent modification dates, and the number of observations marked for deletion.

Two significant procedures have been added to version 6.06: REPORT and SQL.

REPORT. As its name implies, REPORT is a report-writing procedure. It is used interactively to design the report layout, specify variables to use, indicate subtotals and grand totals, sort the dataset, and save the report layout definition for later reference. REPORT does not have the complete control over customized layouts found in DATA steps using FILE and PUT statements. It may be an ideal choice, however, for reports whose complexity falls between the PRINT procedure and DATA step reporting.

SQL. The SQL procedure is an implementation of Structured Query Language (SQL), a popular relational database query and update tool. Use of SQL assumes knowledge of basic principles of relational database theory! Do not assume, particularly in complex SQL statements, that SAS in SQL will behave the way you would expect "regular" SAS to act. SQL is a very powerful tool, probably not appropriate for beginners.

Using PROC SQL, you can use the query language to combine, create, and delete SAS datasets, take subsets of variables and observations, and create new variables. All this can be done without conventional DATA steps, SORTs, and PRINTs, and usually in a fraction of the lines of code.

SQL statements can be stored for later use. These *views* allow users to refer to the result of an SQL query just as they would a SAS dataset. For example, suppose an SQL query that combined two datasets, selected several columns from the result, and calculated a new variable were stored as OUT.QUERY1. A user could print the result of that query by simply entering

```
proc print data=out.query1;
```

There is no need for the user to see the underlying SQL or SAS code used to generate the dataset that will be printed.

Answers

Chapter 2

2.1. a. One DATA step, two PROCs, eight statements.

b. Since the DATA step has an INPUT statement it probably reads (INPUTs) data, probably from a file called 'mast.fin(fy83)'. PRINT probably prints the data and MEANS probably computes mean values for the variables.

c. Probably only part of it would work. The DATA step creates a dataset, in this case, dataset ONE. If the PRINT of the dataset came before the dataset was created, the PROC would not have anything to work with. MEANS would probably run successfully.

d. `data one;infile 'mast.fin(fy83)';input sic $5. earn 5.`
`prof 5.;run;`
`proc print data=one;run;proc means data=one;run;`
This is not at all easy to read and not very easy to change, even with a good program editor. The saving in space (three lines versus the original program's 10) is not worth the cost in legibility.

2.2. 15 columns/variables. You could describe the location by column number: the first variable is always in columns 1 and 2, the second in columns 3 through about 15, and so on. Or you could tell SAS to read numbers and letters until it found a blank, then store the value just read as the first variable's value, then begin reading the second variable when it finds a nonblank character.

2.3. `NOTE: Copyright(c) 1985,86,87 SAS Institute Inc., Cary, NC`
`        27512-8000, U.S.A.`
`NOTE: SAS (r) Proprietary Software Release 6.03`
`        Licensed to D**2 Systems, Site 12345678.`
`  1     data test;`
`  2     length message $25;`
`  3     message = 'This is a test program!';`
`  4     run;`
`NOTE: The data set WORK.TEST has 1 observations and 1`
`    variables.`

```
NOTE: The DATA statement used 6.00 seconds.
    5
    6    proc print data=test;
    7    run;
NOTE: The PROCEDURE PRINT used 4.00 seconds.
NOTE: SAS Institute Inc., SAS Circle, PO Box 8000, Cary,
    NC 27512-8000
---------------------- New page of output --------------
SAS
OBS            MESSAGE
1      This is a test program!
----------------------- End of output -----------------
```

The Log is the portion of the output beginning and ending with NOTEs. The procedure output is the second page of the listing (the line beginning with SAS). The first NOTEs are legalese and the name of the site (D**2 Systems) where the program was run. Other NOTEs appear to be information about how long the program steps took to execute and how large the datasets were. The program is reproduced in the Log and has line numbers added (there were seven lines in this program). The resource messages are probably useful when you want to ensure that the dataset created is the "correct" size. The line numbers are probably especially helpful when you are writing a very long program and an error or other message identifies a problem by the line number (it is easier to find a reference to "line 12" than "variable X").

2.4. The names immediately suggest what the variables represent: HIGHTEMP probably means high temperature; LOWTEMP, low temperature; POP80, population (probably total population) in 1980, and so on. No cross-references to a codebook or other sources are necessary.

2.5. a. The first notes are legalese stating SAS Institute's rights to the program. The notes after line 12 tell where the data are coming from and how many records were in the dataset. The "missing values" note says that since either or both of the variables used in line 11 had missing data, the result (the "generated" value) is also missing. The next note tells about dataset size: dataset WORK.STATES has 51 records (same as the raw data input) and 10 variables. This took 9 seconds to run. The PRINT and MEANS procedures took 11 and 5 seconds to run.
  b. The OBS column is added. This is the number of the observation in the dataset.
  c. Colorado population density: 28
  DC per capita income: 8960
  DC-Alaska population density: 10181 versus 1
  d. Variable names are identified by the column heading "Variable." Statistics columns are labeled using the statistic name: N (count of observations), minimum, maximum, and so on.

2.6. Line 11: Shorten name to maximum of eight characters. Line 12: Reduce number of decimal places. Line 22: Insert an equals sign (=). Other messages do not help diagnosis of the errors.

2.7. a. PRINT is easier to read because commas are inserted in the numbers. Having the state name at the beginning of each line probably makes identification of the observation easier.
  b. MEANS is easier to read as well. The variable label is printed beside the variable name: it is easy to tell what the variable represents (although names at least *suggest* meaning). By specifying the statistics of interest (N, NMISS, and so on in the PROC statement), we see only the statistics we are interested in—extraneous information is eliminated.

Chapter 3

3.1. There are three statements in each. The first two would be easiest to write since they are simply continuous text: no new lines or indentations are required. The last two would be easiest to read: the statements are clearly separated. The last is easiest to alter: to add a new variable you would simply have to insert a line where appropriate (and not have to worry about inserting text *within* a line).

3.2. There are five different types of statements that serve some purpose in the program (DATA, INFILE, INPUT, * -comment, and IF). The ;; represents two "null," or do-nothing, statements, so strictly speaking the correct answer is six.

3.3. There are nine statements in the program (nine semicolons in places that constitute the end of a statement). Refer to the note about null statements in the answer to Exercise 3.2.

3.4. a. Seven comments, five embedded, two comment statements
   b. `*`
   ```
   | Homework 3
   |
   | Listing of EEC sample data (drawn from 1985-1990 tapes).
   | Read and list.
   |
   | Written 10/2/90 (revised 10/5/90)
   *;
   ```
   c. Turn off using comment statements: `*proc print data=one;`
   `*title 'Listing of test data';`
   Turn off using embedded comments: `/*proc print data=one;`
   `title 'Listing of test data'; */`

3.5. Illegal:
   _____(Nine characters is too long. Maximum is eight)
   area code (Embedded blanks are not allowed.)
   -count (hyphen [-] is not valid in variable names)
   prev_pregs (nine characters is too long. Maximum is eight)
   1on1 (Name cannot begin with a number)

3.6. a. apgar1m
   b. mh_gs_f90
   c. q2   *or*   age_measl
   d. q15   *or*   age

3.7. Using BRIDGES as an example, new names might be SPANTYPE   NAME COUNTRY   LEN_MTR   and   YR_COMP.

3.8. a. Seven variables: x1-x6, avg
   b. x1, x2, x3, x4, x5, x6, avg
   c. (1) x1, x2, x3, x4, x5, x6
      (2) x1, x2, x3
      (3) x4, x5, x6
      (4) x4, x5, x6
      (5) x1, x2, x3, avg
      (6) List is invalid (unless there is a variable named _NUMERIC): missing the _ after _NUMERIC.
      (7) x1, x2, x3, avg
      (8) No variables are identified (all are character).
      (9) List is invalid: variables START and END are not defined.
      (10) List is invalid: variables x01 and x06 are not defined even though they look like x1 and x6.

3.9. a. Character: contains nonnumeric text
   b. Character: contains nonnumeric text
   c. Character: even though it looks numeric it will not be used arithmetically, so it can be stored as text that "looks numeric."
   d. Character: same reason as in part c
   e. Numeric: may be handled in arithmetic operations

f. Numeric: same reason as in part e

g. Numeric: the result of an arithmetic operation (change divided by 1980 population multiplied by 100)

h. Character: contains nonnumeric text

3.10. Four periods = four missing values.

3.11. No missing values would be produced since there would be "complete" data. Convert . to 0 since you could argue that someone without a particular account type has, in a sense, a balance of $0 for that account. Leave .'s as is to get an accurate count of nonmissing account types.

3.12. In most cases you would probably not want to enter the numbers as missing. It might be desirable to be able to distinguish the reason for nonresponse. Collapsing the −9's and the −8's into "." loses this distinction.

3.13. Equal pairs are in parts a, b, and d. The pair in part c look identical but are not since one constant is numeric and the other is character.

3.14. Constants that are not valid are part a (no commas allowed), part e (strictly speaking this is valid, but the practice is strongly discouraged), and part g (negative sign must be within the quotation marks).

3.15. a. 14

b. 4.66

c. -8

d. -4.5

e. 33

f. .

g. False, 0

h. True, 1

i. False, 0

j. True, 1

k. True, 1

l. False, 0

m. True, 1

n. False, 0

o. True, 1

Chapter 4

4.1. b (cannot begin with a number); c (exceeds maximum of eight characters).

4.2. a. FRINFO88, FROSH88, INFOFR88, STATS88

b. STATE, STATEDAT

c. STDEMOG

4.3. Possible changes: BRIDGES to LONGBRDG, CMI to WELLNESS, COASTAL to COASTDAT or SHORLINE, COUNTRYS to RESOURCE, LIFESPAN to EXPCTNCE, NATLPARK to PARKDATA, PARKPRES to NATLPRES, PRES to USPRES or PRESDATA, STATES to STPROFIL or STATEDAT.

4.4. a. `filename in 'c:\test.raw';`
    `infile in;`

b. `infile 'c:\test.raw';`

4.5. b (filename specifies INDATA but infile specifies a different name, DATAIN)

4.6. For the LIFESPAN dataset:

a. `input age total wm wf om of;`

b. `input age 1-2 total 4-7 wm 9-12 wf 14-17`
       `om 19-22 of 24-27;`

c. `input age 1-2 total 4-7 wm wf om of;`
   For the NATLPARK dataset:

a. `input park $ state $ coast $ yrestab acres;`

b. `input park $ 1-20 state $ 22-23 coast $ 25`
`        yrestab 27-30 acres 31-39;`
c. `input park st coast yr 27-30 acres 31-39;`

4.7. column ruler
```
—+—1—+—
10 1 2 3    89.8
100 1 2  .   890
5 10 2 3    7.3
```

4.8. d (Once BASE is read, SAS tries to read CODE_NUM beginning in the next nonblank field. It *should* read it from columns 4 through 8. Instead it will read it from columns 25 through 29.)

Chapter 5

5.1. a. (1) `var units us_equiv;`
(2) `var country currency;`
(3) `var country us_equiv;`
(4) `var country us_equiv currency units;`
b. `format country currency $20. us_equiv 5.2 units comma7.;`
c. `label country  = 'Country name'`
`      currency = 'Name of local currency'`
`      units    = '1st class postage, local units'`
`      us_equiv = '1st class postage, $US'   ;`

5.2. MEANS
```
    title 'Mortality ... '
    footnote 'Program in ... '
```
PRINT
```
    title 'Mortality ... '
    title2 'Preliminary rates ... '
    footnote 'Program in ... '
```
FREQ
```
    title 'Mortality ...'
    footnote 'Program in ... '
    footnote2 'Background data ... '
```
PRINT
```
    title 'Demographic ... '
```

5.3. a.

REGION	INDEX
e	2.0
e	2.0
e1	3.1
s	4.2
w	8.0
n	8.0
w	8.1

b.

REGION	INDEX
e	2.0
e	2.0
e1	3.1
n	8.0
s	4.2
w	8.0
w	8.1

c.

REGION	INDEX
w	8.1
w	8.0
s	4.2

```
         n      8.0
         e1     3.1
         e      2.0
         e      2.0
```

d. REGION   INDEX
```
         w      8.0
         w      8.1
         s      4.2
         n      8.0
         e1     3.1
         e      2.0
         e      2.0
```

e. REGION   INDEX
```
         w      8.1
         w      8.0
         s      4.2
         n      8.0
         e1     3.1
         e      2.0
         e      2.0
```

5.4. a.
```
proc sort data=result80;
   by lab_id;
```
b.
```
proc sort data=result80 out=res80srt;
   by lab_id;
```
c.
```
proc sort data=states nodup;
   by region pop_grp;
```

5.5. a. The name of the OUT= dataset needs to be specified.
b. Missing BY statement.

5.6. a.
```
proc print data=rates90 n;
   id yrdiag yr_surv;
   sum _numeric_;
```
b.
```
proc print data=sales d noobs u;
   pageby region;
   by region;
```
c.
```
proc print data=hoop split='*';
   id team year;
```
d.
```
proc print;
   var _numeric_;
   sum _numeric_;
   sumby name;
   by name;
```

5.7. a. Missing BY statement.
b. Only one character should be specified with the SPLIT option.
c. The variable SECTOR in the SUMBY statement is not in the BY statement's variable list.

Chapter 6

6.1.
```
proc means data=coastal n min max mean std; run;
```

6.2. a.
```
proc means mean sum std nmiss n;
   class district;
```
b.
```
proc univariate data=sales;
   var net gross;
```
c.
```
proc univariate data=sales;
```

```
               id saleper;
               var net gross;
          d. proc means data=sales n min max range mean;
             var net gross;
```

6.3. **a.** In the PROC statement replace .2 with 2, STANDARD with STD, VARIANCE with VAR, and COUNT with N.

b. PCTILE is not a PROC option. SALEPER appears in both the variable list and CLASS statement.

c. Missing dataset specification. Although not an error, it is best to explicitly specify the dataset to analyze with the DATA= option.

6.4. **a.** `proc freq data=hydro order=freq;`
`tables lakeid depthgrp;`

b. `proc freq data=hydro;`
`tables lakeid / nocum;`

c. `proc freq data=hydro;`
`tables depthgrp / missing;`

6.5. **a.** Missing dataset specification. Although not an error, it is best to explicitly specify the dataset to analyze with the DATA= option. VAR should be TABLES.

b. REVERSE is not a PROC statement option. A variable list in the TABLES statement is missing.

c. The TABLES statement is *not* in error if there are, in fact, variables named MISSING and NOCUM. Otherwise, there should be a variable list, a slash (/), and the table *options* MISSING and NOCUM. See also the note about the DATA option in the answer to Exercise 6.5a.

6.6. **a.** `proc chart data=sales;`
`hbar prod / type=mean;`

b. `proc chart data=sales;`
`vbar prod / type=mean group=dist sumvar=gsale;`

c. `proc chart data=sales;`
`vbar region / type=freq sumvar=prod;`

d. `proc chart data=sales;`
`vbar prod / sumvar=gsale type=sum group=region`
`              nozeros ascending;`

e. `            proc chart data=sales;`
`hbar dist / symbol='x';`

f. `proc chart data=sales;`
`hbar dist / symbol='x' midpoints = 'N' 'S' 'MW';`

6.7. **a.** NOZEROES should be spelled NOZEROS.

b. Variable DIST is both the analysis variable *and* the GROUP variable.

c. Only one of the DISCRETE and LEVELS options should be specified.

Chapter 7

7.1. Created: d
Replaced: a, b, c

7.2. **a.** 5
b. .
c. 13
d. .
e. 1 (v3 = . is true, or 1, and v2 = 2 is false, or 0. 1 + 0 = 1)

7.3. **a.** There should not be a blank between 2. and 471 (assuming hectares should be multiplied by the constant 2.471).

b. The underscore (_) should be an arithmetic operator (probably a +, since the variable name suggests that it represents a weighted average).

c. This is correct as is.

    d. This is correct as is. The parentheses, however, are not needed.

    e. This is correct as is. If parentheses were being used to highlight the calculation's logic, however, the statement would be better written as:

```
newscore = (score1*1.05) + (score2*.98);
```

7.4. a. 5, 3, 4

    b. 50, 40, 2

    c. 7, 0, 2

    d. 8.5, 12, -0.5

    e. 9.1666, 11.25, 1.4

    f. .5, 5, 5

    g. ., ., .

    h. ., ., .

7.5. a. MINAVG is the smaller of two means: those for GRP1 through GRP5 and GRP10 through GRP15.

    b. MAXMIN is the largest minimum value of three pairs of variables (P1 and P2, P3 and P4, P5 and P6).

    c. FEWEST is the smaller of two counts of nonmissing values: counts of COH70_1 through COH70_10 and COH80_1 through COH80_10.

7.6. a. 1, 0, ., 2, 3, 4, 6

    b. −1, 0, ., 2, 0, 1, 0

    c. ., ., ., ., ., .

    d. 0, 0, ., 0, 0, 0

7.7. a. −2; −2; 0; .; 1; 1; 1; 3

    b. −3; −3; 0; .; 0; 0; 1; 2

    c. −2; −2; 0; .; 0; 0; 1; 2

    d. −3; −2; 0; .; 0; 1; 1; 3

    e. −2; −2; 0; .; 0; 0; 2; 2

    f. −2 : 5; −2 : 5; 0; 0; 0.5; 1; 3

    g. 0; 0; 0; .; 0; 0; 0; 0

Chapter 8

8.1. a. ' ', 'm', ' '

    b. ' ', 'm', ' '

    c. ' ', 'm', ' '

    d. '2', '1', '2'

    e. ' ', 't', ' '

    f. 'n', 'y', 'n'

8.2. a.
```
if          a > 2 & b = 3 then c = 1;
    else if a > 2 | b = 3 then c = 2;
    else                   c = .;
```

    b.
```
if b < 4 & (1 <= a <= 5) then c = 0;
    else if b**2 > 16 & a in (1,2,4,5) then c = 1;
```

    c.
```
if b >= 15 then c = 1;
    else        c = 0;
```

    d.
```
if          b >= 15 then c = 1;
    else if b ^= 0   then c = 0;
```

8.3.
```
if type = 'f' then do;
    style = 'g';
    ok = 1;
    end;
    else if type = 'm' then do;
        style = 'b';
        end;
    else do;
```

```
                    style = ' ';
                    ok = 0;
                    end;
```

8.4.
```
if year <= 1969 then do;
    early = 1;
    late  = 0;
    end;
    else do;
       early = 0;
       late  = 1;
       end;
```

8.5.
```
COHORT values:  'n'   'n'   ' '   'y'   'n'
INT values:      .     .     .     2     .
```

8.6. The DO-group that is a condition of the first ELSE statement is missing an END statement.

8.7. **a.** Insert line (6.5): output;
   b. Insert line (3.5) or (6.5): if keyval <= 100 then delete; Line (3.5) eliminates the unnecessary observations immediately. It does not calculate TOT, AVG, and VALID only to possibly not use them [this is what happens when you insert at (7)]. Inserting at (7) may be preferable since it would be in the same location in the program as a conditional OUTPUT (i.e., they are both in the same place and you know immediately where to look. This argument gets stronger as the DATA step grows longer).
   c. Insert at (3.5): if keyval <= 100 then return; Insert at (6.5): output;
   d. (1) If the RETURN is executed, the observation will have missing values for TOT, AVG, and VALID.
   (2) All observations will have missing values for TOT, AVG, and VALID.
   (3) All observations in dataset TEST will have TOT, AVG, and VALID calculated (any missing values will be a result of S1–S5's values rather than the location of the DELETE statement).

8.8. **a.**

STATUS	T1	T2	INDEX
4	0	1	.
.	0	0	.

   b.

STATUS	T1	T2	INDEX
4	0	1	1
.	0	0	0

Parts c through e are identical to part b.

Chapter 9

9.1. **a.** After line (3): format rating 3.1;
   b. After line (3): format salary dollar8.;
   c. After line (3): format rating 3.; After line (5): format rating 6.2;

9.2.

a.	^-21	^^^0	^^^5	^^19	^120
b.	^-21	0.00	5.20	19.0	^120
c.	-021	0.00	5.20	19.0	0120
d.	^^^^^-21	^^^^^^^0	^^^^^^^5	^^^^^^19	^^ ^^^120
e.	^^-21.30	^^^^0.00	^^^^5.20	^^^19.00	^^120.10
f.	-0021.30	00000.00	00005.20	00019.00	00120.10

9.3.
```
PCTIN Z5.2
FEMPOV COMMA7.
TRASWRK COMMA9.
PCAPINC DOLLAR7.
EMPRES Z5.3
```

9.4. TEMP is read in the DATA step as *numeric* but is displayed in the PRINT with a format preceded by a dollar sign ($), indicating a *character* format.

9.5. `format sal1-sal10 comma8. id $char20. delta1 delta2;`

```
format sal1-sal10 comma8. id $char20.;
format delta1 delta2;
```

9.6. a. `ATTRIB STATE LABEL='State FIPS code' FORMAT=$2.`
```
            NAME LABEL= 'Name of state' FORMAT=$20.
            POP LABEL='Population, 1980, FORMAT=12.
            _5YRPLUS LABEL='# persons>4yrs old, 1980'
                    FORMAT=12.
            _65PLUS LABEL='# persons > 64 yrs old, 1980'
                    FORMAT=12.
            PCTIN LABEL='% 5yr old, diff. st., 1975'
                  FORMAT=6.3
            PCTCROWD LABEL='% occ units > 1 per/room'
                    FORMAT=6.3
    ;
```
b. `LABEL  STATE = 'State FIPS code'`
```
            NAME = 'Name of state'
            POP ='Population, 1980'
            _5YRPLUS = '# persons > 4 yrs old, 1980'
            _65PLUS = '# persons > 64 yrs old, 1980'
            PCTIN = '% 5 yr old, diff. st., 1975'
            PCTCROWD = '% occ units > 1 per/room'
            ;
    FORMAT STATE $2. NAME $20.
            POP _5YRPLUS _65PLUS 12.
            PCTIN PCTCROWD 6.3;
```

9.7. a. After (1) or after (2) and after (5) or after (6): label o2up = 'Oxygen uptake' ;
b. Add SPLIT='*' to the PROC statement to print the label

`'Oxygen*uptake'`

9.8. `DROP RUN;`
Variable RUN is not needed since we know from the program that only values of 'a' and 'A' are in the dataset. The dataset name (RUNA) also suggests it a-only contents.

`DROP RUN DEF1-DEF3;`

Drop RUN for the reason explained above. You might want to drop DEF1–DEF3 if only their average is of interest (consider, though, the backtracking you would have to do if at some point you want to find out how many of DEF1 through DEF3 were used in calculating AVG).

9.9. LEV1 and LEV2 are dropped from dataset ONE but used in the TABLES statement in the FREQ procedure.

9.10. `drop m1-m3 d1-d3;`
`keep batch stat;`

Chapter 10 **10.1.** `libname diss 'c:\diss\dissdata';`
`data diss.temps;`
10.2. `libname datain 'c:\diss\dissdata';`
`proc print data=datain.temps;`
10.3. `libname datain 'c:\diss\dissdata';`
`data datain.subset;`
`set  datain.temps;`
`if reg = 'E' then output;`

10.4.
```
libname datain 'c:\diss\dissdata';
     proc print data=datain.temps(keep=state high low);
```

10.5.
```
libname datain 'c:\diss\dissdata';
     proc print data=datain.temps(keep=state high low
          rename=(high=highest low=lowest));
```

10.6.
```
libname datain 'c:\diss\dissdata';
     data datain.temps (label='Preliminary survey: instrument 1');
```

10.7.
```
libname datain 'c:\diss\dissdata';
```
   a. `proc print data=datain.temps(obs=10);`
   b. `proc print data=datain.temps(firstobs=20 obs=30);`
   c. `proc print data=datain.temps(firstobs=20);`

10.8.
   a. All variables in TWO are dropped. The procedure has no variables to print.
   b. Since labels may be attached to a dataset only when it is written, the label will not be added to ONE. The attempt to label is ignored and does not produce an error!
   c. LIBNAME DATAOUT is not defined: there is no LIBNAME statement identifying DATAOUT before the DATA step.

10.9.
```
libname x '[master.trial]';
     proc contents data=x.trial0 position;
```

10.10. Information lost includes dataset size (number of observations), dataset label, variable type, length (discussed in Chapter 11), format, and label. This information may not be important if all you are interested in obtaining is the position and/or names of the variables. The SHORT/POSITION listing would not be sufficient if you needed to know the size of the dataset, the data type (character or numeric) of a variable, or other, more detailed information about the variables (such as their label or format).

10.11.
```
proc datasets library=datahere;
```

10.12.
   a.
```
delete ver2 ver3;
        save ver1;
```
   b. `change ver1=current;`
   c. `change ver3=oldest ver1=newest ver2=ver1;`
   d.
```
contents data=ver1 short;
        contents data=ver1 position;
```
   e. `contents data=_all_;`
   f. (1) `modify ver1(label='Study grp 5-recent down. data')`
      (2)
```
modify ver1;
           format deficit comma5. surplus;
```
      (3)
```
modify ver1;
           rename q1=q191 q2=q291;
```
   g.
```
copy out=there;
        select ver1 ver2;
```
   h. `copy out=there move;`

10.13.
   a. Correct as is.
   b. NATIONS is deleted, then modified.
   c. The destination of the COPY must be different than the origin. In this case both are the same (HERE).

Chapter 11   **11.1. a.** To PRE: none
                              To POST: 4
                         b. To PRE: 1, 2, 3, 5
                              To POST: 1, 2, 3, 4, 5
                         c. To PRE: 1, 1, 2, 2, 3, 3, 4, 4, 5, 5
                              To POST: 1, 2, 3, 4, 4, 5

11.2. a. Each dataset in the OUTPUT statements is a one-level name. Each name in the DATA statement is two levels.
b. This is correct as is. Each observation will be written to each dataset in the DATA statement.

11.3. a. `length default=3;`
b. `length name_f name_m name_l $20;`
c. `length cityname $30 zip 4;`

11.4. `data trials2;`
`length site labid lev1-lev20 8;`
`set trials;`

11.5. a. `array test(20);`
`array test(20) test1-test20;`
b. `array mth(3) $8 jan feb mar ('January', 'February',`
`'March');`

11.6. a. This statement is correct.
b. TEST was set up with 20 elements but there are 21 elements defined in the variable list.
c. INIT's elements are declared character but are given initial values that are numeric.
d. SAS will assign variable names to the elements, attempting to create variable VALUES100 at the end of the array. This name is nine characters long. The longest legal variables name is eight characters.
e. Q must be referred to in the function as Q(*).

11.7. Change the ending value of the loop's index variable to 19.

11.8. `data testit;`
`do year = 1990 to 1995;`
`  do month = 1 to 12;`
`      flux = ranuni(0);`
`      output;`
`  end;`
`end;`
`run;`

11.9. a. `do i = 1 to 20;`
b. `do i = 2 to 20 by 2;`
c. `do i = 1 to 20 by 3;`
d. `do i = 1 to 10, 15, 20;`

11.10. `array indics{20} q1 - q20;`
`array avgs{18}  avg1-avg18;`
`do i = 2 to 19;`
`   avgs{i} = mean(indics(i-1), indics(i), indics(i+1));`
`end;`

11.11. a. Even though the loop is only executed 13 times it will eventually create a subscript error. When I is 21 there will be a reference to element 21 of array X, which has only 20 elements.
b. This is correct. The subscripting problem of part a is avoided by using variable N as a separate counter and subscript.
c. The loop will not execute as it was probably intended. The default increment is 1, so I will be .07, then 1.07, which is larger than the ending value of .08.

Thus the loop executes only once. Appropriate loop index increments would be values such as .005 or .001;.

11.12. a. 75
     b. 10
     c. 45
     d. 5

11.13. a. `if nbad > 0 then abort;`
     b. `if nbad > 0 then abort abend;`
     c. `if nbad > 0 then stop;`

11.14.
```
input state $ 1-2 name $ 3-32
      / crowd 68-73 pctcrowd 74-78 .3
      /// ;
input state $2. name $20.
      / @68 crowd 6. pctcrowd 5.3
      /// ;
input state $2. name $20.
      #2 @68 crowd 6. pctcrowd 5.3
      #5 ;
```

11.15. A caret (ˆ) indicates a blank column in the input record:

```
ˆˆˆˆdefˆˆaˆbˆc
  gˆˆhˆiˆj
```

11.16. a.

X	Y
1	1.1
4	4.1
5.	5.1

     b.

X	Y
1	1.1
2	2.1
3	3.1
4	4.1
5	5.1

     c.

X
1
4
5
5.1

11.17. `input (v1-v12)(3. 2. 4.1);`

11.18. a. If TYPE is 'm' the program reads NREPS from columns 2 and 3. It then reads NREPS lines from the raw data, assigning values to YEAR and RATE and writing the observation to output dataset ALL.
     b. In ALL: type, nreps, id, year, rate, and i. Duplicated: type, nreps, id.
     c.

TYPE	NREPS	ID	I	YEAR	RATE
m	3	US	1	87	5.4
m	3	US	2	88	5.15
m	3	US	3	89	5.0
m	2	it	1	88	6.0
m	2	it	2	89	5.8

11.19. a.

V1	V2	V3	V4
1	2	3	4
6	.	.	.

     b. Identical to part a.

    c. All values are missing (one observation output). No valid delimiters are found in any line (no blanks, only commas).

11.20. a. `infile rawdata firstobs=10 obs=20;`

    b. `infile rawdata n=3 firstobs=30 obs=44;`

11.21. a. The first and last observations are identical. Only one record from DATAHERE will be used in the DATA step.

    b. The options tell SAS to begin reading *after* the end of the observations to use (the beginning record number cannot be larger than the ending record number).

11.22. a.
```
V1   V2   V3
 1    2    3
 4    .    .
 5    6    .
```
    b.
```
V1   V2   V3
 1    2    3
 4    5    6
```
    c.
```
V1   V2   V3
 1    2    3
 4    .    .
```

Chapter 12

12.1. date, nonumber, center

12.2. PRINT: nocenter, nonumber
FREQ: nocenter, nonumber

12.3. a. `options ps=60 ls=130 nocenter nonumber;`

    b. `options missing='x' ps=22 ls=78;`

    c. `options obs=20;`

    d. `options firstobs=20 obs=81 replace errors=50;`

12.4. `ALL2: 20 observations`
`ALL3: 16 observations`
`ALL4: 12 observations`
`ALL5: 3 observations`

12.5. There is no "correct" answer. Simply run the job and examine the settings of familiar options.

Chapter 13

13.1. a. Concatenation

    b. Matched merge; merge

    c. Update

    d. Interleave

    e. Merge

13.2. a. Interleave, matched merge, update

    b. All

    c. All

    d. Merge, matched merge

    e. Interleave, matched merge, update

13.3. a. All of A's observations, then all of B's.

    b. All of A's observations for a level of ID, then all of B's for the same level. Repeat through the datasets.

    c. Place A and B side by side.

    d. Like part b, only placing observations next to each other instead of "on top of" each other.

    e. Update A's values with those in B.

13.4. **a.**

AREA	SALEPROF	SALE
n	230	.
n	210	.
ne	340	.
s	170	.
n	.	220
e	.	110
ne	.	210
ma	.	180
s	.	120

b.

AREA	SALEPROF
n	230
n	210
ne	340
s	170
n	220
e	110
ne	210
ma	180
s	120

c.

AREA	SALE	SALEPROF
n	220	.
e	110	.
ne	210	.
ma	180	.
s	120	.
n	.	230
n	.	210
ne	.	340
s	.	170

d.

AREA	SALEPROF	SALE
n	230	.
n	210	.
ne	340	.
s	170	.
n	230	.
n	210	.
ne	340	.
s	170	.
n	.	220
e	.	110
ne	.	210
ma	.	180
s	.	120

e.

REGION	SALEPROF	AREA	SALE
n	230		.
n	210		.
ne	340		.
s	170		.
	.	n	220
	.	e	110
	.	ne	210
	.	ma	180
	.	s	120

f.

AREA	SALEPROF	SALE	WHICH
n	230	.	1st

n	210	.	1st
ne	340	.	1st
s	170	.	1st
n	.	220	2nd
e	.	110	2nd
ne	.	210	2nd
ma	.	180	2nd
s	.	120	2nd

13.5. a.

LOT	NPASS	PCTFAIL
10	80	.
10	81	.
10	.	0.01
11	72	.
11	.	0.01
11	.	0.05
15	70	.
15	.	0.10
20	.	0.08
20	.	0.09

b.

LOT	PCTFAIL	NPASS
10	0.01	.
10	.	80
10	.	81
11	0.01	.
11	0.05	.
11	.	72
15	0.10	.
15	.	70
20	0.08	.
20	0.09	.

c.

LOT	NPASS	PCTFAIL
20	.	0.08
20	.	0.09
15	70	.
15	.	0.10
11	72	.
11	.	0.01
11	.	0.05
10	80	.
10	81	.
10	.	0.01

13.6. a.

ID	AVG	MEAN
200	93	93
170	82	92
180	81	94
210	88	98

b.

ID	MEAN	AVG
200	93	93
180	92	82
170	94	81
210	98	88

c.

ID	AVG	MEAN
170	81	92
180	82	94
200	93	93
210	88	98

d.
ID	MEAN	WHERE
200	93	one
170	92	one
180	94	one
210	98	one

13.7. a.
ID	X	Y
10	.	1
15	.	2
15	.	3
19	5	3
19	8	3
21	2	.

b.
ID	X	Y
10	8	1
15	4	2
15	.	3
19	1	3
19	8	3
21	2	.

c.
ID	X	Y
10	.	1
15	.	2
15	.	3
19	5	3
19	8	3

d.
ID	X
10	8
15	4
15	.
19	1
19	8
21	2

e.
ID	Z	Y	X
10	8	1	.
15	4	2	.
15	.	3	.
19	1	3	5
19	1	3	8
21	.	.	2

13.8. a. `if in_a + in_b + in_c = 3 then output;`
  b. `if in_a = 1 & in_b = 0 & in_c = 0 then output;`
  c. `if in_a = 1 & in_b = 1 & in_c = 0 then output;`
  d. `if in_a + in_b + in_c = 2 then output;`
  e. `if in_a + in_b + in_c = 1 then output;`

13.9. a. SECTOR is numeric in P2, character in P1. The new dataset has conflicting data types for SECTOR.
  b. This is correct only if both raw datasets (those read in the two DATA steps) are sorted by LAB and GRP.
  c. Since the only way RATE can be computed is for Q1 to be 0 and Q2 to be 1, you will always have a zero-divide condition when computing RATE: you will be consistently dividing by 0. Perhaps Q1 is not the appropriate divisor.

13.10.
```
data notevery;
    merge input.year1(in=y1)
          input.year2(in=y2)
          input.year3(in=y3);
          input.year4(in=y4);
    by loc id;
    pattern = 1000*y1 + 100*y2 + 10*y3 + y4;
    if pattern ^= 1111 then output;
    run;
```

13.11.
```
proc sort data=pres;
    by name;
    run;


    proc sort data=parkpres(rename=(presname=name));
    by name;
    run;


    data pres;
    * Notice the dataset is sorted by name, not chronologically.
      To put back in its original order sort by INAUG.;
    merge pres parkpres(keep=name _presprk _presacr);
    by name;
    run;
```
A problem you may have with merging by a character variable is that the format of the variable may differ from dataset to dataset. For example, Lyndon Johnson may be "Johnson, L." in one dataset and "L. Johnson" in another. Even though they seem to be identical, for the purposes of comparison during the merge they are different. Another problem may be case: one dataset may use mixed case while another may use only uppercase. Again, even though the data may look identical they will not merge correctly. To get around these problems it is probably best to assign an identification number to each president in both datasets and merge by this number.

13.12. a.

ID	X	Y
10	8	a4
20	.	b4
25	15	1g
30	8	1g
35	21	

b.

ID	X	Y
20	.	b4
25	15	1g
30	8	1g

13.13. a. Only two dataset names may be specified in the UPDATE dataset list.
   b. This is syntactically correct but may not be the logically correct action. Transactions usually update the master, not vice versa. SAS does not treat this as an error and will perform the update.

13.14. a. Variable STCODE is not in each dataset (assuming variable STATE was, in fact, the only state id variable in TWO). The merge will fail.
   b. You cannot merge NOTSORTED datasets.
   c. Although syntactically correct, it is hard to imagine why you would update a dataset by itself.
   d. The datasets are not sorted identically. If the intent was to combine them, pairing observations for similar ID levels, the merge will probably not work.
   e. BATCH is character in ONE and numeric in TWO (notice the COMMA5. format in the PRINT procedure). The merge will fail because the data types are not identical.

Chapter 14

14.1. **a.** picture drcr low -< 0 = '99,999.)' prefix='('
　　　　　　　　0 - high = '99,999' ;
b. value $pt 'A'-'J','a'-'j' = 'Part one'
　　　　　　other　　　　　 = 'Part two' ;
c. value scale 0 - 100 = "Low" (mult=.01)
　　　　　　other　 = "Other" (mult=.01) ;
d. picture num other = '9,999,999' prefix= '->' fill = '*';
e. value drcr low -< 0 = 'deficit'
　　　　　　0　　　　 = 'balanced'
　　　　　　0 >- high = 'surplus'　 ;
f. value bar 0 -< 10 = '**'
　　　　　 10-< 20 = '****'
　　　　　 20 - 30 = '******' ;

14.2. **a.** SAMPLEA: -2, ., 0, high, up to 1, high
SAMPLEB: -2, ., 0, 1 (high), up to 1, 2 (high)
b. format sample samplea.　sampleb splitsmp.;

14.3. **a.** The OTHER category does not necessarily include only numerics: punctuation marks and other special characters may be included. The grouping is correct, of course, if the data are limited to only alphabetic and numeric characters.
b. This is correct but would be better presented if the OTHER category were presented last.
c. Correct.
d. No range is specified.

14.4. **a.** In a microcomputer environment the program would resemble the following:
```
libname library 'd:\fmts';
proc format library=library;
value rankfmt  ... range sepcifications ... ;
```
b. Continuing the example in part a:
```
libname library 'd:\fmts';
proc format library=library fmtlib;
```
c. Continuing the example in part a:
```
libname library 'd:\fmts';
proc print data=test;
format rank rankfmt.;
```

Chapter 15

15.1. **a.** if put(batch, $bgrp.) = '1' then output grp1;
b. btch_grp = put(batch, $bgrp.);
c. value $bgrp '01A' - '50Z' = '1'
　　　　　　 '510' - '99Z' = '2'
　　　　　　 other　　　 = '.' ;

btch_num = input(put(batch, $bgrp.), 1.);

if　　　　'01A' <= batch <= '50Z' then btch_num = 1;
else if　 '510' <= batch <= '99Z' then btch_num = 2;

15.2. Values would be formatted only for store numbers 100, 110–125, 140–169, 170, 180, and 200–250. All other store numbers would remain ungrouped. Thus instead of categories 1 through 5, a FREQ procedure table would have categories 1 through 4 and other values such as 101, 102, and so on.

15.3. Although they are equivalent in their effect on the output dataset, they are *not* equivalent when you look at dataset contents. In the first example the recoded value is transient, available only during the IF statement, while in the second segment it is stored as a variable (GRP) that is kept in the dataset. Although GRP is always "NE" in the dataset, it may be needed if you combine the dataset with other (non-"NE") datasets. So, the segments are *not* equivalent.

15.4. MGR would not be sent to either dataset, nor would 1bc. The IF statements did not accommodate positions other than infield and outfield. MGR does not match mgr, so even if the value did match one in the list, as did '1bc,' it would not be handled in an IF statement.

15.5.
```
proc format;
value $brtype  'c' = 'Cantilever'   's' = 'Suspension'
               'a' = 'Arch';
run;

data bridges;
set bridges;
length br_style $10;
br_style = put(type, $brtype.);
run;
```

15.6.
```
proc format;
value $brtype  'c' = 'Cantilever'   's' = 'Suspension'
               'a' = 'Arch';
run;

proc means;
class type;
var length;
format type $brtype.;
run;
```
Use the formatted rather than the unformatted value to get a clearer understanding of the category's value: "Arch" is more readily understood than "a," for example.

15.7.
```
picture urban   low -< 30 = '999 (Low)'
                30  -< 70 = '999 (Medium)'
                70  -high = '999 (High)' ;
```
PRINT output will show both the number and the category: rather than printing "Low," the procedure displays "24 (Low)." The picture format will also prevent grouping into the three original categories: there will be as many categories in the table as there are unique values of the variable being formatted (i.e., if there are 24 distinct values of percent urban, there will be 24 categories in the frequency distribution).

15.8. a.
```
proc format;
value cont 100-199 = 'Asia'      200-299 = 'Australasia'
           300-399 = 'N. America'  400-499 = 'S. America'
           500-599 = 'Africa'      600-699 = 'Europe'  ;
run;

proc chart data=countrys;
vbar country / type=mean sumvar=energy;
format country cont.;
run;
```
Without the FORMAT statement in the CHART procedure, each bar would represent a country rather than a continent. There would be as many bars as there are countries.

b.
```
value cont
  100-199 = '[2] Asia'
  200-299 = '[3] Australasia'
  300-399 = '[5] N. America'
  400-499 = '[6] S. America'
  500-599 = '[1] Africa'
  600-699 = '[4] Europe';
run;
```

Chapter 16

16.1. a. In a banking application, you may manipulate term deposit data. Date-oriented variables include date of deposit, maturity date, and date terminated (since it usually will not be exactly on the maturity date). You could also use the information to compute exposures at different time frames: for example, what volume of deposits matures in three months?

    b. To calculate the exposures mentioned in part a, you need to take a date such as the Monday of the current week, read each deposit's maturity date, and calculate how many days' difference there is between the two dates. This gives you the basis for assigning the periods (three, six months, one year, and so on.). For example:

```
if mature - today <= 90 then do;
   _3mth = 1;
   _3mthdep = deposit;
   end;
if mature - today <= 180 then do;
   _6mth = 1;
   _6mthdep = deposit;
   end;
```

    You could use the MEANS procedure to add the values of _3MTH and _6MTH to determine how many deposits fall due in those time frames and add _3MTHDEP and _6MTHDEP to see the volume of deposits.

    c. The variable represents a date that occurred 5,423 days after January 1, 1960.

16.2. a. MMDDYY or DDMMYY
    b. DATE
    c. YYQ
    d. YYQ
    e. DDMMYY
    f. MMDDYY
    g. DDMMYY

16.3. The first line is in a format acceptable to DATE: day number is followed by month name and year number. The second line, however, is already in SAS date format: SAS cannot find a month name and thus does not know how to separate the four digits into day, month, and year numbers.

16.4. a. `if bday < '01jan50'd then output;`
    b. `if '01jan70'd <= empstart <= '31dec79'd then do;`
    c. `'01jan92'd - '31mar92' = '92, qtr=1'`
    d. `do incremnt = '01jan90'd to '31dec95'd by 7;`

16.5. a. No slashes (/) should be in the constant.
    b. The *d* should follow the quoted date.
    c. A *d* should follow the quoted date.

16.6. a. DURW is the number of weeks between a starting date and an ending date.
    b. DURQ is the number of quarters between a starting date and an ending date. It is rounded up to the nearest integer value.
    c. Same as part b except not rounded.
    d. DUR is the number of weeks between a starting date and an ending date, rounded to the nearest half week.
    e. TOT_T is the sum of two durations: ST1 to END1 and ST2 to END2.
    f. CURR is the number of years from STDATE to the current date.
    g. Same as part f except rounded to the nearest integer.

16.7. `elapse = mdy(m,d,y) - today();`

16.8.
```
date1 = mdy(m,d,y);
elapse = date1 - today();
juldate = put(date1,julian5.);
```

16.9. a. `if weekday(workday) = 1 | weekday(workday) = 7 then delete;`
    b. `if qtr(due) = 1 then when = 'q1';`
    c.
```
q = qtr(divdue);
if        q = 4 then q = 1;
   else if q = 1 then q = 2;
   else if q = 2 then q = 3;
   else if q = 3 then q = 4;
```
    d.
```
data test;
do date = '01jan92'd to '31dec92'd;
   output;
end;
run;
```
    e.
```
day = day(appt);
mth = month(appt);
year = year(appt);
if mth = 12 then do;
   mth = 1;
   year = year+1;
   end
next = mdy(mth,day,year);
```

16.10. a. 14
    b. 0
    c. 4.66
    d. 01may90
    e. 01jul90
    f. 01jan91

16.11. a. False
    b. True

16.12. a. MONYY
    b. DATE
    c. DATE
    d. WEEKDATE
    e. YYMMDD
    f. YYQ
    g. WORDDATE

16.13. a. 2
    b. 61
    c. 61
    d. 2
    e. 2
    f. 1

Chapter 17

17.1. a. SAMPLE=1
    b. REGION='N'
    c. REGION='N', SAMPLE=1
    d. by _type_ region sample;
    e. REGION='A', SAMPLE=3
    f. REGION and SAMPLE will be missing when the level of summary (noted in _TYPE_ collapses that level. When _TYPE_ is 2, for instance, SAMPLE is missing because 2's summarize regions across sample groups. There cannot be any distinction among samples, so sample is missing.

17.2. a. _R1 is the sum of VAR statement variable R1.
    b. R1 through R3 contain the average, or mean, of VAR statement variables R1 through R3.

   c. COUNT1 through COUNT3 contain the number of nonmissing values of R1 through R3. MEAN1 is the mean of VAR statement variable R1. STDR1 through STDR3 contain the standard deviations of R1 through R3.

17.3. a. `minid(winpct(city))=locity;`
    b. `maxid(tscore(shrtname city))=maxname maxcity;`
    c. `minid(winpct(city), tscore(city))=lowpct lowscore;`

17.4. a.
```
proc means data=cars nway;
class year rating;
var mpg curbwgt;
output out=summ n(mpg curbwgt)=nmpg nwgt
                sum(mpg curbwgt)=summpg sumwgt;
```
    b.
```
proc means data=cars;
class year rating;
var mpg curbwgt;
output out=summ mean=avgmpg avgwgt sum=summpg sumwgt;
minid(mpg(model))=guzzle;
```
    c. `output out=summout.summ(drop=_freq_) mean=;`

17.5. COMP contains variables from the original dataset (MASTER) plus PCT. PCT is an observation's percent of the YEAR's average: 108, for example, means that the observation is 8% above the dataset average.

17.6. a. NGEN is the count of all nonmissing observations, and _FREQ_ is the number of all observations, missing and nonmissing.
    b. "wc"
    c. 326.20 miles. The values agree because no observation can be simultaneously on the East Coast and the West Coast. You cannot have some OCEAN observations on the West Coast ("wc") and some on the East Coast ("ec"). The CLASS variables in the example are somewhat redundant: OCEAN is simply a finer-grained version of COAST.

17.7. a.
```
proc standard data=invntory mean=1 std=0
      out=stdinvt;
```
    b.
```
proc standard data=invntory mean=1 std=0
      out=stdinvt replace;
```
    c.
```
proc standard data=invntory mean=1 std=0
      out=stdinvt replace noprint;
```
    d.
```
proc standard data=invntory out=stdinvt replace;
by partnum;
```

17.8. a. No replacement, mean, or standard deviations are specified.
    b. Correct as is.

17.9. COMPARE, as its name suggests, contains nonstandardized TOTSCOREs, TOTSCORE standardized across the entire dataset (T_ALL), and TOTSCORE standardized within levels of BY variable FY (T_BY). Variable SCORE is 'nonmiss' if the original value of TOTSCORE is nonmissing. This is a way to identify observations where T_ALL or T_BY are due to the REPLACE option being in effect.

17.10. a.
```
proc rank data=records groups=10 out=recrank;
var _88 _89 _90 _91;
ranks dec88 dec89 dec90 dec91;
```
    b.
```
proc rank data=records groups=10 out=recrank;
var _88 _89 _90 _91;
```
    c.
```
proc rank data=records groups=100 out=recrank;
var _88 _89 _90 _91;
```

d. `proc rank data=records groups=2 ties=high;`
  `var _88 _89 _90 _91;`
  `by league;`
  Higher numbers for each variable will be 1. (Ranks begin at 0, so two ranks will have values of 0 and 1.)

17.11. `proc rank data=leagues out=leagrank n=2;`
  `var nonconf;`
  `ranks half;`

Chapter 18

18.1. NAME $25, ADD1 through ADD 4 $20, STYLE 4; ERA $4

18.2. Parts c, d, and e are true.

18.3.

List 1	List 2	List 3
abc	def^g	abc
a	b^c^d	a^b^c
ab	cd	ab^cd
abcde		abcde
abcde	ghijk	abcde

18.4. **a.** `holding = trim(name) || '(' || trim(rating) || ')';`
  b. `if rating =: 'a' | rating =: 'A' then`
        `judgment = 'BLUE CHIP: ' || name;`
      `else judgment = 'SOME RISK: ' || name;`
  c. It is not sorted if using the ASCII character set/collating sequence. It is in sort order if using the EBCDIC character set.
  d.

	ASCII	EBCDIC
(1)	1	6
(2)	6	5
(3)	4	3
(4)	4	3
(5)	5	1
(6)	6	6

18.5. `if length(zipcode) > 5 then _9digit = 1;`
  or

  `if index(zipcode, '-') > 0 then _9digit = 1;`

  `if substr(zipcode,6,1) = '-' then _9digit = 1;`

18.6. `if reverse(mid_name) =: '.' then do;`
  or
  `if substr(mid_name,length(mid_name),1) = '.' then do;`

18.7. `length first last $15 shortnam $20;`
  `first = scan(fullname,1,' ');`
  `last  = scan(fullname,3,' ');`
  `shortnam = trim(first) || ' ' || trim(last);`

18.8. `substr(line,1,1)=upcase(substr(line,1,1);`

18.9. `LEFT:      header = left(header);`
  `RIGHT:     header = right(header);`
  `CENTERED:  length thead $40;`
             `* LEN_HEAD: location of last nonblank char. ;`
             `len_head = length(header);`
             `* VER_HEAD: location of first nonblank char. ;`

```
                      ver_head = verify(header,' ');
                      * New starting location of the nonblank portion;
                      start = (40 - (len_head - ver_head)) / 2 ;
                      * THEAD: nonblank portion of HEADER ;
                      thead = substr(header, ver_head, len_head);
                      header = ' ';
                      substr(header, start, len_head) = thead;
```

18.10.
```
     orig_len = length(line);        * Original line length;
     no_q_len = length(compress(line,'?'));  * Length without ?'s;
     n_of_q = orig_len - no_q_len; * Difference in lengths is the
                                     number of ?'s;
```

Chapter 19

19.1. a. `retain rate1-rate10;`
    b. `retain rate1-rate5 -1 rate6-rate10 0;`
    c. `retain type 'Rental' basis;`

19.2. a. The r0 length is 3; r1 through r5 lengths are each 4.
    b. `r0=., r1-r5='<- ?'`
    c. Variables from IN.DEBUG come first, then INIT, R1-R5, and R0.

19.3. a. LOWAREAS appears to be counting the number of areas that are below median income (it is RETAINed and is incremented by 1 each time MEDIAN_Y is below 12,500). However, it is reset to 0 in *each* observation. Thus all you will see in the dataset is a series of observations whose values are 0 and 1 rather than a count that continually increases. The assignment statement LOWAREAS = 0 should be deleted.
    b. Assuming the intent of LOWAREAS is the same as in part a, LOWAREAS is reset when it should not be. The ELSE sets LOWAREAS to 0 rather than letting it accumulate across observations. The result here, as in part a, is observations with 0 and 1 values rather than a continually increasing count of lower-income observations.

19.4. a. 7
    b. 7
    c. 3
    d. 2
    e. 10
    f. 7
    g. 4
    h. 3

19.5. Parts b and f, and parts c and h are equivalent.

19.6. a. Parts c and h
    b. Part g
    c. Part d

19.7. There is no BY statement. FIRST. and, had it been used, LAST. have no meaning if they do not identify BY-statement variables.

19.8. a.
```
     array q{35} q1--q28y;
     do i = 1 to 35;
         if q(i) = 97 | q(i) = 98 | q(i) = 99 then q(i)=.;
     end;
```
    b.
```
     array q{35} q1--q28y;
     do i = 1 to 35;
         if        q(i) = 97 then q(i) = .a;
             else if q(i) = 98 then q(i) = .b;
             else if q(i) = 99 then q(i) = .c;
     end;
```

19.9.

V1	Category	Count
	.	3
	d	2
	r	1
	1	3

V2	Category	Count
	.	1
	d	4
	1	1
	8	2
	9	1

V3	Category	Count
	.	2
	d	3
	u	1
	1	1
	2	2

19.10. All except f are true. G's character-numeric conversion should be avoided.

19.11.
```
if           .a <= screen <= .m then missrnge = -1;
    else if .n <= screen <= .z then missrnge = -2;
```

19.12.
```
data terms;
   set  in.tnotes;
   * Delete the RETAIN statement and the conditional calculation
     of CURRENT ;
   current = today();
   daysleft = term - current;
```
This program consumes more resources because the calculation is done for every observation rather than only once, at the beginning (_N_ = 1) of the DATA step. Since the date will not change during the DATA step, it is pointless to recalculate it for each observation.

19.13.
```
complobs = _n_ ;
```

19.14. TOT's assignment statement could be moved out of the EOF DO-group and be placed immediately after the IF and ELSE statements computing _EXCESS1 and _EXCESS2. This would be wasteful of computer resources, however, since TOT is meaningful only at the *end* of the datasets' processing. There is no need to calculate it for every observation since its operands (_EXCESS1 and _EXCESS2) are being RETAINed. These variables will have the appropriate counts at the end of the DATA step and need to be added for TOT only once.

19.15. SUMM has one observation. The observation is output when the DATA step is processing the observation number (_N_) equal to what the SET statement identified as the number of observations in the dataset (COUNT).

19.16. If any of the observations have more than 10 missing values for variables V1 through V90, the dataset will have 0 observations (the STOP statement will be executed before the OUTPUT statement). This may or may not be what was intended.

19.17.
```
infile rawdatain length=l;
   if l = 40 then input dirname $char40.;
      else if l=27 then input filename $ 1-8 exten $ 10-12
      size 14-21;
```

19.18.
```
infile cards ls=72;
```

19.19.
```
data people hholds;
   infile survey length=linelen;
   retain hh_id state county n_in_hh hh_inc;
   input @45 rectype $1. @;
```

```
      if rectype = 'H' then do;
         input @1 hh_id $7. state $2. county $3. @25 n_in_hh 2.
               @43 hh_inc 7. ;
         output hholds;
         end;
      else if rectype = 'P' then do;
            input @1 hh_id $7. personid $2. (age income marstat)
               (3. 6. $1.);
            output people;
        end;
      else delete;
   run;
```

19.20. `input name $10. grade $ &  alt : comma6. mean_alt : comma5.`
`disc : date7. ;`

19.21. After the IF ABS(DIFF)... statement, insert: `else deviate = 'N';`

The value may be something other than 'N': the point is that you should be sure not to let values of DEVIATE be accidentally RETAINed.

19.22. a. This program will not work as shown. Insert OUTPUT statements at the bottom of each DO-loop. As it is written now, dataset ONE will contain only one observation (the 50th record read from file ref IN2).

b.
```
do i = 1 to 100;
   if i <= 50 then do;
      infile in1 ... ;
      input ... ;
      end;
   else do;
         infile in2 ... ;
         input ... ;
         end;
   end;
```

19.23. Dataset POINTER contains the first and last observation numbers for each BRANCH in dataset ACCTS.

19.24. The program uses the POINT option and the start-end information in POINTER to directly access records of interest. Dataset SELECT contains branch numbers of interest (100, 120, and so on). SELECT's list of branches and POINTER's branch-observation numbers are merged. If a BRANCH value is common to both datasets (IN variables PTR and SEL both equal 1), use the beginning and ending observation numbers to directly access ACCTS. This technique takes a little "up front" effort to set up. It can, however, save great amounts of computer resources, especially if the ratio of needed records to total records is low. Dataset indexing techniques in version 6.06 of the SAS System would force a reevaluation of this strategy.

Chapter 20   **20.1. a.** `file rpt ls=130 notitles header=pagetop;`
b. `file log line=_line column=_col ls=78;`
c. `file rptout1 print linesleft=_ll ps=55;`

20.2. a. `put name $char20. @23 (score1-score5) (comma10.);`
b. `put _all_;`
c. `put 'Line just read ==> ' _infile_;`

d. `put name $20. / @25 'Scores' (test1-test5) (5.2);`

e. `put @indent name $20. +2 dept $20.;`

20.3. a. The underscores beginning in column 1 will *replace* rather than *underline* the constant "Dept."

b. This is correct only if _page actually is a variable. More likely, it is _page_ misspelled.

c. You can write the results of function calls but not the functions themselves. The following statement will work correctly:

```
stars = repeat("*", 80);
put _infile_ / stars;
```

d. There is no RETURN; statement between the first PUT statement and the PUT labelled by TOP. The page heading will be printed for every observation.

20.4. a.
```
libname append 'd:\data';
data _null_;
set append.natlpark;
if _n_ <= 20 then put _all_;
run;
```

b.
```
libname append 'd:\data';
data _null_;
set append.natlpark;
file 'd:\natlpark.raw' notitles;
* The following are some of the ways to write the PUT
  statement ;
put (park st coast yrestab acres)($20. $2. $1. 4. 9.);
put (park--acres)($20. $2. $1. 4. 9.);
put park $20. st $2. coast $1. yrestab 4. acres 9.;
put park $ 1-20 st $ 21-22 coast $ 23 yrestab 24-27
    acres 28-36;
run;
```

c.
```
libname append 'd:\data';
```

```
proc sort data=append.natlpark out=temp;
by st;
run;
```

```
data _null_;
set temp;
by st;
file print notitles;
if first.st then put st;
* This PUT statement is not the only correct one: any PUT
  using indentation (here, 10 columns) and separating the
  columns (here, 5 spaces) is acceptable. ;
put @10 park $20. +5 acres comma12.;
run;
```

d.
```
data _null_;
set temp;
by st;
file print notitles header=topofpg;
if first.st then do;
   put st;
   totacres = 0;  * Total for state ;
   end;
put @10 park $20. +5 acres comma12.;
totacres + acres;
* If at the end of a state and the beginning flag is 0,
  the state has more than one park.;
```

```
        if last.st = 1 & first.st = 0 then do;
          put @10 'Total for State' @35 totacres comma12.;
          end;
        return;
        topofpg: put 'National Park Data' // ;
                 return;
```

e.
```
data _null_;
  set temp;
  by st;
  file print notitles header=topofpg;
  if first.st then do;
    put st;
    totacres = 0;  * Total for state ;
    end;
  put @8 park $20. +1 acres comma12.;
  totacres + acres;
  if last.st = 1 & first.st = 0 then do;
    put @8 'Total for State' @33 totacres comma12.;
    end;
  return;
  topofpg: put 'National Park Data' // ;
       'State  <---- Park Name ---> <-- Acres ->';
       return;
```

f.
```
libname append 'd:\data';

proc sort data=append.natlpark out=sortpark;
  by st;
  run;

proc means data=append.natlpark nway noprint;
  class st;
  var acres;
  * Keep only the state name and the names of the largest and
    smallest parks. ;
  output out=summ(drop=_type_ _freq_) n=count
       minid(acres(park))=minpark
       maxid(acres(park))=maxpark;
  run;

data _null_;
  merge sortpark summ;
  by st;
  length status $10;
  retain min '(smallest)' max '(largest)';
  retain only1 ' ';  * FLAG: 1 and only 1 state ;
  file print notitles header=topofpg;
  if first.st then do;
    put st;
    if last.st then only1 = 'T';
       else          only1 = 'F';
    totacres = 0;  * Total for state ;
    end;
  * If only 1 park min/max are not defined. Otherwise,
    compare park name to state-level min/max park name and
    set STATUS.;
    if only1 = 'T'  then status = '';
       else if park = minpark then status = min;
       else if park = maxpark then status = max;
       else                    status = ' ';
```

```
              put @8 park $20. +1 acres comma12. +3 status;
              totacres + acres;
              if only1 = 'F' then do;
                 put @8 'Total for State' @33 totacres comma12.;
                 end;
              return;
              topofpg: put 'National Park Data' // ;
                           'State  <---- Park Name ---> <-- Acres ->';
                           return;
```

Chapter 21   **21.1. a.** `vaxis = 1 to 21 by 2;`
                 b. `vaxis = '01mar91'd to '01apr92'd by 'month';`
                 c. `vaxis = 0 to 50 by 5;`
                 d. `vaxis = 1 to 11 by 2, 12 to 20 by 2;`

21.2. a.
```
proc plot nolegend;
   plot nreps*sales/box;
```
b.
```
proc plot vpct=33 data=capital;
   plot nreps*(sales profits netchg) / hzero;
   format sales profits netchg dollar10.;
```
c.
```
proc plot uniform data=capital;
   plot nreps*sales;
   by region;
```
d. `plot nreps*sales='S' nreps*profits='P';`
e.
```
plot nreps*sales='S' nreps*profits='P'
        / href=100000 500000 hrefchar='|'
        / vref=250 vrefchar='-' ;
```

21.3. a. `y1*x y2*x y3*x`
b. `y*x1 y*x2 y*x3`
c. `y1*y2 y1*y3 y2*y3`
d. `y1*x y2*x`
e. `y1*x y2*x`
f. `y1*x1 y1*x2 y1*x3 y2*x1 y2*x2 y2*x3`
g. `y*x`

21.4. a. Nothing is wrong with the syntax. However, the axis scaling required for INPTILE (0 through 100) would "flatten" the narrow range of INCZSCOR (-3 to 3): you would not get a clear idea of INCZSCOR's distribution.
b. Again, the syntax is correct but the display is not useful. The 0 to 100 specification should be used for INPTILE, not INCZSCOR. If run with these axis specifications, the plot would exclude negative INCZSCOR values and cluster the positive ones into a corner of the plot.

21.5. a. `plot y*x / haxis=0 to 50 by 10;`
b. `plot y*x / vaxis=-5 to 0 by 1   haxis=0 to 50 by 5;`
c. `plot y*x / vaxis=-10 to 10 haxis=50 to 100 by 10;`

21.6.
```
proc plot;
plot y1*x;
title "Measure 1 by Income";
run;
plot y2*x;
title "Measure 2 by Income";
quit;
```

21.7. Region "M" (Midwest) tends to have a fairly high portion of its land devoted to farming (variable FARMPCT) and lower values of in-migration (PCTIN). Western states ("W") other than those on the Pacific coast have high FARMPCT values and high in-migration. Southern states ("S") cluster near the middle of both variables' ranges. Northern states ("N") have low farm rates and low- to middle-level in-migration. Finally, the Pacific states ("P") are broadly distributed, with generally mid-level farmland percentages and a fairly wide range of in-migration values (the variety should not be surprising considering this group takes in Alaska, Hawaii, and California, among others).

A simple way to clarify the patterns is to sort the dataset by variable REGION, then plot BY REGION, keeping axis specifications identical in each plot. The plots might have reference lines at the median and/or mean FARMPCT and PCTIN values. These lines would quickly show how a region's states are placed relative to the rest of the country.

21.8. Thirty observations could not be identified with unique coordinates on the plot—points are, in effect, piled on top of each other. This is not an error; just PLOT indicating its inability to produce a clear picture of the distribution. You could change HAXIS and VAXIS options to plot only a portion of the horizontal or vertical axes. You could also use the system options LINESIZE and PAGESIZE to increase the display space. Finally, you could use PROC PLOT statement's VPCT and HPCT options to spread the plot over several pages.

21.9. Enter VPCT=33 in the PROC statement. The new PLOT specifications look like this:

```
proc plot vpct=33 data=in.states ;
plot pcturb*pcapinc / haxis=2000 to 12000 by 2000
     hpos=110 box;
plot pcturb*crimert / box hpos=110;
plot pcturb*crimert='c' pcturb*pcapinc='$' /
     box hpos=110;
```

The last PLOT statement could be rewritten as

```
plot pcturb*(crimert pcapinc) / overlay;
```

but the "c" and "$" distinction would be lost.

21.10.
```
proc univariate data=in.states;
var popsqmi;
id name;
run;
```

Chapter 22   **22.1. a.** 39%; 20/51

b. 28

c. 6; 11% of the total

d. Fifty-one observations were used in the analysis. This figure also represents the sum of the row (28 + 23) and column (26 + 25) marginal totals. The 100.00 represents the sum of the row and column marginal percents: the row percents represented by the counts 28 and 23, 54.9 and 45.10, add up to 100.00. The same is true for the column percents: 50.98 + 49.02 add to 100.00.

e. Nominal and ordinal scales do not require grouping unless you need to collapse the categories (e.g., make nine districts into two "super-districts"). The table in the example would be useless from a statistical point of view because the observations would be spread too sparsely among the cells. FREQ will print a warning about reliability and interpretability if the cell size drops below 5.

22.2. a. `tables a*b;`

b. `tables c*a*b;`

c. `tables d*a*(b c);`

     d. `tables a*(b c d);`
     e. `tables (b c d)*a;`
     f. `tables a*(b c) f*d*e;`
     g. `tables (a d)*(b c);`

22.3. **a.** `nopercent norow nocol`
     b. `nopercent norow nocol deviation expected`
     c. `list`
     d. `missing`
     e. `missprint`
     f. `all`
     g. `chisq`

22.4. **a.** LOST: row and column percentages
        GAINED: cumulative counts and percentages
     b. Use the SPARSE option in the TABLES statement.

22.5. **a.** `proc ttest data=humanres;`
        `class region;`
        `var salary socsec fica;`
     b. `proc ttest data=humanres;`
        `class earn_grp;`
        `by year;`

22.6. **a.** Unless there are only two nonmissing values of DECADE in EXPOSE, the CLASS variable will not be dichotomous.
     b. The CLASS and BY variables are identical (TYPE). By definition, then, there will be only one nonmissing value of TYPE in each BY-group. CLASS variable TYPE will not be dichotomous.

22.7.
```
proc format;
value density low  -100  = '<= 100/sq mi'
               100 <-high = '> 100/sq mi'  ;
run;

proc plot data=in.states;
* Get a visual feel for how the observations are
  distributed. Draw a reference line where the DENSITY
  format splits the population density values. ;
plot crimert * popsqmi / href=100;

proc chart data=in.states;
* Graph average crime rate for each group of POPSQMI. ;
vbar popsqmi / type=mean sumvar=crimert;
* Graph the count of the number of observations in each group
of POPSQMI. ;
vbar popsqmi / type=freq;
format popsqmi density.;
```

22.8. **a.** Highest values: PCTCOLL-PCAPINC (.676), PCTIN-PCTMFG (−.663), PCTURB-PCAPINC (.657). Lowest values: BTUS-PCTFARM (.083), BTUS-PCTURB (−.144), BTUS-PCTCOLL(−.039)
     b. PCTCOLL-PCAPINC: As educational attainment increases (in this case, measured by percent of those over 25 years old completing college), so does per capita income.
        PCTIN-PCTMFG: There is an inverse relationship between concentration of manufacturing jobs and in-migration. The states with high proportions of manufacturing-sector positions are experiencing low amounts of in-migration.

PCTURB-PCAPINC: As the concentration of urban population increases, so does per capita income. Most higher-paying service-sector positions require a "critical mass" of population to be feasible, so this result should not be surprising.

22.9. **a.** `proc corr data=in.perform nosimple;`
   b. `proc corr data=in.perform best=5 rank;`
   c. `proc corr data=in.perform kendall spearman nomiss;`
      `var rate80q1--rate85q4;`
   d. `proc corr data=in.perform noprint outp=pcorr;`
      `var assess01-assess25 overall;`
   e. `proc corr data=in.perform sscp cov;`
      `var assess01-assess25;`
      `with rate80q1--rate85q4;`
   f. `proc corr data=in.perform spearman;`
      `var deprnk1-deprnk5;`
      `with perank1-perank10;`
      `by district;`

22.10. **a.** The matrix is not square (the WITH option was used). Regression procedures require a square, symmetric matrix. Inclusion of all required variables in the VAR statement and elimination of the WITH statement would produce an acceptable matrix.
   b. The WITH statement is not necessary, since it specifies what would be produced by default (a square matrix using BMASS1 to BMASS10).
   c. The CORR procedure will run correctly, but in the regression procedure that follows, the matrix LAB1P will not have a required dependent variable (DECAY). DECAY should be added to CORR's VAR statement.

22.11. **a.** PCAPINC and PCTCOLL each appear five times.
   b. Add NOMISS to the PROC statement.

22.12. **a.** 57,684
   b. PCTURB variance: 224.9
      PCTCOLL variance: 11.11
   c. 50

Chapter 23   **23.1.** **a.** `proc reg data=states;`
      `model pctin = pctfarm;`
   b. `proc reg data=states;`
      `model pctin = pctfarm pctmfg pcapinc;`
   c. Add the following to part b:
      `test pctmfg pcapinc;`
   d. `proc reg data=states;`
      `* VAR statement lists ALL variables that will be used in`
      `  ALL models.;`
      `var pctin pctfarm pctmfg pcapinc;`
      `model pctin = pctfarm;`
      `title 'Reduced Model';`
      `run;`
      `model pctin = pctfarm pctmfg pcapinc;`
      `title 'Full Model';`
      `run;`
      `test pctmfg pcapinc;`
      `title 'Test contribution of PCTMFG and PCAPINC';`
      `quit;`
   e. `data temp;`
      `set states;`

```
* REGION has five levels so four dummy variables are
  created. The omitted category "Pacific" is the reference
  level.;
if region = 'South' then r_south = 1;
  else                     r_south = 0;
if region = "N'east" then r_heast = 1;
  else                     r_neast = 0;
if region = 'Midwest' then r_mwest = 1;
  else                     r_mwest = 0;
if region = 'West' then r_west = 1;
  else                     r_west = 0;
```

23.2. a. Outliers may be detected by scatterplots of the dependent variable with each independent variable (PLOT procedure). Specify the R option in the MODEL statement (REG procedure) to identify observations with high/low residual values. (Residuals do not help identify extreme *individual* variables when there are multiple independent variables.)

b. Specification errors may be detected by plotting residual values against the predicted value of the dependent variable (PLOT procedure). The pattern may indicate that another variable should be added to the model or, if the relationship is nonlinear, that an independent variable should be transformed or removed from the model.

c. Heteroscedasticity may be revealed by plotting residual and predicted values (PLOT procedure). Solutions include transformation of the dependent variable (in the DATA step) or use of the WEIGHT statement in REG to perform "weighted least squares" (create the weights in a DATA step).

d. Autocorrelation may be detected by the DW option in the MODEL statement. If the observations are ordered by time period, the MODEL statement's R option will reveal orderly variation in the size of the residuals. The amount of autocorrelation may be reduced by adding a new (theoretically justifiable) independent variable to the model.

e. Collinearity is detected by the VIF and COLLINOINT options in the MODEL statement. These options may indicate which variable(s) to drop from the model. The CORR procedure run on variables in the model will reveal highly correlated independent variables. This is another way to identify the source of the collinearity if it is caused by *pairs* of variables.

23.3.
```
proc corr data=states noprint outp=pearmtrx;
var pctin pctfarm pctmfg pcapinc;
run;

proc reg data=pearmtrx;
model pctin = pctfarm pctmfg pcapinc;
quit;
```
Execution time of the REG using the correlation matrix will be somewhat faster since the procedure does not have to compute a correlation matrix and other required statistics. The time saving will become more pronounced as the size of the individual-level dataset increases.

23.4.
```
proc reg data=states noprint;
model pctin = pctfarm pctmfg pcapinc;
output out=diag p=predict r=resid;
quit;

proc plot data=diag;
plot resid*predict / vref=0;
quit;
```

23.5. The first reference to X3 in a MODEL statement is *after* the first RUN. Use a VAR statement with *all* variables prior to the first RUN. The X1*X2 interaction specification is not allowed. This should be a variable computed in a DATA step (e.g., INTER = X1 * X2).

23.6. **a.** The NOTE in the Log indicates 51 observations were used in the calculations.

   b. Adjusted R-square is .6903.

   c. CRIMERT mean is 5409.29 (i.e., the "Dep Mean" in the listing).

   d. crimert = −1188 + 69.1(pcturb) + 117.88(pctcoll).

23.7. **a.** Missings will be included. Specify ELSE as:
```
else if eastwest ^= ' ' then dumy_reg = 0;
```

   b.
```
if          eastwest = 'e' then east = 1;
   else if eastwest ^= ' ' then east = 0;
if          eastwest = 'w' then west = 1;
   else if eastwest ^= ' ' then west = 0;
```

   c. No (p value is only .84)

23.8. **a.** Use VAR because a MODEL *after* the first RUN statement will use terms not included in a MODEL *before* the first RUN.

   b. There may be a colinearity problem.

23.9. **a.** Largest residuals: DC, Florida, Delaware, Arizona
   Smallest residuals: Colorado, California

   b. Compute the "Residual" column by subtracting "Predict Value" from "Dep Var CRIMERT." Compute "Student Residual" by dividing "Residual" by "Std Err Residual."

23.10. **a.** Use NOPRINT when the PROC is being run only to produce an output dataset.

   b. The "funneling" suggests heteroscedasticity.

   c. CRIMERT, POPSQMI, PCTCOLL, and PCTURB had these formats in analysis dataset STATES. The formats "travelled" with the variables when DIAGS was created. (Notice that these are system formats—if user-written formats were assigned, SAS would expect the default format library to have the formats defined. See Chapters 14 and 15 for details about creating and using user-written formats.)

Chapter 24

24.1. **a.** Level of indicator variable

TESTGRP	RDU	QDR	XYC
RDU	1	0	0
QDR	0	1	0
XYC	0	0	1

   b. $INCOME_i = a + b_1 RDU_{1i} + b_2 QDR_{2i} + b_3 XYC_{3i} + e_i$

24.2. **a.** Contrast 1: Contrast North and South, disregarding levels of Midwest and West. Only North and South are of interest, so zero out the other levels.
   Contrast 2: Compare the North and Midwest taken together to the South and West taken together. The 1's and −1's suggest that within each group (North-Midwest and South-West) the regions are equal.
   Contrast 3: The average of non-North regions is compared to the North. The non-North coefficients are equal, suggesting the means for regions within the group are equal.

   b. New contrasts 1   − .5   − .25   − .25 compares the North to non-North regions as Contrast 3 did in part a. The differing coefficients, however, suggest the South's mean is twice that of the Midwest and West.

24.3. **a.** One-way ANOVA

   b. Two-way ANOVA

   c. Repeated measures

   d. Repeated measures

   e. One-way

   f. MANOVA

24.4. **a.** mass0 = lab;

   b. mass0 = lab style;

   c. mass0 = lab style lab*style;

    d. mass0-mass3 = lab style;

24.5. a. t3

    b. repeated t 4 contrast(1);

      or

      repeated t contrast(1);

24.6. a. Dep.

var.	LAB	REP
L1	1	1
L2	1	2
L3	1	3
L4	1	4
L5	1	5
L6	2	1
L7	2	2
L8	2	3
L9	2	4
L10	2	5

    b. Dep.

var.	REP	LAB
L1	1	1
L2	1	2
L3	2	1
L4	2	2
L5	3	1
L6	3	2
L7	4	1
L8	4	2
L9	5	1
L10	5	2

24.7. contrast 'N v. non-North' region 1 − .5 − .25 − .25;

24.8. a. MODEL should come before first RUN. The MODEL will only be executed when QUIT is entered.

    b. Correct as is.

    c. Correct as is.

    d. Correct as is.

    e. Correct as is.

24.9. a. Run cross-tabulation of the two factors to see how they are distributed. If cell count is unequal, use GLM. Otherwise use either PROC. FREQ statements look similar to this:

```
proc freq data=freq;
tables div*style;
```

    b.
```
proc anova data=fin;
class div style;
model margin = div style;
```

    c.
```
proc anova data=fin;
class div style;
model margin = div | style;
* Could also write the independent variables as:
    div style div*style
;
```

    d.
```
proc anova data=fin;
class div style;
model margin = div style div*style;
means style / t duncan bon;
```

Index

| (or) operator, 42–43, 594
& (and) operator, 42–43, 595
: comparison operator, 348. *See also* Expressions, character
? format modifier, 54, 58
?? format modifier, 54, 58, *59*
@ column pointer, 212–214, 399, *405, 407, 410, 416, 420, 421*
@ line holder, 218–220, 399
@@ line holder, 217–218
+ column pointer, 212–214, 399, *405, 407, 410, 416, 420, 421*
# line pointer, 210–211, 399
/ line pointer, 210–211, 399

ABEND option, ABORT statement, 209
ABS function, 133, *134*
ABORT statement, 209–210
  compared to STOP statement, 209
Absolute value, 133, *134*
Adding variables
  to datasets, by calculations. *See* Assignment statements
  to datasets, by combining datasets. *See* Combining datasets
  within observations, functions for, *See* Functions, numeric, SUM
  across observations, procedures for. *See* MEANS procedure; SUMBY option, PRINT procedure; UNIVARIATE procedure
ALL
  keyword, CONTENTS procedure, 180–181, *182*

special variable name, 400, *420, 421*
ALL option, FREQ procedure, 452, *459*
ALPHA
  ANOVA procedure, 523, 525
  GLM procedure, 527
Analysis of variance, 517–522. *See also* ANOVA procedure; GLM procedure
  basic issues, 517–519
  contrasts, 519, 520
  interactions, 519–520
  main effects models, 519
  MANOVA, 521
  means and comparison tests, 525
  overspecified models, 518
  repeated measures, 521
  SAS algorithms for, 518
  selecting SAS procedures for, 522
  unbalanced designs, 520
And (&) operator, 42–43, 594
ANOVA, 517–522
ANOVA procedure, 522–526, 532–533, *534–541. See also* Analysis of variance; GLM procedure
  means and comparison tests, 525
  specifying interaction terms, *536*
  statement order, 530
  syntax, 522–526
  used interactively, 530
Arguments, 128
Arrays. *See* ARRAY statement; INPUT statement, reading arrays; PUT statement, writing arrays

ARRAY statement, 199–202
  compared to IF statements, 202, 584
  placement of, 583
  reasons for use, 199
  syntax 200–201
  version 6.06 changes, 635
ASCENDING option, CHART procedure, 107, *121*
Assignment statements, 125–127. *See also* Functions
  examples of use, 127, 584
  impact of missing values in, 126
  sequence of, 126
  syntax of, 125–126
ASSIST add-on product, 634
ATTRIB statement, 69–70, 163. *See also* FORMAT statement; LABEL statement, placement of, 583
Average. *See* Mean

Bar charts. *See* CHART procedure
Batch processing, 594. *See also* Interactive processing; Running a program
BEST format, 159, *161*
BEST option, CORR procedure, 471–472, *478*
Bonferroni *t* tests. *See* BON option
BON option
  ANOVA procedure, 523, 525, *537*
  GLM procedure, 527
BOX option, PLOT procedure, 430, 432, *440, 441*
Box plots, 89

Note: Italicized numbers indicate pages on which exhibits appear.